Lecture Notes in Computer Science 3011

Commenced Publication in 1973
Founding and Former Series Editors:
Gerhard Goos, Juris Hartmanis, and Jan van Leeuwen

T0222719

Springer
Berlin
Heidelberg
New York
Hong Kong
London
Milan
Paris
Tokyo

Jean-Charles Régin   Michel Rueher (Eds.)

Integration of AI and OR Techniques in Constraint Programming for Combinatorial Optimization Problems

First International Conference, CPAIOR 2004
Nice, France, April 20-22, 2004
Proceedings

 Springer

Volume Editors

Jean-Charles Régin
ILOG
Les Taissounières HB2, 1681 route des Dolines, 06560 Valbonne, France
E-mail: regin@ilog.fr

Michel Rueher
Université de Nice - Sophia Antipolis
Projet COPRIN, I3S/CNRS/INRIA, ESSI
BP 145, 06903 Sophia Antipolis, France
E-mail: Michel.Rueher@sophia.inria.fr

Library of Congress Control Number: 2004103824

CR Subject Classification (1998): G.1.6, G.1, G.2.1, F.2.2, I.2, J.1

ISSN 0302-9743
ISBN 3-540-21836-X Springer-Verlag Berlin Heidelberg New York

Springer-Verlag is a part of Springer Science+Business Media

springeronline.com

© Springer-Verlag Berlin Heidelberg 2004
Printed in Germany

Typesetting: Camera-ready by author, data conversion by DA-TeX Gerd Blumenstein
Printed on acid-free paper   SPIN: 10997383   06/3142   5 4 3 2 1 0

Preface

This volume contains the proceedings of the First International Conference on Integration of AI and OR Techniques in Constraint Programming for Combinatorial Optimisation Problems. This new conference follows the series of CP-AI-OR International Workshops on Integration of AI and OR Techniques in Constraint Programming for Combinatorial Optimisation Problems held in Ferrara (1999), Paderborn (2000), Ashford (2001), Le Croisic (2002), and Montreal (2003). The success of the previous workshops has demonstrated that CP-AI-OR is becoming a major forum for exchanging ideas and methodologies from both fields. The aim of this new conference is to bring together researchers from AI and OR, and to give them the opportunity to show how the integration of techniques from AI and OR can lead to interesting results on large scale and complex problems.

The integration of techniques from Artificial Intelligence and Operations Research has provided effective algorithms for tackling complex and large scale combinatorial problems with significant improvements in terms of efficiency, scalability and optimality. The benefit of this integration has been shown in applications such as hoist scheduling, rostering, dynamic scheduling and vehicle routing. At the programming and modelling levels, most constraint languages embed OR techniques to reason about collections of constraints, so-called global constraints. Some languages also provide support for hybridization allowing the programmer to build new integrated algorithms. The resulting multi-paradigm programming framework combines the flexibility and modelling facilities of Constraint Programming with the special purpose and efficient methods from Operations Research.

CP-AI-OR 2004 was intended primarily as a forum to focus on the integration of the approaches of CP, AI, and OR technologies. A secondary aim was to provide an opportunity for researchers in one area to learn about techniques in others. 56 papers were submitted in response to the call for papers. After the reviewing period and some online discussions, the program committee met physically at Nice on January 30 and 31. The program committee decided to accept 23 technical papers and 7 short papers. Short papers present interesting recent results or novel thought-provoking ideas that are not quite ready for a regular full-length paper. Both types of papers were reviewed rigorously and held to a very high standard.

CP-AI-OR 2004 has been fortunate to attract outstanding invited talks. Heinrich Braun and Thomas Kasper discussed about the challenges of optimisation problems in supply chain management. Ignacio Grossmann proposed an hybrid framework that uses mathematical and constraint programming for the scheduling of batch chemical processes. Michel Minoux told us about strengthened relaxations for some CP-Resistant combinatorial problems and their potential usefulness.

We wish to thank our generous sponsors who allowed us to offer substantial allowances to students attending the conference in order to cover their expenses. We extend our gratitude to the outstanding program committee who worked very hard under tight deadlines. We are deeply grateful to Claude Michel who worked in the trenches in preparing the CP meeting at Nice and who dealt with all the difficult organisation aspects of this conference.

April 2004

Jean-Charles Régin
Michel Rueher

Organization

CPAIOR 2004 was organized by INRIA (Institut National de Recherche en Informatique et en Automatique) Sophia Antipolis.

Executive Committee

Conference and Program Chair: Jean-Charles Régin (Ilog SA) and
Michel Rueher (University of Nice-Sophia
Antipolis)
Organization Chair: Claude Michel (University of Nice-Sophia
Antipolis)

Program Committee

Abderrahmane Aggoun, Cosytec, France
Philippe Baptiste, Ecole Polytechnique, France
Roman Bartak, Charles University, Czech Republic
Chris Beck, Cork Constraint Computation Center, Ireland
Mats Carlsson, SICS, Sweden
Alain Colmerauer, University of Marseille, France
Hani El Sakkout, Parc Technologies, UK
Bernard Gendron, CRT and University of Montreal, Canada
Carmen Gervet, IC-Parc, UK
Carla Gomes, Cornell University, USA
Narendra Jussien, Ecole des Mines de Nantes, France
Stefan Karisch, Carmen Systems, Canada
François Laburthe, Bouygues, France
Olivier Lhomme, ILOG, France
Michela Milano, University of Bologna, Italy
George Nemhauser, University of Georgia Tech, USA
Gilles Pesant, CRT and Ecole Polytechnique de Montreal, Canada
Jean-Charles Regin (chair), ILOG, France
Michel Rueher (chair), University of Nice-Sophia Antipolis, France
Christian Schulte, KTH, Sweden
Meinolf Sellmann, Cornell University, USA
Sven Thiel, Max Planck Institute, Germany
Gilles Trombettoni, University of Nice-Sophia Antipolis, France
Michael Trick, Carnegie Mellon University, USA
Pascal van Hentenryck, Brown University, USA
Mark Wallace, IC-Parc, UK

Referees

Abderrahmane Aggoun
Philippe Baptiste
Roman Bartak
Chris Beck
Nicolas Beldiceanu
Pascal Brisset
Mats Carlsson
Alain Colmerauer
Miguel Constantino
Romuald Debruyne
Hani El Sakkout
Andrew Eremin
Marco Gavanelli
Bernard Gendron
Carmen Gervet
Carla Gomes
Idir Gouachi

Brahim Hnich
Narendra Jussien
Stefan Karisch
Irit Katriel
Yahia Lebbah
François Laburthe
Olivier Lhomme
Vassilis Liatsos
Andrea Lodi
Michela Milano
George Nemhauser
Bertrand Neveu
Stefano Novello
Gilles Pesant
Nikolai Pisaruk
Steven Prestwich
Philippe Refalo

Guillaume Rochart
Louis-Martin Rousseau
Christian Schulte
Meinolf Sellmann
Paul Shaw
Josh Singer
Helmut Simonis
Neil Yorke-Smith
Francis Sourd
Sven Thiel
Gilles Trombettoni
Michael Trick
Pascal van Hentenryck
Willem Jan van Hoeve
Mark Wallace
Jean-Paul Watson

Sponsors

Bouygues, France
Conseil Général des Alpes Maritimes
Conseil Régional de Provence-Alpes-Côte d'Azur
Cosytec S.A., France
Carmen System
CoLogNET
ESSI (École Supérieure en Sciences Informatiques), Sophia Antipolis, France
I3S/CNRS–Université de Nice-Sophia Antipolis, France
IISI (The Intelligent Information Systems Institute), USA
Ilog S.A., Paris

Table of Contents

Invited Paper

Technical Papers

Short Papers

Using MILP and CP for the Scheduling of Batch Chemical Processes

Christos T. Maravelias and Ignacio E. Grossmann

Department of Chemical Engineering
Carnegie Mellon University, Pittsburgh, PA 15213, USA
{ctm,ig0c}@andrew.cmu.edu

Abstract. A hybrid framework that uses Mathematical and Constraint Programming for the scheduling of batch chemical processes is proposed. Mathematical programming is used for the high-level optimization decisions (number and type of tasks, and assignment of equipment units to tasks), and Constraint Programming is used for the low-level sequencing decisions. The original problem is decomposed into an MILP master problem and a CP subproblem. The master MILP is a relaxation of the original problem, and given a relaxed solution, the CP subproblem checks whether there is a feasible solution and generates integer cuts. The proposed framework is based on the hybrid algorithm of Maravelias and Grossmann ([1],[2]), and can be used for different objective functions and different plant configurations. In this paper we present the simplifications and enhancements that allow us to use the proposed framework in a variety of problems, and report computational results.

1 Introduction

Scheduling of operations is a common and very important problem in the chemical industry. While related problems have been extensively studied in the Operations Research literature (see [3]), this area has only been addressed recently in process systems engineering (see [4], [5], [6], [7] for reviews).

In terms of plant configurations, problems in chemical industry can be classified into four major categories. In multiple-unit or single-stage plants (Figure 1.a), there are N orders to be scheduled in M units. In flow-shop multi-stage plants (Figure 1.b), there are N orders to be processed in K stages, following the same order. Each stage $k \in \{1, ... K\}$ consists of M_k units, and each order must be processed by one unit in each stage. In general multi-stage plants (Figure 1.c), each order must be processed in all stages but not in the same order. Multipurpose batch plants (Figure 1.d), finally, can be viewed as a generalization of all previous configurations, where batch splitting and mixing, as well as recycle streams are present. It should be noted that the original work in process scheduling concentrated mostly on flow-shop and general multi-stage batch plants (see [3], [5]). The study of general multipurpose plants was largely promoted by the work of Kondili et al. [8].

Preemption is usually not allowed, and utility constraints (e.g. manpower, cooling water, etc.), and release/due times may be present in all configurations. Different storage policies, such as unlimited intermediate storage (UIS), finite intermediate

J.-C. Régin and M. Rueher (Eds.): CPAIOR 2004, LNCS 3011, pp. 1-20, 2004.
© Springer-Verlag Berlin Heidelberg 2004

storage (FIS), no intermediate storage (NIS) and zero-wait (ZW), are used in multi-purpose and multi-stage plants, while in singe-stage plants it is usually assumed that unlimited storage is available. Furthermore, the batch size of a task may be variable, which in many cases leads to variable duration and utility requirements. In terms of objective functions, the most common ones: maximization of production over a fixed time horizon, minimization of makespan for given demand and minimization of cost for given demand with due dates.

A feature that makes scheduling problems in chemical industry hard to solve is that usually the type and number of tasks (jobs) are not uniquely defined, and moreover, a specific task can be performed in more than one unit. These problems are hard to solve because of the large number of different solutions, and have not been studied extensively. Problems where the number and type of tasks are fixed and each task can be assigned to only one machine, on the other hand, have been extensively studied in OR community and efficient algorithms exist for many of these problems.

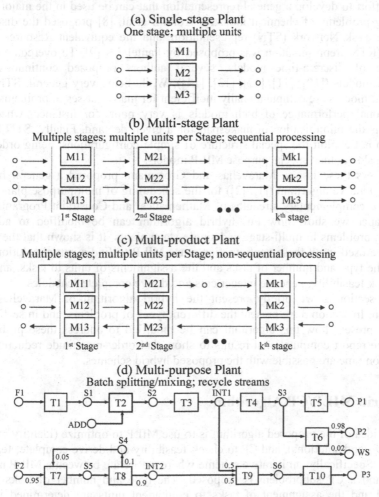

Fig. 1. Common plant configurations in chemical industry

Due to the different plant configurations and process specifications, a wide variety of optimization models, mainly MILP models, have been proposed in the process systems engineering literature. In multipurpose batch plants, for instance, where the level of inventories and the level of resource consumption should be monitored and constrained, the time horizon is partitioned into a sufficiently large number of periods, and variables and constraints are defined for each time period. In single-stage plants, on the other hand, where no batch splitting, mixing and recycle streams are allowed, batch sizes are usually assumed to be constant and this, in turn means that mass balance equations need not be included in the formulation. If, in addition, there are no utility constraints the time horizon is not partitioned into common time periods, and hence, assignment binaries are indexed by tasks and unit time slots, which are generally fewer than time periods. When the number of tasks and the assignments of tasks to units are known, assignments binaries and constraints are dropped, and sequencing binaries are used instead.

In an effort to develop a general representation that can be used in the majority of scheduling problems of chemical industry, Kondili et al. [8] proposed the discrete-time State Task Network (STN) representation, and the equivalent Resource Task Network (RTN) representation was proposed by Pantelides [9]. To overcome some limitations of discrete-time models, several authors proposed continuous-time STN/RTN models ([10], [11], [12], [13], [14]). While being very general, STN and RTN-based models are computationally inefficient for many classes of problems. The computational performance of both models is very poor, for instance, when the objective is the minimization of makespan for a given demand. Finally, STN-based models do not exploit the special structure of simple configurations, being orders of magnitude slower than special purpose MILP models.

To address these issues, Maravelias and Grossmann proposed a general hybrid MILP/CP iterative algorithm ([1], [2]) for the scheduling of multipurpose plants that exploits the complementary strengths of Mathematical and Constraint Programming. In this paper we show how this hybrid algorithm can be modified to address scheduling problems in multi-stage and single-stage plants. It is shown that the same idea can be used in all these configurations: use MILP to find a partial solution that includes the type and number of tasks and the assignments of units to tasks, and use CP to check feasibility, generate integer cuts and derive complete schedules.

In the section 2 we briefly present the hybrid algorithm of Maravelias and Grossmann. In section 3 we present the different types of problems, and in sections 4 and 5 we present how this algorithm can be modified to address these problems. Finally, we report computational results to show that order-magnitude reductions in computation time are possible with the proposed hybrid schemes.

2 Hybrid MILP/CP Algorithm

The main idea of the proposed algorithm is to use MILP to optimize (identify partial, potentially good solutions), and CP to check feasibility and derive complete, feasible schedules. Specifically, an iterative scheme where we iterate between a MILP master problem and a CP subproblem is proposed. The type and number of tasks to be performed and the assignment of tasks to equipment units are determined in the master MILP problem, while the CP subproblem is used to derive a feasible schedule for the partial solution obtained by the master problem. At each iteration, one or more integer cuts are added to the master problem to exclude infeasible or previously obtained assignments. For a maximization problem, the relaxed master problem provides an upper bound and the subproblem, when feasible, provides a lower bound. The algorithm terminates when the bounds converge. To enhance the performance of the algorithm, preprocessing is used to determine Earliest Start Times (*EST*) and Shortest Tails (*ST*) of both tasks ($i \in I$) and units ($j \in J$). Strong integer cuts that are added a priori in the cut-pool of the master problem are also developed during preprocessing. A simplified flow diagram of the proposed algorithm for the maximization of profit is shown in Figure 2.

The proposed decomposition can also be applied as a branch-and-cut algorithm, where the master problem is viewed as a relaxation of the original problem. When a relaxed solution is obtained, the CP solver is called to obtain a complete solution and generate cuts that are added in the MILP relaxation.

2.1 Master Problem

For the master MILP problem an aggregated STN representation with no time periods has been used. Resource constraints and big-M time matching constraints (that lead to poor LP relaxations) have been eliminated and only assignment, batch size and mass balance constraints are included. Mass balance constraints are expressed once (for the total amounts) at the end of the scheduling horizon. Integer cuts are added to exclude previously found or infeasible integer solutions of the master problem.

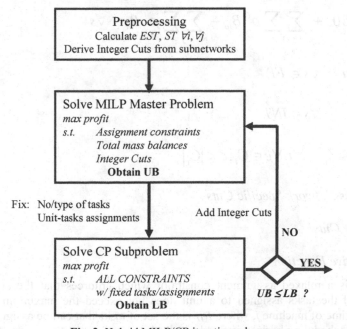

Fig. 2. Hybrid MILP/CP iterative scheme

To decouple units from tasks we use the following rule. If a task i can be performed in both units j and j', then two tasks i (performed in unit j) and i' (performed in unit j') are defined. Thus, by choosing which tasks are performed we also make assignment decisions. For each task i we postulate a set of copies, i.e. an upper bound on the number of batches of task i that can be carried out in any feasible solution.

$$|C_i| = \lfloor H / D_i^{MIN} \rfloor \quad \forall i$$

where H is an upper bound on the length of the time horizon and D_i^{MIN} is the minimum duration of task i. In practice, however, a smaller number of copies can be used based on our knowledge of the process network,

For each copy c of task i we define the binary Z_{ic}, which is equal to 1 if the cth copy of task i is carried out. We also define its duration D_{ic} and batch size B_{ic}. For each state, we define its inventory level S_s at the end of the scheduling horizon. The master MILP problem (MP) consists of constraints (1) to (10):

$$\sum_{i \in I(j)} \sum_c D_{ic} \leq (MS - ST_j) - EST_j \quad \forall j \tag{1}$$

$$D_{ic} = \alpha_i Z_{ic} + \beta_i B_{ic} \quad \forall i, \forall c \in C_i \tag{2}$$

$$B_i^{MIN} Z_{ic} \leq B_{ic} \leq B_i^{MAX} Z_{ic} \quad \forall i, \forall c \in C_i \tag{3}$$

$$S_s = S0_s + \sum_{i \in O(s)} \sum_c \rho_{is}^O B_{ic} - \sum_{i \in I(s)} \sum_c \rho_{is}^I B_{ic} \quad \forall s \tag{4}$$

$$S_s \geq d_s \quad \forall s \in FP \tag{5}$$

$$S_s \leq C_s \quad \forall s \in INT \tag{6}$$

$$Z_{ic+1} \leq Z_{ic} \quad \forall i, \forall c \in C_i, c < |C_i| \tag{7}$$

Process Network Specific Cuts $\tag{8}$

Integer Cuts $\tag{9}$

Objective Function $\tag{10}$

Eq. (1) is a relaxed assignment constraint which enforces that the sum of the durations of the tasks assigned to a unit does not exceed the maximum available processing time of machine j, where $I(j)$ is the set of tasks that can be assigned to unit j. When the scheduling horizon is fixed (maximization of production over a fixed time horizon or problems with deadlines) MS is a parameter equal to the fixed time horizon

H. When the objective is the minimization of the makespan, *MS* is a variable equal to the makespan. Parameter EST_j represents the earliest time that any task can be assigned to start on unit *j*, while parameter ST_j represents the shortest time needed for any material processed in unit *j* to be transformed into a final product. Both EST_j and ST_j are calculated during preprocessing. The duration of copy *c* of task *i* is a function of its batch size, as in eq. (2), and the batch size of copy *c* of task *i* is bounded through eq. (3). The amount of state *s* at the end of the time horizon S_s is calculated by (4) to be equal to the initial amount $S0_s$ plus the amount produced, minus the amount consumed, where $\rho_{is}{}^I$ and $\rho_{is}{}^O$ are the mass fractions for consumption and production, respectively, of state *s* by task *i*, and *O(s)/I(s)* is the set of tasks producing/consuming state *s*. In eq. (5), the inventory S_s of the final product $s \in FP$ must be greater than the demand d_s, while in eq. (6) the inventory of intermediate $s \in INT$ must be less than the capacity C_s of the storage tank of state *s*. Eq. (7) is used to eliminate symmetric assignments, while eq. (8) is used to model special characteristics of a process network (see [1], [2] for details). At a specific iteration *k*, constraints in (9) include all the integer cuts that have been added during preprocessing and in previous iterations. Various objective functions can be accommodated. The master problem (MP) is a relaxation of the original problem because it does not account for the interconnections between tasks and states and does not enforce feasibility throughout the time horizon. Hence, the assignments obtained by (MP) may be infeasible in reality. The feasibility check and the derivation of a complete feasible schedule, if possible, are performed by the CP subproblem.

2.2 CP Subproblem

The modeling language of ILOG's OPL Studio 3.5 ([15], [16]) has been used for the modeling of the CP subproblem. For each equipment unit *j* we define a unary resource *Unit[j]* and for each resource *r* (e.g. cooling water) we define a discrete resource *Utility[r]* with a maximum capacity $R_r{}^{MAX}$. Furthermore, for each state *s* we define a reservoir *State[s]* with capacity C_s and initial level $S0_s$. For each binary Z_{ic} that is equal to 1 in the current optimal solution of the master problem (i.e. copy *c* of task *i* is carried out) we define an activity *Task[i,c]* with duration D_{ic}. We also define a dummy activity *MS* with zero duration and no resource requirements. We also define $|D|$ activities, *Order[d]*, with zero duration, where *D* is the set of orders for final products, *D(s)* is the set of orders for state *s* $(D = \cup_s D(s))$, and for each $d \in D$, AD_d is the amount due and TD_d is the due date. The CP subproblem consists of constraints (11) to (26):

$$B_i^{MIN} \leq B_{ic} \leq B_i^{MAX} \quad \forall i, \forall c \mid Z_{ic} = 1 \tag{11}$$

$$D_{ic} = \alpha_i + \beta_i B_{ic} \quad \forall i, \forall c \mid Z_{ic} = 1 \tag{12}$$

$$Task[i,c] \ requires \ Unit[j] \quad \forall j, \forall i \in I(j), \forall c \mid Z_{ic} = 1 \tag{13}$$

$$B_{ics}^I = \rho_{is}^I B_{ic} \quad \forall s, \forall i, \forall c \mid Z_{ic} = 1 \tag{14}$$

$$B_{ics}^{O} = \rho_{is}^{O} B_{ic} \quad \forall s, \forall i, \forall c \,|\, Z_{ic} = 1 \tag{15}$$

$$Task[i,c] \; consumes \; B_{ics}^{I} \; State[s] \quad \forall s, \; \forall i, \; \forall c \,|\, Z_{ic}=1 \tag{16}$$

$$Task[i,c] \; produces \; B_{ics}^{O} \; State[s] \quad \forall s, \; \forall i, \; \forall c \,|\, Z_{ic}=1 \tag{17}$$

$$R_{icr} = \gamma_{ir} + \delta_{ir} B_{ic} \quad \forall r, \forall i, \forall c \,|\, Z_{ic} = 1 \tag{18}$$

$$Task[i,c] \; requires \; R_{icr} \; Utility[r] \quad \forall i, \; \forall c \,|\, Z_{ic}=1 \tag{19}$$

$$\sum_{i} \sum_{c\,|\,Z_{ic}=1} B_{ics}^{O} \geq d_{s} \quad \forall s \in FP \tag{20}$$

$$Order[d].start = TD_{d} \quad \forall d \tag{21}$$

$$Order[d] \; consumes \; AD_{d} \; State[s] \quad \forall s, \; \forall d \in D(s) \tag{22}$$

$$Task[i,c].end \leq MS.start \quad \forall i, \; \forall c \,|\, Z_{ic}=1 \tag{23}$$

$$Task[i,c] \; precedes \; Task[i,c+1] \quad \forall i, \; \forall c \in C_{i}, c<|C| \tag{24}$$

Special Network Specific Constraints $\tag{25}$

(Optional) Objective Function $\tag{26}$

The batch size of activity *Task[i,c]* is bounded by eq. (11) and its duration is calculated via eq. (12). Constraint (13) enforces tasks in *I(j)* to be assigned to unary resource *Unit[j]*. The amount B_{ics}^{I}/B_{ics}^{O} of reservoir *State[s]* consumed/ produced by activity *Task[i,c]* is calculated by eq. (14)/(15), and the consumption/production of B_{ics}^{I}/B_{ics}^{O} units of reservoir *State[s]* by *Task[i,c]* is enforced by eq. (16)/(17). The amount R_{icr} of discrete resource *Utility[r]* required by activity *Task[i,c]*, throughout its execution, is calculated in eq. (18), and the consumption of R_{icr} units of discrete resource *Utility[r]* by activity *Task[i,c]* is enforced by eq. (19). The condition that the amount of final products should meet the demand is enforced by (20), where d_{s} is the total demand for state *s*. Parameter d_{s} is either given (in the case of fixed demand with no due dates) or calculated as a sum of AD_{d} for all $d \in D(s)$. Each order is executed at its due time (eq. (21)), and the amount delivered is equal to the amount due (eq. (22)). In eq. (23) the end time of all activities is restricted to be smaller than the start of activity *MS*, and *MS* is, (a) fixed finish time when the time horizon is fixed, and (b) a variable finish time when the objective is the minimization of makespan. Constraint (24) is a symmetry-breaking constraint that reduces the number of possible configurations by imposing a sequence between copies of the same task. Constraints that describe some special features of the process network are included in (25). Depending on the nature of the problem (constant vs. variable processing times) and the objective function, we may want to solve the CP subproblem as one feasibility problem, as successive feasibility problems or as an optimization problem. If the CP is an optimization problem we add constraint (26). Details can be found in [2].

2.3 Preprocessing

The performance of the proposed model depends on how fast we solve models (MP) and (SP), and the number of iterations needed to generate solutions and prove optimality. It is crucial, therefore, to exclude infeasible or suboptimal solutions as soon as possible. Preprocessing enhances the performance of the algorithm by (a) tightening existing constraints, and (b) creating strong cuts that are added in the cut-pool of the master problem and are used to eliminate a priori a number of potential configurations. Parameters EST_j and ST_j that are used to tighten constraint (1) are calculated in preprocessing (see [1] and [2] for details). Depending on the characteristics of the problem, different preprocessing can be performed. As will be shown in section 5, for instance, integer cuts can be generated in problems with release and due times.

2.4 Integer Cuts

Another way to reduce the number of iterations is by generating integer cuts that forbid more than one infeasible or suboptimal solutions. While a simple, "no-good" integer cut can always be added at each iteration, in some cases stronger integer cuts that exclude more than one assignment can also be added. The form of the integer cuts is problem specific. Two new classes of integer cuts were proposed in [2] for multi-purpose batch plants. In general, the quality of the integer cuts is crucial for the effectiveness of the proposed scheme, and how to generate effective cuts is an open question.

3   Scheduling Problems

As explained in section 1, a wide variety of scheduling problems appear in chemical industry. While models (MP) and (CP) are very general, they do not exploit the special structure of these problems. A very interesting feature of models (MP) and (CP), however, is that they can be readily *reduced* to models that accurately describe other plant configurations. The reduced models result from models (MP) and (CP) by removing some of the constraints, changing the notation and in some cases adding constraints that describe details of a specific problem and tighten the formulations. Note, however, that no new constructs, variables or types of constraints need to be defined. Another advantage of this *reduction* is that the form and functionality of the two models remain practically the same: the decisions about the number and type of tasks and the assignment of tasks to units (or some of these decisions) are made by the master problem, while the subproblem is used to check feasibility, yield complete schedules and generate integer cuts.

The exact form of the two problems depends on the configuration of the plant, the objective function and the special characteristics of the instance at hand (demand, release/due times, etc.). In terms of plants configurations, the most common are the ones shown in Figure 1. Note that the flow-shop plant is a special case of the multi-stage plant and that the single-stage plant is a special case of the flow-shop plant. In terms of objective functions, the most common are the maximization of production or profit over a fixed time horizon, the minimization of makespan for given demand and the minimization of cost for given demand with due dates. To give an example, if the objective is to maximize the production or profit over a fixed time horizon, the type and number of tasks are unknown, whereas if the objective is to minimize the makespan for a fixed demand expressed in orders for which no batch splitting and mixing is allowed, the type and number of tasks is fixed and only the assignment of units to tasks is determined by the master problem.

A characteristic that differentiates scheduling problems significantly is how the demand for final products is expressed. If it is expressed in terms of fixed quantities, called the *fixed-demand* problem, more than one batches can be combined to meet the demand and thus batch mixing and splitting is allowed. If it is expressed in terms of orders, called the *order* problem, no batch mixing and splitting is allowed throughout the production. Note that in some cases a *fixed-demand* problem can be reduced to an *order* problem by pre-calculating how many and what type of batches are needed to meet the demand. When the demand is expressed in amounts, the number of tasks to be performed is unknown. When the demand is expressed in terms of orders, the number of tasks is fixed and a number of simplifications can be applied. In multipurpose batch plants demand is usually expressed in fixed amounts of final products, while in multi- and single-stage plants demand can be expressed either as fixed amounts or as orders. Next, we present a general *fixed-demand* formulation for the multi-stage plant and reduced *order* formulations for the multi- and single-stage plant (the *fixed-demand* formulation for the single-stage plant is a special case of the multi-stage formulation). When the objective is the maximization of production over a fixed time horizon, the number of tasks is unknown, and thus the *fixed-demand* formulation, without the demand satisfaction constraints, is used.

4   Master Problem Reductions

4.1 Multi-stage Plant: Fixed Demand

In the master problem we use a reduced formulation to determine the number and type of tasks as well as the assignment of units to tasks. A task i corresponds to the processing of a chemical s at a unit j of a stage k. Compared to the multi-purpose plant, in multi-product plants each task consumes and produces only one state and thus the mass fractions for consumption and production in equation (5) are equal to 1:

$$S_s = S0_s + \sum_{i \in O(s)} \sum_c B_{ic} - \sum_{i \in I(s)} \sum_c B_{ic} \quad \forall s \tag{27}$$

All other constraints remain the same. The master problem for the general multi-stage plant consists of equations (1) – (4), (27) and (6) – (10). When units operate at constant batch-size, processing times are parameters and constraints (3) and (4) are dropped, and B_{ic} is equal to $Z_{ic}B_i$, where B_i is the fixed batch-size of task i.

4.2 Multi-stage Plant: Demand in Orders

When demand is expressed in terms of orders, the number of batches is fixed: each order has to be processed at each stage exactly once, i.e. the number of tasks that take place is $|O|*|K|$, where $|O|$ and $|K|$ is the number of orders and stages respectively. Moreover, each order will be processed once at each stage in one of the units of this stage. Thus, we can drop index c for copies, and replace the tuple (i,c) by the triplet (o,k,j), where o, k and j are the indices for orders, stages and units.

Since demand is expressed in orders, constraint (6) is dropped, and since an order corresponds to a certain amount of a final product, batch-sizes are fixed and constraints (2) and (3) are also dropped. Since no batch splitting and mixing is allowed, there are no intermediate chemicals at the end of the horizon. Moreover, we can assume that appropriate storage is available for the finished goods, and thus we can also drop constraints (4) and (5). Since there are no copies of the same task, constraint (7) is dropped, and since tasks are indexed by (o,k,j) instead of (i,c), constraint (1) is written as in (28). The master problem for the multi-stage plant when the demand is expressed in orders consists of equations (8) – (10), (28) and (29).

$$\sum_{j \in J(k)} Z_{okj} = 1 \quad \forall o, \forall k \tag{28}$$

$$\sum_{o} Z_{okj} D_{okj} \leq (MS - ST_j) - EST_j \quad \forall k, \forall j \in J(k) \tag{29}$$

Binary Z_{okj} is equal to 1 if order o is assigned to unit j of stage k (i.e. $j \in J(k)$), and D_{okj} is the fixed duration of task (o,k,j). Constraint (28) enforces that each order is processed at exactly one unit of stage k. Constraint (29) ensures that the sum of processing times of the durations of tasks assigned to unit $j \in J(k)$ does not exceed the maximum processing time available on unit j. Constraints in (8) can include forbidden and processing paths (i.e. if order o cannot be processed in unit $j2 \in J(k+1)$ if previously processed in unit $j1 \in J(k)$ then $Z_{o,j2,k+1} \leq 1 - Z_{o,j1,k}$).

4.3 Single-stage Plant: Demand in Orders

The single-stage plant is a special case of the multi-stage plant. Each order has to be processed only in one stage, and thus the index k is dropped. The master problem consists of equations (8) – (10), (30) and (31).

$$\sum_{j} Z_{oj} = 1 \quad \forall o \tag{30}$$

$$\sum_{o} Z_{oj} D_{oj} \leq (MS - ST_j) - EST_j \quad \forall j \tag{31}$$

5   Subproblem Reductions

In the reduced formulation for the master problem we do not take into account utility constraints. In the subproblem, however, we use constraints (18) and (19) to model utility restrictions. Utility constraints are always the same, i.e. independent of the plant configuration. Furthermore, constraints (20) – (22) are used for the satisfaction of the demand and they are also independent of the plant configuration. Hence, we will refer to utility constraints (18) – (19), and demand satisfaction constraints (20) – (22) as follows:

$\quad$ *Utility Constraints* $\hfill$ (32)

$\quad$ *Demand Satisfaction Constraints* $\hfill$ (33)

$\quad$ Hence, the CP subproblem for the multipurpose batch plant consists of constraints · (11) – (17), (23) – (26) and (32) – (33).

5.1 Multistage Plant: Fixed-Demand

In multi-stage plants there are no recycle streams and all mass fractions are equal to 1, so we can eliminate variables B^I_{ics} and B^O_{ics}, drop constraints (14) and (15) and use constraints (34) and (35) instead of constraints (16) and (17), respectively:

$\quad$ *Task[i,c] consumes B_{cs} State[s]* $\forall s \in SI(i)$, $\forall i$, $\forall c | Z_{ic}=1$ $\hfill$ (34)

$\quad$ *Task[i,c] produces B_{cs} State[s]* $\forall s \in SO(i)$, $\forall i$, $\forall c | Z_{ic}=1$ $\hfill$ (35)

where $SI(i)$ and $SO(i)$ are the sets of the input and output states, respectively, of task i.

$\quad$ The CP subproblem for the fixed-demand multipurpose batch plant consists of equations (11) – (13), (23) – (26), and (32) – (35).

5.2 Multipurpose Plants: Demand in Orders

When the demand is expressed in terms of orders, the number of tasks is fixed, and thus we can drop the index c for copies, and replace index i by the triplet (o,k,j), as in the master problem. This implies that constraint (24) is dropped, and that constraints (13) and (23) are rewritten as in (36) and (37), respectively:

$\quad$ *Task[o,k,j] requires Unit[j]* $\forall o$, $\forall k$, $\forall j \in J(k) | Z_{okj}=1$ $\hfill$ (36)

$\quad$ *Task[o,k,j].end $\leq MS$* $\forall o$, $\forall k$, $\forall j \in J(k) | Z_{okj}=1$ $\hfill$ (37)

$\quad$ Furthermore, the batch-sizes are fixed, so we can remove constraints (11) and (12), and re-write utility constraint (31) as in (38), and constraints (34) – (35) as in (39) and (40), respectively. Since the demand is expressed in orders, any feasible solution of the problem will satisfy the demand, so constraint (33) is not needed.

$\quad$ *Task[o,k,j] requires $Z_{okj}R_{rokj}$ Utility[r]* $\forall r$, $\forall o$, $\forall k$, $\forall j \in J(k) | Z_{okj}=1$ $\hfill$ (38)

$$Task[o,k,j] \text{ } consumes \text{ } B_o \text{ } State[o,SI(o,k)] \quad \forall o, \forall k, \forall j \in J(k) | Z_{okj} = 1 \tag{39}$$

$$Task[o,k,j] \text{ } produces \text{ } B_o \text{ } State[o,SO(o,k)] \quad \forall o, \forall k, \forall j \in J(k) | Z_{okj} = 1 \tag{40}$$

where B_o is the size of order o, R_{rokj} is the amount of utility r required by order o at stage k when processed at unit $j \in J(k)$, and $SI(o,k)$ and $SO(o,k)$ are the sets of input and output states, respectively, of order o at stage k. The CP subproblem consists of equations (25) and (36) – (40).

The redundant constraint (41) that enforces a sequence between tasks of the same order can be added, where $S(o,k)$ is an index set that for order o gives the stage that follows stage k. Note that the sequence between tasks of the same order is also enforced by constraints (39) and (40): a task in stage k can only be performed if the input state is available.

$$Task[o,k,j] \text{ } preceeds \text{ } Task[o,S(o,k),j'] \quad \forall o, \forall k, \forall j \in J(k), \forall j' \in J(S(o,k)) | Z_{okj} = 1 \wedge Z_{o,S(o,k),j'} = 1 \tag{41}$$

If we assume that appropriate dedicated storage is available for all intermediate states, we need not use the construct State and we can drop constraints (39) and (40). In that case, constraint (41) is necessary to impose the sequencing among tasks of the same order. In this case, the CP subproblem consists of constraints (25), (36) – (38) and (41).

5.3 Single-stage Plant: Demand in Orders

In single-stage plants each order has to be processed in only one stage and storage is usually not taken into account. Thus, index k for stages is dropped, the construct *State* is not used and sequencing constraints are not needed. The CP model consists of constraints (25) and (42) – (44), where binary Z_{oj} is 1 if order o is assigned to unit j.

$$Task[o,j] \text{ } requires \text{ } Unit[j] \quad \forall o, \forall j | Z_{oj} = 1 \tag{42}$$

$$Task[o,j].end \leq MS \quad \forall o, \forall j | Z_{oj} = 1 \tag{43}$$

$$Task[o,j] \text{ } requires \text{ } R_{ojr} \text{ } Utility[r] \quad \forall o, \forall j | Z_{oj} = 1 \tag{44}$$

6   Remarks

6.1 Integration with Other Algorithms

An interesting feature of the proposed decomposition framework is that it can be used as a general platform for the integration of two different solution paradigms. In general, the master problem is a relaxation of the original problem. A solution of this relaxed problem defines a subspace that is searched by the subproblem. The main idea of the proposed decomposition, thus, is to use a solution paradigm that is good at identifying promising partial solutions (i.e. use MILP for optimization) and a solution technique that is good at searching the constrained subspace (i.e. use CP for scheduling problems with fixed tasks and fixed assignments).

In general, however, any two solution techniques can be combined, provided that the solution of the master problem can be translated into a meaningful problem for the subproblem. This is particularly useful when the subproblem corresponds to a problem that has been extensively studied by the OR community, and there are efficient algorithms for its solution. When the demand is expressed in orders the subproblem of multi-stage plants (i.e. the reduced problem where the assignments are fixed), for instance, is equivalent to the widely studied job-shop scheduling problem. Thus, for the solution of the subproblem we can use a problem-specific algorithm instead of CP. For the multi-stage problem, specifically, we developed an iterative scheme where we use the Shifting Bottleneck procedure ([17], [18]) (SBP) for the solution of the job-shop problem that arises when the assignment of tasks to specific units is fixed by the master problem.

6.2 Preprocessing and Integer Cuts

Pre-processing algorithms that exploit the special structure of the problem or the specific instance at hand can be used to tighten the models presented above. For the minimization of processing cost of single-stage plants with orders that have release and due times, for instance, we developed a very efficient pre-processing algorithm that generates cover cuts that can be added to the cut pool of the master problem a priori. These cover cuts are generated from knapsack constraints of the form,

$$\sum_o d_{oj} Z_{oj} \leq \max_{o \in O*} \{d_o\} - \min_{o \in O*} \{r_o\} \quad \forall j \qquad (45)$$

where r_o and d_o are the release and due time of order o, and $O*$ is a subset of orders. The proposed pre-processing algorithm was applied in a set of instances studied in [19] and [20] reducing the computational effort by one order of magnitude.

The effectiveness of the proposed framework depends also on the "quality" of the integer cuts. Good integer cuts include only the binary variables that are *responsible* for the infeasibility of a solution, and it is usually very difficult to generate, mainly because the *source* of infeasibility is usually not *revealed* when the CP subproblem is found infeasible. Depending on the configuration of the plant and the algorithm that is used for the solution of the subproblem, it might be possible to derive effective cuts. In Example 2, we present how we used SBP to derive strong integer cuts, based on the fact that successive one-machine problems are solved. This allowed us to detect one-machine infeasible assignments, which is not always possible with CP.

7   Examples and Computational Results

We are currently testing the proposed framework for various plant configurations and objective functions. While the computational performance varies significantly and the hybrid approach does not always outperform other methods, we have found that if problem-specific information is exploited, through pre-processing and the generation of strong integer cuts, the proposed algorithm can be significantly faster than standalone MILP or CP models. In Example 1 we show how pre-processing can be used to eliminate solutions of the master problem, while in Example 2 we show the advantages of the integration of Mathematical Programming with a heuristic method. Example 3, finally, is a multipurpose batch plant with batch splitting and mixing.

7.1 Example 1: Minimization of Processing Cost in Single-stage Plant

Here we study the problem reported in Jain and Grossmann [19]: there are N jobs to be processed in M machines. The processing of job i can start after its release time r_i and must finish before its due time d_i. Job i can be processed in any machine j. The processing cost and the processing time of job i in machine j are C_{ij} and d_{ij} respectively, and the objective is to minimize the total processing cost. The original master problem consists of constraints (8) – (10) and (30) – (31). For this problem, specifically, we can develop a pre-processing algorithm that uses the release and due time data and generates a set of valid inequalities that are a priori added in the cut pool of the master problem.

To illustrate consider the example of Table 1, where three jobs 1, 2 and 3 have to be scheduled on two machines A and B. The processing time of all jobs in both machines is 2 hours, but the processing cost in machine A is much lower, and thus, the objective *favors* an assignment where all jobs are assigned to machine A, if feasible.

Constraint (31) for machine A reads:

$$2Z_{1A} + 2Z_{2A} + 2Z_{3A} \leq max~\{6,3,4\} - min~\{0,1,2\} \Rightarrow 2Z_{1A} + 2Z_{2A} + 2Z_{3A} \leq 6$$

Constraint (31), as well as all other constraints of (MP), is satisfied if $Z_{1A} = Z_{2A} = Z_{3A} = 1$, although, such an assignment is infeasible because job 2 has to start at $t=1$ which means that there is not enough time for job 1 to be performed before job 2 and there is not enough time for jobs 1 and 3 to be performed after job 2. This observation led us to develop a pre-processing algorithm that considers subsets of jobs and checks whether the jobs of these subsets can all be assigned on the same machine, using a knapsack inequality of the form of constraint (45). If the knapsack inequality is violated, the violated knapsack constraint or cover cuts ([21], [22]) of the violated knapsack constraint are added in the master problem (MP). In the example of Table 1, the knapsack inequality for the subset of jobs 2 and 3 reads:

$$2Z_{2A} + 2Z_{3A} \leq max~\{3,4\} - min~\{1,2\} \Rightarrow 2Z_{2A} + 2Z_{3A} \leq 4-1 = 3$$

Since the above constraint is violated, we can add the following cover inequality, which forbids the simultaneous assignment of jobs 2 and 3 in machine A:

$$Z_{2A} + Z_{3A} \leq 1$$

Table 1. Motivating Example 1

I	d_{iA}	d_{iB}	C_{iA}	C_{iB}	r_i	d_i
1	2	2	2	5	0	6
2	2	2	3	6	1	3
3	2	2	3	7	2	4

Table 2: Pre-processing algorithm – Phase I

forall $(i \in I,\ i' \in I|\ r_i \le r_{i'} \wedge d_i \le d_{i'})$
    $S = \text{union } (k \in I|\ r_i \le r_k \wedge d_k \le d_{i'})$
    forall $(j \in J)$
       if $(\Sigma_{k \in S}\ d_{kj} > d_{i'} - r_i)$ then
          Add Knapsack $:= \Sigma_{k \in S}\ d_{kj}\ Z_{kj} \le d_{i'} - r_i$
          or add Cover $:= \Sigma_{k \in S}\ Z_{kj} \le |S| - 1$

A straightforward implementation of a pre-processing algorithm that generates some of the violated knapsack constraints or cover cuts is presented in Table 2. While a more efficient algorithm that takes into account the ordering of tasks by ascending release time and descending due time can be developed, the computational time required for this step is negligible and thus we implemented as is. The constraints generated are a priori added in the cut pool of (MP). The algorithm can be further enhanced if all cover inequalities of a given knapsack are generated.

Another interesting case is the one illustrated through the example of Table 3.

In this example, the knapsack constraints for sets $\{1,2,3\}$, $\{1,2\}$, $\{1,3\}$ and $\{2,3\}$ are the following:

$$2Z_{1A} + 2Z_{2A} + 2Z_{3A} \le max\ \{4,3,6\} - min\ \{0,1,3\} = 6 \tag{46}$$

$$2Z_{1A} + 2Z_{2A} \le max\ \{4,3\} - min\ \{0,1\} \Rightarrow 2Z_{1A} + 2Z_{2A} \le 4 \tag{47}$$

$$2Z_{1A} + 2Z_{3A} \le max\ \{4,6\} - min\ \{0,3\} \Rightarrow 2Z_{1A} + 2Z_{3A} \le 6 \tag{48}$$

$$2Z_{2A} + 2Z_{3A} \le max\ \{3,6\} - min\ \{1,3\} \Rightarrow 2Z_{2A} + 2Z_{3A} \le 5 \tag{49}$$

All knapsacks are satisfied if all binaries are 1, which means that none of these assignments can be excluded by the pre-processing described in Table 2. However, in any assignment job 2 must start at $t=1$ and finish at $t=3$, which means that job 1 cannot be assigned in the same machine as job 2 because its due time is at $t=4$. This assignment could have been excluded if we had adjusted the RHS of (47) to account for the fact that both the maximum due time and the minimum release time for subset $\{1,2\}$ correspond to the same job, namely job 1. Whenever this is the case, the RHS must be adjusted as follows:

$$\sum_{i \in S} d_{ij} Z_{ij} \le d_{i*} - r_{i*} - min\{d_{i*} - max_{i \in S \setminus \{i*\}}, min_{i \in S \setminus \{i*\}} - r_{i*}\} \tag{50}$$

where i* is the task in subset S with the smallest release and largest due time.

Table 3. Motivating Example 2

i	d_{iA}	d_{iB}	C_{iA}	C_{iB}	r_i	d_i
1	2	2	2	5	0	4
2	2	2	3	6	1	3
3	2	2	3	7	3	6

Table 4. Pre-processing algorithm – Phase II

---

forall ($i \in I$)
$\quad$ S = union ($k \in I |\ r_i \leq r_k \wedge d_k \leq d_i$)
$\quad$ forall ($j \in J$)
$\qquad A := min_{k \in S \setminus \{i\}}\{r_k\} - r_i\ ;\quad B := d_i - max_{k \in S \setminus \{i\}}\{d_k\}$
$\qquad C := min\{A, B\}$
$\qquad$ if ($\Sigma_{k \in S}\ d_{kj} > d_{i'} - r_i$) then
$\qquad\qquad$ Add Knapsack := $\Sigma_{k \in S}\ d_{kj}\ Z_{kj} \leq d_{i'} - r_i - C$
$\qquad\qquad$ or add Cover := $\Sigma_{k \in S}\ Z_{kj} \leq |S| - 1$

For the subset {1,2}, constraint (50) gives knapsack inequality (51), which in turn gives the cover cut (52). Both inequalities exclude the infeasible assignment $Z_{1A} = 1$ AND $Z_{2A} = 1$.

$$2Z_{1A} + 2Z_{2A} \leq 4 - 0 - min\{4 - max\{3\}, min\{1\} - 0\} = 4 - 1 \Rightarrow 2Z_{1A} + 2Z_{2A} \leq 3 \qquad (51)$$

$$Z_{1A} + Z_{2A} \leq 1 \qquad (52)$$

The pre-processing routine that generates cuts for the exclusion of infeasible assignments that exhibit the feature described above is given in Table 4 (Phase II). The two phases of the preprocessing and the iterative hybrid MILP/CP algorithm were implemented in OPL Studio 3.6, on a PIII PC at 1GHz. For the solution of the master MILP model (MP) we used CPLEX 8.0.

The proposed method was tested in a set of problems by Jain and Grossmann ([19]). The authors showed that standalone MILP and CP models are not computationally efficient and proposed an iterative MILP/CP algorithm. Bockmayr and Pisaruk [20] proposed a branch-and-cut scheme where CP is used for the derivation of integer cuts. Computational results of standalone MILP and CP approaches, of the MILP/CP hybrid schemes of Jain and Grossmann (J & G) and Bockmayr and Pisaruk (B & P) and the proposed approach are given in Table 5.

As shown, all hybrid schemes are more efficient that the standalone MILP and CP models. Using the proposed pre-processing we were able to solve all problems in less than three CPU seconds. Note that the pre-processing algorithm generates a large number of integer cuts that are added in the cut pool of the master problem, and thus very few iterations are needed.

Table 5. Computational Results

| Problem | | Obj | MILP | CP | J & G | | B & P | | Proposed Approach | | |
M	N		Time	Time	Iter's	Time	Nodes	Time	Iter's	Cuts	Time
3	12	101	220	3.8	31	12.7	78	0.5	2	108	0.6
3	12	83	1.8	0.4	1	0.5	149	0.9	1	20	0.2
5	15	115	180.4	553.5	18	5.1	144	0.7	3	303	1.5
5	15	102	61.8	9.3	1	4.3	177	0.9	1	113	0.2
5	20	158	>20000	68854	31	41.5	924	4.0	2	581	2.7
5	20	140	106.3	2673.9	6	162	38303	245.8	1	302	0.3

7.2 Example 2: Minimization of Processing Cost in Multi-stage Plant

In the multi-stage problem addressed in [23] there are N orders with release and due times to be processed in K stages, where each stage k has M_k machines. The assignment of a job to a machine has a processing cost and a processing time and the objective is to find the assignment with the minimum cost that satisfies the release and due times of jobs, subject to forbidden job-machine assignments and production paths. For the solution of this problem the authors proposed a hybrid MILP/CP scheme. While the decomposition into a master MILP and a subproblem CP model has the advantages discussed above, in this specific problem it was not clear how to develop effective integer cuts. The results with only the "no-good" cuts were poor, and the authors had to relax the CP subproblem (i.e. allow for violation of the due dates) and use the information about the late orders to develop two classes of heuristic cuts that enhanced the performance of their algorithm.

It became clear, thus, that having an efficient algorithm for the solution of the flow-shop subproblem with fixed assignments which can additionally provide us with information that can be used for the generation of strong cuts would be very useful. Such an algorithm is the SBP algorithm because in its first stage checks whether a feasible schedule can be obtained given the assignments for each single machine; i.e. it detects infeasibilities that are due to the assignments on a single machine. Thus, if infeasibility is detected at this stage, we can add a strong integer cut that forbids the current assignment on this machine. The MILP/SBP integration was implemented in Mosel 1.2, using XPRESS-MP 14.2 for the solution of the master MILP and the SBP for the solution of the subproblem, on a 1GHz Pentium III. The computational results of standalone MILP and CP models, of the MILP/CP hybrid scheme of Harjunkoski and Grossmann ([23]) and the proposed MILP/SBP heuristic are shown in Table 6. Note that the SBP is a heuristic algorithm for the solution of job-shop problems, which means that a feasible partial solution of the master problem may be found infeasible by the SBP. In such a case, the iterative algorithm will not terminate when the optimal solution is found, yielding a suboptimal solution. Thus, the proposed scheme is a heuristic. However, note that for all ten instances the proposed scheme obtained the optimal solution.

Table 6. Computational Results for Multi-stage Plants

	Obj	MIP CPU s	CP CPU s	MIP/CP: CUT H Iter's/Cuts	CPU s	MIP/SBH Iter's/Cuts	CPU s
P1D1	39	0.1	0.1	5/4	0.1	2/1	0.2
P1D2	112	0.1	0.1	3/2	0.1	2/1	0.2
P2D1	153	4.4	0.1	16/19	0.7	4/4	1.3
P2D2	188	0.8	0.3	4/3	0.1	1/0	0.2
P3D1	56	446.6	4.2	29/45	8.7	26/32	16.9
P3D2	1113	0.4	1375.0	5/5	0.4	2/2	1.1
P4D1	149	27.3	447.2	17/22	28.8	12/15	7.8
P4D2	946	1.4	359.0	20/26	11.0	6/14	9.8
P5D1	111	4041.7	293.0	43/56	318.4	19/25	17.2
P5D2	704	2767.8	712.6	2/2	13.7	2/1	0.4

While the computational times are not directly comparable due to differences in the software and hardware, note that the MILP/SBP integration requires fewer iterations and cuts. As shown, when additional integer cuts are used, problems that were not solvable in 1,000 CPU seconds are solved in 8 to 320 seconds, and the number of iterations is significantly reduced.

7.3 Example 3: Minimization of Makespan in Multi-purpose Plant

Finally, we present computational results from Maravelias and Grossmann [2] for the batch plant shown in Figure 3. The objective is to find the schedule of minimum makespan for the production of 5 tons of products P1, P2, P3 and P4. When formulated as a continuous time MILP model [2] with 10 time points, this problem was intractable as it could not be solved in 10 hours of CPU-time with CPLEX7.5. With the proposed hybrid scheme, assuming that we can have at most 4 copies of each task, the optimal solution of 15 hours is found in 5 iterations. The assignment that gives the optimal solution is found in the first iteration with a lower bound of 14 hours. Successive feasibility CP problems are solved for this assignment, and a feasible schedule with a makespan of 15 hours is found in the second subproblem. The subsequent master problems give solutions with a lower bound on the makespan equal to 14 hours, but none of these assignments yields a feasible schedule with makespan shorter than 15 hours. The fifth MILP is infeasible, which means that there are no more assignments that can meet the given demand. The total computational time is 1.80 CPU seconds, from which 0.03 seconds are spent in preprocessing, 0.70 seconds are spent for the master problem (approximately 0.14 sec for each MILP), and 1.07 seconds are spent for all the CP subproblems.

Generally, the modeling of multipurpose batch plants is a complex task, and the solution of the resulting MILP models is hard. Existing models are moderately effective when the objective function is the maximization of profit or production over a fixed time horizon, but very slow when the objective is the minimization of makespan. To our experience, the proposed hybrid scheme seems to be moderately effective for the maximization of profit and very effective for the minimization of makespan, enabling us to solve problems that were previously unsolvable.

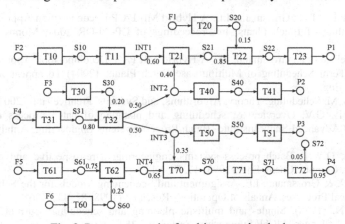

Fig. 3. Process network of multipurpose batch plant

8   Conclusions

A general decomposition framework, which uses Mathematical and Constraint Programming techniques, for the solution of scheduling problems in chemical industry is presented. The number and type of tasks performed, as well as the assignment of units to tasks are determined by the master MILP model, while the CP subproblem is used to check feasibility and derive complete schedules. The advantage of the proposed framework is that it can be readily applied to many classes of problems. Furthermore, the underlying decomposition idea can also be used to integrate Mathematical Programming with scheduling algorithms other than CP.

The computational efficiency of the proposed model varies significantly. There is good evidence, however, that for some classes of problems it outperforms existing methods. In general, if the structure of the problem at hand is exploited by efficient preprocessing and the generation of strong integer cuts, it is expected that hybrid schemes will be more effective because they combine the complementary strengths of two solution techniques. Finally, we showed how the special structure or characteristics of a problem can be exploited by simple additions (pre-processing in Example 1) and modifications (solution of subproblem using SBP in Example 2) of the proposed scheme. Furthermore, in Example 3 the proposed hybrid scheme was able to solve in few seconds a complex scheduling problem that proved to be intractable when solved as an MILP problem.

Acknowledgements

We would like to thank Dr. Alkis Vazacopoulos for making the SBP code available to us, and his helpful comments on how to generate cuts from the SBP. The authors would also like to acknowledge financial support from the National Science Foundation under Grant ACI-0121497.

References

[1]   Maravelias, T.C.; Grossmann, I.E. A Hybrid MILP/CP Decomposition Approach for the Scheduling of Batch Plants. In Proceedings of CP-AI-OR 2003, Montreal, Canada (2003).

[2]   Maravelias, C.T.; Grossmann, I.E. A Hybrid MILP/CP Decomposition Approach for the Short Term Scheduling of Multipurpose Batch Plants. (2004) To appear in Comput. Chem. Eng.

[3]   Pinedo, M. Scheduling: Theory, Algorithms, and Systems. Prentice Hall (2001).

[4]   Reklaitis, G.V. Overview of scheduling and planning of batch process operations, NATO Advanced Study Institute- Batch Process Systems Engineering, Antalya, Turkey (1992).

[5]   Rippin, D.W.T. Batch process systems engineering: a retrospective and prospective review. Comput. Chem. Eng. (1993) 17, S1-S13.

[6]   Pinto, J. & Grossmann, I.E. Assignment and Sequencing Models for the Scheduling of Chemical Processes. Annals of Operations Research (1998) 81, 433-466.

[7]   Shah, N. (1998). Single- and multisite planning and scheduling: Current status and future challenges. AIChE Symposium Series, 94, 75.

[8]   Kondili, E.; Pantelides, C. C.; Sargent, R. A General Algorithm for Short-Term Scheduling of Batch Operations – I. MILP Formulation. Comput. Chem. Eng. (1993) 17, 211-227.

[9]   Pantelides, C. C. Unified Frameworks for the Optimal Process Planning and Scheduling. In Proceedings on the Second Conference on Foundations of Computer Aided Operations. (1994) 253-274.

[10]   Schilling, G.; Pantelides, C. C. A Simple Continuous-Time Process Scheduling Formulation and a Novel Solution Algorithm. Comput. Chem. Eng. (1996) 20, S1221-1226.

[11]   Zhang, X.; Sargent, R. W. H. The Optimal Operation of Mixed Production Facilities – General Formulation and Some Approaches for the Solution. Comput. Chem. Eng. (1996) 20, 897-904.

[12]   Mockus, L.; Reklaitis, G.V. Continuous Time Representation Approach to Batch and Continuous Process Scheduling. 1. MINLP Formulation. Ind. Eng. Chem. Res. (1999) 38, 197-203.

[13]   Castro, P.; Barbosa-Povoa, A. P. F. D.; Matos, H. An Improved RTN Continuous-Time Formulation for the Short-term Scheduling of Multipurpose Batch Plants. Ind. Eng. Chem. Res. (2001) 40, 2059-2068.

[14]   Maravelias, C.T.; Grossmann, I.E. A New General Continuous-Time State Task Network Formulation for the Short-Term Scheduling of Multipurpose Batch Plants. Ind. Eng. Chem. Res. (2003) 42 (13), 3056-3074.

[15]   ILOG OPL Studio 3.5: The Optimization Language, ILOG Inc. (2001).

[16]   ILOG OPL Studio 3.5: The User's Manual, ILOG Inc. (2001).

[17]   Adams, J.; Balas, E.; Zawack, D. The Shifting Bottleneck Procedure for Job Shop Scheduling. Management Science (1988) 34, 391-401.

[18]   Balas, E.; Vazacopoulos,A. Guided Local Search with Shifting Bottleneck for Job Shop Scheduling, Management Science (1998) 44(2), 262-275.

[19]   Jain, V; Grossmann, I. E. Resource-constrained Scheduling of Tests in New Product Development. Ind. Eng. Chem. Res. (1999) 38, 3013-3026.

[20]   Bockmayr, A.; Pisaruk, N. Detecting Infeasibility and Generating Cuts for MIP Using CP. In Proceedings of CP-AI-OR 2003, Montreal, Canada (2003).

[21]   Balas, E. Facets of the Knapsack Polytope. Mathematical Programming (1975) 8, 146-164.

[22]   Wolsey, L.A. Faces for a Linear Inequality in 0-1 Variables. Mathematical Programming (1975) 8, 165-178.

[23]   Harjunkoski, I.; Grossmann, I.E. Decomposition Techniques for Multistage Scheduling Problems Using Mixed-Integer and Constrained Programming Methods. Comput. Chem. Eng. (2002) 26, 1533-1552.

SIMPL: A System for Integrating Optimization Techniques*

Ionuţ Aron, John N. Hooker, and Tallys H. Yunes

Graduate School of Industrial Administration
Carnegie Mellon University, Pittsburgh, PA 15213-3890, USA
{iaron,jh38,thy}@andrew.cmu.edu

Abstract. In recent years, the Constraint Programming (CP) and Operations Research (OR) communities have explored the advantages of combining CP and OR techniques to formulate and solve combinatorial optimization problems. These advantages include a more versatile modeling framework and the ability to combine complementary strengths of the two solution technologies. This research has reached a stage at which further development would benefit from a general-purpose modeling and solution system. We introduce here a system for integrated modeling and solution called SIMPL. Our approach is to view CP and OR techniques as special cases of a single method rather than as separate methods to be combined. This overarching method consists of an infer-relax-restrict cycle in which CP and OR techniques may interact at any stage. We describe the main features of SIMPL and illustrate its usage with examples.

1 Introduction

In recent years, the Constraint Programming (CP) and Operations Research (OR) communities have explored the advantages of combining CP and OR techniques to formulate and solve combinatorial optimization problems. These advantages include a more versatile modeling framework and the ability to combine complementary strengths of the two solution technologies. Examples of existing programming languages that provide mechanisms for combining CP and OR techniques are ECL^{i}PS^e [,], OPL [ ] and Mosel [ ].

Hybrid methods tend to be most effective when CP and OR techniques interact closely at the micro level throughout the search process. To achieve this one must often write special-purpose code, which slows research and discourages broader application of integrated methods. We address this situation by introducing here a system for integrated modeling and solution called SIMPL (Programming Language for Solving Integrated Models). The SIMPL modeling language formulates problems in such a way as to reveal problem structure to the solver. The solver executes a search algorithm that invokes CP and OR techniques as needed, based on problem characteristics.

---

* This work has been supported by the National Science Foundation under grant ACI-0121497 and by the William Larimer Mellon Fellowship.

J.-C. Régin and M. Rueher (Eds.): CPAIOR 2004, LNCS 3011, pp. 21– , 2004.

The design of such a system presents a significant research problem in itself, since it must be flexible enough to accommodate a wide range of integration methods and yet structured enough to allow high-level implementation of specific applications. Our approach, which is based partly on a proposal in [,], is to view CP and OR techniques as special cases of a single method rather than as separate methods to be combined. This overarching method consists of an infer-relax-restrict cycle in which CP and OR techniques may interact at any stage.

This paper is organized as follows. In Sect. , we briefly review some of the fundamental ideas related to the combination of CP and OR that are relevant to the development of SIMPL. We describe the main concepts behind SIMPL in Sect.   and talk about implementation details in Sect. . Section   presents a few examples of how to model optimization problems in SIMPL, explaining the syntax and semantics of the language. Finally, Sect.   outlines some additional features provided by SIMPL, and Sect.   discusses directions for future work.

2   Previous Work

A comprehensive survey of the literature on the cooperation of logic-based, Constraint Programming (CP) and Operations Research (OR) methods can be found in [ ]. Some of the concepts that are most relevant to the work presented here are: decomposition approaches (e.g. Benders [ ]) that solve parts of the problem with different techniques [, , , , ,]; allowing different models/solvers to exchange information [ ]; using linear programming to reduce the domains of variables or to fix them to certain values [, ,]; automatic reformulation of global constraints as systems of linear inequalities [ ]; continuous relaxations of global constraints and disjunctions of linear systems [, , , , , , ,]; understanding the generation of cutting planes as a form of logical inference [,]; strengthening the problem formulation by embedding the generation of valid cutting planes into CP constraints [ ]; maintaining the continuous relaxation of a constraint updated when the domains of its variables change [ ]; and using global constraints as a key component in the intersection of CP and OR [ ].

Ideally, one would like to incorporate all of the above techniques into a single modeling and solving environment, in a clean and generic way. Additionally, this environment should be flexible enough to accommodate improvements and modifications with as little extra work as possible. In the next sections, we present the concepts behind SIMPL that aim at achieving those objectives.

3   SIMPL Concepts

We first review the underlying solution algorithm and then indicate how the problem formulation helps to determine how particular problems are solved.

3.1   The Solver

SIMPL solves problems by enumerating problem restrictions. (A restriction is the result of adding constraints to the problem.) Each node of a classical branch-

and-bound tree, for example, can be viewed as a problem restriction defined by fixing certain variables or reducing their domains. Local search methods fit into the same scheme, since they examine a sequence of neighborhoods, each of which is the feasible set of a problem restriction. Thus SIMPL implements both exact and heuristic methods within the same architecture.

The search proceeds by looping through an infer-relax-restrict cycle: it infers new constraints from the current problem restriction, then formulates and solves relaxations of the augmented problem restriction, and finally moves to another problem restriction to be processed in the same way. The user specifies the overall search procedure from a number of options, such as depth-first branching, local search, or Benders decomposition. The stages in greater detail are as follows.

Infer. New constraints are deduced from the original ones and added to the current problem restriction. For instance, a filtering algorithm can be viewed as inferring indomain constraints that reduce the size of variable domains. A cutting plane algorithm can generate inequality constraints that tighten the continuous relaxation of the problem as well as enhance interval propagation.

Relax. One or more relaxations of the current problem restriction are formulated and solved by specialized solvers. For instance, continuous relaxations of some or all of the constraints can be collected to form a relaxation of the entire problem, which is solved by a linear or nonlinear programming subroutine. The role of relaxations is to help direct the search, as described in the next step.

Restrict. The relaxations provide information that dictates which new restrictions are generated before moving to the next restriction. In a tree search, for example, SIMPL creates new restrictions by branching on a constraint that is violated by the solution of the current relaxation. If several constraints are violated, one is selected according to user- or system-specified priorities (see Sect.   ). Relaxations can also influence which restriction is processed next, for instance by providing a bound that prunes a branch-and-bound tree.

If desired, an inner infer-relax loop can be executed repeatedly before moving to the next problem restriction, since the solution of the relaxation may indicate further useful inferences that can be drawn (post-relaxation inference). An example would be separating cuts, which are cutting planes that "cut off" the solution of the relaxation (see Sect.   ).

The best-known classical solution methods are special cases of the infer-relax-restrict procedure:

- In a typical *CP solver*, the inference stage consists primarily of domain reduction. The relaxation stage builds a (weak) relaxation simply by collecting the reduced domains into a constraint store. New problem restrictions are created by splitting a domain in the relaxation.
- In a *branch-and-bound solver* for integer programming, the inference stage can be viewed as "preprocessing" that takes place at the root node and possibly at subsequent nodes. The relaxation stage drops the integrality constraints and solves the resulting problem with a linear or perhaps nonlinear

programming solver. New problem restrictions are created by branching on an integrality constraint; that is, by branching on a variable with a fractional value in the solution of the relaxation.

- A *local search* procedure typically chooses the next solution to be examined from a neighborhood of the current solution. Thus local search can be regarded as enumerating a sequence of problem restrictions, since each neighborhood is the feasible set of a problem restriction. The "relaxation" of the problem restriction is normally the problem restriction itself, but need not be. The restriction may be solved to optimality by an exhaustive search of the neighborhood, as in tabu search (where the tabu list is part of the restriction). Alternatively, a suboptimal solution may suffice, as in simulated annealing, which selects a random element of the neighborhood.
- In *Branch-and-Infer* [ ], the relaxation stage is not present and branching corresponds to creating new problem restrictions.

An important advantage of SIMPL is that it can create new infer-relax-restrict procedures that suit the problem at hand. One example is a hybrid algorithm, introduced in [,   ], that is obtained through a generalization of Benders decomposition. It has provided some of the most impressive speedups achieved by hybrid methods [,   ,   ,   ,   ]. A Benders algorithm distinguishes a set of primary variables that, when fixed, result in an easily-solved subproblem. Solution of an "inference dual" of the subproblem yields a Benders cut, which is added to a master problem containing only the primary variables. Solution of the master problem fixes the primary variables to another value, and the process continues until the optimal values of the master problem and subproblem converge. In typical applications, the master problem is an integer programming problem and the subproblem a CP problem. This method fits nicely into the infer-relax-restrict paradigm, since the subproblems are problem restrictions and master problems are relaxations. The solution of the relaxation guides the search by defining the next subproblem.

The choice of constraints in a SIMPL model can result in novel combinations of CP, OR and other techniques. This is accomplished as described in Sect.   .

3.2   Modeling

SIMPL is designed so that the problem formulation itself determines to a large extent how CP, OR, and other techniques interact. The basic idea is to view each constraint as invoking specialized procedures that exploit the structure of that particular constraint. Since some of these procedures may be from CP and some from OR, the two approaches interact in a manner that is dictated by which constraints appear in the problem.

This idea of associating constraints with procedures already serves as a powerful device for exploiting problem substructure in CP, where a constraint typically activates a specialized filtering algorithm. SIMPL extends the idea by associating each constraint with procedures in all three stages of the search. Each constraint can (a) activate inference procedures, (b) contribute constraints to one or more relaxations, and (c) generate further problem restrictions if the search branches on that particular constraint.

If a group of constraints exhibit a common structure—such as a set of linear inequalities, flow balance equations, or logical formulas in conjunctive normal form—they are identified as such so that the solver can take advantage of their structure. For instance, a resolution-based inference procedure might be applied to the logical formulas.

The existing CP literature typically provides inference procedures (filters) only for CP-style global constraints, and the OR literature provides relaxations (cutting planes) only for structured groups of linear inequalities. This poses the research problem of finding specialized relaxations for global constraints and specialized filters for structured linear systems. Some initial results along this line are surveyed in [ ].

Some examples should clarify these ideas. The global constraint *element* is important for implementing variable indices. Conventional CP solvers associate *element* with a specialized filtering algorithm, but useful linear relaxations, based on OR-style polyhedral analysis, have recently been proposed as well [ ]. Thus each *element* constraint can activate a domain reduction algorithm in the inference stage and generate linear inequalities, for addition to a continuous relaxation, in the relaxation stage. If the search branches on a violated *element* constraint, then new problem restrictions are generated in a way that makes sense when that particular constraint is violated.

The popular *all-different* and *cumulative* constraints are similar in that they also have well-known filters [,] and were recently provided with linear relaxations [,]. These relaxations are somewhat weak and may not be useful, but the user always has the option of turning off or on the available filters and relaxations, perhaps depending on the current depth in the search tree.

Extensive polyhedral analysis of the traveling salesman problem in the OR literature [,] provides an effective linear relaxation of the *cycle* constraint. In fact, SIMPL has the potential to make better use of the traditional OR literature than commercial OR solvers. Structured groups of inequalities can be represented by global constraints that trigger the generation of specialized cutting planes, many of which go unused in today's general-purpose solvers.

4   From Concepts to Implementation

SIMPL is implemented in C++ as a collection of object classes, as shown in Fig. .

This makes it easy to add new components to the system by making only localized changes that are transparent to the other components. Examples of components that can be included are: new constraints, different relaxations for existing constraints, new solvers, improved inference algorithms, new branching modules and selection modules, alternative representations of domains of variables, etc. The next sections describe some of these components in detail.

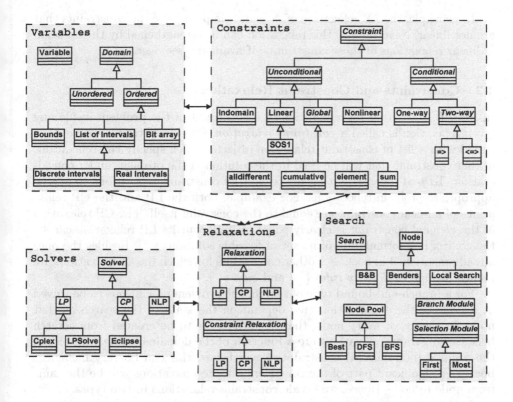

Fig. 1. Main components of SIMPL

4.1 Multiple Problem Relaxations

Each iteration in the solution of an optimization problem P examines a *restriction* N of P. In a tree search, for example, N is the problem restriction at the current node of the tree. Since solving N can be hard, we usually solve a *relaxation* N_R of N, or possibly several relaxations.

In an integrated CP-IP modeling system, the linear constraints in the hybrid formulation are posted to a Linear Programming (LP) solver, and some (or all) of them may be posted to a CP solver as well. The CP solver also handles the constraints that cannot be directly posted to the LP solver (e.g. global constraints). Notice that each solver only deals with a relaxation of the original problem P (i.e. a subset of its constraints). In this example, each problem restriction N has two relaxations: an *LP relaxation* and a *CP relaxation*. Extending this idea to more than two kinds of relaxations is straightforward.

---

[1] In general, we say that problem Q_R is a relaxation of problem Q if the feasible region of Q_R contains the feasible region of Q.

In principle, the LP relaxation of N could simply ignore the constraints that are not linear. Nevertheless, this relaxation can be strengthened by the addition of linear relaxations of those constraints, if available (see Sect.   ).

4.2   Constraints and Constraint Relaxations

In SIMPL, the actual representation of a constraint of the problem inside any given relaxation is called a *constraint relaxation*. Every constraint can be associated with a list of constraint relaxation objects, which specify the relaxations of that constraint that will be used in the solution of the problem under consideration. To *post* a constraint means to add its constraint relaxations to all the appropriate problem relaxations. For example, both the LP and the CP relaxations of a linear constraint are equal to the constraint itself. The CP relaxation of the *element* constraint is clearly equal to itself, but its LP relaxation can be the convex hull formulation of its set of feasible solutions [ ]. Besides the ones already mentioned in Sect.   , other constraints for which linear relaxations are known include *cardinality rules* [ ] and *sum* [ ].

For a branch-and-bound type of search, the problem relaxations to be solved at a node of the enumeration tree depend on the state of the search at that node. In theory, at every node, the relaxations are to be created from scratch because constraint relaxations are a function of the domains of the variables of the original (non-relaxed) constraint. Nevertheless, this can be very inefficient because a significant part of the constraints in the relaxations will be the same from node to node. Hence, we divide constraint relaxations in two types:

Static: those that change very little (in structure) when the domains of its variables change (e.g. relaxations of linear constraints are equal to themselves, perhaps with some variables removed due to fixing);

Volatile: those that radically change when variable domains change (e.g. some linear relaxations of global constraints).

To update the problem relaxations when we move from one node in the search tree to another, it suffices to recompute volatile constraint relaxations only. This kind of update is not necessary for the purpose of creating valid relaxations, but it is clearly beneficial from the viewpoint of obtaining stronger bounds.

```
procedure Search(A)
    If A ≠ ∅ and stopping criteria not met
        N := A.getNextNode()
        N.explore()
        A.addNodes(N.generateRestrictions())
        Search(A)
```

Fig. 2. The main search loop in SIMPL

```
1. Pre-relaxation inference
2. Repeat
3.     Solve relaxations
4.     Post-relaxation inference
5. Until (no changes) or (iteration limit)
```

Fig. 3. The node exploration loop in branch-and-bound

4.3   Search

The main search loop in SIMPL is implemented as shown in Fig. . Here, N is again the current problem restriction, and A is the current list of restrictions waiting to be processed. Depending on how A, N and their subroutines are defined, we can have different types of search, as mentioned in Sect.   . The routine N.explore() implements the infer-relax sequence. The routine N.generateRestrictions() creates new restrictions, and A.addNodes() adds them to A. Routine A.getNextNode() implements a mechanism for selecting the next restriction, such as depth-first, breadth-first or best bound.

In tree search, N is the problem restriction that corresponds to the current node, and A is the set of open nodes. In local search, N is the restriction that defines the current neighborhood, and A is the singleton containing the restriction that defines the next neighborhood to be searched. In Benders decomposition, N is the current subproblem and A is the singleton containing the next subproblem to be solved. In the case of Benders, the role of N.explore() is to infer Benders cuts from the current subproblem, add them to the master problem, and solve the master problem. N.generateRestrictions() uses the solution of the master problem to create the next subproblem.

In the sequel, we will restrict our attention to branch-and-bound search.

Node Exploration. Figure   describes the behavior of N.explore() for a branch-and-bound type of search. Steps 1 and 4 are inference steps where we try to use the information from each relaxation present in the model to the most profitable extent. Section   provides further details about the types of inference used in those steps. The whole loop can be repeated multiple times, as long as domains of variables keep changing because of step 4, and the maximum number of iterations has not been reached. This process of re-solving relaxations and looking for further inferences behaves similarly to a fix point calculation.

Branching. SIMPL implements a tree search by branching on constraints. This scheme is considerably more powerful and generic than branching on variables alone. If branching is needed, it is because some constraint of the problem is violated and that constraint should "know" what to do. This knowledge is embedded in the so called *branching module* of that constraint. For example, if a variable $x \in \{0,1\}$ has a fractional value in the current LP, its indomain constraint I_x is violated. The branching module of I_x will then output two constraints: $x \in \{0\}$ and $x \in \{1\}$, meaning that two subproblems should be created by the inclusion of those two new constraints. In this sense, branching on the variable x can be interpreted as branching on I_x. In general, a branching module returns a sequence of *sets* of constraints $C_1, \ldots, C_k$. This sequence means that k subproblems should be created, and subproblem i can be constructed from the current problem by the inclusion of all constraints present in the set C_i. There is no restriction on the types of constraints that can be part of the sets C_i.

Clearly, there may be more than one constraint violated by the solution of the current set of problem relaxations. A *selection module* is the entity responsible for selecting, from a given set of constraints, the one on which to branch next. Some possible criteria for selection are picking the first constraint found to be violated or the one with the largest degree of violation.

4.4   Inference

We now take a closer look at the inference steps of the node exploration loop in Fig. . In step 1 (pre-relaxation inference), one may have domain reductions or the generation of new implied constraints (see [ ]), which may have been triggered by the latest branching decisions. If the model includes a set of propositional logic formulas, this step can also execute some form of resolution algorithm to infer new resolvents. In step 4 (post-relaxation inference), other types of inference may take place, such as fixing variables by reduced cost or the generation of cutting planes. After that, it is possible to implement some kind of primal heuristic or to try extending the current solution to a feasible solution in a more formal way, as advocated in Sect. 9.1.3 of [ ].

Since post-relaxation domain reductions are associated with particular relaxations, the reduced domains that result are likely to differ across relaxations. Therefore, at the end of the inference steps, a synchronization step must be executed to propagate domain reductions across different relaxations. This is shown in Fig. . In step 6, D_v^r denotes the domain of v inside relaxation r, and D_v works as a temporary domain for variable v, where changes are centralized. The initial value of D_v is the current domain of variable v. By implementing the changes in the domains via the addition of indomain constraints (step 8), those changes will be transparently undone when the search moves to a different part of the enumeration tree. Similarly, those changes are guaranteed to be redone if the search returns to descendents of the current node at a later stage.

```
1.  V := ∅
2.  For each problem relaxation r
3.        V_r := variables with changed domains in r
4.        V := V ∪ V_r
5.        For each v ∈ V_r
6.              D_v := D_v ∩ D_v^r
7.  For each v ∈ V
8.        Post constraint v ∈ D_v
```

Fig. 4. Synchronizing domains of variables across multiple relaxations

5 SIMPL Examples

SIMPL's syntax is inspired by OPL [ ], but it includes many new features.

Apart from the resolution algorithm used in Sect.   , SIMPL is currently able to run all the examples presented in this section. Problem descriptions and formulations were taken from Chapter 2 of [ ].

5.1 A Hybrid Knapsack Problem

Let us consider the following integer knapsack problem with a side constraint.

$$\min \ 5x_1 + 8x_2 + 4x_3$$
$$\text{subject to } 3x_1 + 5x_2 + 2x_3 \geq 30$$
$$\text{all-different}(x_1, x_2, x_3)$$
$$x_j \in \{1, 2, 3, 4\}, \text{ for all } j$$

To handle the *all-different* constraint, a pure MIP model would need auxiliary binary variables: $y_{ij} = 1$ if and only if $x_i = j$. A SIMPL model for the above problem is shown in Fig. . The model starts with a DECLARATIONS section in which constants and variables are defined. The objective function is defined in line 06. Notice that the range over which the index i takes its values need not be explicitly stated. In the CONSTRAINTS section, the two constraints of the problem are named **totweight** and **distinct**, and their definitions show up in lines 09 and 12, respectively. The RELAXATION statements in lines 10 and 13 indicate the relaxations to which those constraints should be posted. The linear constraint will be present in both the LP and the CP relaxations, whereas the **alldiff** constraint will only be present in the CP relaxation. In the SEARCH section, line 15 indicates we will do branch-and-bound (BB) with depth-first search (DEPTH). The BRANCHING statement in line 16 says that we will branch on the first of the x variables that is not integer (remember from Sect.   that branching on a variable means branching on its indomain constraint).

```
01. DECLARATIONS
02.     n = 3; cost[1..n] = [5,8,4]; weight[1..n] = [3,5,2]; limit = 30;
03.     DISCRETE RANGE xRange = 1 TO 4;
04.     x[1..n] IN xRange;
05. OBJECTIVE
06.     MIN SUM i OF cost[i]*x[i]
07. CONSTRAINTS
08.     totweight MEANS {
09.         SUM i OF weight[i]*x[i] >= limit
10.         RELAXATION = {LP, CS} }
11.     distinct MEANS {
12.         alldiff(x)
13.         RELAXATION = {CS} }
14. SEARCH
15.     TYPE = {BB:DEPTH}
16.     BRANCHING = {x:FIRST}
```

Fig. 5. SIMPL model for the Hybrid Knapsack Problem

Initially, bounds consistency maintenance in the CP solver removes value 1 from the domain of x_2 and the solution of the LP relaxation is $x = (2\frac{2}{3}, 4, 1)$. After branching on $x_1 \leq 2$, bounds consistency determines that $x_1 \geq 2$, $x_2 \geq 4$ and $x_3 \geq 2$. At this point, the `alldiff` constraint produces further domain reduction, yielding the feasible solution $(2, 4, 3)$. Notice that no LP relaxation had to be solved at this node. In a similar fashion, the CP solver may be able to detect infeasibility even before the linear relaxation has to be solved.

5.2   A Lot Sizing Problem

A total of P products must be manufactured over T days on a single machine of limited capacity C, at most one product each day. When manufacture of a given product i begins, it may proceed for several days, and there is a minimum run length r_i. Given a demand d_{it} for each product i on each day t, it is usually necessary to hold part of the production of a day for later use, at a unit cost of h_{it}. Changing from product i to product j implies a setup cost q_{ij}. Frequent changeovers allow for less holding cost but incur more setup cost. The objective is to minimize the sum of the two types of costs while satisfying demand.

Let $y_t = i$ if and only if product i is chosen to be produced on day t, and let x_{it} be the quantity of product i produced on day t. In addition, let u_t, v_t and s_{it} represent, respectively, for day t, the holding cost, the changeover cost and the ending stock of product i. Figure   exhibits a SIMPL model for this problem. We have omitted the data that initializes matrices d, h, q and r. We have also left out the statements that set $y_0 = 0$ and $s_{i0} = 0$ for $i \in \{1, \ldots, P\}$.

In line 07, we use the predefined continuous range **nonegative**. Notice the presence of a new section called RELAXATIONS, whose role in this example is to define the default relaxations to be used. As a consequence, the absence of

```
01. DECLARATIONS
02.     P = 5; T = 10; C = 50;
03.     d[1..P,1..T] = ; h[1..P,1..T] = ; q[0..P,1..P] = ; r[1..P] = ;
04.     CONTINUOUS RANGE xRange = 0 TO C;
05.     DISCRETE RANGE yRange = 0 TO P;
06.     x[1..P,1..T] IN xRange; y[0..T] IN yRange;
07.     u[1..T], v[1..T], s[1..P,0..T] IN nonegative;
08. OBJECTIVE
09.     MIN SUM t OF u[t] + v[t]
10. RELAXATIONS
11.     LP, CS
12. CONSTRAINTS
13.     holding MEANS { u[t] >= SUM i OF h[i,t]*s[i,t] FORALL t }
14.     setup MEANS { v[t] >= q[y[t-1],y[t]] FORALL t }
15.     stock MEANS { s[i,t-1] + x[i,t] = d[i,t] + s[i,t] FORALL i,t }
16.     linkyx MEANS { y[t] <> i -> x[i,t] = 0 FORALL i,t }
17.     minrun MEANS {
18.         y[t-1] <> i and y[t] = i ->
19.         (y[t+k] = i FORALL k IN 1 TO r[i]-1) FORALL i,t }
20. SEARCH
21.     TYPE = {BB:BEST}
22.     BRANCHING = {setup:MOST}
```

Fig. 6. SIMPL model for the Lot Sizing Problem

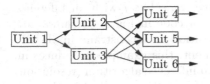

```
build MEANS {
  y1 => (y2 or y3), y3 => y4, y2 => y1,
  y3 => (y5 or y6), y2 => (y4 or y5),
  y4 => (y2 or y3), y2 => y6, y3 => y1,
  y5 => (y2 or y3), y6 => (y2 or y3)
  RELAXATION = {LP, CS}
  INFERENCE = {RESOLUTION} }
```

Fig. 7. (a) Network superstructure   (b) The INFERENCE statement in SIMPL

RELAXATION statements in the declaration of constraints means that all constraints will be posted to both the LP and CS relaxations. The **holding** and **stock** constraints define, respectively, holding costs and stock levels in the usual way. The **setup** constrains make use of variable indexing to obtain the desired meaning for the v_t variables. The CS relaxation of these constraints uses *element* constraints, and the LP relaxation uses the corresponding linear relaxation of *element*. The symbol -> in lines 16 and 18 implements a one-way link constraint of the form $A \to B$ (see [ ]). This means that whenever condition A is true, B is imposed as a constraint of the model, but we do not worry about the contrapositive. Condition A may be a more complicated logical statement and B can be any collection of arbitrary constraints. There are also two-way link constraints such as "implies" (=>) and "if and only if" (<=>) available in SIMPL. Here, the **linkyx** constraints ensure that x_{it} can only be positive if $y_t = i$, and the **minrun** constraints make production last the required minimum length. The statements in lines 21 and 22 define a branch-and-bound search with best-bound node selection, and branching on the most violated of the **setup** constraints, respectively.

5.3   Processing Network Design

This problem consists of designing a chemical processing network. In practice one usually starts with a network that contains all the processing units and connecting links that could eventually be built (i.e. a superstructure). The goal is to select a subset of units that deliver the required outputs while minimizing installation and processing costs. The discrete element of the problem is the choice of units, and the continuous element comes in determining the volume of flow between units. Let us consider the simplified superstructure in Fig. (a). Unit 1 receives raw material, and units 4, 5 and 6 generate finished products. The output of unit 1 is then processed by unit 2 and/or 3, and their outputs undergo further processing. For the purposes of this example, we will concentrate on the selection of units, which is amenable to the following type of logical reasoning. Let the propositional variable y_i be true when unit i is installed and false otherwise. From Fig. (a), it is clearly useless to install unit 1 unless one installs unit 2 or unit 3. This condition can be written as $y_1 \Rightarrow (y_2 \vee y_3)$. Other rules of this kind can be derived in a similar way. SIMPL can take advantage of the presence of such rules in three ways: it can relax logical propositions into linear constraints; it can use the propositions *individually* as two-way link constraints (see Sect.   ); and it can use the propositions *collectively* with an inference algorithm to deduce stronger facts. The piece of code in Fig. (b) shows how one would group this collection of logical propositions as a constraint in SIMPL. In addition to the known RELAXATION statement, this example introduces an INFERENCE statement whose role is to attach an inference algorithm (resolution) to the given group of constraints. This algorithm will be invoked in the pre-relaxation inference step, as described in Sect.   . Newly inferred resolvents can be added to the problem relaxations and may help the solution process.

5.4   Benders Decomposition

Recall from Sect.   that Benders decomposition is a special case of SIMPL's search mechanism. Syntatically, to implement Benders decomposition the user only needs to include the keyword MASTER in the RELAXATION statement of each constraint that is meant to be part of the master problem (remaining constraints go to the subproblem), and declare TYPE = {BENDERS} in the SEARCH section. As is done for linear relaxations of global constraints, Benders cuts are generated by an algorithm that resides inside each individual constraint. At present, we are in the process of implementing the class *Benders* in the diagram of Fig. .

6   Other SIMPL Features

Supported Solvers. Currently, SIMPL can interface with CPLEX [ ] and LP_SOLVE [ ] as LP solvers, and with ECL^{i}PS^e [ ] as a CP solver. Adding a new solver to SIMPL is an easy task and amounts to implementing an interface to that solver's callable library, as usual. The rest of the system does not need to be changed or recompiled. One of the next steps in the development of SIMPL is the inclusion of a solver to handle non-linear constraints.

Application Programming Interface. Although SIMPL is currently a purely declarative language, it will eventually include more powerful (imperative) search constructs, such as loops and conditional statements. Meanwhile, it is possible to implement more elaborate algorithms that take advantage of SIMPL's paradigm via its Application Programming Interface (API). This API can be compiled into any customized C++ code and works similarly to other callable libraries available for commercial solvers like CPLEX or XPRESS [ ].

Search Tree Visualization. Once a SIMPL model finishes running, it is possible to visualize the search tree by using Leipert's VBC Tool package [ ]. Nodes in the tree are colored red, green, black and blue to mean, respectively, pruned by infeasibility, pruned by local optimality, pruned by bound and branched on.

7   Conclusions and Future Work

In this paper we introduce a system for dealing with integrated models called SIMPL. The main contribution of SIMPL is to provide a user-friendly framework that generalizes many of the ways of combining Constraint Programming (CP) and Operations Research (OR) techniques when solving optimization problems. Although there exist other general-purpose systems that offer some form of hybrid modeling and solver cooperation, they do not incorporate various important features available in SIMPL.

The implementation of specialized hybrid algorithms can be a very cumbersome task. It often involves getting acquainted with the specifics of more than one type of solver (e.g. LP, CP, NLP), as well as a significant amount of computer programming, which includes coordinating the exchange of information among solvers. Clearly, a general purpose code is built at the expense of performance. Rather than defeating state-of-the-art implementations of cooperative approaches that are tailored to specific problems, SIMPL's objective is to be a generic and easy-to-use platform for the development and empirical evaluation of new ideas in the field of hybrid CP-OR algorithms.

As SIMPL is still under development, many new features and improvements to its functionality are the subject of ongoing efforts. Examples of such enhancements are: increasing the vocabulary of the language with new types of constraints; augmenting the inference capabilities of the system with the generation of cutting planes; broadening the application areas of the system by supporting other types of solvers; and providing a more powerful control over search. Finally, SIMPL is currently being used to test integrated models for a few practical optimization problems such as the lot-sizing problem of Sect.   .

References

[1] E. Balas.   Disjunctive programming: Properties of the convex hull of feasible points. *Discrete Applied Mathematics*, 89:3–44, 1998.

[2] P. Baptiste, C. Le Pape, and W. Nuijten. *Constraint-Based Scheduling*. Kluwer, 2001.

[3] J. F. Benders. Partitioning procedures for solving mixed-variables programming problems. *Numerische Mathematik*, 4:238–252, 1962.

[4] H. Beringer and B. de Backer. Combinatorial problem solving in constraint logic programming with cooperating solvers. In C. Beierle and L. Plümer, editors, *Logic Programming: Formal Methods and Practical Applications*. Elsevier Science, 1995.

[5] M. Berkelaar. LP_SOLVE. Available from ftp://ftp.ics.ele.tue.nl/pub/lp_solve/.

[6] A. Bockmayr and F. Eisenbrand. Combining logic and optimization in cutting plane theory. In H. Kirchner and C. Ringeissen, editors, *Proceedings of the Third International Workshop on Frontiers of Combining Systems (FroCos)*, *LNAI* 1794, 1–17. Springer-Verlag, 2000.

[7] A. Bockmayr and T. Kasper. Branch and infer: A unifying framework for integer and finite domain constraint programming. *INFORMS Journal on Computing*, 10(3):287–300, 1998.   ,

[8] Y. Colombani and S. Heipcke. *Mosel: An Overview.* Dash Optimization, 2002.

[9] Dash Optimization. XPRESS-MP. http://www.dashoptimization.com.

[10] A. Eremin and M. Wallace. Hybrid Benders decomposition algorithms in con-
     straint logic programming. In Toby Walsh, editor, *Proceedings of the Seventh
     International Conference on Principles and Practice of Constraint Programming
     (CP)*, LNCS 2239, 1–15. Springer-Verlag, 2001.   ,

[11] F. Focacci, A. Lodi, and M. Milano. Cost-based domain filtering. In J. Jaffar, ed-
     itor, *Proceedings of the Fifth International Conference on Principles and Practice
     of Constraint Programming (CP)*, LNCS 1713, 189–203. Springer-Verlag, 1999.

[12] F. Focacci, A. Lodi, and M. Milano. Cutting planes in constraint programming:
     A hybrid approach. In R. Dechter, editor, *Proceedings of the Sixth International
     Conference on Principles and Practice of Constraint Programming (CP)*, LNCS
     1894, 187–201. Springer-Verlag, 2000.

[13] G. Gutin and A. P. Punnen, editors. *Traveling Salesman Problem and Its Varia-
     tions.* Kluwer, 2002.

[14] J. N. Hooker. *Logic-Based Methods for Optimization.* Wiley-Interscience Series in
     Discrete Mathematics and Optimization, 2000.   ,   ,   ,   ,

[15] J. N. Hooker. Logic, optimization and constraint programming. *INFORMS Jour-
     nal on Computing*, 14(4):295–321, 2002.   ,

[16] J. N. Hooker. A framework for integrating solution methods. In H. K. Bhargava
     and M. Ye, editors, *Computational Modeling and Problem Solving in the Net-
     worked World*, pages 3–30. Kluwer, 2003. Plenary talk at the Eighth INFORMS
     Computing Society Conference (ICS).   ,

[17] J. N. Hooker. Logic-based benders decomposition for planning and scheduling.
     Manuscript, GSIA, Carnegie Mellon University, 2003.

[18] J. N. Hooker and M. A. Osorio. Mixed logical/linear programming. *Discrete Ap-
     plied Mathematics*, 96–97(1–3):395–442, 1999.   ,   ,

[19] J. N. Hooker and G. Ottosson. Logic-based benders decomposition. *Mathematical
     Programming*, 96:33–60, 2003.   ,

[20] J. N. Hooker, G. Ottosson, E. Thorsteinsson, and H.-J. Kim. On integrating
     constraint propagation and linear programming for combinatorial optimization.
     In *Proceedings of the 16th National Conference on Artificial Intelligence*, pages
     136–141. MIT Press, 1999.

[21] J. N. Hooker and H. Yan. Logic circuit verification by Benders decomposition. In
     V. Saraswat and P. Van Hentenryck, eds., *Principles and Practice of Constraint
     Programming: The Newport Papers*, MIT Press (Cambridge, MA, 1995) 267-288.
     ,

[22] J. N. Hooker and H. Yan. A relaxation for the cumulative constraint. In
     P. Van Hentenryck, editor, *Proceedings of the Eighth International Conference on
     Principles and Practice of Constraint Programming (CP)*, LNCS 2470, 686–690.
     Springer-Verlag, 2002.   ,

[23] ILOG S. A. The CPLEX mixed integer linear programming and barrier optimizer.
     http://www.ilog.com/products/cplex/.

[24] V. Jain and I. E. Grossmann. Algorithms for hybrid MILP/CP models for a class
     of optimization problems. *INFORMS Journal on Computing*, 13(4):258–276, 2001.
     ,

[25] E. L. Lawler, J. K. Lenstra, A. H. G. Rinnooy Kan, and D. B. Shmoys. *The Trav-
     eling Salesman Problem: A Guided Tour of Combinatorial Optimization.* John
     Wiley & Sons, 1985.

[26] S. Leipert. The tree interface version 1.0: A tool for drawing trees. Available at
     http://www.informatik.uni-koeln.de/old-ls_juenger/projects/vbctool.html.

[27] M. Milano, G. Ottosson, P. Refalo, and E. S. Thorsteinsson. The role of integer programming techniques in constraint programming's global constraints. *INFORMS Journal on Computing*, 14(4):387–402, 2002.

[28] G. Ottosson, E. S. Thorsteinsson, and J. N. Hooker. Mixed global constraints and inference in hybrid CLP-IP solvers. In *CP'99 Post Conference Workshop on Large Scale Combinatorial Optimization and Constraints*, pages 57–78, 1999.

[29] P. Refalo. Tight cooperation and its application in piecewise linear optimization. In J. Jaffar, editor, *Proceedings of the Fifth International Conference on Principles and Practice of Constraint Programming (CP)*, LNCS 1713, 375–389. Springer-Verlag, 1999.

[30] P. Refalo. Linear formulation of constraint programming models and hybrid solvers. In R. Dechter, editor, *Proceedings of the Sixth International Conference on Principles and Practice of Constraint Programming (CP)*, LNCS 1894, 369–383. Springer-Verlag, 2000.

[31] J.-C. Régin. A filtering algorithm for constraints of difference in CSPs. In *Proceedings of the National Conference on Artificial Intelligence*, pages 362–367, 1994.

[32] R. Rodošek, M. Wallace, and M. T. Hajian. A new approach to integrating mixed integer programming and constraint logic programming. *Annals of Operations Research*, 86:63–87, 1999.   ,

[33] E. S. Thorsteinsson. Branch-and-Check: A hybrid framework integrating mixed integer programming and constraint logic programming. In Toby Walsh, editor, *Proceedings of the Seventh International Conference on Principles and Practice of Constraint Programming (CP)*, LNCS 2239, 16–30. Springer-Verlag, 2001.

[34] P. Van Hentenryck. *The OPL Optimization Programming Language*. MIT Press, 1999.   ,

[35] M. Wallace, S. Novello, and J. Schimpf. ECL^{i}PS^e: A platform for constraint logic programming. *ICL Systems Journal*, 12:159–200, 1997.   ,

[36] H. P. Williams and H. Yan. Representations of the all_different predicate of constraint satisfaction in integer programming. *INFORMS Journal on Computing*, 13(2):96–103, 2001.   ,

[37] H. Yan and J. N. Hooker. Tight representations of logical constraints as cardinality rules. *Mathematical Programming*, 85:363–377, 1999.   ,

[38] T. H. Yunes. On the sum constraint: Relaxation and applications. In P. Van Hentenryck, editor, *Proceedings of the Eighth International Conference on Principles and Practice of Constraint Programming (CP)*, LNCS 2470, 80–92. Springer-Verlag, 2002.   ,

A New Exact Solution Algorithm
for the Job Shop Problem
with Sequence-Dependent Setup Times

Christian Artigues[1], Sana Belmokhtar[2], and Dominique Feillet[1]

[1] Laboratoire d'Informatique d'Avignon, FRE CNRS 2487
Université d'Avignon, Agroparc BP 1228, 84911 Avignon Cedex 9, France
[2] MSGI-G2I, Ecole des Mines de Saint Etienne 158
cours Fauriel, 42023 Saint Etienne Cedex 2, France

Abstract. We propose a new solution approach for the Job Shop Problem with Sequence Dependent Setup Times (SDST-JSP). The problem consists in scheduling jobs, each job being a set of elementary operations to be processed on different machines. The objective pursued is to minimize the completion time of the set of operations. We investigate a relaxation of the problem related to the traveling salesman problem with time windows (TSPTW). Our approach is based on a Branch and Bound procedure, including constraint propagation and using this relaxation. It compares favorably over the best available approaches from the literature on a set of benchmark instances.

1 Introduction

In this work, we consider the Job Shop Problem with Sequence Dependent Setup Times (SDST-JSP). The Job Shop Problem is used for modeling problems where a set of jobs, consisting of a sequence of elementary operations to be executed on distinct machines, has to be scheduled. This problem is widely investigated in the literature and many efficient approaches exist for its resolution (see, e.g., Blazewicz et al. [5], Nowicki and Smutnicki [13] or Vaessens et al. [15]). The SDST-JSP is a variant problem where machines have to be reconfigured between two successive operations. The most common objective is to minimize the completion time of the set of operations, i.e., the so-called makespan.

The SDST-JSP is trivially NP-hard in the strong sense since it admits the Job Shop Problem as a special case, which is also NP-hard in the strong sense. However, despite its similarities with the Job Shop Problem, few works have been devoted to its solution. Some heuristic solution approaches have been proposed in the literature (Choi and Choi [8], Artigues and Buscaylet[3], Artigues et al [2]), but no exact solution algorithm strictly addresses this problem. Actually, Brucker and Thiele [6] describe a Branch and Bound algorithm for the General-Shop Problem (GSP), which admits the SDST-JSP as a special case. Also, Focacci et al. [9] propose a Branch and Bound scheme for a variant problem where the machines used for operations are not fixed but to be chosen in a given subset.

J.-C. Régin and M. Rueher (Eds.): CPAIOR 2004, LNCS 3011, pp. 37–49, 2004.

In this paper, we also propose a Branch and Bound procedure using constraint propagation techniques. A relaxation is introduced, based on the search of feasible solutions for Traveling Salesman Problem with Time Windows (TSPTW) instances. Note that our approach presents many similarities with Focacci *et al.* [ ]'s one, except that Focacci *et al.* [ ] prefer to relax subtour constraints and tackle assignment problems. Indeed assignment problems can be solved in polynomial time, while the TSPTW is NP-hard in the strong sense. The assignment problem relaxation is embedded into a global constraint associated with a reduced cost-based filtering algorithm. In our algorithm the TSPTW is considered without relaxation but a long term memory is introduced for limiting computing times. It memorizes TSPTW instances solved throughout the search tree and permit the quick obtaining of feasible solutions in many cases. The relaxation is used to compute an initial lower bound and to test feasibility at the nodes of the search in a dichotomy framework, but it does not provide additional domain filtering.

The problem and the model used are described in section . Section   then presents the backbone of the algorithm, i.e., the Branch and Bound scheme. The relaxation is described in section . Numerical experiments conclude the paper in section .

2   Presentation and Mathematical Formulation for the SDST-JSP

We consider a set $J = \{J_1, \ldots, J_n\}$ of n jobs. Jobs have to be processed on a set $M = \{M_1, \ldots, M_m\}$ of m resources (machines). Each job is made up of m operations $O_{i1}, \ldots, O_{im}$, to be processed in this order, without overlapping or preemption. The set of all operations is noted O and its size is $N = n \times m$. Operation $O_{ij} \in O$ requires machine $m_{ij} \in M$ and necessitates processing time $p_{ij} \geq 0$. Machines admit at most one operation at a time.

A set of setup types σ is defined, with a matrix $(|\sigma| + 1) \times (|\sigma|)$ of setup times noted s. To each operation O_{ij} is associated setup type $\sigma_{ij} \in \sigma$. A setup time $s_{\sigma_{ij}\sigma_{kl}}$ is then necessary between two consecutive operations O_{ij} and O_{kl} on the same machine. Also, an initial setup times $s_{0\sigma_{ij}}$ is necessary if O_{ij} is the first operation on machine m_{ij}. Finally, we assume that setup times satisfy the triangle inequality.

The objective pursued in this work is the minimization of the makespan, that is the completion time of the last operation processed, even if other objectives could have been considered.

A solution of the SDST-JSP can be represented with the help of the so-called job and machine Gantt's diagrams. An example of such diagrams is given in figure   for some instance of the SDST-JSP with 3 jobs and 3 machines.

On these diagrams, we represent setup times with hatched rectangles. Full rectangles stand for idle times. The two diagrams respectively represent a solution from both machine and job views.

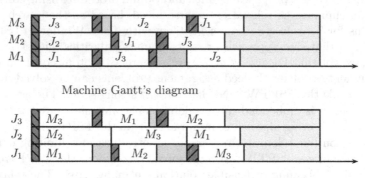

Fig. 1. Gantt's diagrams

Using the constraint programming framework, a simple model can easily be proposed for the SDST-JSP. For each operation O_{ij}, we introduce a variable S_{ij} indicating the starting time of the operation. A variable C_{max} is also introduced to represent the value of the makespan. The model is then:

$$\text{minimize } C_{max} \tag{1}$$

subject to

$$C_{max} \geq S_{im} + p_{im} \quad (i = 1, \ldots, n) \tag{2}$$

$$S_{ij} + p_{ij} \leq S_{i(j+1)} \quad (O_{ij} \in O; j < m) \tag{3}$$

$$(S_{ij} \geq S_{hl} + p_{hl} + s_{\sigma_{hl}\sigma_{ij}}) \text{ or}$$
$$(S_{hl} \geq S_{ij} + p_{ij} + s_{\sigma_{ij}\sigma_{hl}}) \quad (O_{ij}, O_{hl} \in O; m_{ij} = m_{hl}) \tag{4}$$

$$S_{ij} \geq S_{0\sigma_{ij}} \quad (O_{ij} \in O) \tag{5}$$

$$S_{ij} \in \mathbb{N} \quad (O_{ij} \in O) \tag{6}$$

$$C_{max} \in \mathbb{N} \tag{7}$$

Constraints (2) state C_{max} as the makespan in optimal solutions. Constraints (3) represent the precedence constraints between the successive operations of the same job. Disjunctive constraints between operations processed on a same machine are enforced with constraints (4). Setup times between operations on a same machine appear in these constraints, while initial setup times require a last set of constraints (5).

A useful tool for the solution of scheduling problems is the so-called disjunctive graph. This graph provides an efficient representation of the decisions, while limiting the solution space. It is defined as follows.

Let $G = (X, U, E)$ be the disjunctive graph. The set of vertices X is made up of the set of operations plus two dummy vertices representing the beginning and the end of the schedule. Thus, X has $n \times m + 2$ vertices. A set of arcs U and a set of edges E are defined. Arcs in U represent precedence constraints between operations and are called conjunctive arcs. They are weighted with the processing time of the origin vertex of the arc. Edges in E represent disjunctive constraints between operations on a same machine and are called disjunctive arcs. Actually, disjunctive arcs can be interpreted as the union of two exclusive arcs with opposite directions.

By definition, a selection is a state of the disjunctive graph where a direction is chosen for some disjunctive arcs. A selection is said to be complete when every arc has a direction. A complete selection coresponds to a unique semi-active schedule if the resulting graph is acyclic. Once they are directed, disjunctive arcs are weighted with the sum of the processing time of the origin vertex of the arc plus the setup time required between the origin and the destination vertices. Minimizing the makespan then reduces to the search of the longest path in the graph, the makespan being the length of such a path. Hence, the SDST-JSP reduces exactly to the problem of finding a complete acyclic selection for which the longest path is minimum. This standpoint relies on the property that it is possible to consider only semi-active scheduling, that is scheduling where operations are started as soon as possible (when disjunctive and conjunctive constraints are satisfied), to find an optimal solution. An example of a disjunctive graph is presented in figure .

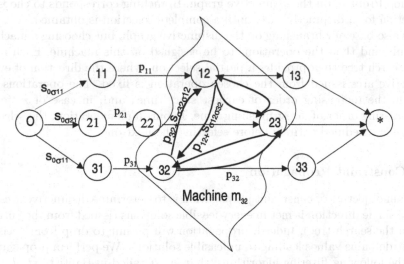

Fig. 2. Disjunctive graph for the SDST-JSP

3   Branch and Bound Scheme

The Branch and Bound technique is an approach based on a non-exhaustive enumeration of the solutions of a problem. Two main features of the technique are the branching scheme and the bounding scheme. Branching replaces a problem (represented by a node of the search tree) with several new problems of reduced sizes (included in the tree as descendant nodes). Bounding permits the pruning of nodes for which it appears that the associated problems do not contain any optimal solution.

Our algorithm is based on a Branch and Bound embedded in a dichotomy framework, the Branch and Bound being limited to the search of a feasible solution. An initial lower bound LB is obtained by the relaxation of precedence constraints ( ) - see section   for details. An initial upper bound UB is obtained using an existing heuristic solution approach. At each step of the dichotomy a value L is chosen in the interval $[LB, UB]$ that sets an upper limit for C_{max}. Depending whether a solution is found or not using the Branch and Bound, UB is fixed to the value of this solution or LB is fixed to L and the algorithm iterates.

In the following subsections, we focus on the Branch and Bound algorithm implemented in the dichotomy framework.

3.1   Branching Scheme

Brucker and Thiele [ ] base their branching scheme for the solution of the General Shop Problem on the disjunctive graph. Branching corresponds to choosing a direction for a disjunctive arc, until a complete selection is obtained.

We also base our branching on the disjunctive graph, but choosing a machine randomly and then the operations to be assigned on this machine. Each node of the search tree then provides a partial selection, where the direction of every disjunctive arcs issued from the chosen operation is fixed. The operations are taken in the increasing order of earliest start times and, in case of a tie, in the increasing order of latest finishing time. In our implementation, the selected machine is scheduled entirely before selecting the next one.

3.2   Constraint Propagation

The main objective of constraint propagation is to determine disjunctive arcs for which a single direction is met in every feasible solutions (issued from the current node of the search tree). Indeed, propagation will permit to drop from decision variable domains values leading to unfeasible solutions. We perform propagation using the following filtering algorithms, that are detailed in Baptiste *et al.* [ ].

Precedence Constraints. Each node of the search tree corresponds to a partial selection in the disjunctive graph $G = (V, U, E)$, where U is the set of conjunctive arcs and disjunctive arcs for which a direction has been chosen. Consistency is

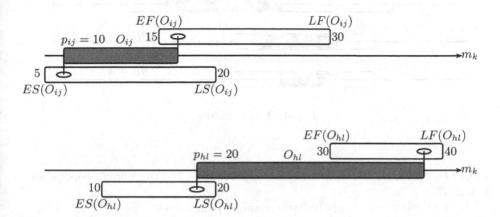

Fig. 3. Disjunctive constraints

ensured at the bounds of the domains of variables S_{ij} (with $O_{ij} \in O$) using constraints $S_{i_2 j_2} \geq S_{i_1 j_1} + w_{i_1 j_1, i_2 j_2}$ $(O_{i_1 j_1}, O_{i_2 j_2}) \in U$, where $w_{i_1 j_1, i_2 j_2}$ is the weight of the arc in the graph. This propagation is performed by computing longest paths in (V, U).

Disjunctive Constraints. Using domains of variables S_{ij} (with $O_{ij} \in O$) and disjunctive constraints between operations on a same machine, it might be possible to deduce that an operation is necessarily processed before another one. Such a deduction corresponds to setting a direction of a disjunctive arc in the disjunctive graph.

For every operation $O_{ij} \in O$, we respectively note ES_{ij}, LS_{ij}, EF_{ij} and LF_{ij} the earliest starting time, latest starting time, earliest finishing time and latest finishing time of the operation, that are directly deduced from S_{ij}. Let O_{ij} and O_{hl} be two operations requiring the same machine. If $EF_{hl} + s_{\sigma_{hl}\sigma_{ij}} > LS_{ij}$, O_{hl} can not be processed before O_{ij}. This possibility is illustrated on figure , where $s_{\sigma_{hl}\sigma_{ij}} = 5$.

Propagation is activated as soon as the bounds of the domain of a variable S_{ij} are changed, for all the operations requiring the same machine m_{ij}.

Edge Finding. This propagation scheme is a generalization of the preceding one. It permits to determine that an operation has to be processed before or after a set of other operations using the same machine. This propagation scheme is illustrated on figure , where operation O_{11} has to be processed after opera-

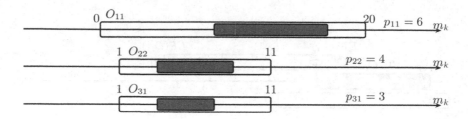

Fig. 4. Edge Finding

tions O_{22} and O_{31}, which cannot be deduced using the preceding propagation scheme.

It is worth mentioning that setup times penalize the effectiveness of this scheme. Indeed, while estimating the finishing time of a set of operations, the last operation of the set is not known and the setup time considered after the last operation of the set is the minimum setup time among all the operations of the set. Hence, some extensions of edge finding have been proposed [6, 16]. However, the classical version of edge finding ignoring setup times is considered in this study.

4   TSPTW Relaxation

In this section, we describe the TSPTW relaxation used to compute an initial lower bound and, at the nodes of the dichotomizing search, for pruning as soon as unfeasibility is detected. This is performed through the relaxation of the precedence constraints in the model (1-7) of section 2. The new problem then decomposes into several TSPTWs.

The TSPTW consists in finding a route visiting one and only one time each vertex of a graph with a minimum cost, a travel cost being associated with the arcs of the graph. The visit of customers (vertices) is constrained to take place within given time windows. In the context of the SDST-JSP, customers stand for operations.

When relaxing precedence constraints in the model (1)-(7), m independent TSPTW instances appear, one for each machine. In the disjunctive graph, this relaxation is characterized by the elimination of the conjunctive arcs and by the appearance of m connected components. Hence, noting $TSPTW_k$ $(k = 1, \ldots, m)$ the TSPTW instance associated to machine k, finding k such that $TSPTW_k$ is unfeasible ensures that the current selection admits no feasible solution.

Figure 5 provides an example of this relaxation scheme with 3 jobs and 3 machines (where arcs issued from vertex 0 are omitted)

SDST-JSP

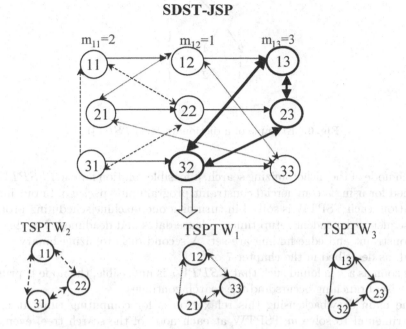

Fig. 5. Relaxation of the SDST-JSP into m TSPTWs

For every k in $\{1,\ldots,m\}$, we address the solution of $TSPTW_k$ with the following model. Let $O_k \subset O$ be the subset of operations to be considered in subproblem $TSPTW_k$ (operations on machine M_k). Variables S_{ij} of model ( )-( )are used for operations $O_{ij} \in O_k$. Note that the domain of these variables is reduced through the branching decisions and the constraint propagation, which corresponds to time windows. We introduce a new variable $Tour_k$ for the objective function.

$$\text{minimize } Tour_k \tag{8}$$

subject to

$$Tour_k \geq S_{ij} + p_{ij} \; (O_{ij} \in O_k) \tag{9}$$

$$S_{ij} \geq s_{0\sigma_{ij}} \; (O_{ij} \in O_k) \tag{10}$$

$$S_{ij} \geq S_{hl} + p_{hl} + s_{\sigma_{hl}\sigma_{ij}}$$
$$\text{or } S_{hl} \geq S_{ij} + p_{ij} + s_{\sigma_{ij}\sigma_{hl}} \; (O_{ij}, O_{hl} \in O_k; m_{ij} = m_{hl} = m_k) \tag{11}$$

$$S_{ij} \in \mathbb{N} \; (O_{ij} \in O) \tag{12}$$

$$Tour_k \in \mathbb{N} \tag{13}$$

Before running the dichotomy, all $TSPTW_k$ are solved to optimality to compute a lower bound equal to the greatest optimal solution among the m problems.

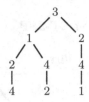

Fig. 6. Example of a dictionary for a $TSPTW_k$

At each node of the dichotomizing search, a feasible solution for each $TSPTW_k$ is searched for using a commercial constraint programming package. In our implementation, each TSPTW is solved in turn as a one-machine scheduling problem with sequence-dependent setup times, release dates and deadlines. The disjunctive constraint and edge finding are set. A second dichotomizing binary search is used, as described in the chapter 7 of [ ].

As soon as a k is found such that $TSPTW_k$ is unfeasible, the node is pruned. Otherwise, branching occurs and the search continues.

The main drawback using this relaxation is for computing time. It might be detrimental to solve m TSPTW at each node of the search tree, even if it provides a strong tool for pruning. In order to limit this drawback, we propose to memorize solutions of $TSPTW_k$ obtained so far during the search. Indeed, one might have to solve instances of the $TSPTW$ for which only some time windows have changed and for which sequences of customers previously found feasible might still be feasible.

Each time a solution is found, the ordered sequence of visited customers is memorized as a word in a dictionary. Then, when solving a $TSPTW_k$, the dictionary associated to machine M_k is first scanned to determine whether one of the sequences memorized is feasible for the current instance. If not, constraint programming is applied as explained above.

For each subproblem $TSPTW_k$, the dictionary is defined as a forest, as illustrated in figure . On this example, the dictionary contains sequences $J_3 \prec J_1 \prec J_2 \prec J_4$, $J_3 \prec J_1 \prec J_4 \prec J_2$ and $J_3 \prec J_2 \prec J_4 \prec J_1$. This structure is convenient as well for adding new words than for checking words for feasibility. For this last case especially, a simple depth first search can be implemented. When nodes are attained, earliest arrival times are computed, which permits to eventually backtrack. Hence, sequences beginning with the same unfeasible subsequence can be discarded simultaneously.

5 Computational Results

In this section, we present computational experiments conducted to evaluate the quality of our approach. For this purpose, we use benchmark instances from Brucker and Thiele [ ]. These instances are issued from the classical Adams

Table 1. Bound value at the root of the search tree for instances $5 \times 10 \times 5$

Instance	BT96	LB	UB
t2-ps01	756	796	818
t2-ps02	705	715	829
t2-ps03	658	678	782
t2-ps04	627	647	745
t2-ps05	653	671	704

Table 2. Bound value at the root of the search tree for instances $5 \times 15 \times 5$

Instance	BT96	LB	UB	CPU LB
t2-ps06	986	996	1026	1
t2-ps07	940	927	1033	0.9
t2-ps08	913	923	1002	3
t2-ps09	1001	1012	1060	19
t2-ps10	1008	1018	1036	141

et al.'s [ ] instances devoted to the Job Shop Problem, introducing setup. Each instance is characterized by a number of machines, a number of jobs to be scheduled and a number of setup types for the operations. These three parameters define a triplet with the format machines $\times$ jobs $\times$ type. Computational experiments are realized on two sets of $5 \times 10 \times 5$ and $5 \times 15 \times 5$ instances. For comparison purpose, we use also two large $5 \times 20 \times 10$ instances.

Algorithms are implemented in C++ on a Pentium IV 2 GHz computer. ILOG Scheduler 5.2 [ ] and ILOG Solver 5.2 [ ] are used for constraint propagation and Branch and Bound.

Tables   and   compare the value of the bound we obtained at the root of the tree with the value obtained by Brucker and Thiele [ ], respectively for the sets of $5 \times 10 \times 5$ and $5 \times 15 \times 5$ instances. Note that Brucker and Thiele [ ] compute a lower bound by relaxing the GSP into a one machine problem and by using Jackson preemptive scheduling principle (JPS) and results from Carlier and Pinson [ ]. Columns of these tables indicate successively the name of the instance, the value of Brucker and Thiele's bound, the value of our bound and an upper bound provided by Artigues *et al.* [ ]. These two last values define the initial interval for dichotomy. In table , the computing time of the lower bound is given in seconds. These tables show that solving TSPTWs provide better bounds than the approach of Brucker and Thiele [ ] for most of the instances. Although Focacci *et al* [ ] do not explicitly provide a lower bound, the Assignment Problem solved in [ ] is a relaxation of the TSPTW considered in the present paper and would consequently give a weaker (but faster) bound.

Table 3. Makespan value for instances $5 \times 10 \times 5$

Instance	BT		FNL		ABF		ABF*
	Cmax	CPU (sec.)	Cmax	CPU (sec.)	Cmax	CPU (sec.)	
t2-ps01	**798**	502.1	798	>1800	**798**	522	1168.5
t2-ps02	**784**	157.8	784	88.4	**784**	7.7	14.8
t2-ps03	**749**	1897.7	749	144.2	**749**	47.8	111.7
t2-ps04	**730**	188.8	730	388.3	**730**	34.3	93.6
t2-ps05	**691**	770.1	691	30.4	**691**	30.2	136.7

Table   compare the value of the makespan computed with our approach (ABF) with the value obtained by Brucker and Thiele [ ] (BT) and Focacci *et al* [ ] (FNL) for the smallest instances. A time limit of 7200 seconds is set for the computations for BT and ABF whereas a time limit of 1800 seconds is set for FNL. Columns indicate the name of the instance and, for each approach, the value obtained with the computing times. A last column ABF* indicates computing times when dictionaries are not used to memorize TSPTW solutions. Times are indicated in seconds. Times indicated for Brucker and Thiele [ ]'s results are obtained on a Sun 4/20 station. Times indicated for Focacci *et al* [ ]'s results are obtained on a Pentium II 200 MHz. Proven optimal solutions are indicated in bold. The table show that all $5 \times 10 \times 5$ instances are solved by our method in less than 500 seconds. The benefit of using the dictionary is clear. The comparison with BT and FNL approaches is not easy according to the considerable difference between the speeds of the computers. However, applying factors of 3 and 10 for CPU time comparisons with FNL and BT, respectively, our approach remains competitive.

Table   compare the results of our approach with BT on the $5 \times 15 \times 5$ instances. The results of Focacci *et al* [ ] are not available on these instances. We also indicate the current value of the lower bound when the algorithm is stopped, for our approach. For $5 \times 15 \times 5$ instances, all the results from Brucker and Thiele [ ] are improved. Two new instances of this size are closed for the first time (t2-ps10 and t2-ps07), with additional CPU time requirements for the latter one. A comparison with BT is here even more difficult for two reasons. First, it would be necessary to test the BT Branch and Bound on a recent computer. Second, as shown in table   all upper bounds given in a few seconds by the fast heuristic proposed in [ ] are better than the ones provided by BT after 2 hours of computation. Hence, a better heuristic should be used before running the BT algorithm.

To compare our approach on large instances with the one of Brucker and Thiele [ ] and Focacci *et al* [ ], we give in table   the results on the two SDST-JSP $5 \times 20 \times 10$ used in [ ]. The limits of our approach is reached since after 2 hours the dichotomizing search is unable to increase the initial lower bound displayed in column ABF/LB and does not improve the initial upper bound

Table 4. Makespan and lower bound value for instances $5 \times 15 \times 5$

Instance	BT		ABF		
	Cmax	CPU (sec.)	Cmax	BInf	CPU (sec.)
t2-ps06	1056	7200	1026	996	7200
t2-ps07	1087	7200	970 (**970**)	966 (970)	7200 (16650)
t2-ps08	1096	7200	1002	923	7200
t2-ps09	1119	7200	1060	1037	7200
t2-ps10	1058	7200	**1018**	1018	498

Table 5. Makespan and lower bound value for instances $5 \times 20 \times 10$

Instance	BT			FNL		ABF		
	Cmax	CPU (sec.)	LB	Cmax	CPU (sec.)	Cmax	CPU (sec.)	LB
t2-ps12	1528	7200	1139	1448	120	1319	7200	1159
t2-ps13	1549	7200	1250	1658	120	1439	7200	1250

displayed in column ABF/Cmax. However the initial upper bound is here again surprisingly better than the results obtained by FNL and BT and the proposed lower bound still improves the BT lower bound.

6 Conclusion

In this paper, we propose a new algorithm for the exact solution of the SDST-JSP. The main framework of the algorithm is a Branch and Bound strategy embedding constraint propagation at each node of the search tree. In order to prune nodes effectively, we introduce a global constraint by relaxing precedence constraints between operations. Filtering algorithm for this global constraint reduces to solving a TSPTW instance for each machine. The principle is to detect infeasible TSPTW instances, indicating unfeasibility for the global problem associated to the current node of the search tree. In order to improve computing time, we use a long term memory by storing feasible solutions of TSPTW instances found so far. This permits to limit significantly the negative impact of the global constraint on computing time.

This approach is competitive with the approaches of Brucker and Thiele [ ] and Focacci *et al* [ ] and permits to close some new instances. Possible improvements in the future are the introduction of a Lagrangian relaxation type penalization of the precedence constraints relaxed within the bounding scheme and the implementation of more efficient algorithms for the solution of TSPTW instances (as the ones proposed in Pesant *et al.* [ ] or Focacci and Lodi [ ]).

Acknowledgements

The authors would like to thank Sophie Demassey for her help as a LaTex specialist. We would like also to thank the anonymous referees for their helpful comments.

References

[1] Adams J., Balas E. and Zawack D. (1988) The Shifting Bottleneck Procedure for Job-Shop Scheduling. Manag Sci 34:391-401.

[2] Artigues C., Lopez P. and Ayache P. D. (2003), Schedule generation schemes and priority rules for the job-shop problem with sequence dependent setup times: Dominance properties and computational analysis. to appear in Annals of Operations Research.   ,   ,

[3] Artigues C. and Buscaylet F. (2003), A fast tabu search method for the job-shop problem with sequence-dependent setup times. Proceedings of the Metaheuristic International Conference MIC'2003, Kyoto 2003.

[4] Baptiste P., Le Pape C. and Nuijten W. (2001), Constrained-Based scheduling, Applying constraint programming to Scheduling Problems, Kluwer's International Series, 19-37.

[5] Blazewicz J. ,Domschke W. and Pesch E. (1996), The job shop scheduling problem: Conventional and new solution techniques. European Journal of Operational Research, 93(1):1–33.

[6] Brucker P. and Thiele O. (1996), A branch and bound method for the general-shop problem with sequence-dependent setup times, Operations Research Spektrum, 18, 145-161.   ,   ,   ,   ,   ,   ,

[7] Carlier J. and Pinson E. (1989), An algorithm for solving the Job-Shop Problem. Management Science 35, 164-176.

[8] Choi I.-N. and Choi D.-S. (2002), A local search algorithm for jobshop scheduling problems with alternative operations ans sequence-dependent setups, Computers and Industrial Engineering 42, 43-58.

[9] Focacci F., Laborie P. and Nuijten W. (2000), Solving scheduling problems with setup times and alternative resources. In Fifth International Conference on Artificial Intelligence Planning and Scheduling, Breckenbridge, Colorado, 92-101.   ,
   ,   ,   ,

[10] Focacci F. and Lodi, A. (1999), Solving TSP with Time Windows with Constraints, International Conference on Logic Programming 515-529.

[11] ILOG SOLVER 5.2 User's Manuel (2000), ILOG.

[12] ILOG SCHEDULER 5.2 User's Manuel (2000), ILOG.

[13] Nowicki E. and Smutnicki C. (1996), A fast taboo search algorithm for the job shop problem, Management Science 42, 797-813.   ,

[14] Pesant G., Gendreau M., Potvin J-Y. and Rousseau J-M. (1999), An Exact Constraint Logic Programming Algorithm for the Traveling Salesman Problem with Time Windows, European Journal of Operational Research 117, 253-263.

[15] Vaessens R. J. M., Aarts E. H. L. and Lenstra J. K. (1996), Job Shop Scheduling by Local Search, INFORMS Journal on Computing 8, 302-317.

[16] Vilím P. and Barták, R. (2002), Filtering Algorithms for Batch Processing with Sequence Dependent Setup Times, In Ghallab, Hertzberg, and Traverso (eds.) Proceedings of The Sixth International Conference on Artificial Intelligence Planning and Scheduling (AIPS 2002). AAAI Press, Toulouse, p. 312-320.

Simple Rules for Low-Knowledge Algorithm Selection*

J. Christopher Beck and Eugene C. Freuder

Cork Constraint Computation Centre
Department of Computer Science, University College Cork, Cork, Ireland
{c.beck,e.freuder}@4c.ucc.ie

Abstract. This paper addresses the question of selecting an algorithm from a predefined set that will have the best performance on a scheduling problem instance. Our goal is to reduce the expertise needed to apply constraint technology. Therefore, we investigate simple rules that make predictions based on limited problem instance knowledge. Our results indicate that it is possible to achieve superior performance over choosing the algorithm that performs best on average on the problem set. The results hold over a variety of different run lengths and on different types of scheduling problems and algorithms. We argue that low-knowledge approaches are important in reducing expertise required to exploit optimization technology.

1 Introduction

Using constraint technology still requires significant expertise. A critical area of research if we are to achieve large scale adoption is the reduction of the skill required to use the technology. In this paper, we adopt a low-knowledge approach to automating algorithm selection for scheduling problems. Specifically, given an overall time limit T to find the best solution possible to a problem instance, we run a set of algorithms during a short "prediction" phase. Based on the quality of the solutions returned by each algorithm, we choose one of the algorithms to run for the remainder of T. A low-knowledge approach is important in actually reducing the expertise required rather than simply shifting it to another portion of the algorithm selection process. Empirical analysis on two types of scheduling problems, disjoint algorithm sets, and a range of time limits demonstrates that such an approach consistently achieves performance no worse than choosing the best pure algorithm and furthermore can achieve performance significantly better.

The contributions of this paper are the introduction of a low-knowledge approach to algorithm selection, the demonstration that such an approach can achieve performance better than the best pure algorithm, and the analysis of the empirical results to characterize limits on the performance of any on-line algorithm selection technique.

---

* This work has received support from Science Foundation Ireland under Grant 00/PI.1/C075 and ILOG, SA.

J.-C. Régin and M. Rueher (Eds.): CPAIOR 2004, LNCS 3011, pp. 50–   , 2004.
© Springer-Verlag Berlin Heidelberg 2004

2   The Algorithm Selection Problem

The algorithm selection problem consists of choosing the best algorithm from a predefined set to run on a problem instance [ ]. In AI, the algorithm selection problem has been addressed by building detailed, high-knowledge models of the performance of algorithms on specific types of problems. Such models are generally limited to the problem classes and algorithms for which they were developed. For example, Leyton-Brown et al. [ ] have developed strong selection techniques for combinatorial auction algorithms that take into account 35 problem features based on four different representations. Other work applying machine learning techniques to algorithm generation [ ] and algorithm parameterization [, ,] is also knowledge intensive, developing models specialized for particular problems and/or search algorithms and algorithm components.

Our motivation for this work is to lessen the expertise necessary to use optimization technology. While existing algorithm selection techniques have shown impressive results, their knowledge-intensive nature means that domain and algorithm expertise is necessary to develop the models. The overall requirement for expertise has not been reduced: it has been shifted from algorithm selection to predictive model building. It could still be argued that the expertise will have been reduced if the predictive model can be applied to different types of problems. Unfortunately, so far, the performance of a predictive model tends to be inversely proportional to its generality: while models accounting for over 99% of the variance in search cost exist, they are not only algorithm and problem specific, but also problem instance specific [ ]. While the model building *approach* is general, the requirement for expertise remains: an in-depth study of the domain and of different problem representations is necessary to identify features that are predictive of algorithm performance. To avoid shifting the expertise to model building, we examine models that require no expertise to build. The feature used for prediction is the solution quality over a short period of time.

The distinction between low- and high-knowledge (or knowledge-intensive) approaches focuses on the number, specificity, and computational complexity of the measurements of a problem instance required to build a model. A low-knowledge approach has very few, inexpensive metrics, applicable to a very wide range of algorithms and problem types. A high-knowledge approach has more metrics, that are more expensive to compute, and are more specific to particular problems and algorithms. This distinction is independent of the model building approach. In particular, sophisticated model building techniques based on machine learning techniques are consistent with low-knowledge approaches.

3   On-Line Scenario and Prediction Techniques

We use the following on-line scenario: a problem instance is presented to a scheduling system and that system has a fixed CPU time of T seconds to return a solution. We assume that the system designer has been given a learning set of problem instances at implementation time and that these instances are representative of the problems that will be later presented. We assume that there exists a set of algorithms, A, that can be applied to the problems in question. Algorithm selection may be done off-line by, for example, using the learning set to identify the best pure algorithm overall and running that on each problem instance. Alternatively, algorithm selection can be done on-line, choosing the algorithm only after the problem instance is presented. In the latter case, the time to make the selection must be taken into account. To quantify this, let t_p represent the prediction time and t_r the subsequent time allocated to run the chosen pure technique. It is required that $T = t_p + t_r$.

For the low-knowledge techniques investigated here, each pure algorithm, $a \in A$, is run for a fixed number of CPU seconds, t, on the problem instance. The results of each run are then used to select the algorithm that will achieve the best performance given the time remaining. We require that $t_p = |A| \times t$. The learning set is used to identify t^* which is the value of t that leads to the best system performance.

Three simple prediction rules each with three variations are investigated:

- **pcost** - Selection is based on the cost of the best solution found by each algorithm. The three variations are: $pcost\_min(t)$: the algorithm that has found the minimum cost solution over all algorithms by time t is selected; $pcost\_mean(t)$: the algorithm with the minimum mean of the best solutions (sampled at 10 second intervals) is selected; $pcost\_median(t)$: identical to $pcost\_mean(t)$ except the median is used in place of the mean.
- **pslope** - Selection is based on the change in the cost of the best solutions found at 10 second intervals. The three variations are: $pslope\_min(t)$: the algorithm that has the minimum slope between $t-10$ and t seconds is selected; $pslope\_mean(t)$: the algorithm with minimum mean slope for each pair of consecutive 10 second intervals is selected; $pslope\_median(t)$: identical to $pslope\_mean(t)$ except the median is used in place of the mean.
- **pextrap** - Selection is based on the extrapolation of the current cost and slope to a predicted cost at T. As above, the three variations are: $pextrap\_min(t)$: the best solutions for an algorithm at time t and $t - 10$ are used to define a line which is used to extrapolate the cost at time T; the algorithm that has the minimum extrapolated cost is chosen; $pextrap\_mean(t)$ the algorithm with the minimum mean extrapolated cost over each interval of 10 seconds from 20 seconds to t seconds is selected; $pextrap\_median(t)$ identical to $pextrap\_mean(t)$ except the median is used in place of the mean.

For all rules ties are broken by selecting the algorithm with the best mean solution quality on the learning set at time T. The sampling interval was arbitrarily set to 10 seconds as it allowed time for a reasonable amount of search.

4   Initial Experiment

Our initial experiment divides a set of problem instances into a learning set and a test set, uses the learning set to identify t^* for each prediction rule and variation, and then applies each rule variation using $t_p = |A| \times t^*$ to the test problems.

4.1   Problem Sets and Algorithms

Three sets of 20×20 job shop scheduling (JSP) problems are used. A total of 100 problem instances in each set were generated and 20 problems per set were chosen as the learning set. The rest were placed in the test set. The problem sets have different structure based on the generation of the activity durations.

- **Rand**: Durations are drawn randomly with uniform probability from the interval [1, 99].
- **MC**: Durations are drawn randomly from a normal distribution. The distributions for activities on different machines are independent. The durations are, therefore, machine-correlated (MC).
- **JC**: Durations are drawn randomly from a normal distribution. The distributions for different jobs are independent. Analogously to the MC set, these problems are job-correlated (JC).

These different problem structures have been studied for flow-shop scheduling [ ] but not for job shop scheduling. They were chosen based on the intuition that the different structures may differentially favor one pure algorithm and therefore the algorithms would exhibit different relative performance on the different sets. Such a variation is necessary for on-line prediction to be useful: if one algorithm dominates on all problems, the off-line selection of that algorithm will be optimal.

Three pure algorithms are used. These were chosen out of a set of eight algorithms because they have generally comparable behavior on the learning set. The other techniques investigated performed much worse (sometimes by an order of magnitude) on every problem. The three algorithms are:

- **tabu-tsab**: a sophisticated tabu search due to Nowicki & Smutnicki [ ]. The neighborhood is based on swapping pairs of adjacent activities on a subset of a randomly selected critical path. An important aspect of tabu-tsab is the use of an evolving set of the five best solutions found. Search returns to one of these solutions and moves in a different direction after a fixed number (1000 in our experiments) of iterations without improvement.
- **texture**: a constructive search technique using texture-based heuristics [ ], strong constraint propagation [,   ], and bounded chronological backtracking. The bound on the backtracking follows the optimal, zero-knowledge pattern of 1, 1, 2, 1, 1, 2, 4, ... [ ]. The texture-based heuristic identifies a resource and time point with maximum competition among the the activities and chooses a pair of unordered activities, branching on the two possible orders. The heuristic is randomized by specifying that the resource and time point is chosen with uniform probability from the top 10% most critical resources and time points.

- **settimes**: a constructive search technique using the SetTimes heuristic [ ], the same propagation as texture, and slice-based search [ ], a type of discrepancy-based search. The heuristic chronologically builds a schedule by examining all activities with minimal start time, breaking ties with minimal end time, and then breaking further ties arbitrarily. The discrepancy bound follows the pattern: 2, 4, 6, 8, ...

4.2   Experimental Details

For these experiments, the overall time limit, T, is 1200 CPU seconds. Each pure algorithm is run for T seconds with results being logged whenever a better solution is found. This design lets us process the results to examine the effect of different settings for the prediction time, t_p, and different values for $T \leq 1200$. As noted, the number of algorithms, $|A|$, is 3.

   To evaluate the prediction rules, we process the data as follows. Given a pure algorithm time of $t = t_p/|A|$ seconds, we examine the best makespans found by each algorithm on problem instance k up to t seconds. Based on the prediction rule, one algorithm, a^*, is chosen. We then examine the best makespan found by a^* for k at $t + t_r$ where $t_r = T - t \times |A|$. This evaluation means that we are assuming that each pure algorithm can be run for t CPU seconds, one can be chosen, and that chosen one can continue from where it left off.

4.3   Results

Learning Set. Table   displays the fraction of learning problems in each subset and overall for which each algorithm found the best solution. It also shows the mean relative error (MRE), a measure of the mean extent to which an algorithm finds solutions worse than the best known solutions. MRE is defined as follows:

$$MRE(a, K) = \frac{\sum_{k \in K} \frac{c(a,k) - c^*(k)}{c^*(k)}}{|K|} \qquad (1)$$

   Where:

- K is a set of problem instances
- $c(a, k)$ is the lowest cost solution found by algorithm a on k
- $c^*(k)$ is the lowest cost solution known k.

   Tabu-tsab finds the best solution for slightly more problems than texture and produces the lowest MRE. These differences are slight as in 50% of the problems, texture finds the best solution. As expected, there are significant differences among the problems sets: while tabu-tsab clearly dominates in the MC problem set, the results are more uniform for the random problem set and texture is clearly superior in the JC set.

   The best variation of each prediction rule are shown in Figure . This graph presents the relative MRE (RMRE) found by the prediction rule as we varied the prediction time, $t \in \{10, 20, \ldots, 400\}$. The RMRE displayed for each prediction

Table 1. Fraction of problems in each learning problem set for which the best solution was found by each algorithm (Frac. Best) and their mean relative error (MRE)

	MC		Rand		JC		All	
	Frac. Best	MRE	Frac. Best	MRE	Frac. Best	MRE	Best	MRE
tabu-tsab	0.7	0.00365	0.6	0.00459	0.3	0.00688	0.53	0.00504
texture	0.2	0.01504	0.4	0.00779	0.9	0.00092	0.5	0.00792
settimes	0.1	0.03752	0	0.03681	0.5	0.00826	0.2	0.02753

rule, p, is the MRE relative to the MRE of best pure algorithm, in this case tabu-tsab, calculated as follows: $RMRE(p(t)) = MRE(p(t))/MRE(tabu\text{-}tsab)$. Values below $y = 1$ represent performance better than the best pure algorithm. For example, the RMRE for $pextrap\_mean(50)$ is 0.736: the MRE achieved by the $pextrap\_mean$ rule at $t = 50$ is 73.6% that achieved by tabu-tsab.

It is possible that the pure algorithms have such similar performance that any prediction rule would perform well. Therefore, two additional "straw men" prediction rules are included in Figure : $pcost\_max(t)$ and $pcost\_rand(t)$. The algorithm whose best solution at t is the maximum over all algorithms is chosen in the former and a random algorithm is chosen in the latter. These two techniques perform substantially worse than the real prediction rules, lending support to the claim that the observed results are not due to a floor effect.

The best performance for each prediction rule is seen with $pcost\_min(110)$, $pslope\_mean(50)$, and $pextrap\_mean(120)$. The differences between the MRE of each prediction rule and tabu-tsab are not statistically significant.

Test Set. Table displays the fraction of the problems in the test set for which each algorithm found the best solution (Fraction Best) and the MRE for each pure algorithm and for the best variation and prediction time, t^*, of each prediction rule. On the basis of the fraction of best solutions, all prediction rules are worse than the best pure algorithm (texture) however none of these differences are statistically significant. Based on MRE, while tabu-tsab and texture are very closely matched, settimes performs significantly worse and each prediction rule performs better than each pure algorithm. Statistically, however, only $pcost\_min(110)$ achieves performance that is significantly better than the best pure algorithm. In fact, $pcost\_min(t)$ is robust to changes in t as a difference at the same level of significance is found for all $t \in \{80, \ldots, 260\}$.

A Static Prediction Technique. The existence of widely differing pure algorithm performance on the different problem subsets (Table) suggests that a high-knowledge, static prediction technique could be built based on categorizing a problem instance into one of the subsets and then using the algorithm that

---

[1] All statistical results in this paper are measured using a randomized paired-t test [ ] and a significance level of $p \leq 0.005$.

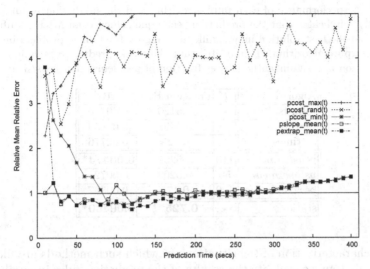

Fig. 1. The performance of the best variation of each prediction rule at prediction time $t \in \{10, 20, \ldots, 400\}$ on the JSP learning set. The graph shows the MRE relative to the MRE achieved by the best pure algorithm, tabu-tsab

performed best on that subset in the learning phase. The *static* prediction technique uses texture on the JC problems and tabu-tsab on the other two sets. The results for static presented in Table   make two strong assumptions: the mapping of a problem instance to a subset is both infallible and takes no CPU time. These assumptions both favor the static technique over the low-knowledge prediction techniques. The results indicate that the static technique outperforms all the other prediction techniques and the pure algorithms in terms of the fraction of problems solved and does the same as *pcost_min* on MRE.

The static technique is knowledge-intensive: one has to know to look for the specific duration structure before algorithm performance correlations can be developed. Therefore, we are not interested specifically in the static technique. It is included to demonstrate that a high-knowledge technique, even under idealized assumptions, may not significantly out-perform a low-knowledge technique.

5   Investigations of Generality

Our initial experiment demonstrates that, at least for the problem sets, algorithms, and time limit used, it is possible to use low-knowledge prediction and simple rules to do algorithm selection. Furthermore *pcost_min(t)* achieves MRE performance that is significantly better than the best pure algorithm and comparable to an idealized high-knowledge approach. A number of questions are raised with respect to the generality of these results. How sensitive are the results to different choices of parameters such as the overall time limit? Can such simple rules be successfully applied to other problems and algorithms? Can we

Table 2. The performance of each pure algorithm and the prediction techniques on the test set. '*' indicates that the prediction technique achieved an MRE significantly lower or found the best solution in a significantly higher fraction of problem instances than the best pure algorithm. '‡' indicates that the static prediction technique found the best solution in a significantly greater fraction of problems than *pcost_min*

Algorithm	t^*	Fraction Best	MRE
tabu-tsab	-	0.5125	0.00790
texture	-	0.5458	0.00859
settimes	-	0.125	0.02776
pcost_min	110	0.5292	**0.00474***
pslope_mean	50	0.5208	0.00726
pextrap_mean	120	0.475	0.00577
static	-	**0.725*‡**	**0.00460***

develop a characterization of the situations in which such methods are likely to be successful? Can we evaluate the results of the prediction rules in an absolute sense and therefore provide intuitions as to the likelihood that more sophisticated prediction techniques may be able to improve upon them? In this section, we will address these questions.

5.1 Other Time Limits

In all experiments presented above, the overall CPU time limit, T, was 1200 seconds. Table   reports a series of experiments with $T \in \{100, 200, \ldots, 1200\}$. For each time limit, we repeated the experiment: t^*, the prediction time with the lowest MRE on the learning set for the best variation of each prediction rule, was identified, each problem in the test set was solved with each prediction rule using its t^* value, and the MRE was compared against the best pure algorithm. There were no significant differences between the MRE of the best pure technique and those of the prediction rules across all the T values on the learning set. The results for the test set are displayed in the final four columns. For time limits $500 \leq T \leq 1200$, $pcost\_min(t^*)$ performs significantly better than the best pure technique. For $T = 100$ the best pure technique (texture) has a significantly lower MRE than $pcost\_min(t^*)$ and $pslope\_min(t^*)$. For $T = 100$, the static technique is able to find significantly lower RMREs than $pcost\_min$. No other time limits showed any difference between static and $pcost\_min$. These results indicate that the results using $T = 1200$ are relatively robust to different T values.

5.2 Other Problems

Earliness/tardiness scheduling problems (ETSPs) define a set of jobs to be scheduled on a set of resources such that each job has an associated due date and costs associated with finishing the last activity in a job before or after that due date. The activities within jobs are completely ordered and the resources can only execute a single activity at any time. Three ETSP algorithms are used here:

Table 3. The results of the best variations of the prediction rules relative to the best pure technique for different run-time limits for the JSP problems. '*' indicates that the prediction rule achieved an RMRE significantly lower than the best pure algorithm, '†' indicates that the best pure technique is significantly better than the prediction rule, and '‡' indicates a time limit where static is significantly better than $pcost\_min$

Time	Learning Set						Test Set			
	pcost		pslope		pextrap		pcost	pslope	pextrap	static
Limit	RMRE	t^*	RMRE	t^*	RMRE	t^*	RMRE			
100	1.114	20	1.139	20	1.094	20	**1.110†**	**1.134†**	1.062	**0.821*‡**
200	1.138	40	1.004	30	0.992	30	1.001	1.103	0.990	**0.793***
300	1.048	60	0.967	30	0.935	30	0.925	1.067	1.012	**0.782***
400	0.969	90	0.879	30	0.895	50	0.914	1.039	0.912	**0.783***
500	0.849	90	0.761	30	0.730	50	**0.814***	1.043	0.920	**0.776***
600	0.815	110	0.751	50	0.740	50	**0.799***	1.024	0.964	**0.738***
700	0.824	110	0.683	50	0.683	50	**0.752***	0.926	0.882	**0.707***
800	0.772	100	0.668	50	0.668	50	**0.683***	0.921	0.877	**0.689***
900	0.748	110	0.702	30	0.679	120	**0.650***	0.977	0.772	**0.660 ***
1000	0.681	90	0.663	50	0.646	120	**0.625***	0.878	0.635	**0.633***
1100	0.680	90	0.671	50	0.600	120	**0.630***	0.909	0.724	**0.618***
1200	0.754	110	0.736	50	0.642	120	**0.600***	0.919	0.730	**0.583***

- **hls**: a hybrid local search algorithm combining tabu search with linear programming.
- **mip**: a pure mixed-integer programming approach using the default search heuristics in CPLEX 7.2 with an emphasis on good solutions over optimal.
- **probeplus**: a probe-based algorithm combining linear programming and constraint programming search.

Details of these algorithms, problems sets, and results can be found in Beck & Refalo [ ].

We divided the 90 ETSP problems into a learning set of 36 problems and a test set of 54 problems. The experimental design is identical to our first experiment. In particular, the overall time, $T = 1200$, and the number of algorithms, $|A| = 3$.

Instead of makespan minimization, the optimization criteria on ETSPs is the minimization of weighted earliness/tardiness cost. It is possible for problems to have a optimal cost of 0 and a number of the easier problem instances do. Therefore, MRE is not well-formed as it would require a division by 0. Instead, we calculate the normalized cost (NC) for each problem and use the mean normalized cost (MNC) as one of our evaluation criteria. NC is commonly used in work that has applied genetic algorithms to ETSPs [ ]. In that literature, the cost of a solution is divided by the sum of the durations of all activities in the problem weighted by the earliness/tardiness cost of each job. In our problems, the earliness and tardiness weights for a single job are independent. Therefore,

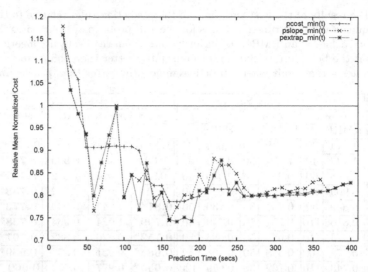

Fig. 2. The performance of the best variations of the three prediction rules at different prediction times on the ETSP learning set. The graph plots the mean normalized costs of each rule at each t value relative to the mean normalized cost achieved by the best pure algorithm

we have modified this normalization to weight the duration sum with the mean of the two cost weights. The NC for algorithm a on problem instance k is

$$NC(a, k) = \frac{c(a, k)}{\sum_{j \in Jobs(k)} (\frac{ec_j + tc_j}{2} \times \sum_{a \in Job_j} dur_a)} \qquad (2)$$

Where:

- $c(a, k)$ is the lowest cost for algorithm a on problem instance k
- $Jobs(k)$ is the set of jobs in problem instance k
- Job_j is the set of activities in job j
- ec_j and tc_j are respectively the earliness and tardiness costs for job j

Figure   presents the MNC of the three best prediction rule variations (relative to the best pure technique, hls for the learning set) with $t \in \{20, 30, \ldots, 400\}$. The plot is analogous to Figure . For each prediction rule the "min" variations results in the best performance with the following t^* values: $pcost\_min(160)$, $pslope\_min(160)$, and $pextrap\_min(170)$. As with the JSP problem set, however, none of these results are significantly different from those found by hls on the learning set.

Table   presents the fraction of the test problems for which each pure and prediction-based technique found the best solution and the MNC over all problem instances in the test set. The prediction rules perform very well on both measures. However, none of them achieve performance on either measure that is significantly different from the best pure technique. The pure technique that

Table 4. The mean normalized cost (MNC) for each pure technique and the best prediction rules on the ETSP test set. None of the prediction rules achieve a significantly different MNC or fraction best than the best pure technique

Algorithm	t^*	Fraction Best	MNC
mip	-	0.4259	0.02571
hls	-	0.6481	0.02908
probeplus	-	0.5	0.02645
pcost_min	160	0.6296	0.01721
pslope_min	160	0.7407	0.01753
pextrap_min	170	0.6667	0.01825

achieves the best solution on the highest number of problem instances (hls) is worst on the basis of MNC. The reverse is also true, as mip finds the lowest MNC but finds the best solution on the fewest number of instances.

5.3   Characterizations of Prediction Techniques

Clearly, two interacting factors determine the performance of the prediction rules tested in this paper and, indeed, any on-line prediction technique: the accuracy of prediction and the computation time required to make the prediction.

We expect prediction accuracy to increase as t_p is increased since more computation time will result in better data regarding algorithm performance. Furthermore, since we have a fixed time limit, the larger t_p, the closer it is to this time limit and the less far into the future we are required to predict. To evaluate the data underlying the accuracy of predictions for the *pcost* rule, in Figure   we present the mean Spearman's rank correlation coefficient between t and $t + t_r$ for the learning sets of both the JSP and the ETSP problems. For a problem instance, k, and prediction time, t, we rank each of the pure algorithms in ascending order of the best makespan found by time t. We then create the same ranking at time $t + t_r$, the total run-time of the chosen algorithm. The correlation between these rankings is calculated using Spearman's rank correlation coefficient and the mean coefficient over all the problems in the set is plotted. It is reasonable to expect that the accuracy of $pcost\_min(t)$ depends on the extent to which the algorithm ranking at time t is correlated with that at $t + t_r$. We can see in the graph that the lower the value of t, the lower the correlation and, therefore, the lower the accuracy of the predictions. Both from the graph and from the reasoning above, to achieve a greater accuracy, prediction should be as late as possible.

For $t = 10$ the JSP rankings are negatively correlated. The appropriate heuristic for choosing a pure algorithm at $t = 10$ is to choose the algorithm whose best makespan is *largest*. This is exactly $pcost\_max(t)$ plotted in Figure   and in that graph, $pcost\_max(10)$ does indeed perform better than $pcost\_min(10)$.

The second factor is the time required to measure the instance and make the prediction. In an on-line context, more time spent predicting means less

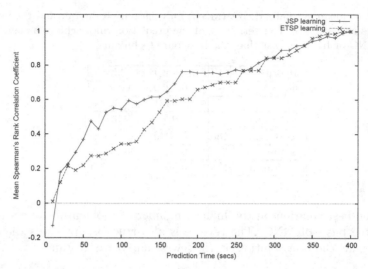

Fig. 3. The mean Spearman's rank correlation coefficient between rankings of the pure algorithms at prediction time, t, and $t + t_r$ for problems in the JSP learning set and the ETSP learning set

spent solving the problem with the chosen algorithm. If $T = 1200$ and $|A| = 3$, then $t = 200$ means that 600 seconds have expired when the algorithm choice is made. Only 600 additional seconds are available to run the chosen algorithm. This has a large implication for performance of prediction-based techniques. This is illustrated in Figure . The $pperf(t)$ plot is the MRE of a perfect prediction on the test set. For example, for $t = 200$, the effective run time of the chosen technique is 800 seconds: 200 seconds during the prediction phase and then the remaining 600 seconds. The perfect MRE for $t = 200$ therefore is found using the lowest makespan found by any pure technique by time 800 and calculating the error compared to the best known makespan. When t is very small, the MRE of $pperf(t)$ is very small too. This reflects the fact that the pure algorithms do not find large improvements extremely late in the run. As the t increases however, the best case MRE increases: the time used in prediction instead of solving results in worse performance even with perfect prediction.

These graphs demonstrate the trade-off inherent for any on-line prediction technique: for accuracy the prediction time should be as late as possible but to allow time for the pure algorithm to run after it is chosen, it should be as early as possible. While the correlation graph presents data specific to a prediction rule used in this paper, we expect a similar graph of accuracy vs. the prediction time for all prediction techniques. The perfect prediction graphs clearly have a general interpretation since, by definition, no prediction technique can achieve better performance.

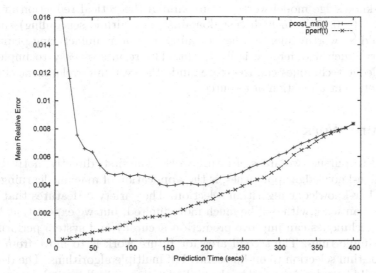

Fig. 4. The MRE on the JSP test set for *pcost_min*(*t*) and when we assume perfect prediction

6   Discussion

We have shown that low-knowledge metrics of pure algorithm behavior can be used to form a system that performs as well, and sometimes better, than the best pure algorithm. If our goal was to win the algorithmic "track meet" and publish better results, our results are not spectacular. However, our goal was not to build a better algorithm through applying our expertise. Our goal was to exploit existing techniques with minimal expertise. From that perspective, the fact that applying simple rules to an instance-specific prediction phase is able to outperform the best pure algorithm is significant. We believe this study serves as a proof-of-concept of low-knowledge approaches and indicates that they are an important area of study in their own right.

Beyond the importance of low-knowledge approaches to reduce expertise, a prosaic reason to develop these approaches is that they can provide guidance in deciding whether the effort and expense of applying human expertise is worthwhile. Figure   shows that at a prediction time of $t = 100$ the mean r-value for *pcost_min*(*t*) on the JSP learning set is 0.543. This is a relatively low correlation, providing support for the idea that a more informed approach can significantly increase prediction accuracy. On the other hand, if we expected the on-line computation required for a high-knowledge approach to take more time (e.g., $t = 250$), the return on an investment in a high-knowledge approach seems less likely: the mean r-value is 0.775 so there is less room for improvement in prediction accuracy. Similarly, Figure   shows that at a prediction time of $t = 100$ the MRE of *pcost_min* on the JSP test set is 0.0046. Based on the *pperf*(*t*) plot, any predictive approach can only reduce this MRE to 0.0013. Is the development of a high-knowledge model worth the maximum theoretical reduction in MRE from 0.46% to 0.13%? In high cost domains (e.g., airline scheduling) such an effort would be worthwhile. In other domains (e.g., a manufacturing plant with uncertainty) such a difference is irrelevant. The results of easy to implement low-knowledge techniques can therefore guide the system development effort in the more efficient allocation of resources.

7   Future Work

We intend to pursue two areas of future work. The first, directly motivated by existing high-knowledge approaches, is the application of machine learning techniques to low-knowledge algorithm selection. The variety of features that these techniques can work with will be much more limited, but we expect that better grounded techniques can improve prediction accuracy and system performance over the simple rules. The second area for future work is to move from "one-shot" algorithm selection to on-line control of multiple algorithms. The decision making could be extended to allow the ability to dynamically switch among pure algorithms based on algorithm behavior.

Another consideration is the types of problems that are appropriate for prediction techniques or control-level reasoning. A real system is typically faced with a series of changing problems to solve: a scheduling problem gradually changes as new orders arrive and existing orders are completed. As the problem or algorithm characteristics change, prediction-based techniques may have the flexibility to appropriately change the pure algorithms that are applied.

8   Conclusion

We have shown that a low-knowledge approach based on simple rules can be used to select a pure algorithm for a given problem instance and that these rules can lead to performance that is as good, and sometimes better, than the best pure algorithm. We have argued that while we expect high-knowledge approaches will result in better performance, low-knowledge techniques are important from the perspective of reducing the expertise required to use optimization technology and have a useful role in guiding the expert in deciding when high-knowledge approaches are likely to be worthwhile.

References

[1] Rice, J.: The algorithm selection problem. Advances in Computers **15** (1976) 65–118

[2] Leyton-Brown, K., Nudelman, E., Shoham, Y.: Learning the empirical hardness of optimization problems: The case of combinatorial auctions. In: Proceedings of the Eighth International Conference on Principles and Practice of Constraint Programming (CP02). (2002) 556–572

[3] Minton, S.: Automatically configuring constraint satisfaction programs: A case study. CONSTRAINTS **1** (1996) 7–43

[4] Horvitz, E., Ruan, Y., Gomes, C., Kautz, H., Selman, B., Chickering, M.: A bayesian approach to tacking hard computational problems. In: Proceedings of the Seventeenth Conference on uncertainty and Artificial Intelligence (UAI-2001). (2001) 235–244

[5] Kautz, H., Horvitz, E., Ruan, Y., Gomes, C., Selman, B.: Dynamic restart policies. In: Proceedings of the Eighteenth National Conference on Artifiical Intelligence (AAAI-02). (2002) 674–681

[6] Ruan, Y., Horvitz, E., Kautz, H.: Restart policies with dependence among runs: A dynamic programming approach. In: Proceedings of the Eighth International Conference on Principles and Practice of Constraint Programming (CP-2002), Springer-Verlag (2002) 573–586

[7] Watson, J. P.: Empirical Modeling and Analysis of Local Search Algorithms for the Job-Shop Scheduling Problem. PhD thesis, Dept. of Computer Science, Colorado State University (2003)

[8] Watson, J. P., Barbulescu, L., Whitley, L., Howe, A.: Constrasting structured and random permutation flow-shop scheduling problems: search-space topology and algorithm performance. INFORMS Journal on Computing **14** (2002)

[9] Nowicki, E., Smutnicki, C.: A fast taboo search algorithm for the job shop problem. Management Science **42** (1996) 797–813

[10] Beck, J. C., Fox, M. S.: Dynamic problem structure analysis as a basis for constraint-directed scheduling heuristics. Artificial Intelligence **117** (2000) 31–81

[11] Nuijten, W. P. M.: Time and resource constrained scheduling: a constraint satisfaction approach. PhD thesis, Department of Mathematics and Computing Science, Eindhoven University of Technology (1994)

[12] Laborie, P.: Algorithms for propagating resource constraints in AI planning and scheduling: Existing approaches and new results. Artificial Intelligence **143** (2003) 151–188

[13] Luby, M., Sinclair, A., Zuckerman, D.: Optimal speedup of Las Vegas algorithms. Information Processing Letters **47** (1993) 173–180

[14] Scheduler: ILOG Scheduler 5.2 User's Manual and Reference Manual. ILOG, S. A. (2001)

[15] Beck, J. C., Perron, L.: Discrepancy-bounded depth first search. In: Proceedings of the Second International Workshop on Integration of AI and OR Technologies for Combinatorial Optimization Problems (CPAIOR'00). (2000)

[16] Cohen, P. R.: Empirical Methods for Artificial Intelligence. The MIT Press, Cambridge, Mass. (1995)

[17] Beck, J. C., Refalo, P.: Combining local search and linear programming to solve earliness/tardiness scheduling problems. In: Proceedings of the Fourth International Workshop on Integration of AI and OR Techniques in Constraint Programming for Combinatorial Optimization Problems (CPAIOR'02). (2002)

[18] Vazquez, M., Whitley, L. D.: A comparision of genetic algorithms for the dynamic job shop scheduling problem. In: Proceedings of the Genetic and Evolutionary Computation Conference (GECCO-2000), Morgan Kaufmann (2000) 1011–1018

Filtering Algorithms for the *Same* Constraint

Nicolas Beldiceanu[1], Irit Katriel[2]*, and Sven Thiel[2]**

[1] Ecole des Mines de Nantes, Nantes, France
nicolas.beldiceanu@emn.fr
[2] Max-Planck-Institut für Informatik, Saarbrücken, Germany
{irit,sthiel}@mpi-sb.mpg.de

Abstract. We define the *Same* and *UsedBy* constraints. *UsedBy* takes two sets of variables X and Z such that $|X| \geq |Z|$ and assigns values to them such that the multiset of values assigned to the variables in Z is contained in the multiset of values assigned to the variables in X. *Same* is the special case of *UsedBy* in which $|X| = |Z|$. In this paper we show algorithms that achieve arc consistency and bound consistency for the *Same* constraint and in its extended version we generalize them for the *UsedBy* constraint.

1 Introduction

As a motivating example, we consider simple scheduling problems of the following type. The organization Doctors Without Borders [ ] has a list of doctors and a list of nurses, each of whom volunteered to go on one rescue mission in the next year. Each volunteer specifies a list of possible dates and each mission should include one doctor and one nurse. The task is to produce a list of pairs such that each pair includes a doctor and a nurse who are available on the same date and each volunteer appears in exactly one pair. Since the list of potential rescue missions at any given date is infinite, it does not matter how the doctor-nurse pairs are distributed among the different dates.

We model such a problem by the $Same(X = \{x_1,\ldots,x_n\}, Z = \{z_1,\ldots,z_n\})$ constraint which is defined on two sets X and Z of distinct variables such that $|X| = |Z|$ and each $v \in X \cup Z$ has a domain $D(v)$. A solution is an assignment of values to the variables such that the value assigned to each variable belongs to its domain and the multiset of values assigned to the variables of X is identical to the multiset of values assigned to the variables of Z.

This problem can be generalized to the case in which there are more nurses than doctors and the task is to create a list of pairs as above, with the requirement that every doctor appears in exactly one pair and every nurse in at most one pair (naturally, not all of the nurses will be paired). For this version we use the general case of the $UsedBy(X = \{x_1,\ldots,x_n\}, Z = \{z_1,\ldots,z_m\})$ constraint where $|X| = n \geq m = |Z|$ and a solution is an assignment of values to the variables such that the multiset of values assigned to the variables of Z is contained in the multiset of values assigned to the variables of X.

In this paper we show filtering algorithms for the *Same* constraint and in the full version [ ] we generalize them to solve the *UsedBy* constraint. Let Y be the union of

---

* Partially supported by DFG grant SA 933/1-1.
** Partially supported by the EU IST Programme, IST-1999-14186 (ALCOM-FT).

the domains of the variables in $X \cup Z$ and let $n' = |Y|$. The arc consistency algorithms run in time $O(n^2 n')$ and the bound consistency algorithms in time $O(n' \alpha(n', n') + n \log n)$, where α is the inverse of Ackermann's function .

The general approach we take resembles the flow-based filtering algorithms for the *AllDifferent* [,] and *Global Cardinality* (*GCC*) [,] constraints: We construct a bipartite variable-value graph, find a single solution in it and compute the strongly connected components (SCCs) of the residual graph. We show that an edge is consistent iff both of its endpoints are in the same SCC.

The main difference compared to the previous constraints that were solved by the flow-based approach is that we now have three sets of nodes. One set for each set of variables and a third set for the values. This significantly complicates the bound consistency algorithms, in particular the SCC computation compared to the corresponding stage in the *AllDifferent* and *GCC* cases. Our contribution is therefore not only in providing a solution to these constraints but also in showing that the ideas that appear in the previous algorithms can be extended to much more complex variable-value graphs.

In Section   we define the variable-value graph for the *Same* constraint and characterize the solutions to the constraint in terms of subsets of the edges in this graph. In Section   we show the arc consistency algorithm and in Section   we show the bound consistency algorithm. Source code for the bound consistency algorithm is available by request from the authors.

2   The *Same* Constraint

We represent the *Same* constraint as a bipartite graph $B = (X \cup Z, Y, E)$, which we call the *variable-value graph*, where $E = \{\{v, y\} \mid v \in X \cup Z \land y \in Y \land y \in Dom(v)\}$. That is, the nodes on one side represent the variables and the nodes on the other represent the values and every variable is connected by an edge to all values in its domain. We now characterize the set of all solutions to the constraint in terms of subsets of edges of B.

Definition 1. *Let $M \subseteq E$ be a set of edges of B. For any node $v \in X \cup Y \cup Z$, let $N_M(v)$ be the set of nodes which are neighbors of v in $B' = (X \cup Z, Y, M)$. We say that M is a* parity matching *(PM) in B iff $\forall_{v \in X \cup Z} |N_M(v)| = 1$ and $\forall_{y \in Y} |N_M(y) \cap X| = |N_M(y) \cap Z|$.*

Lemma 1. *There is a one to one correspondence between the solutions to the Same constraint and the PMs in B.*

Proof. Given a PM M in B, we can construct the solution

$$Same(\{N_M(x_1), \ldots, N_M(x_n)\}, \{N_M(z_1), \ldots, N_M(z_n)\}).$$

Since $|N_M(v)| = 1$ for all $v \in X \cup Z$, all of the assignments are well defined. In addition, for each edge (v, y) in B, and in particular in M, $y \in Dom(v)$. Finally, since $|N_M(y) \cap X| = |N_M(y) \cap Z|$ for all $y \in Y$, each value is assigned the same number of times to variables of X and Z. Hence, the constraint is satisfied.

Given a solution $Same(\{y(x_1), \ldots, y(x_n)\}, \{y(z_1), \ldots, y(z_n)\})$ where $y(v)$ is the value

---

[1] For all practical purposes, $\alpha(n', n')$ can be regarded as a small constant.

Table 1. Domains of the variables for our example

j	$D(x_j)$	$D(z_j)$
1	{1,2}	{2,3}
2	{3,4}	{4,5}
3	{4,5,6}	{4,5}

assigned to v, define $M = \{\{v,y\}|v \in X \cup Z \wedge y = y(v)\}$. Since $y(v) \in Dom(v)$ for all v, we have $M \subseteq E$. Since $y(v)$ is determined for all variables, $|N_M(v)| = 1$ for all $v \in X \cup Z$ and since each $y \in Y$ appears the same number of times in $\{y(x_1),\ldots,y(x_n)\}$ and in $\{y(z_1),\ldots,y(z_n)\}$), $|N_M(y) \cap X| = |N_M(y) \cap Z|$, so M is a PM.   □

In the next sections we show the filtering algorithms for *Same*, first arc consistency and then bound consistency. We will illustrate them with the aid of the following example. $|X| = |Z| = 3$, $|Y| = 6$ and the domains of the variables of $X \cup Z$ are as in Table .

3   Arc Consistency

Inspired by Régin [ ], we convert the graph B into a capacitated and directed graph $B = (V,E)$, as follows. We direct the edges from X to Y and from Y to Z and assign a capacity requirement of $[0,1]$ to each of these edges. We add two nodes s and t, an edge with capacity $[1,1]$ from s to each $v \in X$ and from each $v \in Z$ to t and an edge with capacity $[n,n]$ from t to s (see Figure). A flow in B is feasible iff there is a flow of value n on the arc from t to s. This implies that one unit of flow goes through every node in $X \cup Z$. By flow conservation, every node in Y is connected by edges that carry flow to the same number of nodes from X and from Z. The correspondence between PMs in B and feasible flows in B should be obvious.

The algorithm uses Ford and Fulkerson's augmenting paths method to find a feasible flow f in B. If there is no such flow it reports that the constraint is not satisfiable. Otherwise, it removes the nodes s and t and builds the *residual graph* $B_f = (V_f, E_f)$ where $V_f = X \cup Y \cup Z$ and $E_f = E \cup \{(v,u)|u,v \in V \wedge (u,v) \in E \wedge f(e) = 1\}$. That is, all edges appear in their original orientation and the edges that carry flow appear also in

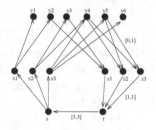

Fig. 1. The capacitated graph for the example in Table

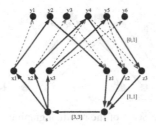

Fig. 2. A feasible flow in the graph of Figure   (left) and the corresponding residual graph (right)

reverse direction (see Figure   ). We now show that B_f can be used to determine which of the edges of B are consistent.

Lemma 2. *An edge $e = (u,v) \in B_f$ is consistent iff u and v belong to the same SCC.*

Proof. Suppose that u and v are in the same SCC. Let $M = \{e'|f(e') = 1\}$ be the PM that corresponds to the flow f. If $e \in M$ then e participates in the solution M and is therefore consistent. Assume that $e \notin M$. Then there is a cycle P in B_f that uses e. Starting at any y node on the cycle, number its nodes: $y_0, v_0, y_0, v_0, \ldots, y_{p-1}, v_{p-1}$ where $y_i \in Y$ and $v_i \in X \cup Z$ for all $0 \leq i < p$. Let $M' = M \oplus P$ be the symmetric difference between the PM M and the cycle P. That is, $M' = \{e'|e' \in M \wedge e' \in P \wedge e' \notin M \cap P\}$. If we look at $y_i \in P$ and its two neighbors, we have four possibilities: If $v_{(i-1) \bmod p}, v_i \in X$ or $v_{(i-1) \bmod p}, v_i \in Z$ then in M' each of $v_{(i-1) \bmod p}, v_i$ remains matched and y_i is matched with the same number of nodes from each of X and Z as it was in M'.
The other two options are that $v_{(i-1) \bmod p} \in X$ and $v_i \in Z$ or $v_{(i-1) \bmod p} \in Z$ and $v_i \in X$. Then the edges $(v_{(i-1) \bmod p}, y_i)$ and (y_i, v_i) are either both in M or both in M'. To see that this is true, note that the cycle can only enter a node $v \in X$ by an edge in M and leave it by an edge which is not in M. On the other hand, it can only enter a node $v \in Z$ by an edge which is not in M and leave it by an edge in M. This implies that either $v_{(i-1) \bmod p} \in Z$ and $v_i \in X$ and both edges are in M or $v_{(i-1) \bmod p} \in X$ and $v_i \in Z$ and both edges are in M'. In either case y_i either lost or gained a neighbor from each of X and Z, so it is still matched with the same number of nodes from each of these sets.

Each of $v_{(i-1) \bmod p}, v_i$ is adjacent on P to one edge from M and one edge which is not in M. Hence, in M' it is still matched with exactly one y-node. We get that M' is a PM that contains e, so e is consistent.

It remains to show that an edge e which is not in an SCC of B_f (which implies $f(e) = 0$) is not consistent. Let $C_1, \ldots, C_k$ be the SCCs of B_f. Since all edges in M appear in both directions, any edge between two SCCs is not in M. For any node v, let $C(v)$ be the SCC that v belongs to and let $E_{in} = \{(u,v)|C(u) = C(v)\}$ be the edges for which both endpoints are in the same SCC. Assume that an edge $e \in E \setminus E_{in}$ is consistent and let M' be a PM such that $e \in M'$. Consider the graph $B'_f = (V_f, E_{in} \cup M')$. That is, B'_f contains all edges within SCCs plus the edges between SCCs which are in M'. If we shrink each SCC of B'_f into a single node, we get a DAG (directed acyclic graph). Let D_e be the connected component of this DAG which contains e and let C be a root of D_e, i.e., there are only outgoing edges from C. Let E_{xy} be the $x \to y$ edges and E_{yz} the

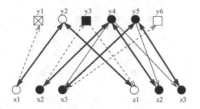

Fig. 3. The SCCs of the residual graph of Figure

$y \to z$ edges out of C. Since C is not an isolated node, $|E_{xy}| + |E_{yz}| \geq 1$. Let Y_C be the set of y nodes in the SCC represented by C. Then $|M'(Y_C) \cap X| = |M(Y_C) \cap X| - |E_{xy}|$ and $|M'(Y_C) \cap Z| = |M(Y_C) \cap Z| + |E_{yz}|$, contradicting the assumption that M' is a PM.   □

Figure   shows the SCCs of B_f for the example in Table . The nodes of each SCC have a distinct shape and the inconsistent edges are dashed.

Let $|E|$ denote the number of edges in B and recall that $n = |X| = |Z|$ and $n' = |Y|$. Clearly, $|E| = O(nn')$. The running time of the algorithm is dominated by the time required to find a flow, which is $O(n|E|) = O(n^2 n')$ [,].

4   Bound Consistency

The bound consistency algorithm does the same as the arc consistency algorithm, but achieves faster running time by exploiting the simpler structure of the graph B: As the domain of every $v \in X \cup Z$ is an interval, B is *convex*, which means that the neighborhood of every variable node is a consecutive sequence of value nodes. We will show that in a convex graph we can find a PM in time $O(n' + n \log n)$ and compute the SCCs of the residual graph in time $O(n' \alpha(n', n))$.

4.1   Finding a Parity Matching

Figure   shows the algorithm for finding a PM in the graph B. It uses two priority queues, P_x for the nodes in X and P_z for the nodes in Z. In both queues the nodes are sorted by the upper endpoints of their domains.

For any $v \in X \cup Z$, let $\underline{D}(v)$ and $\overline{D}(v)$ denote the lower and upper endpoints of $D(v)$, respectively. The algorithm traverses the value nodes from y_1 to $y_{n'}$ and for each y_i inserts to the respective queue all variable nodes $v \in X \cup Z$ with $\underline{D}(v) = i$. It then checks whether there is a node in one of the queues (the node with minimum priority) whose domain ends at i. If so, it tries to match this node and a node from the other queue with y_i. If the other queue is empty, it declares that there are no PMs in B. The PM obtained by the algorithm for the example in Table   corresponds to the flow shown in Figure .

Lemma 3. *If there is a PM in B then the algorithm in Figure   finds one.*

(* Assumption: X and Z are sorted according to $\underline{D}$. *)
$P_x \leftarrow []$ (* priority queue containing x nodes sorted by $\overline{D}$ *)
$P_z \leftarrow []$ (* priority queue containing z nodes sorted by $\overline{D}$ *)
$j \leftarrow 0$
for $i = 1$ **to** n' **do**
    forall x_h with $\underline{D}(x_h) = i$ **do** P_x.Insert x_h
    forall z_h with $\underline{D}(z_h) = i$ **do** P_z.Insert z_h
    (* Assume that MinPriority of an empty queue is ∞ *)
    while P_x.MinPriority $= i$ or P_z.MinPriority $= i$ **do**
        if P_x.IsEmpty or P_z.IsEmpty **then** report failure
        $j \leftarrow j + 1$
        $x \leftarrow P_x$.ExtractMin; match x with y_i
        $z \leftarrow P_z$.ExtractMin; match z with y_i
    end while
endfor
if P_x and P_z are both empty **then** report success
else report failure

Fig. 4. Algorithm to find a parity matching in a convex graph

Proof. We show by induction on i that if there is a PM M then for all $0 \leq i \leq n'$, there is a PM M_i in B which matches $\{y_1, \ldots, y_i\}$ with the same matching mates as the algorithm.

For $i = 0$, the claim holds with $M_0 = M$. For larger i, given a PM M, we can assume by the induction hypothesis that there is a PM M_{i-1} that matches the nodes in $\{y_1, \ldots, y_{i-1}\}$ with the same matching mates as the algorithm. We show how to construct M_i from M_{i-1}. As long as the matching mates of y_i are not the same as the ones determined by the algorithm, perform one of the following transformations:

If y_i is matched with a pair x_j, z_k such that neither one of x_j and z_k was matched with y_i by the algorithm, then since x_j and z_k were not matched by the algorithm with any of $\{y_1, \ldots, y_{i-1}\}$, they both remained in the queues after iteration i, which implies $\overline{D}(x_j) > i$ and $\overline{D}(z_k) > i$. In M_i we match both of them with y_{i+1}.

The other option is that the algorithm matched y_i with a pair x_j, z_k which are not both matched with y_i in M_i. Since the algorithm extracted them from the queues, we know that at least one of them has upper domain endpoint equal to i. Assume w.l.o.g. that $\overline{D}(x_j) = i$. Since M_{i-1} agrees with the algorithm on the matching mates of $\{y_1, \ldots, y_{i-1}\}$, we get that in M_{i-1}, x_j is matched with y_i and z_k is matched with $y_{i'}$ for some $i' > i$. Hence, there is some other node $z_{k'} \in Z$ which is matched with y_i in M_{i-1} and which was matched by the algorithm with $y_{i''}$ for some $i'' > i$. When the algorithm extracted z_k from the queue, $z_{k'}$ was in the queue because it is a neighbor of y_i. Hence, $\overline{D}(z_{k'}) \geq \overline{D}(z_k)$ so we can exchange z_k and $z_{k'}$. That is, we can match z_k with y_i and $z_{k'}$ with $y_{i'}$.

Each time we apply one of the transformations above to M_{i-1}, we decrease the number of differences between the set of matching mates of y_i in M_{i-1} and in the matching generated by the first i iterations of the algorithm. So we can continue until we obtain a matching M_i that agrees with the algorithm on the matching mates of $\{y_1, \ldots, y_i\}$. □

Lemma 4. *If the algorithm in Figure   reports success then it constructs a PM in B.*

Proof. If the algorithm reports success then P_x and P_z are empty at the end, which means that all nodes in $X \cup Z$ were extracted and matched with y nodes. In addition, since the algorithm did not report failure during the extractions, whenever $v \in X \cup Z$ was matched with y_i, $\underline{D}(v) \leq i \leq \overline{D}(v)$. For all $1 \leq i \leq n'$, whenever y_i is matched with some node from X it is also matched with a node from Z, and vice versa. Hence, we get that the matching that was constructed is a PM. □

4.2   Finding Strongly Connected Components

Having found a PM in B, which we can interpret as a flow in B, we next wish to find the SCCs of B_f (cf. Figure). Mehlhorn and Thiel [ ] gave an algorithm that does this in the residual graph of the *AllDifferent* constraint in time $O(n)$ plus the time required for sorting the variables according to the lower endpoints of their domains. Katriel and Thiel [ ] enhanced this algorithm for the *GCC* constraint, in which a value node can be matched with more than one variable node. For our graph, we need to construct a new algorithm that can handle the distinction between the nodes in X and in Z and the more involved structure of the graph.

As in [,], the algorithm in Figure   begins with n' initial components, each containing a node $y_i \in Y$ and its matching mates (if any). It then merges these components into larger ones. While the algorithm used for the *AllDifferent* graph can do this in one pass over the y nodes from $y_1, \ldots, y_{n'}$, our algorithm makes two such passes for reasons that will be explained in the following. The first pass resembles the SCC algorithm for the *AllDifferent* graph. It traverses the y nodes from y_1 to $y_{n'}$ and uses a stack to merge components which are strongly connected and are adjacent to each other. It maintains a list *Comp* of completed components and a stack *CS* of temporary components. The components in both *Comp* and *CS* are not guaranteed to be SCCs of B_f. They are strongly connected but may not be maximal. However, a component in *Comp* is completed with respect to the first pass, while the components in *CS* may still be merged with unexplored components and with other components in *CS*.

Let B_f^i be the graph induced by $\{y_1, \ldots, y_i\}$ and their matching mates. The first pass begins with the empty graph B_f^0 and in iteration i moves from the graph B_f^{i-1} to B_f^i, as follows. As long as the topmost component in *CS* does not reach any $y_{i'}$ with $i' > i$ by a single edge, this component is popped from *CS* and appended to *Comp*. Then the algorithm creates a new component C with y_i and its matching mates. It repeatedly checks if C and the topmost component in *CS* reach each other by a single edge in each direction. If so, it pops the component from *CS* and merges it into C. Finally, it pushes C onto *CS* and proceeds to the next iteration.

The reason that this pass is enough for the *AllDifferent* graph but not for ours is that in our case the outgoing edges of a y-node do not fulfil the convexity criteria: It could be that there is an edge from y_i to a z node which is matched with $y_{i'}$ for some $i' > i + 1$ while there is no edge from y_i to any of y_{i+1}'s matching mates. In the *AllDifferent* case,

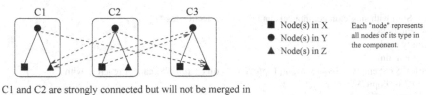

C1 and C2 are strongly connected but will not be merged in
the first pass because C2 (and not C1) is the topmost
component in CS when C3 is created and pushed onto CS.

Fig. 5. Example of components that will not be merged by the first pass of the SCC algorithm

this could not happen: If a matching mate of y_i can reach any $y_{i'}$ by a single edge then convexity implies that it can reach all y nodes between y_i and $y_{i'}$. This means that in our graph, there could be two components C, C' in CS such that C reaches C' by a single edge and C' reaches C by a single edge, but this is not detected when the second of them was inserted into CS because the first was not the topmost in the stack (see, e.g., Figure). The second pass, which merges such components, will be described later.

In the following, whenever we speak of a component C of B_f, we refer to a set of nodes such that for every node in C, all of its matching mates are also in C. This means that a component C is strongly connected, but may not be maximal. We say that a component C *reaches* a component C' if there is a path in B_f from a node in C to a node in C'. In addition, two components C and C' are *linked* if there is an edge from C to C' and there is an edge from C' to C. They are *linked by $x \to y$ (linked by $y \to z$) edges* if the edges in both directions are $x \to y$ ($y \to z$) edges.

The pseudo-code in Figure   uses the following shortcuts. In the full version of this paper [ ] we show how to implement them with linear-time preprocessing. Let C, C' be two components.

- *MinY[C] (MaxY[C])* is the minimum (maximum) index of a y node in C.
- *ReachesRight[C]* is the largest index i such that y_i or one of its matching mates can be reached by a single edge from a node in C.
- *xyLeftLinks[C] (yzLeftLinks[C])* is true iff C is linked with some component to its left by $x \to y$ ($y \to z$) edges.
- *Linked[C,C']* is true iff C and C' are linked.
- *xyLinks[C,C'] (yzLinks[C,C'])* is true iff C and C' are linked by $x \to y$ ($y \to z$) edges.

The following lemmas examine the components that are generated by the first pass of the algorithm, first with respect to their order and connectivity in CS and then with respect to the SCCs of B_f that they compose. They will help us to show that the SCCs of B_f, when viewed as combinations of components that are generated by the first pass of the algorithm, have a relatively simple structure which enables to identify them in the second pass.

Lemma 5. *Let $CS = < C_1, C_2, \cdots >$ be the components in CS at the end of iteration i of the first pass (ordered from bottom to top). Then for all κ, $MaxY[C_\kappa] < MinY[C_{\kappa+1}]$. In other words, no component is nested in another and the components appear in CS in increasing order of the indices of their y nodes.*

(* Pass 1: Start with singleton components and merge adjacent ones *)
$Comp \leftarrow$ empty list
$CS \leftarrow$ empty stack
for $i = 1$ **to** n' **do**
    while CS not empty $\wedge ReachesRight(\text{Top}(CS)) < i$ **do** (* Top(CS) cannot reach $y_{i'}$ with $i' \geq i$. *)
        $C' \leftarrow \text{Pop}(CS)$
        append C' to $Comp$
    end while
    $C \leftarrow \{y_i\} \cup N_M(y_i)$
    while CS not empty $\wedge Linked(TOP(CS), C)$
        $C' \leftarrow \text{Pop}(CS)$
        $C \leftarrow C' \circ C$ (* Merge components *)
    end while
    push C onto CS
endfor
while CS not empty **do**
    $C' \leftarrow \text{Pop}(CS)$
    append C' to $Comp$
end while
(* Pass 2: merge non-adjacent components *)
$SCCs \leftarrow$ empty list
$CS \leftarrow$ empty stack
for $i = 1$ **to** $|Comp|$ **do** (* Traverse the components of $Comp$ by $MinY[C]$ order *)
    while CS not empty $\wedge ReachesRight(\text{Top}(CS)) < MinY[C_i]$ **do** (* Top(CS) cannot reach $C_{i'}$ with $i' \geq i$. *)
        $C \leftarrow \text{Pop}(CS)$
        if $xyLeftLinks[C]$ **then** push C onto $CSxy$
        else if $yzLeftLinks[C]$ **then** push C onto $CSyz$
        else append C to $SCCs$
        if $xyLinks[\text{Top}(CS), \text{Top}(CSxy)]$ **then**
            $C \leftarrow \text{Pop}(CS)$
            $C' \leftarrow \text{Pop}(CSxy)$
            $C \leftarrow C' \circ C$ (* Merge components *)
            while $Linked[C, \text{Top}(CS)]$ **do**
                $C' \leftarrow \text{Pop}(CS)$
                $C \leftarrow C' \circ C$ (* Merge components *)
            end while
            push C onto CS
        end if
        if $yzLinks[\text{Top}(CS), \text{Top}(CSyz)]$ **then**
            $C \leftarrow \text{Pop}(CS)$
            $C' \leftarrow \text{Pop}(CSyz)$
            $C \leftarrow C' \circ C$ (* Merge components *)
            while $Linked[C, \text{Top}(CS)]$ **do**
                $C' \leftarrow \text{Pop}(CS)$
                $C \leftarrow C' \circ C$ (* Merge components *)
            end while
            push C onto CS
        end if
    end while
    push C_i onto CS
endfor
return $SCCs$

Fig. 6. Algorithm to find the SCCs of the residual graph

Proof. Initially, CS is empty and the claim holds. Assume that it holds after iteration $i - 1$ and consider the changes made to the stack during iteration i. First some of the topmost components are popped from the stack; this does not affect the correctness of the claim. Then some of the topmost components are merged with each other and with the new component C and the result is pushed to the top of the stack. By the induction hypothesis, all components that were popped and merged contain y nodes with larger indices than the components that remained in CS. In addition, all y nodes in CS after iteration $i - 1$ have indices smaller than i. So the claim holds after iteration i.   □

Lemma 6. *Let $CS = < C_1, C_2, \cdots >$ be as in Lemma . Then for all κ, C_κ and $C_{\kappa+1}$ are not linked.*

Proof. Again, the claim clearly holds for the empty stack. Assume that it is true after iteration $i - 1$. By the induction hypothesis, the claim holds for every adjacent pair of components that remained in CS after popping the completed components. If the new component C that is pushed onto CS is linked with the component C' which is immediately below it, then the algorithm would have popped C' and merged it with C. Hence, the claim holds at the end of iteration i.   □

Lemma 7. *If $CS = < C_1, C_2, \cdots >$ is as in Lemma and $\{C_{i_1}, \ldots, C_{i_\kappa}\}$ is a maximal set of components in CS which belong to the same SCC of B_f such that $i_1 \leq \cdots \leq i_\kappa$ then $C_{i_{\kappa-1}}$ and C_{i_κ} are linked.*

Proof. Assume that there is such a set of components $\{C_{i_1}, \ldots, C_{i_\kappa}\}$ in CS where $C_{i_{\kappa-1}}$, C_{i_κ} are not linked.

Case 1: C_{i_κ} does not reach $C_{i_{\kappa-1}}$ by a single edge. Then it must reach a component C_{i_j} with $i_j < i_{\kappa-1}$ by a single edge. This edge must be a $y \to z$ edge because otherwise convexity and Lemma would imply that it also reaches $C_{i_{\kappa-1}}$ by an $x \to y$ edge, in contradiction to our assumption. Let z_γ be the target of this edge and y_β be its matching mate. Assume that β is maximal among β' such that $y_{\beta'}$ is in one of the components $C_{i_1}, \ldots, C_{i_{\kappa-2}}$ and it has a matching mate z_γ which is reachable from C_{i_κ} by a $y \to z$ edge. There is a path from z_γ to $C_{i_{\kappa-1}}$. If the path includes an $x \to y$ edge from a matching mate x_α of y_β to $y_{\beta'}$ with $\beta' > \beta$, then $\overline{D}(x_\alpha) > \beta$ and $\overline{D}(z_\gamma) > \beta$ so the algorithm in Figure could not have matched x_α and z_γ with y_β, a contradiction. If the path includes a $y \to z$ edge from $y_{\beta'}$ with $\beta' \leq \beta$ to a node $z_{\gamma'}$ which is matched with $y_{\beta''}$ where $\beta'' > \beta$, then $z_{\gamma'}$ was in P_z when z_γ was extracted, so by convexity and Lemma it is reachable from C_{i_κ} by a $y \to z$ edge, in contradiction to the maximality of β. We get that the path must go from z_γ to the left and then bypass y_β by an $x \to y$ edge $(x_{\alpha'}, y_{\beta''})$ such that $x_{\alpha'}$ is matched with $y_{\beta'}$ and $\beta' < \beta < \beta''$. We can assume w.l.o.g. that the path from z_γ to $x_{\alpha'}$ does not go to the left of $x_{\alpha'}$; if it does then one can show that either Lemma is violated or there also exists a path that shortcuts the part that goes to the left of $x_{\alpha'}$. If the path ends with a $y \to z$ edge into a matching mate z_γ of $y_{\beta'}$ followed by the matching edges $(z_\gamma, y_{\beta'}) \circ (y_{\beta'}, x_{\alpha'})$ then $\overline{D}(x_{\alpha'}) > \beta'$ and $\overline{D}(z_\gamma) > \beta'$ so the algorithm in Figure could not have matched $x_{\alpha'}$ and z_γ with $y_{\beta'}$, a contradiction. So the path ends with an $x \to y$ edge $(x_{\alpha'''}, y_{\beta'})$ followed by the edge $(y_{\beta'}, x_{\alpha'})$, where $x_{\alpha'''}$ is matched with $y_{\beta'''}$ for some $\beta' < \beta''' \leq \beta$. $x_{\alpha'''}$ was in P_x when $x_{\alpha'}$ was extracted, so it also reaches $y_{\beta''}$ by an $x \to y$ edge. By continuing backwards along the path and applying the same considerations, we get that there is an $x \to y$ edge from x_α to $y_{\beta''}$, hence again $\overline{D}(x_\alpha) > \beta$ and $\overline{D}(z_\gamma) > \beta$, so the algorithm in Figure   could not have matched x_α and z_γ with y_β, a contradiction.

Case 2: $C_{i_{\kappa-1}}$ does not reach C_{i_κ} by a single edge. Then C_{i_κ} is reached from another component C_{i_j} by an $x \to y$ edge $(x_\alpha, y_{\beta'})$. Assume that x_α is matched with y_β and that β is maximal among β'' such that $y_{\beta''}$ is in one of the components $\{C_{i_1}, \ldots, C_{i_{\kappa-2}}\}$ and has a matching mate $x_{\alpha''}$ which reaches C_{i_κ} by an $x \to y$ edge. There is a path from $C_{i_{\kappa-1}}$ to y_β. If there is an edge $(x_{\alpha''}, y_\beta)$ from $x_{\alpha''}$ which is matched with $y_{\beta''}$ for some $\beta < \beta'' < \beta'$, then by convexity and Lemma , $x_{\alpha''}$ reaches C_{i_κ} by an $x \to y$ edge, in contradiction to the maximality of β. If there is an edge $(y_{\beta''}, z_\gamma)$ such that z_γ is matched with y_β and $\beta'' > \beta$ then $\overline{D}(x_\alpha) > \beta$ and $\overline{D}(z_\gamma) > \beta$ so the algorithm in Figure   could not have matched x_α and z_γ with y_β, a contradiction. We get that the path must bypass y_β by a $y \to z$ edge $(y_{\beta''}, z_{\gamma''})$ such that $z_{\gamma''}$ is matched with $y_{\beta'''}$ for some $\beta''' < \beta < \beta''$, and then return from $z_{\gamma''}$ to y_β. With arguments similar to the ones used in case 1, we get that the path from $z_{\gamma''}$ to y_β must consist of $y \to z$ edges, which implies that there is a $y \to z$ edge from $y_{\beta''}$ to z_γ, hence again $\overline{D}(x_\alpha) > \beta$ and $\overline{D}(z_\gamma) > \beta$ so the algorithm in Figure   could not have matched x_α and z_γ with y_β, a contradiction. □

Corollary 1. *Let $CS = < C_1, C_2, \cdots >$ and $\{C_{i_1}, \ldots, C_{i_\kappa}\}$ be as in Lemma . Then $C_{i_{\kappa-1}}$ and C_{i_κ} are either linked by $x \to y$ edges or linked by $y \to z$ edges.*

Proof. Lemma   guarantees that $C_{i_{\kappa-1}}$ and C_{i_κ} are linked. Assume that these edges are not of the same type. That is, one is an $x \to y$ edge and the other is a $y \to z$ edge. Then by convexity and Lemma   we get that either $C_{i_{\kappa-1}}$ and $C_{i_{\kappa-1}+1}$ or $C_{i_\kappa-1}$ and C_{i_κ} are linked, and this contradicts Lemma . □

Lemma 8. *Let $CS = < C_1, C_2, \cdots >$ and $\{C_{i_1}, \ldots, C_{i_\kappa}\}$ be as in Lemma   and assume that there is at least one unexplored node which is in the same SCC of B_f as $\{C_{i_1}, \ldots, C_{i_\kappa}\}$. Then there is an edge from C_{i_κ} to an unexplored node.*

Proof. Assume the converse. Then there is an edge from one of the components in $\{C_{i_1}, \ldots, C_{i_{\kappa-1}}\}$ to an unexplored node. Let j be maximal such that $C_j \in \{C_{i_1}, \ldots, C_{i_{\kappa-1}}\}$ has an edge to an unexplored node. If this is a $y \to z$ edge then by convexity and Lemma C_{i_κ} is also connected by a $y \to z$ edge to an unexplored node. So it must be an $x \to y$ edge. Let $x_j \in C_j$ be its source. With arguments similar to the ones used in the proof of Lemma   we can show that there cannot be a path from C_{i_κ} back to x_j. This contradicts the assumption that C_{i_κ} and C_j are in the same SCC of B_f. □

Corollary 2. *If $\{C_1, C_2, \ldots, C_{p_1}\}$ are the components found by the first pass of the algorithm and $\{C_{i_1}, \ldots, C_{i_\kappa}\}$ is a maximal subset of these components which are strongly connected between them such that $MinY[C_{i_1}] \leq \cdots \leq MinY[C_{i_\kappa}]$. Then (1) No component in $\{C_{i_1}, \ldots, C_{i_\kappa}\}$ is nested in another. That is, for all $j, j' \in \{i_1, \ldots, i_\kappa\}$ such that $j < j'$, $MaxY[C_j] < MinY[C_{j'}]$. (2) $C_{i_\kappa-1}$ and C_{i_κ} are linked. Furthermore, they are either linked by $x \to y$ edges or linked by $y \to z$ edges.*

Proof. Assume that there is a component C_j in the set which is nested in another component $C_{j'}$. Then $C_{j'}$ consists of nodes which are both to the left and to the right of C_j. This means that at some point, the algorithm merged these nodes into one component.

At this point in time, the topmost component in CS consisted only of nodes which are to the left of C_j and it was merged with a component that contains only nodes to the right of C_j. By Lemma , this means that C_j was popped before this iteration, but this contradicts Lemma , so we have shown that (1) holds.

By Lemma , we know that none of $\{C_{i_1},\ldots,C_{i_{\kappa-1}}\}$ were popped from CS before C_{i_κ} was pushed onto it. At a certain iteration, C_{i_κ} was pushed onto CS and stayed there at least until the next iteration. This, together with Corollary , implies (2). $\qquad\square$

To sum up, the first pass partitions the nodes into components such that the components that compose an SCC of B_f are not nested and the two rightmost components of each SCC are linked by edges of the same type. The second pass merges components that belong to the same SCC. It starts with an empty stack CS and an empty list $SCCs$ and traverses the components found in the first pass by increasing order of $MinY[C]$. When considering a new component C, it first pops from CS all topmost components that cannot reach C or beyond it. For each such component C', it first checks if C' is linked with a component to its left by $x \rightarrow y$ edges. If so, there are components in CS that it needs to be merged with. So the algorithm pushes C' to a second stack $CSxy$. Otherwise, it checks if C' is linked with a component to its left by $y \rightarrow z$ edges and if so, pushes it to a third stack $CSyz$. Otherwise, it appends C' to $SCCs$ because C' is not linked with any component in CS and it does not reach unexplored components.

Before popping the next component, it checks whether the topmost component in CS and the topmost component in $CSxy$ are linked by $x \rightarrow y$ edges. If so, it pops each from its stack and merges them. It then repeatedly checks if the merged component is linked with the topmost component on CS. If so, the two are merged. Finally, the component which is the result of the merges is pushed back onto CS. The algorithm then checks whether the topmost component in CS and the topmost component in $CSyz$ are linked by $y \rightarrow z$ edges and if so, handles this in a similar way.

In the remaining part of this section we show that this algorithm finds the SCCs of B_f. Denote the set of components found in the first pass of the algorithm by $Comp = \{C_1, C_2, \ldots, C_{p_1}\}$, such that for all $1 \le i < p_1$, $MaxY[C_i] < MinY[C_{i+1}]$. Since each of these components is strongly connected but is not necessarily an SCC of B_f, the components in $Comp$ are partitioned into sets of components such that the components of each set compose an SCC of B_f.

Definition 2. *A subset $S = \{C_{i_1}, \ldots, C_{i_\kappa}\}$ of Comp is an SCC set if the union of the components in S is an SCC of B_f. In the following we will use this notation while assuming that $MinY[C_{i_1}] < \cdots < MinY[C_{i_\kappa}]$.*

Definition 3. *Let $S_1 = \{C_{i_1}, \ldots, C_{i_\kappa}\}$ and $S_2 = \{C_{j_1}, \ldots, C_{j_{\kappa'}}\}$ be distinct SCC sets. Then S_1 is nested in S_2 if there exists $j \in \{j_1, \ldots, j_{\kappa'-1}\}$ such that for all $C_i \in S_1$, $MinY[C_j] < MinY[C_i] < MinY[C_{j+1}]$. S_1 and S_2 are interleaved if there exist $i, i' \in \{i_1, \ldots, i_{\kappa-1}\}$ and $j, j' \in \{j_1, \ldots, j_{\kappa'-1}\}$ such that $MinY[C_i] < MinY[C_j] < MinY[C_{i'}] < MinY[C_{j'}]$.*

The SCC sets of $Comp$ can be interleaved in one another, but in the following lemma we show that this can only occur in a restricted form. This will help us to show that the second pass of Algorithm identifies all SCC sets in $Comp$.

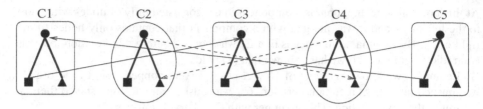

Fig. 7. Two interleaved SCC sets

Lemma 9. *Let $S_1 = \{C_{i_1}, \ldots, C_{i_\kappa}\}$ and $S_2 = \{C_{j_1}, \ldots, C_{j_{\kappa'}}\}$ be interleaved SCC sets. If $MinY[C_i] < MinY[C_j] < MinY[C_{i+1}] < MinY[C_{j+1}]$ then there are links between $S_i^l = \{C_1, \ldots, C_i\}$ and $S_i^h = \{C_{i+1}, \ldots, C_{i_\kappa}\}$ and there are links between $S_j^l = \{C_1, \ldots, C_j\}$ and $S_j^h = \{C_{j+1}, \ldots, C_{j_{\kappa'}}\}$. These links can be of one of two forms: (1) S_i^l and S_i^h are linked by $x \rightarrow y$ edges; S_j^l and S_j^h are linked by $y \rightarrow z$ edges. (2) S_i^l and S_i^h are linked by $y \rightarrow z$ edges; S_j^l and S_j^h are linked by $x \rightarrow y$ edges.*

Proof. We show that any other option is not possible. If S_i^l and S_i^h are linked by an $x \rightarrow y$ edge from S_i^l to S_i^h and a $y \rightarrow z$ edge from S_i^h to S_i^l then by convexity and Lemma , S_i^l and S_j^l are also linked, which means that the components of S_1 and S_2 belong to the same SCC of B_f, contradicting the assumption that S_1 and S_2 are distinct SCC sets. The same follows if S_i^l and S_i^h are linked by a $y \rightarrow z$ edge from S_i^l to S_i^h and an $x \rightarrow y$ edge from S_i^h to S_i^l and if S_j^l and S_j^h are linked by an $x \rightarrow y$ edge in one direction and a $y \rightarrow z$ edge in the other.

Assume that each of the pairs S_i^l, S_i^h and S_j^l, S_j^h is linked by $x \rightarrow y$ edges in both directions. Then by convexity and Lemma , S_j^l and S_i^h are also linked by $x \rightarrow y$ edges and again the components of S_1 and S_2 are in the same SCC of B_f. The same holds if the pairs S_i^l, S_i^h and S_j^l, S_j^h are both linked by $y \rightarrow z$ edges. $\square$

Figure shows an example of two interleaved SCC sets $S_1 = \{C_1, C_3, C_5\}$ and $S_2 = \{C_2, C_4\}$. For clarity, some of the edges were not drawn but the reader should assume that all edges that are implied by convexity exist in the graph. The SCCs of S_1 are linked by $x \rightarrow y$ edges and the SCCs of S_2 are linked by $y \rightarrow z$ edges.

The following resemble Lemmas and but refer to the second pass.

Lemma 10. *Let $CS = <C_1, C_2, \cdots >$ be the components in CS at the end of iteration i of the second pass (ordered from bottom to top). Then for all κ, $MaxY[C_\kappa] < MinY[C_{\kappa+1}]$.*

Proof. At the beginning CS is empty and the claim clearly holds. Assume that the claim holds after iteration $i - 1$ and consider the changes made to the stack during iteration i. When a topmost component of the stack is popped it does not affect the correctness of the claim. When a new component is pushed onto the stack it is the result of merging some of the topmost components in the stack with components from the temporary stacks $CSxy$ and $CSyz$. Since the temporary stacks contain only components with higher y nodes than CS, the merged component contains indices which are all higher than what is in components which are below it in CS. $\square$

Lemma 11. *Let* $S = \{C_{i_1}, \ldots, C_{i_\kappa}\}$ *be an SCC set. Then for all* $\ell \in \{1, \ldots, \kappa - 1\}$, C_{i_ℓ} *reaches at least one of* $\{C_{i_{\ell+1}}, \ldots, C_{i_\kappa}\}$ *by a single edge.*

Proof. By Lemmas   and   , the components of S appear in CS in the second pass in the same order in which they appear in the first pass. This implies that after a certain iteration of the second pass, $\{C_{i_1}, \ldots, C_{i_\ell}\}$ are in CS and $\{C_{i_{\ell+1}}, \ldots, C_{i_\kappa}\}$ are unexplored. Since there is an edge from one of $\{C_{i_1}, \ldots, C_{i_\ell}\}$ to one of $\{C_{i_{\ell+1}}, \ldots, C_{i_\kappa}\}$, we can show the claim by arguments which are similar to the ones used in the proof of Lemma   .   □

We can now show that the second pass of the algorithm identifies the SCCs of B_f.

Lemma 12. *Let* $S = \{C_{i_1}, \ldots, C_{i_\kappa}\}$ *be an SCC set. Then in the second pass of the algorithm, the components of S will be merged.*

Proof. Let S be an SCC set such that all SCC sets which are nested in S were merged. We show by induction that all components of S are merged in the second pass. That is, we show that for all i from $i_{\kappa-1}$ to i_1, $C_i, \ldots, C_{i_\kappa}$ will be merged. For $i = i_{\kappa-1}$, we know by Corollary   that $C_{i_{\kappa-1}}$ and C_{i_κ} are linked by either $x \to y$ edges in both directions (case 1) or $y \to z$ edges in both directions (case 2). By Lemma   , we know that $C_{i_{\kappa-1}}$ is not popped from CS before C_{i_κ} is pushed onto it. When C_{i_κ} is popped from CS for the first time, it is pushed onto the stack CS' where CS' is $CSxy$ (case 1) or $CSyz$ (case 2). Assume that when $C_{i_{\kappa-1}}$ became the topmost component in CS, C_{i_κ} was not the topmost component in CS'. If it was popped from CS' before that time, this is because a component above $C_{i_{\kappa-1}}$ in CS is linked with it. Since this component is not in S, this contradicts the assumption that S is an SCC set. On the other hand, if there was another component C' above C_{i_κ} in CS', then by Lemma   this is because C' is linked by $x \to y$ edges (case 1) or $y \to z$ edges (case 2) with another component C'' which was above $C_{i_{\kappa-1}}$ in CS. If C' and C'' were not merged, we get that there must have been a component above C' in CS' when C'' was the topmost component on CS. Applying the same argument recursively, we get that the number of components is infinite. Hence, C' and C'' were merged before $C_{i_{\kappa-1}}$ was the topmost component in CS, so C' could not have been above C_{i_κ} in CS'.

Assume that $C_{i_j+1}, \ldots, C_{i_\kappa}$ were merged by the algorithm into a larger component C. By Lemma   we know that C_{i_j} reaches C by a single edge and this implies that it was not popped from CS before any of $C_{i_j+1}, \ldots, C_{i_\kappa}$. In addition, since C and C_{i_j} are in the same SCC, there is a path from C to C_{i_j}. This path does not go through components that are between C and C_{i_j} in the stack because that would place these components in the same SCC as C and C_{i_j}, contradicting the assumption that S is an SCC set. Assume that the path begins with an $x \to y$ edge from C to C_{i_j} or a component below it in CS. Then by convexity there is also an $x \to y$ edge from C to C_{i_j}. If the edge from C_{i_j} to C is a $y \to z$ edge then by convexity and Lemma   , C is linked with C_{i_j} and all components which are above it in the stack and will be merged with them by the algorithm. If this edge is an $x \to y$ edge then C will be pushed onto $CSxy$ and as in the base case, it will be merged with C_{i_j} when the later will become the topmost component in CS.

If the path from C to C_{i_j} begins with a $y \to z$ edge from C to a component below C_{i_j} then arguments similar to the ones used in the proof of Lemma   imply that C also reaches C_{i_j} by a $y \to z$ edge. If the edge from C_{i_j} to C is an $x \to y$ edge then again by convexity and Lemma C_{i_j} and all components above it in CS are linked with C and will be merged with it. If it is a $y \to z$ edge then C will be pushed onto $CSyz$ and will be merged with C_{i_j} when the later will become the topmost component in CS.   □

4.3 Complexity Analysis

The PM algorithm traverses the y nodes and performs $2n$ priority queues operations. This takes time $O(n' + n\log n)$. In the SCC computation everything takes linear time except for maintaining the component list. We wish to be able to merge components and to find which component a node belongs to. For this we use a Union-Find[ ] data structure over the y nodes, on which we perform $O(n')$ operations that take a total of $O(n'\alpha(n', n'))$ time. Narrowing the domains of the variables can then be done in time $O(n + n')$ as in [ ]. So the total running time is $O(n'\alpha(n', n') + n\log n)$.

References

[1] N. Beldiceanu, I. Katriel, and S. Thiel. Filtering algorithms for the Same and UsedBy constraints. Research Report MPI-I-2004-1-001, Max-Planck-Institut für Informatik, Saarbrücken, Germany, 2004.   ,

[2] H. N. Gabow and R. E. Tarjan. A Linear-Time Algorithm for a Special Case of Disjoint Set Union. *Journal of Computer and System Sciences*, 30(2):209–221, 1985.

[3] L. R. Ford Jr. and D. R. Fulkerson. *Flows in Networks*. Princeton University Press, 1962.

[4] I. Katriel and S. Thiel. Fast bound consistency for the global cardinality constraint. In *Proceedings of the 9th International Conference on Principles and Practice of Constraint Programming (CP 2003)*, volume 2833 of *LNCS*, pages 437–451, 2003.   ,   ,

[5] K. Mehlhorn and S. Thiel. Faster Algorithms for Bound-Consistency of the Sortedness and the Alldifferent Constraint. In *Proceedings of the 6th International Conference on Principles and Practice of Constraint Programming (CP 2000)*, volume 1894 of *LNCS*, pages 306–319, 2000.   ,

[6] J.-C. Régin. A filtering algorithm for constraints of difference in CSPs. In *Proceedings of the 12th National Conference on Artificial Intelligence (AAAI-94)*, pages 362–367, 1994.

[7] J.-C. Régin. Generalized Arc-Consistency for Global Cardinality Constraint. In *Proceedings of the 13th National Conference on Artificial Intelligence (AAAI-96)*, pages 209–215, 1996.
,   ,

[8] Médecins sans Frontières. http://www.doctorswithoutborders.org/.

Cost Evaluation of Soft Global Constraints

Nicolas Beldiceanu and Thierry Petit

LINA FRE CNRS 2729
École des Mines de Nantes, 4 rue Alfred Kastler, 44307 Nantes Cedex 3, France
{nicolas.beldiceanu,thierry.petit}@emn.fr

Abstract. This paper shows that existing definitions of costs associated with soft global constraints are not sufficient to deal with all the
usual global constraints. We propose more expressive definitions: *refined
variable-based cost*, *object-based cost* and *graph properties based cost*. For
the first two ones we provide ad-hoc algorithms to compute the cost
from a complete assignment of values to variables. A representative set
of global constraints is investigated. Such algorithms are generally not
straightforward and some of them are even NP-Hard. Then we present
the major feature of the graph properties based cost: a systematic way
for evaluating the cost with a polynomial complexity.

1 Introduction

Within the context of disjunctive scheduling, Baptiste et al. [ ] have extended
the concept of global constraint to over-constrained problems. Such soft global
constraints involve a cost variable used to quantify the violation. This concept
has been initially investigated in a generic way in [ ]. However, the question of
violation costs associated with global constraints still contains many challenging
issues:

- As it has been shown for the `AllDifferent` constraint in [ ] there is no
  one single good way to relax a global constraint and the choice is essentially
  application dependent. Moreover, existing definitions of costs associated with
  soft global constraints are not sufficient to deal with all the usual global
  constraints.
- Defining how to relax a global constraint is not sufficient from an operational
  point of view. One has also to at least provide an algorithm to compute the
  cost that has been defined. As we will see this is usually not straightforward.
- Finally we would like to address the two previous issues in a systematic way
  so that each global constraint is not considered as a special case.

Our main motivation is to deal with over-constrained problems. Further, providing global constraints with generic definitions of costs and algorithms to compute them constitute a necessary step to unify constraint programming and local
search techniques [ ]. It would allow to use systematically global constraints as
higher level abstractions for describing the problem, together with local search
techniques. This again requires evaluating the constraint violation cost.

J.-C. Régin and M. Rueher (Eds.): CPAIOR 2004, LNCS 3011, pp. 80– , 2004.
© Springer-Verlag Berlin Heidelberg 2004

1.1   Context

A Constraint Network is a triple $\mathcal{R} = (\mathcal{X}, \mathcal{D}(\mathcal{X}), \mathcal{C})$. $\mathcal{X}$ is a set of variables. Each $x_i \in \mathcal{X}$ has a finite domain $\mathcal{D}_i \in \mathcal{D}(\mathcal{X})$ of values that can be assigned to it. $\mathcal{C}$ is a set of constraints. A constraint $C \in \mathcal{C}$ links a subset of variables $var(C) \subseteq \mathcal{X}$ by defining the allowed combinations of values between them. An assignment of values to a set of variables $\mathcal{Y} \subseteq \mathcal{X}$ is an instantiation $\mathcal{I}(\mathcal{Y})$ of $\mathcal{Y}$ (denoted simply by I when it is not confusing). An instantiation may satisfy or violate a given constraint. The *Constraint Satisfaction Problem* (CSP) consists of finding a complete instantiation $I(\mathcal{X})$ such that $\forall C \in \mathcal{C}$, I satisfies C.

A problem is over-constrained when no instantiation satisfies all constraints (the CSP has no solution). Such problems arise frequently when real-world applications are encoded in terms of CSPs. In this situation, the goal is to find a compromise. Violations are allowed in solutions, providing that such solutions retain a practical interest. A cost is generally associated with each constraint to quantify its degree of violation [ ]. For instance if $x \leq y$ is violated then a natural valuation of the distance from the satisfied state is $(x - y)^2$ [ ].

It is natural to integrate the costs as new variables of the problem [ ]. Then reductions in cost domains can be propagated directly to the other variables, and vice-versa. Furthermore, soft global constraints [,] can be defined. They extend the concept of global constraint [, , ,] to over-constrained problems. The use of constraint-specific filtering algorithms associated with global constraints may improve significantly the solving process as their efficiency can be much higher.

In [ ] two definitions of the cost associated with a global constraint were proposed. The first one is restricted to global constraints which have a primal representation [,],i.e., that can be directly decomposed as a set of binary constraints without adding new variables. The other one is more generic: the *variable-based cost*. It consists of counting the minimum number of variables that should change their value in order to return to a satisfied state. We do not place restrictions w.r.t. constraints. Therefore we focus on the generic definition. Consider the `AllDifferent`($\mathcal{X}$) constraint [ ]. It enforces all values assigned to variables in $\mathcal{X}$ to be pairwise different. For instance if $\mathcal{X} = \{x_1, x_2, x_3, x_4, x_5\}$, the cost of the tuple $(3, 7, 8, 2, 19)$ is 0 (satisfied) while the cost of the tuple $(3, 7, 8, 7, 7)$ is 2 (it is required to change the value of at least two variables to turn it satisfied).

Although the purpose of this definition is to be generic, some important questions remain open in regards to practical applications. The first one is related to the meaning of such a cost. Consider the `NumberOfDistinctValues`(N, $\mathcal{X}$) constraint [,]. It holds if variable N is equal to the number of distinct values assigned to variables in $\mathcal{X}$. For any assignment the maximal possible cost is 1: it is always sufficient to change the value of N. Such a limited cost variation is not satisfying. The second question is related to the complexity of the algorithms which evaluate the cost. Depending on constraints, it may be hard to compute the cost. We aim at providing generic answers to these two questions.

1.2 Contributions

The first contribution of this paper is related to expressiveness. We propose three new definitions of cost. They should allow to deal with most of soft global constraints.

1. **Refined variable-based cost**: we partition variables in two sets to distinguish changeable variables from fixed ones in the cost computation. The motivation is to deal with a common pattern of global constraints where one or several variables express something computed from a set of variables.
2. **Object-based cost**: application-directed constraints deal with high level objects, for instance activities in scheduling. The information provided by a cost related to the minimum number of objects to remove instead of the minimum number of variables to change should be more exploitable.
3. **Graph properties based cost**: we introduce a definition based on the graph properties based representation of global constraints [ ]. The important point of this definition is to have a systematic way to evaluate the cost with a polynomial complexity.

The second contribution is to provide algorithms to compute the cost from a complete instantiation of variables. With respect to the first two definitions we study ad-hoc algorithms for a set of well-known constraints: the `GlobalCardinality` [ ], `OneMachineDisjunctive` [ ], and `NonOverlappingRectangles` [ ]. Concerning the `GlobalCardinality` constraint, algorithms are provided for several types of costs. For the other ones we selected only the cost that makes sense. Through these examples we point out that the cost computation is generally not obvious and may even be NP-Hard. Then, we present a systematic way to evaluate the graph properties based cost with a polynomial complexity according to the number of variables occurring within the global constraint under consideration.

2 Soft Global Constraints

In an over-constrained problem encoded trough a constraint network $\mathcal{R} = (\mathcal{X}, \mathcal{D}(\mathcal{X}), \mathcal{C})$, we denote by $\mathcal{C}_h \subseteq \mathcal{C}$ the set of *hard* constraints that must necessarily be satisfied. $\mathcal{C}_s = \mathcal{C} \setminus \mathcal{C}_h$ is the set of *soft* constraints. Let I be an instantiation of X. If I is a solution then $\forall C \in \mathcal{C}_h$, I satisfies C. We search for the solution that respects as much as possible the set of constraints $\mathcal{C}$. A cost is associated with each constraint. A global objective related to the whole set of costs is usually defined. For instance, the goal can be to minimize the sum of elementary costs. While in Valued and Semi-Ring CSPs [ ] costs are defined trough an external structure (that is, the CSP framework has to be augmented), a model has been presented in [ ] to handle over-constrained problems directly as classical optimization problems. In this model costs are represented by variables. Features of classical constraint solvers can be directly used, notably to define soft global constraints including the cost variable. For sake of clarity, we consider in this paper only positive integer costs .

Definition 1. *Let $C \in \mathcal{C}_s$. The* **cost constraint** *of C, denoted by $\bar{C}$, is a constraint involving at least the cost variable (denoted by cost). This constraint should both express the negation of C and give the semantics of the violation.*

Generally $var(\bar{C}) = var(C) \cup \{cost\}$.

Definition 2. *Let $C \in \mathcal{C}_s$ be a soft constraint. A* **disjunctive relaxation** *of C is a constraint of the form: $[C \wedge [cost = 0]] \vee [\bar{C} \wedge [cost > 0]]$.*

This definition imposes explicitly to value 0 for variable *cost* to be not consistent with a violated state. In many cases 0 should not be consistent with $\bar{C}$. Then the statement $[cost > 0]$ is not necessary. This is the case in the following example.

Example 1. Let $C \in \mathcal{C}_s$ such that $C : [x \leq y]$. We define the cost constraint $\bar{C}$ of C by: $[[x > y] \wedge [cost = (x - y)^2]]$. The corresponding disjunctive relaxation of C is: $[[x \leq y] \wedge [cost = 0]] \vee [[x > y] \wedge [cost = (x - y)^2]]$.

This example illustrates directly what $\bar{C}$ expresses: the negation of C, $[x > y]$, and the semantics of the violation, $[cost = (x - y)^2]$.

From Definition   it is simple to turn an over-constrained problem into a classical optimization problem [ ]. We add the set of cost variables to $\mathcal{X}$ (one per soft constraint) and we replace each $C \in \mathcal{C}_s$ by its disjunctive relaxation, defined as a hard constraint. The optimization criterion involves the whole set of cost variables. It is shown in [ ] that this formulation entails no penalty in terms of space or time complexity compared with the other reference paradigms. In the next sections the term *soft global constraint* will denote the cost constraint of a global constraint. One can also use the term "relaxed version" of a global constraint.

3   Generic Definitions of Violation Costs

This section presents definitions of the semantics of the violation of a global constraint. We formalize the notions which are necessary to handle in a generic way violation costs. Definitions above should allow to relax most global constraints.

3.1   Refined Variable-Based Cost

Definition 3. Variable-Based Violation Cost [ ]
Let C be a global constraint. The cost of the violation of C is defined as the minimum number of values assigned to variables in $var(C)$ that should change in order to make C satisfied.

---

[1] This assumption implies that costs are totally ordered. It is a minor restriction [ ].

The cost is expressed by a variable *cost* such that $max(D(cost)) \leq |\mathcal{X}|+1$, where $|\mathcal{X}|$ denotes the number of variables in $\mathcal{X}$. Given $C \in \mathcal{C}_s$, we will deduce the cost constraint $\bar{C}$ from this definition. Such a definition is theoretically valid for any global constraint. However, in practice, it is not suited to all global constraints. For instance, consider the NumberOfDistinctValues(N, $\mathcal{X}$) constraint which holds if the variable N is equal to the number of distinct values assigned to a set of variables $\mathcal{X}$. If this constraint is violated, by changing the value assigned to N to the effective number of distinct values in $\mathcal{X}$ we make it satisfied. That is, in any case at most one variable has to be changed. The information provided by the cost value is poor. Therefore the user would prefer to have a finer way to evaluate the cost. We propose to count the minimum number of variables to change in $\mathcal{X}$ in order to make the current value assigned to N equal to the effective number of distinct values. In other terms, we isolate N as a variable having a fixed value, that cannot be changed. In this way we can deal with a common pattern of global constraints where one variable expresses something computed from a set of variables. For some constraints a subset of variables has to be fixed instead of one. Therefore Definition 4 is more general.

Definition 4. Refined Variable-Based Violation Cost
Let C be a constraint. Let us partition $var(C)$ in two sets $\mathcal{X}_f$ (fixed variables) and $\mathcal{X}_c$ (changeable variables). The cost of the violation of C can be defined as the minimum number of values assigned to variables in $\mathcal{X}_c$ that should change in order to make C satisfied. If this is not possible the cost is defined as $|\mathcal{X}_c| + 1$.

We may use in the same problem both Definition   and Definition   and aggregate directly the different costs because the two definitions are related to a minimum number of variables to change.

3.2   Object-Based Cost

At the user level, some application-oriented constraints handle objects which are not simple variables. For instance in scheduling, the Cumulative constraint [ ] deals with tasks which are defined by four variables (origin, end, duration, required amount of resource) and a variable equal to the maximum peak of resource utilization. In this case providing the constraint with a violation cost related to variables may give a poor information. A higher level cost directly related to tasks should be more convenient. The definitions are similar to Definitions   and . Instead of taking into account the minimum number of variables to change we consider the minimum number of objects to remove . The interested reader can refer to [ ] where a soft global constraint handling a variant of the One-Machine problem [ ] is described. Dealing with task-based costs in a solver dedicated to scheduling problems makes sense.

---

[2] This definition is valid for global constraints where removing an object makes it easier to satisfy the constraint. This is the case, for instance, for the Cumulative and Diffn [ ] constraints, but not for AtLeast.

3.3   Graph Properties Based Cost

As we will see in section , one of the main difficulty related to soft global constraints is that, even if one defines in a generic way the violation cost, it is usually far from obvious to come up with a polynomial algorithm for evaluating the cost for a specific global constraint. This is a major bottleneck if one wants to embed in a systematic way global constraints within local search or if one wants to allow relaxing global constraints. Therefore this section presents a definition of the cost based on the description of global constraints in terms of graph properties introduced in [ ]. We first need to recall the principle of description of global constraints. For general notions on graphs we invite the reader to refer to [ ].

A global constraint C is represented as an initial directed graph $G_i = (\mathcal{X}_i, U_i)$: to each vertex in $\mathcal{X}_i$ corresponds a variable involved in C, while to each arc e in U_i corresponds a binary constraint involving the variables at both endpoints of e. Unlike what is done in conventional constraints networks [ ] we do not ask anymore all binary constraints to hold. We consider G_i and remove all binary constraints which do not hold. This new graph is denoted by G_f. We enforce a given property on this graph. For instance, we ask for a specific number of strongly connected components.

Let us now explain how to generate G_i. A global constraint has one or more parameters which usually correspond to a domain variable or to a collection of domain variables. Therefore we discuss the process which allows to go from the parameters of a global constraint to G_i. For this purpose we came up with a set of arcs generators described in [ ]. We illustrate with a concrete example this generation process. Consider the `NumberOfDistinctValues`(N, $\mathcal{X}$) constraint where $\mathcal{X} = \{x_1, ..., x_m\}$. We first depict G_i. We then give the binary constraint associated to each arc. Finally we introduce the graph characteristics. The left and right parts of Figure 1 respectively show the initial graph G_i generated for the `NumberOfDistinctValues` constraint with $\mathcal{X} = \{x_1, ..., x_4\}$ as well as the graph G_f associated to the instantiation `NumberOfDistinctValues`($3, \{5, 8, 1, 5\}$).

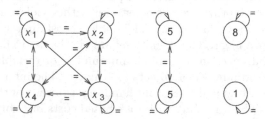

Fig. 1.   Initial graph associated with the `NumberOfDistinctValues`(N, $\mathcal{X}$) constraint where $\mathcal{X} = \{x_1, ..., x_4\}$ and final graph of the ground solution `NumberOfDistinctValues`($3, \{5, 8, 1, 5\}$)

We indicate for each vertex of G_i its corresponding variable. In G_f we show the value assigned to the corresponding variable. We have removed from G_f all the arcs corresponding to the equality constraints which are not satisfied. The NumberOfDistinctValues$(3, \{5, 8, 1, 5\})$ holds since G_f contains three strongly connected components, which can be interpreted as the fact that N is equal to the number of distinct values taken by the variables x_1, x_2, x_3 and x_4.

As shown in [ ], most global constraints can be described as a conjunction of graph properties where each graph property has the form P op I; P is a graph characteristic, I is a fixed integer and op is one of the comparison operator $=, \neq, <, \geq, >, \leq$.

Definition 5. Violation Cost of a Graph Property *Consider the graph property P op I. Let p denotes the effective value of the graph characteristic P on the final graph G_f associated to the instantiated global constraint we consider. Depending on the value of op, the violation cost of P op I is (abs denotes the absolute value): $cost(p, =, I) = \mathbf{abs}(p - I)$, $cost(p, \neq, I) = 1 - min(1, \mathbf{abs}(p - I))$, $cost(p, <, I) = max(0, p + 1 - I)$, $cost(p, \geq, I) = max(0, I - p)$, $cost(p, >, I) = max(0, I + 1 - p)$, $cost(p, \leq, I) = max(0, p - I)$.*

Definition 6. Graph Properties Based Violation Cost *Consider a global constraint defined by a conjunction of graph properties. The violation cost of such a global constraint is the sum of the violation costs of its graph properties.*

The most common graph characteristics used for defining a global constraint are for instance:

- *NVERTEX* the number of vertices of G_f.
- *NARC* the number of arcs of G_f.
- *NSINK* the number of vertices of G_f which don't have any successor.
- *NSOURCE* the number of vertices of G_f which don't have any predecessor.
- *NCC* the number of connected components of G_f.
- *MIN_NCC* the number of vertices of the smallest connected component of G_f.
- *MAX_NCC* the number of vertices of the largest connected component of G_f.
- *NSCC* the number of strongly connected components of G_f.
- *MIN_NSCC* the number of vertices of the smallest strongly connected component of G_f.
- *MAX_NSCC* the number of vertices of the largest strongly connected component of G_f.
- *NTREE* the number of vertices of G_f that do not belong to any circuit and for which at least one successor belongs to a circuit.

As a first concrete illustration of the computation of the cost, consider again the NumberOfDistinctValues$(N, \mathcal{X})$ constraint. N is defined as the number of strongly connected components of the final graph stem from the different variables in $\mathcal{X}$ and from their assigned values. Therefore the violation cost is defined as the absolute value of the difference between the effective number of strongly connected components and the value assigned to N. For instance the cost of NumberOfDistinctValues$(3, \{5, 5, 1, 8, 1, 9, 9, 1, 7\})$ is equal to abs$(5 - 3) = 2$ which can be re-interpreted as the fact that we have two extra distinct values. In section   we highlight the fact that for all global constraints which are representable as a conjunction of graph properties, the graph-based cost can be computed with a polynomial complexity according to the number of variables of the considered global constraint.

In order to further illustrate the applicability of the graph based cost we exemplify its use on three other examples. For the last one we show its ability for defining various violation costs for a given global constraint.

Consider the Change$(N, \mathcal{X})$ [ ] constraint where N is the number of times that the disequality constraint holds between consecutive variables of $\mathcal{X} = \{x_1, ..., x_m\}$. To each variable of $\mathcal{X}$ corresponds a vertex of G_i. To each pair of consecutive variables (x_j, x_{j+1}) of $\mathcal{X}$ corresponds an arc of G_i labelled by the disequality constraint $x_j \neq x_{j+1}$. Since the Change constraint is defined by the graph property $NARC = N$, its violation cost is $cost(NARC, =, N) =$ abs$(NARC - N)$. For instance the cost of Change$(1, \{4, 4, 8, 4, 9, 9\})$ is equal to abs$(3 - 1) = 2$ which can be re-interpreted as the fact that we have two extra disequalities which hold.

Consider now the Cycle constraint [ ] which, unlike the previous examples, is defined by two graph properties. The Cycle$(N, \mathcal{X})$ constraint holds when $\mathcal{X} = \{x_1, ..., x_m\}$ is a permutation with N cycles. To each variable of $\mathcal{X}$ corresponds a vertex of G_i. To each pair of variables (x_j, x_k) of $\mathcal{X}$ corresponds an arc of G_i labelled by the equality constraint $x_j = k$. The Cycle constraint is defined by the conjunction of the following two graph properties $NTREE = 0$ and $NCC = N$. The first one is used in order to avoid having vertices which both do not belong to a circuit and have at least one successor located on a circuit. The second one counts the number of connected components of G_f which, when $NTREE = 0$, is the number of cycles. From these properties we have that the violation cost associated to the Cycle constraint is $cost(NTREE, =, 0) + cost(NCC, =, N) = NTREE + $ abs$(NCC - N)$. For instance the cost of Cycle$(1, \{2, 1, 4, 5, 3, 7, 8, 7\})$ is equal to $1 + $ abs$(3 - 1) = 3$ which can be re-interpreted as the fact that we have one vertex which does not belong to a cycle as well as two extra cycles (see Figure 2.).

A given constraint may have several graph representations, leading to different ways to relax it. For instance consider the AllDifferent$(\mathcal{X})$ constraint where $\mathcal{X} = \{x_1, ..., x_m\}$. The Alldifferent constraint holds when all the variables of $\mathcal{X}$ are assigned to distinct values.

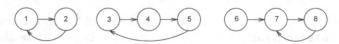

Fig. 2. Cycle$(1, \{2, 1, 4, 5, 3, 7, 8, 7\})$

It may be represented first by an initial directed graph G_i where each node is a distinct variable and between each ordered pair of variables (the order is used to avoid counting twice an elementary violation) there is an arc representing the binary constraint $\neq$. To obtain G_f we remove all arcs where the binary constraint is violated. The `AllDifferent` constraint is defined by the graph property $NARC = m * (m - 1)/2$. This definition leads to the following violation cost $cost(NARC, =, m * (m - 1)/2) = m * (m - 1)/2 - NARC$. This corresponds to the number of binary constraints violated, that is, the *primal graph based cost* introduced in [ ].

A second way to describe the `AllDifferent`$(\mathcal{X})$ constraint is to build a graph where all arcs are defined between pairs of variables like the initial graph of the `NumberOfDistinctValues` constraint (see the left part of Figure 1). The binary constraints will then be $=$ and the graph property $NSCC = m$. The previous graph property enforces G_f to contain m strongly connected components. Since G_f has m vertices this forbid having a strongly connected component with more than one vertex. This second definition of `Alldifferent` leads to the violation cost $cost(NSCC, =, m) = m - NSCC$. This corresponds in fact to the variable-based cost introduced in [ ].

Finally a third way to model the `AllDifferent`$(\mathcal{X})$ constraint is to define the initial graph G_i as in our second model and to use the graph property $MAX\_NSCC = 1$. This imposes the size of the largest strongly connected component of G_f to be equal to 1. This third definition of `Alldifferent` leads to the violation cost $cost(MAX\_NSCC, =, 1) = MAX\_NSCC - 1$. This can be interpreted a the difference between the number of occurrences of the value which occurs the most in $\mathcal{X}$ and 1.

4   Cost Computation Algorithms

This section presents algorithms to compute the cost from a complete assignment of value to variables $I(\mathcal{X})$ for a representative set of global constraints. We focus on variable-based and object-based costs. A systematic way to compute graph-based costs will be provided in the next section. Let us recall the context of this work. When we define a soft global constraint it is necessary to be able:

(1) To compute the cost from a complete assignment. This is the basic step to evaluate how much the constraint is violated.

(2) To compute a lower bound of the cost from a partial assignment. This step is useful to have a consistency condition, based on the bounds of $D(cost)$.

(3) To provide the constraint with a filtering algorithm which deletes values that can not belong to any assignment with an acceptable cost (i.e., $\in D(cost)$). A first algorithm can always be deduced from step (2) by applying for each value v the consistency condition with $D(x) = \{v\}$.

This section focuses on step (1). We show that computing a cost from a complete assignment is generally not obvious and even may be hard in the algorithmic sense. Well-known constraints which were selected differ significantly one another. For a given constraint, several definitions of the cost are considered when it makes sense.

4.1   GlobalCardinality : Variable-Based Cost

In a GlobalCardinality($\mathcal{X}$, T, $\mathcal{N}$) constraint, $\mathcal{X}$ is a set of variables, T an array of values, and $\mathcal{N}$ a set of positive integer variables one-to-one mapped with values in T. This constraint enforces each v in T to be assigned in $\mathcal{X}$ a number of times equal to the value assigned to the variable in $\mathcal{N}$ which represents v.

To turn the GlobalCardinality($\mathcal{X}$, T, $\mathcal{N}$) constraint from a violated state to a satisfied one, it is necessary to change variables (in $\mathcal{X}$ or $\mathcal{N}$) in order to make the cardinalities equal to the corresponding occurrences of values in $\mathcal{X}$. We assume in this section that the maximum value of the domain of a cardinality variable does not exceeds $|\mathcal{X}|$. Given a complete assignment of $\mathcal{X} \cup \mathcal{N}$ we have a possibly empty subset $\mathcal{N}_{false} \subseteq \mathcal{N}$ of cardinalities that do not correspond to the current number of occurrences of their values in the assignment $I(\mathcal{X})$ of $\mathcal{X}$. A way to make the constraint satisfied is to change the value of each variable in $\mathcal{N}_{false}$ to the effective cardinalities relatively to $I(\mathcal{X})$. Unfortunately, as shown by example 2, $|\mathcal{N}_{false}|$ is not necessarily the cost value of Definition .

Example 2. Let $\mathcal{X} = \{x_1, x_2\}$, $D(x_1) = D(x_2) = \{a, b\}$, $T = [a, b]$, and $\mathcal{N} = \{c_a, c_b\}$ with $D(c_a) = D(c_b) = [0, 2]$. The assignment $\mathcal{I}(\mathcal{X}) = \{(x_1, a),$ $(x_2, b), (c_a, 0), (c_b, 2)\}$ violates the GlobalCardinality($\mathcal{X}$, T, $\mathcal{N}$) constraint. $\mathcal{N}_{false} = \{c_a, c_b\}$. We can turn it satisfied by changing the value of x_1 to b, that is the minimum number of variables to change is $cost = 1$. Thus $|\mathcal{N}_{false}| > cost$.

We have to solve the following problem: how many variables should change their value to reduce $\mathcal{N}_{false}$ to the empty set? As a short cut we say that $x \in \mathcal{X}$ "saves" one variable c_v from $\mathcal{N}_{false}$ when by changing x we can turn $c_v \in \mathcal{N}_{false}$ to the effective number of occurrences of v in $I(\mathcal{X})$.

Notation 1 *Let v be a value of $\mathcal{D}(\mathcal{X})$, occ_v its effective number of occurrences in $\mathcal{I}(\mathcal{X})$ and req_v, $req_v \leq |\mathcal{X}|$, the value assigned to its cardinality variable c_v. d_v and s_v respectively denote $\mathbf{abs}(occ_v - req_v)$ and $occ_v - req_v$.*

Lemma 1 *If no x in $\mathcal{X}$ saves one variable in $\mathcal{N}_{false}$ then for any value v assigned to variables in $\mathcal{X}$ such that $c_v \in \mathcal{N}_{false}$ we have $d_v \geq 2$.*

Proof $c_v \in \mathcal{N}_{false} \Rightarrow d_v > 0$. If $d_v = 1$ then by changing one variable it is possible to turn d_v to 0, i.e., c_v to the effective number of occurrences of v in $I(\mathcal{X})$.

Lemma 2 *Changing the value of one variable in $\mathcal{X}$ changes the value assigned to at most two variables in $\mathcal{N}_{false}$ and may save at most those two variables.*

Proof If x changes from v_1 to v_2 only two cardinalities change: the ones of v_1 and v_2.

Lemma 3 $cost \leq |\mathcal{N}_{false}|$

Proof it is possible to change all variables in $\mathcal{N}_{false}$ to make the constraint satisfied, and then obtain a cost equal to $|\mathcal{N}_{false}|$.

Theorem 1. *If no x in $\mathcal{X}$ save variables in $\mathcal{N}_{false}$ then $cost = |\mathcal{N}_{false}|$.*

Proof Hypothesis H: no x in $\mathcal{X}$ can be changed to save c_v in $\mathcal{N}_{false}$. Assume $cost \neq |\mathcal{N}_{false}|$. By Lemma , $cost < |\mathcal{N}_{false}|$. We consider the number of steps to turn $\mathcal{N}_{false}$ to $\emptyset$ by changing one variable at each step, either in $\mathcal{N}_{false}$ or in $\mathcal{X}$. Let the current step be n. Changing $x \in X$ saves 0 and is supposed to lead in a next step $n + k$ to a better result than turning k variables in $\mathcal{N}_{false}$. From Lemma , at $n + 1$ two variables c_{v1} and c_{v2} in $\mathcal{N}_{false}$ have been changed and from H none were saved (distances ≥ 2). Thus by Lemma , at step $n + 2$ at most two variables can be saved in $\mathcal{N}_{false}$ (distance $= 1$). Let us detail the 3 cases at $n + 2$: $\diamond$ 1. Two variables in $\mathcal{N}_{false}$ are saved. From H and Lemma they are c_{v1} and c_{v2} (if not we could have saved the other ones at $n + 1$, contradiction with H). The situation remains equivalent to step n except $\mathcal{N}_{false} \setminus \{c_{v1}, c_{v2}\}$ is considered (we made 2 saves, 2 variables have been changed). $\diamond$ 2. One variable in $\mathcal{N}_{false}$ is saved. It is necessarily c_{v1} or c_{v2}, which is removed from $\mathcal{N}_{false}$. The consequence can be in the better case to have a new distance becoming 1 when saving c_{v1} or c_{v2}. In this case at step $n + 3$ we return to this situation. In this way it will never be possible to compensate the loss of one of step $n + 1$. $\diamond$ 3. No variables can be saved. No $c_v \in \mathcal{N}_{false}$ is such that $d_v < 2$. The situation at $n + 2$ is equivalent to the one of step n and no variables were saved. $\diamond$ Conclusion: no case can lead to a better situation at next steps. The cost is $|\mathcal{N}_{false}|$.

Notation 2 *Given a complete instantiation we define:*
$\mathcal{N}_{false}^{+} = \{c_v \in \mathcal{N}_{false} \text{ such that } s_v > 0\}$
$\mathcal{N}_{false}^{-} = \{c_v \in \mathcal{N}_{false} \text{ such that } s_v < 0\}$
$\mathcal{D}_{false}^{+} = \{d_v \text{ such that } c_v \in \mathcal{N}_{false}^{+}\}$ *ordered by increasing* d_v,
$\mathcal{D}_{false}^{-} = \{d_v \text{ such that } c_v \in \mathcal{N}_{false}^{-}\}$ *ordered by increasing* d_v,
$n_1^{+} = |\{d_v \in \mathcal{D}_{false}^{+} \text{ such that } d_v = 1\}|$,
$n_1^{-} = |\{d_v \in \mathcal{D}_{false}^{-} \text{ such that } d_v = 1\}|$.

Figure 3. represents the two sets $\mathcal{D}_{false}^{+}$ and $\mathcal{D}_{false}^{-}$ for a `GlobalCardinality` constraint where the current instantiation has $|\mathcal{N}_{false}| = 13$ wrong cardinalities. The cost is equal to the minimal number of changes to remove each column, provided that:
− If we change x with initial value v and $d_v = 1$ we remove one column in $\mathcal{D}_{false}^{+}$.
− If we change x to a new value w with $d_w = 1$ we remove one column in $\mathcal{D}_{false}^{-}$.
− From these two first rules, if we change x with initial value v to w and $d_v = d_w = 1$ then we remove two columns in one step.
− Finally, if we change the value of one cardinality c_v to the real number of occurences of v in $\mathcal{X}$ then we remove its corresponding column.

To remove all columns with a minimal number of changes, one can select, while either $d_v = 1$ or $d_w = 1$, a variable x assigned to value v with minimal d_v in $\mathcal{D}_{false}^{+}$ and change its value v to value w with minimal d_w in $\mathcal{D}_{false}^{-}$. In this

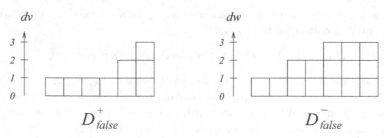

Fig. 3. Graphical representation of the two sets $\mathcal{D}_{false}^{+}$ and $\mathcal{D}_{false}^{-}$ with corresponding distances from real number of occurences of values in an instantiation $I(\mathcal{X})$

way, if the two minima verify $d_v = d_w = 1$ we remove two columns in one step. If not we tend to the most favourable next state. When the two minima d_v and d_w are such that $d_v > 1$ and $d_w > 1$, by Theorem   the remaining minimal number of changes to make is $|\mathcal{N}_{false}|$. From this process, Theorem   gives the exact cost of an instantiation $I(\mathcal{X})$.

Theorem 2. *Let k be the positive integer defined as follows:*

- *If $n_1^{+} \geq n_1^{-}$, k is the smallest possible number of first elements of $\mathcal{D}_{false}^{-}$ satisfying $\sum_{i=1}^{i=k} d_{v_i} > n_1^{+}$ (or $k = |\mathcal{N}_{false}^{-}| + 1$ if $\sum_{i=1}^{i=|\mathcal{N}_{false}^{-}|} d_{v_i} \leq n_1^{+}$).*

- *If $n_1^{+} < n_1^{-}$, k is the smallest possible number of first elements of $\mathcal{D}_{false}^{+}$ satisfying $\sum_{i=1}^{i=k} d_{v_i} > n_1^{-}$ (or $k = |\mathcal{N}_{false}^{+}| + 1$ if $\sum_{i=1}^{i=|\mathcal{N}_{false}^{+}|} d_{v_i} \leq n_1^{-}$).*

The exact cost of the instantiation is $|\mathcal{N}_{false}| - k + 1$.

4.2   GlobalCardinality : Refined Variable-Based Cost (1)

We consider a first refined variable-based definition of the cost where fixed variables are the cardinalities $\mathcal{N}$ and changeable ones are $\mathcal{X}$.

Basic Notions on Flows

Flow theory was originally introduced by Ford and Flukerson [ ]. Let $G = (X, U)$ be a directed graph. An arc (u, v) *leaves* u and *enters* v. $\Gamma^{-}(v)$ is the set of edges entering a vertex v. $\Gamma^{+}(v)$ is the set of edges leaving v. $\Gamma(v) = \Gamma^{-} \cup \Gamma^{+}$. Consider a graph $G = (X, U)$ such that each arc (u, v) is associated with two positive integers $lb(u, v)$ and $ub(u, v)$. $ub(u, v)$ is called the *upper bound capacity* of (u, v) and $lb(u, v)$ the *lower bound capacity*. A *flow* in G is a function f satisfying the two following two conditions: $\diamond$ 1. For any arc (u, v), $f(u, v)$ represents the amount of commodity which flows along the arc. Such a flow is allowed only in the direction of the arc (u, v), that is, from u to v. $\diamond$ 2. A *conservation law* is observed at each of the vertices: $\forall v \in X : \sum_{u \in \Gamma^{-}(v)} f(u, v) = \sum_{w \in \Gamma^{+}(v)} f(v, w)$. $\diamond$

The *feasible flow problem* is the problem of the existence of a flow in G which satisfies the *capacity constraint*, that is: $\forall(u, v) \in U : lb(u, v) \leq f(u, v) \leq ub(u, v)$.

Application to the GlobalCardinality Constraint

We aim to take into account the fact that modifying one $x \in X$ changes two cardinalities. The problem can be formulated as a feasible flow with minimum cost (Min Cost Flow [ ]) on the bipartite variable-values graph defined below.

In this graph we consider for the GlobalCardinality constraint one arc with a specific cost is defined for each pair variable-value for values belonging to each initial domain. We assume that the constraint has intially at least one feasible soltion. Arcs corresponding to the current assignment have cost 0, other have cost 1.

Definition 7. $G_{gcc} = (X \cup D_0(X) \cup \{s, t\}, E)$, where $D_0(X)$ denotes the set of initial domains of X, is defined by:

- s, a vertex such that $\forall x \in X$, $(s, x) \in E$. These arcs have a capacity equal to $[1, 1]$ and are valued by $cost(s, x) = 0$,
- t, a vertex such that $\forall v \in D_0(X)$, $(v, t) \in E$. The capacity of each arc (v, t) is req_v (see Notation in previous subsection). The valuation is $cost(v, t) = 0$,
- $(t, s) \in E$, capacity $[|X|, |X|]$, $cost(s, t) = 0$,
- $\forall (x, v)$, $x \in X$ and $v \in D_0(X)$ one arc $\in E$. Capacity $[0, 1]$, $cost(x, v)$ equal to 0 if v is assigned to x in $I(X)$, 1 otherwise.

computeCost($I(X)$)
        return *the cost of a minimum cost flow in G_{gcc};*

Complexity: from [ ] the complexity is $O(n_1 * m * log(n^2/m) * log(n_1 * maxC))$ where $n_1 = |X|$, $n = |X| + |\cup_i D(x_i), x_i \in X|$, $m = |E|$ and $maxC$ is a maximum cost of an arc, that is in our problem $maxC = 1$.

4.3   GlobalCardinality: Refined Variable-Based Cost (2)

We consider a second refined variable-based definition of the cost where fixed variables are X and changeable ones are the cardinalities N.

computeCost($I(X)$)
        return $|N_{false}|$;

Complexity: $O(1)$, assuming that $|N_{false}|$ is incrementally maintained when assigning variables in X and N.

4.4   OneMachineDisjunctive: Object-Based Refined Cost

The OneMachineDisjunctive(T, M) constraint is defined on a set of tasks T and a makespan variable M. Each task T_i is defined by two variables: a start s_i

and a duration d_i. Its purpose is to enforce all the tasks to be pairwise disjoint and executed before M ($\forall s_i, d_i, s_i + d_i \leq M$). We consider the object-based refined cost where fixed objects $\mathcal{O}_f$ reduce to $\{M\}$ and changeable ones $\mathcal{O}_c$ are the set $\mathcal{T}$.

We search for the minimum number of tasks to remove to make the constraint satisfied. The optimal number can be found by a greedy algorithm derived from [ ]. The principle is to sort tasks by increasing $f_i = s_i + d_i$ and use a greedy algorithm which adds at each step $i + 1$ a task T_{i+1} of minimum end time f_{i+1} after T_i if and only if $s_{i+1} \geq f_i$. This is done until the makespan is reached. In this way a maximal number of tasks are added, as it is proven in [ ] by a simple induction.

Complexity: $O(nlog(n))$ where $n = |\mathcal{T}|$.

4.5   NonOverlappingRectangles: Object-Based Cost

The NonOverlappingRectangles($\mathcal{R}$) constraint holds if the set $\mathcal{R}$ of two dimensional rectangles parallel to axis and defined by their respective left-up origin and sizes is such that all the rectangles do not pairwise overlap.

The cost is the minimum number of rectangles to remove in order to make them disjoint one another. This number is equal to $|\mathcal{R}|$ minus the size of a maximum independent set of a rectangle intersection graph. A rectangle intersection graph is a graph where vertices are rectangles and an edge exists between two rectangles if and only if these rectangles intersect.

Complexity: finding the maximum independent set of a rectangle intersection graph is known to be NP-Hard [   ].

5   Computation of the Graph Properties Based Cost

Computing the cost requires evaluating different graph characteristics on the final graph G_f. This graph is computed in the following way: we discard all the arcs of the initial graph for which the corresponding elementary constraints do not hold. We assume that it is possible to check in polynomial time whether a ground instance of an elementary constraint holds. Since all the characteristics we mentioned in [ ] can be evaluated in a polynomial time according to the number of vertices of G_f, evaluating the cost can also be performed in a polynomial time.

6   Perspectives

This paper presents generic definitions of costs and the related computation algorithms. Such definitions are required to define soft global constraints. We directed the paper to over-constrained problems but our cost definitions are also useful to mix local search based techniques and constraint programming.

6.1   Incremental Algorithms

When a move is made on a given assignment of values to variables it is necessary to quantify its impact on the constraints. As a perspective of this work a detailed study of *incremental* cost computation algorithms should be made, depending on their definition and the moves that are supposed to be performed. For some constraints and some costs this step is easy. This was done in COMET [ ] for the AllDifferent and the GlobalCardinality. An atomic move consists of modifying a given instantiation by changing the value assigned to one variable.

6.2   Filtering Algorithms

The other main perspective is related to the variable-based and object-based definitions of cost: for all the usual global constraints it may be interesting to provide *filtering algorithms*, like the ones proposed for the AllDifferent in [ ]. Finally, with respect to the graph properties based cost, filtering algorithms provided for the different graph characteristics would be directly suitable to soft global constraints.

7   Conclusion

We pointed out that cost definitions presented in [ ] for soft global constraints are not sufficient. We introduced three new definitions. For the first two ones we investigated algorithms to compute the cost of a complete instantiation for a representative set of global constraints. The last definition is based on graph properties representation of constraints [ ]. Its major feature is to come up with a systematic way for evaluating the cost with a polynomial complexity. We discussed the perspectives of this work, notably with respect to combination of the expressive power of global constraints with the local search frameworks.

Acknowledgements

The authors thank the reviewers for the helpful comments they provided. This work also benefited from early discussions with Markus Bohlin.

References

[1] A. Aggoun and N. Beldiceanu. Extending CHIP in order to solve complex scheduling and placement problems. *Mathl. Comput. Modelling*, 17(7):57–73, 1993.

[2] R.K. Ahuja, J.B. Orlin, C. Stein, and R.E. Tarjan. Improved algorithms for bipartite network flow. *SIAM Journal on Computing*, 23:5:906–933, 1994.

[3] J-C. Bajard, H. Common, C. Kenion, D. Krob, J-M. Muller, A. Petit, Y. Robert, and M. Morvan. Exercices d'algorithmique : oraux d'ENS. *International Thomson Publishing, in French*, pages 72–74, 1997.

[4] P. Baptiste, C. Le Pape, and L. Peridy. Global constraints for partial CSPs: A case-study of resource and due date constraints. *Proceedings CP*, pages 87–102, 1998.   ,   ,

[5] N. Beldiceanu. Global constraints as graph properties on a structured network of elementary constraints of the same type. *Proceedings CP*, pages 52–66, 2000.   ,   ,   ,   ,

[6] N. Beldiceanu. Pruning for the minimum constraint family and for the number of distinct values constraint family. *Proceedings CP*, pages 211–224, 2001.

[7] N. Beldiceanu and M. Carlsson. Revisiting the cardinality operator and introducing the cardinality-path constraint family. *Proceedings ICLP*, 2237:59–73, 2001.

[8] N. Beldiceanu and E. Contejean. Introducing global constraints in CHIP. *Journal of Mathematical and Computer Modelling*, 20(12):97–123, 1994.   ,   ,

[9] C. Berge. Graphs and hypergraphs. *Dunod, Paris*, 1970.

[10] C. Bessière and P. Van Hentenryck. To be or not to be... a global constraint. *Proceedings CP*, pages 789–794, 2003.

[11] S. Bistarelli, U. Montanari, F. Rossi, T. Schiex, G. Verfaillie, and H. Fargier. Semiring-based CSPs and valued CSPs: Frameworks, properties, and comparison. *Constraints*, 4:199–240, 1999.   ,

[12] R. Dechter. Constraint networks. *Encyclopedia of Artificial Intelligence*, pages 276–285, 1992.

[13] R. Dechter and J. Pearl. Network-based heuristics for constraint-satisfaction problems. *Artificial Intelligence*, 34:1–38, 1987.

[14] L. Ford and D. Flukerson. Flows in networks. *Princeton University Press*, 1962.

[15] R. J. Fowler, M. S. Paterson, and S. L. Tanimoto. Optimal packing and covering in the plane are np-complete. *Inform. Processing Letters*, 12(3):133–137, 1981.

[16] M. R. Garey and D. S. Johnson. Computers and intractability : A guide to the theory of NP-completeness. *W. H. Freeman and Company*, ISBN 0-7167-1045-5, 1979.

[17] I. Gent, K. Stergiou, and T. Walsh. Decomposable constraints. *Artificial Intelligence*, 123:133–156, 2000.

[18] E. Lawler. Combinatorial optimization: Networks and matroids. *Holt, Rinehart and Winston*, 1976.

[19] F. Pachet and P. Roy. Automatic generation of music programs. *Proceedings CP*, pages 331–345, 1999.

[20] T. Petit, J-C. Régin, and C. Bessière. Meta constraints on violations for over constrained problems. *Proceedings IEEE-ICTAI*, pages 358–365, 2000.   ,   ,

[21] T. Petit, J-C. Régin, and C. Bessière. Specific filtering algorithms for over constrained problems. *Proceedings CP*, pages 451–463, 2001.   ,   ,   ,   ,

[22] J-C. Régin. A filtering algorithm for constraints of difference in CSPs. *Proceedings AAAI*, pages 362–367, 1994.

[23] J-C. Régin. Generalized arc consistency for global cardinality constraint. *Proceedings AAAI*, pages 209–215, 1996.

[24] P. Van Hentenryck and L. Michel. Control abstractions for local search. *Proceedings CP*, pages 66–80, 2003.   ,

SAT-Based Branch & Bound and Optimal Control of Hybrid Dynamical Systems

Alberto Bemporad and Nicolò Giorgetti

Dip. Ingegneria dell'Informazione
University of Siena, via Roma 56, 53100 Siena, Italy
{bemporad,giorgetti}@dii.unisi.it

Abstract. A classical hybrid MIP-CSP approach for solving problems having a logical part and a mixed integer programming part is presented. A Branch and Bound procedure combines an MIP and a SAT solver to determine the optimal solution of a general class of optimization problems. The procedure explores the search tree, by solving at each node a linear relaxation and a satisfiability problem, until all integer variables of the linear relaxation are set to an integer value in the optimal solution. When all integer variables are fixed the procedure switches to the SAT solver which tries to extend the solution taking into account logical constraints. If this is impossible, a "no-good" cut is generated and added to the linear relaxation. We show that the class of problems we consider turns out to be very useful for solving complex optimal control problems for linear hybrid dynamical systems formulated in discrete-time. We describe how to model the "hybrid" dynamics so that the optimal control problem can be solved by the hybrid MIP+SAT solver, and show that the achieved performance is superior to the one achieved by commercial MIP solvers.

1 Introduction

In this paper we consider the general class of *mixed logical/convex* problems:

$$\min_{z,\nu,\mu} f(z) \quad \text{(Convex function)} \tag{1a}$$

$$\text{s.t.} \ g_c(z) \leq 0, \ h_c(z) = 0 \quad \text{(Continuous constraints)} \tag{1b}$$

$$g_m(z,\mu) \leq 0, \ h_m(z,\mu) = 0 \quad \text{(Mixed constraints)} \tag{1c}$$

$$g_L(\nu,\mu) = \text{TRUE} \quad \text{(Logic constraints)} \tag{1d}$$

$$z \in \mathbb{R}^{n_z}, \ \nu \in \{0,1\}^{n_\nu}, \ \mu \in \{0,1\}^{n_\mu},$$

where $g_c : \mathbb{R}^{n_z} \to \mathbb{R}^{q_{gc}}$, $g_m : \mathbb{R}^{n_z+n_\mu} \to \mathbb{R}^{q_{gm}}$ are convex functions, $h_c : \mathbb{R}^{n_z} \to \mathbb{R}^{q_{hc}}$, $h_m : \mathbb{R}^{n_z+n_\mu} \to \mathbb{R}^{q_{hm}}$ are affine functions, and $g_L : \{0,1\}^{n_\nu \times n_\mu} \to \{0,1\}^{n_{CP}}$ is a Boolean function.

An MIP solver provides the solution to ( ) after solving a sequence of relaxed convex problems, typically standard linear or quadratic programs (LP, QP). A potential drawback of MIP is (a) the need for converting the logic constraints

J.-C. Régin and M. Rueher (Eds.): CPAIOR 2004, LNCS 3011, pp. 96–   , 2004.

( ) into mixed-integer inequalities, therefore losing most of the original discrete structure, and (b) the fact that its efficiency mainly relies upon the tightness of the continuous LP/QP relaxations.

Such a drawback is not suffered by techniques for solving constraint satisfaction problems (CSP), i.e., the problem of determining whether a set of constraints over discrete variables can be satisfied. Under the class of CSP solvers we mention constraint logic programming (CLP) [ ] and SAT solvers [ ], the latter specialized for the satisfiability of Boolean formulas.

While CSP methods are superior to MIP approaches for determining if a given problem has a feasible (integer) solution, the main drawback is their inefficiency for solving optimization, as they do not have the ability of MIP approaches to solve continuous relaxations (e.g., linear programming relaxations) of the problem in order to get upper and lower bounds to the optimum value.

For this reason, it seems extremely interesting to integrate the two approaches into one single solver. Some efforts have been done in this direction [, , , ,], showing that such mixed methods have a tremendous performance in solving mathematical programs with continuous (quantitative) and discrete (logical/symbolic) components, compared to MIP or CSP individually. Such successful results have stimulated also industrial interest: ILOG Inc. is currently distributing OPL (Optimization Programming Language), a modeling and programming language which allows the formulation and solution of optimization problems, using both MIP and CSP techniques, combining to some extent the advantages of both approaches; European projects with industrial participants, such as LISCOS [ ], developed and are developing both theoretical insights and software tools for applying the combined approach of MIP and CSP to industrial case studies.

In this paper, we focus on combinations of convex programming (e.g., linear, quadratic, etc.) for optimization over real variables, and of SAT-solvers for determining the satisfiability of Boolean formulas. The main motivation for our study stems from the need for solving complex optimal control problems of theoretical and industrial interest based on "hybrid" dynamical models of processes that exhibit a mixed continuous and discrete nature. Hybrid models are characterized by the interaction of continuous models governed by differential or difference equations, and of logic rules, automata, and other discrete components (switches, selectors, etc.). Hybrid systems can switch between many operating modes where each mode is governed by its own characteristic continuous dynamical laws. Mode transitions may be triggered internally (variables crossing specific thresholds), or externally (discrete commands directly given to the system). The interest in hybrid systems is mainly motivated by the large variety of practical situations where physical processes interact with digital controllers, as for instance in embedded control systems.

Several authors focused on the problem of solving optimal control problems for hybrid systems. For continuous-time hybrid systems, most of the literature either studied necessary conditions for a trajectory to be optimal, or focused on the computation of optimal/suboptimal solutions by means of dynamic programming or the maximum principle [, ,].

The hybrid optimal control problem becomes less complex when the dynamics is expressed in discrete-time, as the main source of complexity becomes the combinatorial (yet finite) number of possible switching sequences. In particular, in [, ,] the authors have solved optimal control problems for discrete-time hybrid systems by transforming the hybrid model into a set of linear equalities and inequalities involving both real and (0-1) variables, so that the optimal control problem can be solved by a mixed-integer programming (MIP) solver.

At the light of the benefits and drawbacks of the previous work in [, ,] for solving control and stability/safety analysis problems for hybrid systems using MIP techniques, we follow a different route that uses the aforementioned approach combining MIP and CSP techniques.

We build up a new modeling approach for hybrid dynamical systems directly tailored to the use of the hybrid MIP+SAT solver for solving optimal control problems, and show its computational advantages over pure MIP methods. A preliminary work in this direction appeared in [ ], where generic constraint logic programming (CLP) was used for handling the discrete part of the optimal control problem.

The paper is organized as follows. In Section   optimal control problems of discrete-time hybrid models are introduced and in Section   are reformulated to the general class ( ). Section   introduces the new solution algorithm for the general class ( ). An example of optimal control problem of a hybrid model showing the benefits of the solution algorithm, compared to pure MIP approaches [,] is shown in section .

2   Motivating Application

2.1   Optimal Control of Discrete-Time Hybrid Systems

Following the ideas in [ ], a hybrid system can be modeled as the interconnection of an automaton (AUT) and a switched affine system (SAS) through an event generator (EG) and a mode selector (MS). The discrete-time hybrid dynamics is described as follows [ ]:

$$(\text{AUT}) \quad x_l(k+1) = f_l(x_l(k), u_l(k), e(k)),$$
$$y_l(k) = g_l(x_l(k), u_l(k), e(k)), \tag{2a}$$
$$(\text{SAS}) \quad x_c(k+1) = A_{i(k)}x_c(k) + B_{i(k)}u_c(k) + f_{i(k)},$$
$$y_c(k) = C_{i(k)}x_c(k) + D_{i(k)}u_c(k) + g_{i(k)}, \tag{2b}$$
$$(\text{EG}) \quad [e_j(k) = 1] \longleftrightarrow [a_j^T x_c(k) + b_j^T u(k) \leq c_j] \tag{2c}$$
$$(\text{MS}) \quad i(k) = \begin{bmatrix} \delta_1 \\ \vdots \\ \delta_s \end{bmatrix} = f_{\text{MS}}(x_l(k), u_l(k), i(k-1)) \tag{2d}$$

The automaton (or finite state machine) describes the logic dynamics of the hybrid system. We will only refer to "synchronous automata", where transitions are clocked and synchronous with the sampling time of the continuous dynamical equations. The dynamics of the automaton evolves according to the logic update functions ( ) where $k \in \mathbb{Z}^+$ is the time index, $x_l \in \mathcal{X}_l \subseteq \{0,1\}^{n_l}$ is the logic state, $u_l \in \mathcal{U}_l \subseteq \{0,1\}^{m_l}$ is the exogenous logic input, $y_l \in \mathcal{Y}_l \subseteq \{0,1\}^{p_l}$ is the logic output, $e \in \mathcal{E} \subseteq \{0,1\}^{n_e}$ is the endogenous input coming from the EG, and $f_l : \mathcal{X}_l \times \mathcal{U}_l \times \mathcal{E} \longrightarrow \mathcal{X}_l$, $g_l : \mathcal{X}_l \times \mathcal{U}_l \times \mathcal{E} \longrightarrow \mathcal{Y}_l$ are deterministic boolean functions.

The SAS describes the continuous dynamics and it is a collection of affine systems ( ) where $x_c \in \mathcal{X}_c \subseteq \mathbb{R}^{n_c}$ is the continuous state vector, $u_c \in \mathcal{U}_c \subseteq \mathbb{R}^{m_c}$ is the exogenous continuous input vector, $y_c \in \mathcal{Y}_c \subseteq \mathbb{R}^{p_c}$ is the continuous output vector, $i(k) \in \mathcal{I} \triangleq \left\{ \begin{bmatrix} 1\,0\,\cdots\,0 \end{bmatrix}^T, \cdots, \begin{bmatrix} 0\,\cdots\,0\,1 \end{bmatrix}^T \right\} \subseteq \{0,1\}^s$ is the "mode" in which the SAS is operating, $\sharp\mathcal{I} = s$ is the number of elements of $\mathcal{I}$, and $\{A_i, B_i, f_i, C_i, D_i, g_i\}_{i \in \mathcal{I}}$ is a collection of matrices of opportune dimensions. The mode $i(k)$ is generated by the mode selector, as described below. A SAS of the form ( ) preserves the value of the state when a switch occurs. Resets can be modeled in the present discrete-time setting as detailed in [ ].

The event generator (EG) is a mathematical object that generates a Boolean vector according to the satisfaction of a set of *threshold events* ( ) where $_j$ denotes the j-th component of the vector, and $a_j \in \mathbb{R}^{n_c}$, $b_j \in \mathbb{R}^{m_c}$, $c_j \in \mathbb{R}$ define the hyperplane in the space of continuous states and inputs.

The mode selector (MS) selects the dynamic mode $i(k) \in \mathcal{I} \subseteq \{0,1\}^s$, also called the *active mode*, of the SAS and it is described by the logic function (   ) where $f_{MS} : \mathcal{X}_l \times \mathcal{U}_l \times \mathcal{I} \longrightarrow \mathcal{I}$ is a Boolean function of the logic state $x_l(k)$, of the logic input $u_l(k)$, and of the active mode $i(k-1)$ at the previous sampling instant. We say that a *mode switch* occurs at step k if $i(k) \neq i(k-1)$. Note that contrarily to continuous time hybrid models, where switches can occur at any time, in our discrete-time setting a mode switch can only occur at sampling instants.

A finite-time optimal control problem for the class of hybrid systems is formulated as follows:

$$\min_{\{x(k+1),u(k)\}_{k=0}^{T-1}} \sum_{k=0}^{T-1} \ell_k(x(k+1) - r_x(k+1), u(k) - r_u(k)) \tag{3a}$$

$$\text{s.t. dynamics }(\ ),(\ ),(\ ),(\ ) \tag{3b}$$

$$h_D(x(0), \{x(k+1), u(k), e(k), i(k)\}_0^{T-1}) \leq 0 \tag{3c}$$

$$h_A(x(0), \{x(k+1), u(k), e(k), i(k)\}_0^{T-1}) \leq 0 \tag{3d}$$

where T is the control horizon, $\ell_k : \mathbb{R}^{n \times m} \to \mathbb{R}$ is a nonnegative convex function, $n = n_c + n_l$, $m = m_c + m_l$, $r_x \in \mathbb{R}^n$, $r_u \in \mathbb{R}^m$ are given reference trajectories to be tracked by the state and input vectors, respectively.

The constraints of the optimal control problem can be classified as *dynamical constraints* ( ), representing the discrete-time hybrid system, *design constraints* ( ), artificial constraints imposed by the designer to fulfill the required specifications, and *ancillary constraints* ( ), an a priori additional and auxiliary information for determining the optimal solution which does not change the solution itself, rather help the solver to find it more easily.

2.2 Problem Reformulation

Problem ( ) can be solved via MILP when the costs ℓ_k are convex piecewise linear functions, for instance $\ell_k(x, u) = \|Q_x x\|_\infty + \|Q_u u\|_\infty$, where Q_x, Q_u are full-rank matrices and $\| \cdot \|_\infty$ denotes the infinity-norm ($\|Qx\|_\infty = \max_{j=1,\ldots,n} |Q^j x|$, where Q^j is the j-th row of Q) [ ], or via MIQP (mixed integer quadratic programming) when $\ell_k(x, u) = x'Q_x x + u'Q_u u$, where Q_x, Q_u are positive (semi)definite matrices [ ]. In this paper we wish to solve problem ( ) by using MIP and SAT techniques in a combined approach, taking advantage of SAT for dealing with the purely logic part of the problem. In order to do this, we need to reformulate the problem in a suitable way.

The automaton and mode selector parts of the hybrid system are described as a set of Boolean constraints so they do not require transformations. The event generator and SAS parts can be equivalently expressed, by adopting the so-called "big-M" technique [ ], as a set of continuous and mixed constraints. Problem ( ) can be cast as the mixed logical/convex program

$$\min_{\substack{\{x(k+1), u(k), \\ w(k), \delta(k)\} \\ k = 0, \ldots, T-1}} \sum_{k=0}^{T-1} \ell_k(x(k+1) - r_x(k+1), u(k) - r_u(k)) \tag{4a}$$

$$\text{s.t. } Ax_c(k) \leq b, \; x_c(k+1) = \sum_{i=1}^{s} w_i(k) \tag{4b}$$

$$M_1 x_c(k) + M_2 u_c(k) + M_3 w(k) \leq M_4 e(k) + M_5 \delta(k) + M_6 \tag{4c}$$

$$g(x_l(k+1), x_l(k), u_l(k), e(k), \delta(k)) = \texttt{TRUE} \tag{4d}$$

$$w(k) = [w_1(k) \ldots w_s(k)]', \; w_i(k) \in \mathbb{R}^{n_c}, \; \delta(k) \in \{0, 1\}^s,$$

where $\{x_c(k+1), u_c(k), w(k)\}_{k=0}^{T-1}$ are the continuous optimization variables, $\{x_l(k+1), u_l(k), \delta(k), e(k)\}_{k=0}^{T-1}$ are the binary optimization variables, $x_c(0)$, $x_l(0)$ is a given initial state, constraints ( ), ( ) represent the EG and SAS parts ( ), ( ), and the purely continuous or mixed constraints from ( ), ( ), while ( ) represents the automaton ( ), the mode selector ( ), possible purely Boolean constraints from ( ), ( ). Matrices M_i, $i = 1 \ldots 6$, are obtained by the big-M translation.

   Problem ( ) belongs to the general class ( ) in which all constraints depend on the state initial condition $[x_c(0)' \; x_l(0)']'$ of the hybrid system. In the hybrid optimal control problem at hand, z collects all the continuous variables ($x_c(k+1)$, $u_c(k)$, $k = 0, \ldots, T-1$), the auxiliary variables needed for expressing the SAS dynamics, possibly slack variables for upper bounding the cost function in ( ) [ ], μ collects the integer variables that appear in mixed constraints ($e(k)$, $\delta_i(k)$, $k = 0, \ldots, T-1$, $i = 1, \ldots, s$), and ν collects the integer variables such as $x_l(k)$, $u_l(k)$ that only appear in logic constraints. Note that in general if the objective function in the the form $f(z, \mu)$ we could consider the new objective function ϵ, $\epsilon \in \mathbb{R}$, and an additional constraint $f(z, \mu) \leq \epsilon$ which is a mixed convex constraint that could be included in ( ).

3   SAT-Based Branch&Bound

3.1   Constraint Satisfaction and Optimization

While optimization is primarily associated with mathematics and engineering, CSP was developed (more recently) in the computer science and artificial intelligence communities. The two fields evolved more or less independently until a few years ago. Yet they have much in common and are applied to solve similar problems. Most importantly for the purposes of this paper, they have complementary strengths, and the last few years have seen growing efforts to combine them [, , , ,].

The recent interaction between CSP and optimization promises to affect both fields. In the following subsections we illustrate an approach for merging them into a single problem-solving technology, in particular by combining convex optimization and satisfiability of Boolean formulas (SAT).

Convex Optimization. Convex optimization is very popular in engineering, economics, and other application domains for solving nontrivial decision problems. Convex optimization includes linear, quadratic, and semidefinite programming, for which several extremely efficient commercial and public domain solvers are nowadays available. An excellent reference to convex optimization is the book by Boyd and Vandenberghe [ ].

SAT Problems. An instance of a satisfiability (SAT) problem is a Boolean formula that has three components:

- A set of n variables: $x_1, x_2, \ldots, x_n$.
- A set of literals. A literal is a variable $(Q = x)$ or a negation of a variable $(Q = \neg x)$.
- A set of m distinct clauses: $C_1, C_2, \ldots, C_m$. Each clause consists of only literals combined by just logical *or* ($\vee$) connectives.

The goal of the satisfiability problem is to determine whether there exists an assignment of truth values to variables that makes the following Conjunctive Normal Form (*CNF*) formula satisfiable:

$$C_1 \wedge C_2 \wedge \ldots \wedge C_m,$$

where $\wedge$ is a logical "and" connective. For a survey on SAT problems and related solvers the reader is referred to [ ].

3.2 A SAT-Based Hybrid Algorithm

The basic ingredients for an integrated approach are (1) a solver for convex problems obtained from relaxations over continuous variables of mixed integer convex programming problems of the form ( )-( )-( ), and (2) a SAT solver for testing the satisfiability of Boolean formulas of the form ( ). The relaxed model is used to obtain a solution that satisfies the constraint sets ( ) and ( ) and optimizes the objective function ( ). The optimal solution of the relaxation may fix some of the (0-1) variables to either 0 or 1. If all the (0-1) variables in the relaxed problem have been assigned (0-1) values, the solution of the relaxation is also a feasible solution of the mixed integer problem. More often, however, some of the (0-1) variables have fractional parts, so that further "branching" and solution of further relaxations is necessary. To accelerate the search of feasible solutions one may use the fixed (0-1) variables to "infer" new information on the other (0-1) variables by solving a SAT problem obtained by constraint ( ). In particular, when an integer solution of μ is found from convex programming, a SAT problem then verifies whether this solution can be completed with an assignment of ν that satisfies ( ).

The basic branch&bound (B&B) strategy for solving mixed integer problems can be extended to the present "hybrid" setting where both convex optimization and SAT solvers are used. In a B&B algorithm, the current best integer solution is updated whenever an integer solution with an even better value of the objective function is found. In the hybrid algorithm at hand an additional SAT problem is solved to ensure that the integer solution obtained for the relaxed problem is feasible for the constraints ( ) and to find an assignment for the other logic variables ν that appear in ( ). It is only in this case that the current best integer solution is updated.

The B&B method requires the solution of a series of convex subproblems obtained by branching on integer variables. Here, the non-integer variable to branch on is chosen by selecting the variable with the largest fractional part (i.e., the one closest to 0.5), and two new convex subproblems are formed with that variable fixed at 0 and at 1, respectively. When an integer feasible solution of the relaxed problem is obtained, a satisfiability problem is solved to complete the solution. The value of the objective function for an integer feasible solution of the whole problem is an upper bound (UB) of the objective function, which may be used to rule out branches where the optimum value attained by the relaxation is larger than the current upper bound.

Let P denote the set of convex and SAT subproblems to be solved. The proposed SAT-based B&B method can be summarized as follows:

1. **Initialization.** $UB = \infty$, $P = \{(p^0, SAT^0)\}$. The convex subproblem p^0 is generated by using ( ),( ), ( ) along with the relaxation $\mu \in [0,1]^{n_\mu}$, and the SAT subproblem SAT^0 is generated by using ( ).
2. **Node selection.** If $P = \emptyset$ then go to   ; otherwise select and remove a (p, SAT) problem from the set P; The criterion for selecting a problem is called *node selection rule*.

3. **Logic inference.** Solve problem SAT. If it is infeasible go to step
4. **Convex reasoning.** Solve the convex problem p, and:
    4.1. If the problem is infeasible or the optimal value of the objective function is greater than UB then go to step
    4.2. If the solution is not integer feasible then go to step
5. **Bounding.** Let $\mu^* \in \{0,1\}^{n_\mu}$ be the integer part of the optimal solution found at step   ; to extend this partial solution, solve the SAT problem finding ν such that $g(\nu, \mu^*) =$TRUE. If the SAT problem is feasible then update UB; otherwise add to the LP problems of the set P the "no-good" cut [ ]

$$\sum_{i \in T^*} \mu_i - \sum_{j \in F^*} \mu_j \le B^* - 1,$$

   where $T^* = \{i | \mu_i^* = 1\}$, $F^* = \{j | \mu_j^* = 0\}$, and $B^* = |T^*|$. Go to step
6. **Branching.** Among all variables that have fractional values, select the one closest to 0.5. Let μ_i be the selected non-integer variable, and generate two subproblems $(p \cup \{\mu_i = 0\},\ SAT\&\{\neg\mu\})$, $(p \cup \{\mu_i = 1\},\ SAT\&\{\mu\})$ and add them to set P; go to step
7. **Termination.** If $UB = \infty$, then the problem is infeasible. Otherwise, the optimal solution is the current value UB.

Remark 1. At each node of the search tree the algorithm executes a three-step procedure: logic inference, solution of the convex relaxation, and branching. The first step and the attempted completion of the solution do not occur in MIP approaches but they are introduced here by the distinction of mixed (0-1) variables μ and pure (0-1) variables ν. The logic inference and the attempted completion steps do not change the correctness and the termination of the algorithm but they improve the performance of the algorithm because of the efficiency of the SAT solver in finding a feasible integer solution.

Remark 2. The class of problems ( ) is similar to the MLLP framework introduced by Hooker in [ ],

$$\min c'x \tag{5a}$$
$$\text{s.t. } p_j(y, h) \rightarrow [A^j(x) \ge a^j],\ j \in J \tag{5b}$$
$$q_i(y, h),\ i \in I, \tag{5c}$$

where $x \in \mathbb{R}^{n_x}$, $y \in \{0,1\}^{n_y}$, $h \in \mathcal{D} \subset \mathbb{Z}^{n_h}$, ( ) is the continuous part, and ( ) is the logic part. If we consider the only y variables as discrete variables and a liner cost function, constraints ( ), ( ) represent the linearization of ( ), and constraints ( ) are equivalent to ( ).
There are however a few differences between frameworks ( ) and ( ). First, the relaxation problem of ( ) is the same for each node in the search tree, while in ( ) the relaxation depends on which left-hand side of ( ) is true. Second, in the class of problems ( ) constraints of type

$$[\mu = 1] \longleftrightarrow [z_1 + z_2 \ge \alpha],$$

can not be introduced and they have to be converted into inequalities, becoming part of constraints ( ). Inference is done only in the logic part, by the SAT solver, and no information is derived by the continuous part. In the MLLP framework, instead, inferences are made in both ways.

Remark 3. The modeling framework ( ) can also be solved by using a combined approach of MIP and CLP [ ]. The role of constraint propagation is obviously to reduce as much as possible the domain sets of the μ variables that appear in the constraints managed by the CLP solver. In this way, the constraint propagation can reduce the search space removing some branches in the search tree that can not have feasible solutions. Moreover the constraint propagation together with choice points can help to find a completion of the solution trying to fix the ν variables.

The SAT solver behaves in a similar way to CP solver. The SAT inference is a feasibility check. If a partial assignment of the μ variables is infeasible for the set of constraints ( ) SAT is able to find the infeasibility easier and more quickly than a CLP solver. SAT solvers are also more efficient for finding a feasible assignment for the ν variables with respect to CLP solvers.

However the efficiency of SAT solvers relies upon the representation of the logic part of the problem. While CLP can be used both with logic formulas and linear constraints, as well as global constraints, SAT turns out to be useful only with Boolean formulas.

4   Numerical Results

In this section we show on an example of hybrid optimal control problem that the hybrid solution technique described in the previous sections has a better performance compared to commercial MIP solvers.

4.1   "Hybrid" Model

Consider a room with two bodies with temperatures T_1, T_2 and let T_{amb} be the room temperature (this example is an extension of the example reported in [ ]). The room is equipped with a heater, close to body 1, delivering thermal power u_{hot} and an air conditioning system, close to body 2, draining thermal power u_{cold}. These are turned on/off according to some rules dictated by the closeness of the two bodies to each device. We want guarantee that the bodies are not cold or hot.

The discrete-time continuous dynamics of each body is described by the difference equation

$$\frac{T_i(k+1) - T_i(k)}{T_s} = -\alpha_i(T_i(k) - T_{amb}) + k_i(u_{hot}(k) - u_{cold}(k)) + cu_e(k), \quad (6)$$

where $i = 1, 2$, α_i, k_i, c are suitable constants, T_s is the sampling time, and $u_e(k)$ is an exogenous input that can be used to deliver or drain thermal power manually (e.g. by opening a window or by changing the water flow from a centralized heating system).

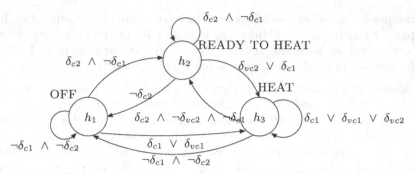

Fig. 1. Automaton regulating the heater

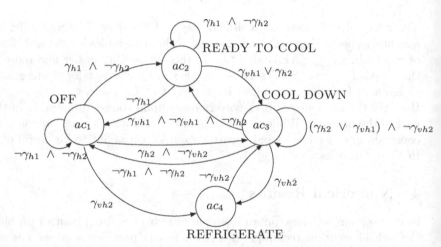

Fig. 2. Air conditioning system automaton

The automaton part of the system is described by the two automata represented in Figures   and , where $\delta_{ci}, \delta_{vci}, \gamma_{hi}$ and γ_{vhi}, for $i = 1, 2$, are logic variables defined as follows

$$[\delta_{vci}(k) = 1] \longleftrightarrow [T_i(k) \leq T_{vci}], \qquad (7a)$$

$$[\delta_{ci}(k) = 1] \longleftrightarrow [T_i(k) \leq T_{ci}], \qquad (7b)$$

$$[\gamma_{hi}(k) = 1] \longleftrightarrow [T_i(k) \geq T_{hi}], \qquad (7c)$$

$$[\gamma_{vhi}(k) = 1] \longleftrightarrow [T_i(k) \geq T_{vhi}], \qquad (7d)$$

and where $T_{vci} \leq T_{ci} \leq T_{hi} \leq T_{vhi}$ are constant thresholds. The automaton for the heater (Figure) sets the heater in the "ready to heat" state if body 2 is cold, and will go in "heat" state if body 2 is very cold. If body 1 is cold or very cold the heater is turned on immediately. The automaton of the air conditioning (A/C) system (Figure) sets the air conditioning system in the "ready to cool" state if body 1 is hot, unless body 2 is cold, in other words, the A/C system is turned on only when body 1 is very hot. However, the draining thermal power is half of the full power. The A/C system is set to the maximum power if the body 2 is very hot but it is immediately switched to half power as soon as body 2 is only hot (due to energy consumptions of the A/C system).

The heater delivers thermal power and the A/C system drains thermal power according to the following rules:

$$u_{\text{hot}} = \begin{cases} u_H & \text{if } h_3 = 1 \\ 0 & \text{otherwise} \end{cases} \qquad u_{\text{cold}} = \begin{cases} u_C & \text{if } ac_4 = 1 \\ \frac{u_C}{2} & \text{if } ac_3 = 1 \\ 0 & \text{otherwise} \end{cases}. \qquad (8)$$

By following the notation of ( ), we have $x_l = [h_1 \ h_2 \ h_3 \ ac_1 \ ac_2 \ ac_3 \ ac_4]' \in \{0,1\}^7$, $u_l = \emptyset$ and $e(k) = [\delta_{vc1} \ \delta_{vc2} \ \delta_{c1} \ \delta_{c2} \ \gamma_{h1} \ \gamma_{h2} \ \gamma_{vh1} \ \gamma_{vh2}]' \in \{0,1\}^8$.

The system has six modes: $(u_{\text{hot}}, u_{\text{cold}}) \in \{(0,0), (u_H, 0), (0, u_C), (0, u_C/2), (u_H, u_C), (u_H, u_C/2)\}$. The mode selector function is defined as follows

$$i(k) = \begin{bmatrix} \neg h_3(k) \wedge \neg ac_4(k) \wedge \neg ac_3(k) \\ h_3(k) \wedge \neg ac_4(k) \wedge \neg ac_3(k) \\ \neg h_3(k) \wedge ac_4(k) \wedge \neg ac_3(k) \\ \neg h_3(k) \wedge \neg ac_4(k) \wedge ac_3(k) \\ h_3(k) \wedge ac_4(k) \wedge \neg ac_3(k) \\ h_3(k) \wedge \neg ac_4(k) \wedge ac_3(k) \end{bmatrix} \in \{0,1\}^6,$$

which only depends on logic states.

The SAS dynamics ( ), i.e., the continuous part of the hybrid system, is translated into a set of inequalities using the Big-M technique, which provides the set of constraints

$$A x_c(k) + B u_c(k) + C w(k) \le D \delta(k) + E, \qquad (9)$$

where $x_c = [T_1 \ T_2]'$, $u_c = u_e$, $w(k) \in \mathbb{R}^3$ contains the auxiliary continuous variables needed to represent the conditions $u_{\text{hot}} = u_H$, $u_{\text{cold}} = u_C$, $u_{\text{cold}} = u_C/2$, and $\delta(k) = [h_3(k) \ ac_3(k) \ ac_4(k)] \in \{0,1\}^3$. Constraints ( ) are obtained by employing the HYSDEL compiler [ ], a dedicated "hybrid" system description language and compiler which translates a description of the problem into the mathematical mixed+logical dynamical (MLD) representation introduced in [ ], a mathematical framework useful for defining optimal control problems as pure MIP problems.

Finally, the event generator is represented by ( ) and ( ). These are translated by HYSDEL into a set of linear inequalities:

$$G'_x x_c(k) + G'_u u_c(k) + D' e(k) \le E', \qquad (10)$$

where $e(k) = [\delta_{vc1} \ \delta_{vc2} \ \delta_{c1} \ \delta_{c2} \ \gamma_{h1} \ \gamma_{h2} \ \gamma_{vh1} \ \gamma_{vh2}]' \in \{0,1\}^8$.

4.2 Optimal Control Problem

The goal is to design an optimal control profile for the continuous input u_e that minimizes $\sum_{k=0}^{T} |T_i(k) - T_{amb}|$ subject to the hybrid dynamics and the following additional constraints:

- Continuous constraints on temperatures to avoid that they assume unacceptable values

$$-10 \leq T_1(k) \leq 50 \qquad\qquad -10 \leq T_2(k) \leq 50. \tag{11a}$$

These constraints may be interpreted as dynamical constraints due to physical limitations of the bodies.
- A continuous constraint on exogenous input to avoid excessive variations:

$$-10 \leq u_e(k) \leq 10. \tag{12}$$

This constraint may be interpreted as a design constraint of the form ( ).

4.3   Results

The above dynamics and constraints are also modeled in HYSDEL [ ] to obtain an MLD model of the hybrid system in order to compare the performance achieved by the hybrid solver with the one obtained by employing a pure MILP approach.

The optimal control problem is defined over horizon of T steps as:

$$\min_{\{x,u,z,\delta,\epsilon_T\}} \sum_{k=0}^{T-1} \epsilon_T(k) \tag{13a}$$

$$\text{s.t. } \epsilon_T(k) \begin{bmatrix} 1 \\ \vdots \\ 1 \end{bmatrix} \geq \pm(T_i(k) - T_{amb}), \tag{13b}$$

$$\text{automata Figures } , , \tag{13c}$$

$$(\ ), (\ ) \tag{13d}$$

$$(\ ), (\ ) \tag{13e}$$

where $\{x,u,z,\delta,\epsilon_T\}=\{x(k),u(k),z(k),\delta(k), \epsilon_T(k)\}_{k=0}^{T-1}$, $\epsilon_T =[\epsilon_{T1}(0), \epsilon_{T2}(0),\ldots,$ $\epsilon_{T1}(T-1), \epsilon_{T2}(T-1)]' \in \mathbb{R}^{2T}$.

Each part of the optimal control problem is managed by either the SAT solver or the LP solver: the cost function ( ), the inequalities ( ), ( ), and the additional constraints ( ) are managed by the LP solver, the logic part ( ) is managed by the SAT solver. Our simulations have been done describing and solving the problem within the Matlab environment and calling, through MEX interfaces, respectively, zCHAFF [ ] for SAT and CPLEX [ ] for LP.

In all our simulations we have adopted depth first search as the *node selection rule*, to reduce the amount of memory used during the search.

For the initial condition $T_1(0) = 5°$ C, $T_2(0) = 2°$ C and for $T_{amb} = 25°$ C we have done simulations for different horizons (the obtained optimal solution is clearly the same both using the SAT-based B&B and the MILP), reported in Table .

We can see that the performance of the SAT-based B&B is always better than the one obtained by using the commercial MILP solver of CPLEX. In Table , we

108   Alberto Bemporad and Nicolò Giorgetti

Table 1. Optimal control solution: comparison among SAT-based B&B, naive MILP, and CPLEX MILP

T	Bool. Vars	SATbB&B (s)	LPs	SATs	CPLEX (s)	LPs	Naive MILP (s)	LPs
5	82	0.09	5	6	0.03	18	0.48	23
10	157	0.18	5	6	0.13	79	3.7150	119
15	232	0.33	5	6	0.42	199	83.69	943
20	307	0.5110	6	8	0.5410	243	109.0870	2181
25	382	0.7620	8	10	0.8210	286	503.0030	3833
30	457	1.0520	9	12	1.0110	333	1072.3	6227
35	532	1.4420	10	13	1.7170	341	> 1200	–
40	607	1.8630	13	16	2.5030	374	> 1200	–
45	682	2.7740	15	20	3.8320	475	> 1200	–

Table 2. Computation time for solving a pure integer feasibility problem: comparison between SAT (zCHAFF) and MILP (CPLEX)

T	Bool. Vars	Constr	SAT (s)	MILP (s)
5	82	460	0	0.02
10	157	920	0.01	0.02
15	232	1380	0.02	0.03
20	307	1840	0.03	0.03
25	382	2300	0.04	0.05
30	457	2760	0.05	0.06
35	532	3220	0.06	0.07
40	607	3680	0.08	0.10
45	682	4140	0.09	0.13

also compare the performance of a "naive MILP" solver, that is obtained from the SAT-based B&B code by simply disabling SAT inference. The main reason is that the SAT B&B algorithm solves a much smaller number of LPs than an MILP solver. The "cuts" performed by the SAT solver, i.e. the infeasible SAT problems, obtained at step 3 of the algorithm turn out very useful to exclude subtrees containing no integer feasible solution, see Figure . Moreover, the time spent for solving the integer feasibility problem at the root node of the search treee described as SAT problem is much smaller than solving a pure integer feasibility problem, see Table . We can also see from Table that the number of feasible SAT solved equals the number of LP solved plus one. This one more SAT is used to complete a feasible solution and it is very useful to further reduce the computation time.

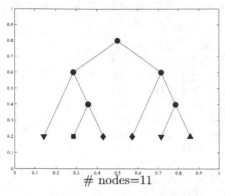

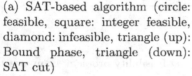

(a) SAT-based algorithm (circle: feasible, square: integer feasible, diamond: infeasible, triangle (up): Bound phase, triangle (down): SAT cut)

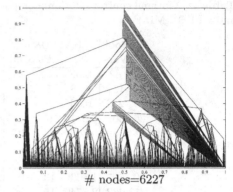

(b) Naive MILP algorithm

Fig. 3. Comparison of the trees generated by the SAT-based and naive MILP algorithms (T=30)

The results were simulated on a PC Pentium IV 1.8 GHz running CPLEX 9.0 and zCHAFF 2003.12.04.

5   Conclusions

In this paper we have proposed a new unifying framework for MIP and CSP techniques based on the integration of convex programming and SAT solvers for solving optimal control problems for discrete-time hybrid systems. The approach consists of a logic-based branch and bound algorithm, whose performance in terms of computation time can be superior in comparison to pure mixed-integer programming techniques, as we have illustrated on an example.

Ongoing research is devoted to the improvement of the logic-based method by including relaxations of the automaton and MS parts of the hybrid system in the convex programming part, to the investigation of alternative relaxations of the SAS dynamics that are tighter than the big-M method, and to the use of SAT solvers for also performing domain reduction (cutting planes).

Acknowledgement

This research was supported by the European Union through project IST-2001-33520 "CC-Computation and Control".

References

[1] K. Marriot and P. J. Stuckey. *Programming with constraints: an introduction.* MIT Press, 1998.

[2] J. Gu, P. W. Purdom, J. Franco, and B. Wah. Algorithms for the satisfiability (SAT) problem: A survey. In *DIMACS Series on Discrete Mathematics and Theoretical Computer Science*, number 35, pages 19–151. American Mathematical Society, 1997. ,

[3] J. Hooker. *Logic-based methods for Optimization.* Wiley-Interscience Series, 2000.
, ,

[4] A. Bockmayr and T. Kasper. Branch and infer: A unifying framework for integer and finite domain constraint programming. *INFORMS Journal on Computing*, 10(3):287–300, Summer 1998. ,

[5] R. Rodosek, M. Wallace, , and M. Hajian. A new approach to integrating mixed integer programming and constraint logic programming. *Annals of Oper. Res.*, 86:63–87, 1997. ,

[6] F. Focacci, A. Lodi, and M. Milano. Cost-based domain filtering. In *J. Jaffar, editor, Principle and Practice of Constraint Programming*, volume 1713 of *Lecture Notes in Computer Science*, pages 189–203. Springer Verlag, 2001.

[7] I. Harjunkoski, V. Jain, , and I. E. Grossmann. Hybrid mixed-integer/constraint logic programming strategies for solving scheduling and combinatorial optimization problems. *Comp. Chem. Eng.*, 24:337–343, 2000.

[8] Large scale integrated supply chain optimisation software. A European Union Funded Project, 2003. http://www.liscos.fc.ul.pt/.

[9] X. Xu and P. J. Antsaklis. An approach to switched systems optimal control based on parameterization of the switching instants. In *Proc. IFAC World Congress*, Barcelona, Spain, 2002.

[10] B. Lincoln and A. Rantzer. Optimizing linear system switching. In *Proc. 40th IEEE Conf. on Decision and Control*, pages 2063–2068, 2001.

[11] F. Borrelli, M. Baotic, A. Bemporad, and M. Morari. An efficient algorithm for computing the state feedback optimal control law for discrete time hybrid systems. In *Proc. American Contr. Conf.*, Denver, Colorado, 2003.

[12] A. Bemporad and M. Morari. Control of systems integrating logic, dynamics, and constraints. *Automatica*, 35(3):407–427, March 1999. , ,

[13] A. Bemporad, F. Borrelli, and M. Morari. Piecewise linear optimal controllers for hybrid systems. In *Proc. American Contr. Conf.*, pages 1190–1194, Chicago, IL, June 2000. ,

[14] F. D. Torrisi and A. Bemporad. HYSDEL - A tool for generating computational hybrid models. *IEEE Transactions on Control Systems Technology*, 12(2), March 2004. , , ,

[15] A. Bemporad and N. Giorgetti. A logic-based hybrid solver for optimal control of hybrid systems. In *Proc. 43th IEEE Conf. On Decision and Control*, Maui, Hawaii, USA, Dec. 2003. ,

[16] H. P. Williams. *Model Building in Mathematical Programming.* John Wiley & Sons, Third Edition, 1993.

[17] G. Ottosson. *Integration of Constraint Programming and Integer Programming for Combinatorial Optimization.* PhD thesis, Computing Science Department, Information Technology, Uppsala University, Sweden, 2000.

[18] E. Tsang. *Foundations of Constraint Satisfaction.* Academic Press, 1993.

[19] S. Boyd and L. Vandenberghe. *Convex Optimization.* In press., 2003.

[20] J. N. Hooker and M. A. Osorio. Mixed logical/linear programming. *Discrete Applied Mathematics*, 96-97(1-3):395–442, 1999.

[21] A. Bemporad. Efficient conversion of mixed logical dynamical systems into an equivalent piecewise affine form. *IEEE Trans. Automatic Control*, 2003. In Press.

[22] M. Moskewicz, C. Madigan, Y. Zhao, L. Zhang, and S. Malik. Chaff: Engineering an efficient sat solver. In *39th Design Automation Conference*, June 2001.

[23] ILOG, Inc. *CPLEX 8.1 User Manual.* Gentilly Cedex, France, 2002.

Solving the Petri Nets Reachability Problem Using the Logical Abstraction Technique and Mathematical Programming

Thomas Bourdeaud'huy[1], Saïd Hanafi[2], and Pascal Yim[1]

[1] L.A.G.I.S., Ecole Centrale de Lille
Cité Scientifique, B.P. 48, 59651 Villeneuve d'Ascq Cedex, France
{thomas.bourdeaud_huy,pascal.yim}@ec-lille.fr
[2] L.A.M.I.H., Université de Valenciennes
59313 Valenciennes Cedex 9, France
said.hanafi@univ-valenciennes.fr

Abstract. This paper focuses on the resolution of the reachability problem in Petri nets, using the *logical abstraction* technique and the mathematical programming paradigm. The proposed approach is based on an implicit exploration of the Petri net reachability graph. This is done by constructing a unique sequence of *partial steps*. This sequence represents *exactly* the total behavior of the net. The logical abstraction technique leads us to solve a constraint satisfaction problem. We also propose different new formulations based on integer and/or binary linear programming. Our models are validated and compared on large data sets, using `Prolog IV` and `Cplex` solvers.

1 Introduction

The operational management of complex systems is characterized, in general, by the existence of a huge number of solutions. Decision-making processes must be implemented in order to find the best results. These processes need suitable modelling tools offering true practical resolution perspectives. Among them, Petri nets (PN) provide a simple graphic model taking into account, in the same formalism, concurrency, parallelism and synchronization.

In this paper, we consider the PN reachability problem. Indeed, it seems very efficient to model discrete dynamic systems in a flexible way, as can be seen from the fact that many operations research problems have been defined using reachability between states (e.g. scheduling problems [ ], railway traffic planning [ ] or car-sequencing problems [ ]). Furthermore, a large number of PN analysis problems such as deadlock freeness or liveness, are equivalent to the reachability problem, or to some of its variants or sub-problems [ ].

Various methods have been suggested to handle the PN reachability problem. In this paper, we study more precisely the PN logical abstraction technique proposed initially by Benasser [ ]. This method consists in developing a unique sequence of *partial steps* corresponding to the total behavior of the system. It

J.-C. Régin and M. Rueher (Eds.): CPAIOR 2004, LNCS 3011, pp. 112–   , 2004.
© Springer-Verlag Berlin Heidelberg 2004

was validated on several examples using logical constraint programming techniques. Numerical results using `Prolog IV` have shown that the method can be more effective than other generic solvers, and can even compete with heuristics dedicated to particular classes of problems [ ]. However, its resolution performance is limited in practice by the incremental search mechanism used: memory overflows appear soon when the size of the problem grows.

We propose an alternative based on a mathematical programming formulation. We model the problem as an integer linear program, then we solve it with a branch-and-bound technique, using the `Cplex` optimization software. We show that the performances achieved are better than those of the initial technique in terms of speed and memory.

The paper is organized as follows. In section 2, after the introduction of the PN terminology, we define the reachability problem to be solved. In section 3, we recall the notions of partial markings and partial steps used in the Petri nets logical abstraction technique, and the main results presented in [ ]. In the section 4, we use firstly the logical abstraction technique to model the PN reachability problem as a constraint satisfaction problem. Then we propose new formulations based on integer linear programming. Computational results on a set of benchmarks are presented and compared in section 5. Finally, as a conclusion, we describe a few promising research directions.

2   Petri Net Reachability Problem

2.1   Petri Net Structure

A Petri net [ ] $(R = (\mathbb{P}, \mathbb{T}, W), m_0)$ can be defined as a *bipartite directed graph* where:

- $\mathbb{P} = \{p_1, \dots, p_M\}$ is a finite set of places, with $|\mathbb{P}| = M$. Places are represented as circles;
- $\mathbb{T} = \{t_1, \dots, t_N\}$ is a finite set of transitions, with $|\mathbb{T}| = N$. Transitions are represented as rectangles;
- $W : \mathbb{P} \times \mathbb{T} \cup \mathbb{T} \times \mathbb{P} \to \mathbb{N}$ is the weighted flow relation representing the arcs. It associates to each pair *(place, transition)* or *(transition, place)* the weight of the corresponding arc in the net;
- $m_0 : \mathbb{P} \to \mathbb{N}$ associates to each place $p \in \mathbb{P}$ an integer $m_0(p)$ called the *marking* of the place p. Markings are represented as full disks called *tokens* inside the place.

The matrices which are defined below (precondition (Pre: $m \times n$), postcondition (Post: $m \times n$) and incidence (C: $m \times n$)) are useful in PN analysis:

- $\forall p \in \mathbb{P}, \forall t \in \mathbb{T}, Pre(p, t) = k \Leftrightarrow W(p, t) = k;$
- $\forall p \in \mathbb{P}, \forall t \in \mathbb{T}, Post(p, t) = k \Leftrightarrow W(t, p) = k;$
- $\forall p \in \mathbb{P}, \forall t \in \mathbb{T}, C(p, t) = Post(p, t) - Pre(p, t).$

An example of Petri net is presented in figure .

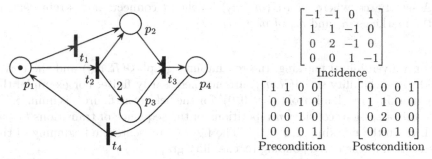

Fig. 1. A Petri net and its Pre, Post and Incidence matrices

2.2   Token Game

In a Petri net, the markings of the places represent the state of the corresponding system at a given moment. This marking can be modified by the firing of transitions. A transition t is *fireable* for a marking m_0 (denoted by $m_0[t\rangle$), when $\forall p \in \mathbb{P}, m_0(p) \geq W(p,t)$. If this condition is satisfied, a new marking m_1 is produced from the marking m_0 (denoted by $m_0[t\rangle m_1$): $\forall p \in \mathbb{P}, m_1(p) = m_0(p) - W(p,t) + W(t,p)$.

The previous equations can be generalized to the firing of *fireable transition sequences*. Given $\sigma = t_{\sigma_1} t_{\sigma_2} \ldots t_{\sigma_r}$ a sequence of transitions of (R, m_0), we define the *firing count vector* $\vec{\sigma}$ associated to σ, where $\vec{\sigma}(t)$ is the number of times the transition t has been fired in σ, $\forall t \in \mathbb{T}$. Formally, $\vec{\sigma} = \sum_{j=1}^{r} \vec{e_{t_{\sigma_j}}}$, where $\vec{e_{t_j}}$ represents the characteristic vector of t_j. Therefore we have:

$$\sigma \text{ is fireable from } m_0 \overset{\text{def}}{\Longleftrightarrow} m_0[\sigma\rangle \Rightarrow Pre.\vec{\sigma} \leq \overrightarrow{M_0} \tag{1}$$

$$m_0[\sigma\rangle m_1 \Rightarrow \overrightarrow{M_1} = \overrightarrow{M_0} + C.\vec{\sigma} \tag{2}$$

The equation ( ) is called the *fundamental (or state) equation*. One will denote by $\mathbb{T}^\infty$ the set of all sequences of elements of $\mathbb{T}$, and $\mathcal{T}(R, m_0)$ the set of all transitions sequences fireable from m_0.

2.3   Reachability Problem

The firing rule can be used to define a *reachability graph* associated with the Petri net. The reachability graph corresponds to the usual formal representation of the behavior of the net. The reachability graph of a net R, denoted by $\mathcal{G}(R, m_0)$, is defined by:

- A set of nodes $\mathcal{A}(R, m_0)$ which represent all the markings reachable by any fireable transition sequence. Formally, $\mathcal{A}(R, m_0) = \{m_f \mid \exists \sigma \in \mathcal{T}(R, m_0)$ s.t. $m_0[\sigma\rangle m_f\}$;

– A set of arcs, where an arc (m_i, m_j) labelled t connects nodes representing the markings m_i and m_j iff $m_i[t\rangle m_j$.

For a given initial marking, the reachability graph $\mathcal{G}(R, m_0)$ and the corresponding reachability set $\mathcal{A}(R, m_0)$ are not necessarily finite. For example, the set of markings reachable from $(1, 0, 0, 0)^t$ for the net of the figure   is infinite. To prove this assertion, consider n repetitions of the sequence of transitions $t_2 t_3 t_4$, which reach the marking $(1, 0, n, 0)^t$. The figure   presents the beginning of the construction of the corresponding reachability graph.

A net is called *bounded* iff $\exists k \in \mathbb{N}, \forall m \in \mathcal{A}(R, m_0), \forall p \in \mathbb{P}, m(p) \leq k$. The reachability set of bounded nets is finite (it is clearly limited by k^M).

The reachability problem is defined as follows: " *Given a Petri net R with the initial marking m_0 and a final marking $m_f \in \mathbb{N}^M$, decide if m_f is reachable from m_0 (i.e. if $m_f \in \mathcal{A}(R, m_0))$* ". To solve this problem, we need to find a fireable sequence of transitions from $\mathcal{T}(R, m_0)$ such that $m_0[\sigma\rangle m_f$. A " naive " approach consists in exploring the reachability graph exhaustively. It has been shown that the reachability problem is decidable [ ]. However it is *EXP-TIME* and *EXP-SPACE hard* in the general case ([ ]).

Practically, it is not possible to explore the reachability graph exhaustively due to the well known problem of *combinatorial explosion*: the size of the state-space may grow exponentially with the size of a system configuration. Many methods have been studied to limit this explosion. Let us mention the three main ones. The first manages the combinatorial explosion without modifying the studied reachability graph. Classical approaches are *graph compressions*, particularly *bdd encoding* [ ]) and *forward checking* [ ]. Some other techniques construct a reduced reachability graph associated to the original, based on some properties to preserve: symmetries [ ], reductions [ ] and partial order (covering step graphs [ ], *stubborn sets* [ ]) are the main approaches. The last ones are based on the state equation: we can distinguish *parametrized analysis* [ ] and *algebraic methods* [ ].

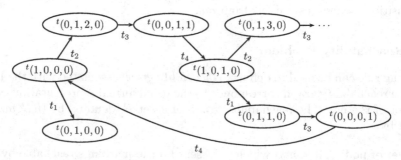

Fig. 2. Reachability graph for the PN fig.   starting $(1, 0, 0, 0)^t$

In the following sections, we propose new approaches to find fireable transitions sequences leading to a target marking. Our methods are based on the *Petri net logical abstraction* and *mathematical programming* techniques.

3   Petri Net Logical Abstraction Technique

3.1   Steps and Steps Sequences

We generalize the notion of transition firing to step firing. A step corresponds to the simultaneous firing of several transitions, the same transition can be fired several times (we call this *reentrance*). We represent a step as a multi-set over transitions, i.e. a set which can contain several copies of the same element, for example $\{2t_1, t_2\}$, which we would note simply $2t_1 + t_2$.

A step $\varphi = \alpha_1 t_1 + \ldots + \alpha_N t_N$, where $\alpha_1, \ldots, \alpha_N \in \mathbb{N}$, is fireable from a marking m iff $\forall p \in \mathbb{P}, m(p) \geq \sum_{j=1}^{N} \alpha_j W(p, t_j)$. The marking must contain enough marks so that each transition of the step may consume its own tokens. We associate a step φ and a characteristic vector $\overrightarrow{\varphi}$ in the classical manner, as a linear combination with positive coefficients of the characteristic vectors of each transition, i.e. $\overrightarrow{\varphi} = \sum_{j=1}^{N} \alpha_j \overrightarrow{e_{t_j}}$.

Equations ( ) and ( ) can be generalized to steps and step sequences. In the following sections, we will use the notations already used previously: $m[\varphi\rangle$, $m_0[\varphi\rangle m_1$, $m_0[\varphi_1 \varphi_2 \ldots \varphi_k\rangle$ and $m_0[\varphi_1 \varphi_2 \ldots \varphi_k\rangle m_k$ to indicate that a step or a step sequence is fireable, and the marking obtained in each case. We denote by $\mathbb{T}^*$ the set of steps built over $\mathbb{T}$, and $\mathcal{S}(\mathbb{P}, \mathbb{T}, W, m_0)$, the set of steps fireable from m_0.

3.2   Partial Steps and Markings

Steps and step sequences capture parallel executions in a unique step fire. In this section, we present briefly the notions of partial steps and markings introduced in [ ]. These new structures will capture the total behavior of the Petri net in a unique sequence of partial steps. Informally, intermediate steps and markings are considered as vectors of variables, which are associated to a formula. Formulae correspond to constraints required for variables so as to guarantee that all possible instantiations will always represent valid concrete steps and markings.

Let $\mathcal{L}$ be a first order logic whose domain is $\mathbb{Z}$, we denote respectively $\mathcal{E}_{\mathcal{L}}$ and $\mathcal{F}_{\mathcal{L}}$ the set of expressions and formulae of $\mathcal{L}$.

A partial marking (resp. partial step) is a pair $MP = (m, F_m)$ (resp. $SP = (\varphi, F_\varphi)$) where:

- $m : \mathbb{P} \mapsto \mathcal{E}_{\mathcal{L}}$ (resp. $\varphi : \mathbb{T} \mapsto \mathcal{E}_{\mathcal{L}}$) is a mapping associating to each place (resp. transition) an expression of the language $\mathcal{L}$;
- F_m (resp. F_φ) is a formula from $\mathcal{L}$.

We denote $\mathcal{SP}(R)$ and $\mathcal{MP}(R)$ the sets of partial steps and markings, respectively. The firing properties can be extended easily to partial steps and markings.

Of course, concrete steps and markings are particular cases of partial ones associated to the truth formula, so that the extension can be made in a natural way.

Let $R = (\mathbb{P}, \mathbb{T}, W)$ be a Petri net, $MP_0 = (m_0, F_{m_0})$ be a partial marking from $\mathcal{MP}(R)$ and $SP_1 = (\varphi_1, F_{\varphi_1})$ be a partial step from $\mathcal{SP}(R)$. The partial step SP_1 is called *fireable* from MP_0 (denoted by $MP_0[SP_1\rangle$) iff the formula $F_{m_0} \wedge F_{\varphi_1} \wedge \left(\bigwedge_{p \in \mathbb{P}} \left(m_0(p) \geq \sum_{t \in T} W(p,t)\varphi_1(t) \right) \right)$ is satisfiable.

Within these conditions, the *firing* of the partial step SP_1 creates a new partial marking $MP_1 = (m_1, F_{m_1})$ (denoted by $MP_0[SP_1\rangle MP_1$) such that:

$$- \quad \forall p \in \mathbb{P}, \ m_1(p) \overset{\text{def}}{=} m_0(p) - \sum_{t \in T} W(p,t)\varphi_1(t) + \sum_{t \in T} W(t,p)\varphi_1(t); \qquad (E_1)$$

$$- \quad F_{m_1} \overset{\text{def}}{\equiv} F_{m_0} \wedge F_{\varphi_1} \wedge \left(\bigwedge_{p \in \mathbb{P}} \left(m_0(p) \geq \sum_{t \in T} W(p,t)\varphi_1(t) \right) \right). \qquad (E_2)$$

3.3   Complete Partial Steps Sequences

With some additional hypothesis on the constraints F_{φ_i}, linked to the partial steps, we can define a *complete* sequence of partial steps. This complete partial steps sequence will be used to find all the concrete steps sequences of a given size.

Let $R = (\mathbb{P}, \mathbb{T}, W)$ be a Petri net. Let $SSP = SP_1 SP_2 \ldots SP_k$ be a partial steps sequence from $\mathcal{SP}(R)$, such that $\forall i \in [\![1, k]\!], SP_i = (\varphi_i, F_{\varphi_i})$. Let $MP_0, \ldots, MP_k$ be partial markings from $\mathcal{MP}(R)$ s.t. $\forall i \in [\![1, k]\!], MP_{i-1}[SP_i\rangle MP_i$. We note $\varphi_{01}, \varphi_{02}, \ldots, \varphi_{0M}$ the variables embedded in the partial marking MP_0, and V_{φ_i} those corresponding to SP_i, for $i \in [\![1, k]\!]$.

If a sequence of partial steps SSP satisfies the following conditions, then it is complete from the marking MP_0:

- $\forall i \in [\![1, k]\!], V_{\varphi_i} = \{\varphi_{i1}, \varphi_{i2}, \ldots, \varphi_{iN}\}$;
- $\forall i \in [\![1, k]\!], \forall j \in [\![1, N]\!], \varphi_i(t_j) = \varphi_{ij}$;
- The symbols of variables φ_{ij}, for $i \in [\![0, k]\!]$ and $j \in [\![1, N]\!]$ are different;
- $\forall i \in [\![1, k]\!], F_{\varphi_i} \equiv \bigwedge_{j=1}^{N} (\varphi_{ij} \geq 0). \qquad\qquad\qquad (E_3)$

Benasser [ ] proved that a complete sequence of partial steps of length k captures exactly all the concrete fireable sequences of steps of the same length. More exactly:

- Any sequence of concrete steps of length k corresponds to an instantiation of a complete sequence of k partial steps;
- Every possible instantiation of a complete sequence of partial steps corresponds to a valid sequence of concrete steps.

On the other hand, from the point of view of the reachable markings, the partial markings produced by the complete sequence of partial steps represent all the markings reachable in at most k steps and only those markings. It is thus equivalent to build the reachability graph generated by the fire of k steps, or to build the complete sequence of partial steps of length k.

4 Resolution of the Reachability Problem

4.1 Formulations

According to the previous section, a complete sequence of partial steps can capture the behavior of a Petri net from the point of view of step sequences as well as of reachable markings. One can use this partial sequence to search for concrete fireable sequences which can produce a given marking m_f from the initial marking m_0, in order to solve the reachability problem defined as follows:

" *Let* $(\mathbb{P}, \mathbb{T}, W, m_0)$ *be a Petri net,* $K \in \mathbb{N}$ *and* m_f *be a marking from* $\mathbb{N}^M$. *Find all steps sequences allowing to reach the marking* m_f *from* $\qquad$ (P_1) *the marking* m_0 *in at most* K *steps* ".

In fact, it is sufficient to use K partial steps, to replace the formula F_{m_K} concerning the last partial marking (m_K, F_{m_K}) by the formula F' defined by:

$$F' \equiv F_{m_K} \wedge (m_K = m_f) \qquad (E_4)$$

and to solve the associated system of equations.

In this way, the exploration of the reachability graph and the resolution of the corresponding reachability problem are reduced to the resolution of a system of equations. The interest of this technique is to avoid the exploration of the branches of the graph which do not lead to the desired final marking.

In table , we present in a condensed way two formulations of the reachability problem. The first column describes the problem in the form of a constraint satisfaction program (CSP). The second column corresponds to a model described as an integer linear program (ILP). We split the vectors of logical expressions and the associated formulae: vectors become vectors of variables, and logical formulae are expressed as constraints using variables defined in these vectors. The corresponding formulations are directly deduced from the equations E_1 to E_4 defined previously.

4.2 Constraint Satisfaction Approach

Benasser [ ] proposed an algorithm to solve the reachability problem using the logical abstraction and constraint programming techniques. This algorithm iterates the number of partial steps used, adding one new step at each iteration, in order to test *all the lengths* of complete sequences of partial steps lower than K.

Table 1. Formulations of the reachability problem using partial steps and markings

Given an integer K and two markings m_0 and $m_f \in \mathbb{N}^M$, let $X = (\varphi_1, \ldots, \varphi_K)^t$ be the vector of variables where $\varphi_i = (\varphi_{i1}, \ldots, \varphi_{in})^t \in \mathcal{E}_{\mathcal{L}}{}^N$, $\forall i \in [1, K]$. The problem consists of finding the values of X such that the following constraints are satisfied:

CSP Formulation	ILP Formulation
Step firability constraints (E_1 and E_2)	

- $F_{m_0} \equiv True$
- $\forall i \in [1, K]$,
  $$F_{m_i} \equiv F_{m_{i-1}} \wedge \bigwedge_{p \in P} \left(m_{i-1}(p) \geq \sum_{t \in T} W(p,t)\, \varphi_i(t) \right).$$
- $\forall i \in [1, K]$,
  $$m_i(p) = m_{i-1}(p) - \sum_{t \in T} W(p,t)\varphi_i(t) + \sum_{t \in T} W(t,p)\varphi_i(t).$$

- $\forall i \in [1, K], \overrightarrow{M_{i-1}}[\overrightarrow{\varphi_i}\rangle$
  $\Leftrightarrow \forall i \in [1, K], Pre.\overrightarrow{\varphi_i} \leq \overrightarrow{M_{i-1}}$
  $\Leftrightarrow \forall i \in [1, K], Pre.\overrightarrow{\varphi_i} \leq \overrightarrow{M_0} + C.\overrightarrow{\varphi_1} + \ldots + C.\overrightarrow{\varphi_{i-1}}$
  $\Leftrightarrow \forall i \in [1, K], -Pre.\overrightarrow{\varphi_i} + C.\overrightarrow{\varphi_1} + \ldots + C.\overrightarrow{\varphi_{i-1}} \geq -\overrightarrow{M_0}$
  $$\Leftrightarrow \begin{bmatrix} -Pre & 0 & \cdots & & 0 \\ C & -Pre & \ddots & (0) & 0 \\ C & C & -Pre & \ddots & \vdots \\ \vdots & (C) & \ddots & \ddots & 0 \\ C & C & \cdots & C & -Pre \end{bmatrix} . X \geq -\overrightarrow{M_0}$$

Positivity and Integrality Constraints (E_3)	

- The langage $\mathcal{L}$ has the domain $\mathbb{Z}$
- $\forall i \in [1, K]$,
  $$F_{\varphi_i} \equiv \bigwedge_{j=1}^{N} (\varphi_{ij} \geq 0)$$

- $\forall i \in [1, K], \forall j \in [1, N], \varphi_{ij} \in \mathbb{Z}$
- $\forall i \in [1, K], \forall j \in [1, N], \varphi_{ij} \geq 0$
  $$\Leftrightarrow \begin{bmatrix} \mathbb{I}_N & \cdots & (0) \\ \vdots & \ddots & \vdots \\ (0) & \cdots & \mathbb{I}_N \end{bmatrix} . X \geq \overrightarrow{0}$$

Final marking reachability constraints (E_4)	

- $m_K = m_f$

- $\overrightarrow{M_K} = \overrightarrow{M_f}$
  $\Leftrightarrow \overrightarrow{M_{K-1}} + C.\overrightarrow{\varphi_K} = \overrightarrow{M_f}$
  $\Leftrightarrow \overrightarrow{M_0} + \sum_{i=1}^{K} C.\overrightarrow{\varphi_i} = \overrightarrow{M_f}$
  $\Leftrightarrow [C\, C \ldots C] . X = \overrightarrow{M_f} - \overrightarrow{M_0}$

This algorithm is *correct* since the sequences found are effectively sequences of steps which produce the desired final marking. It is also *complete* since it can enumerate *all* the solutions of length smaller than a given integer. In each iteration, the algorithm uses a mechanism of linear constraints solving. It has been implemented using the constraint logic programming software `Prolog IV`.

The interest using `Prolog IV` is that its constraints resolution mechanism is *incremental* [ ]. Indeed, it is not necessary to redefine in each iteration the constraints incorporated into the previous stage. The constraints are added in the constraints solver so that it can reuse the results of the previous constraints propagation. The search for the concrete results is made at the end by an *enumeration* of all the possible integer solutions.

The corresponding results are presented in the benchmarks of section 5. As all the lengths of steps are tested in the iterative algorithm, an additional constraint is introduced to exclude empty steps, so as to reduce the search space.

4.3 Mathematical Programming Approach

It is clear that the complexity of the problem grows as the length k of the sequence of steps used increase. In this section, we are interested in finding the smallest value of the parameter k from which a solution exists, denoted by k_{min}. Hence, we define a new reachability problem in the following way:

" *Let (R, m_0) be a Petri net, and m_f be a marking from $\mathbb{N}^M$ reachable from m_0. Find the minimal length, denoted by k_{min}, of a sequence of steps allowing to reach the marking m_f from the marking m_0* ". (P_2)

In this section, we solve the reachability problem P_2 using logical abstraction and *mathematical programming techniques* based on the ILP formulation described in the second column of table .

To solve it, we operate a *jump search*. This technique consists of increasing progressively the size k of the sequences tried (not necessarily by one unit) until we find a solution. The amplitude of jumps is defined so as to find a compromise between the search space growth and the consecutive combinatorial difficulty. It depends on the number of variables added for each new step used and thus directly on the size of the network (N supplementary variables are needed for each new step used). We do not use a dichotomic search because we do not have an upper bound on the k_{min} value.

In contrast with a constraint satisfaction problem, an optimization problem requires the definition of a criterion to be optimized. For our problem P_2, we studied several types of objectives to minimize, among which:

– A function *vanishing identically*: $obj_1(X) = 0$;
– The L_1 *norm of steps* φ_i: $obj_2(X) = \sum\limits_{i=1}^{K} \sum\limits_{j=1}^{N} \varphi_{ij}$;
– The *number of empty steps*: $obj_3(X)$.

To define obj_3, we introduce new binary variables $\alpha_1, \alpha_2, \ldots, \alpha_k$, where $\alpha_i = 0$ if the step i is empty, and $\alpha_i = 1$ otherwise. Furthermore, we incorporate in the ILP model the following additional constraints:

$$\forall i \in [\![1, k]\!], \ 1 + (\alpha_i - 1)B \leq \sum\limits_{j=1}^{N} \varphi_{ij} \leq \alpha_i B;   (E_5)$$

where B is a sufficiently big number, chosen much bigger than the number of transitions and tokens in the net.

We propose also a $0 - 1$ linear programming $(0 - 1LP)$ formulation of the reachability problem. The binary variables used in the $0 - 1LP$ means to " forbid " transition reentrance. The $0 - 1LP$ model is obtained from the ILP formulation by replacing the integrality constraints by

$$\forall i \in [\![1, k]\!], \forall j \in [\![1, \mathrm{N}]\!], \varphi_{ij} \in \{0, 1\}. \tag{E_6}$$

Compared to the ILP model, the usefulness of the $0 - 1LP$ model, is that we can model a notion of *partial order* in the steps by linear constraints. The partial order considered is obtained by integrating into the $0 - 1LP$ model the following constraints:

$$\forall i \in [\![1, K]\!], \mathrm{N}. \sum_{j=1}^{N} \varphi_{ij} \geq \sum_{j=1}^{N} \varphi_{(i+1)j}. \tag{E_7}$$

The constraints E_7 express that the empty steps have to appear *at the end* of the sequence searched.

4.4   Relaxations

In the algorithms described previously to determine k_{min}, we have to check the feasibility of a family of *NP-hard* optimization problems. For all the values $k < k_{min}$, we know that the associated problem has no solution. In this section, we propose to use relaxation techniques in order to decrease the necessary time needed to conclude to the infeasibility of the system of equations.

Relaxation techniques are useful in the field of combinatorial optimization. The principle of these techniques is to replace the complex original problem by one or several simpler ones. A relaxation of an optimization problem P of type maximisation is an optimization problem R such as:

- Each feasible solution for P is also a feasible one for R;
- The value of the objective function of any solution of R is greater than or equal to the value of the objective function of the same solution for P.

One of the useful properties of a relaxation in our context is that *if the relaxed problem does not admit a solution, then the initial problem would not admit it either.* For the values of $k < k_{min}$, it can be sufficient to study a relaxed problem to conclude.

In the literature [ ], there are several techniques of relaxation. We distinguish:

- The *LP-relaxation* which consists in relaxing integrality constraints;
- The *Lagrangean relaxation* which consists in relaxing a set of constraints by integrating them via penalties in the objective function. In our context, we can relax for example the step firability constraints (E_1 and E_2) and/or reachability ones (E_4);

– The *surrogate relaxation* which consists in replacing a set of constraints by
only one constraint which is a linear combination of the relaxed constraints.
In our problem, we can aggregate the step firability constraints.

We have tested these various relaxations, we discuss about LP-relaxation in
section .

5   Computational Results

The numerical experiments were carried out on an Intel Pentium 1Ghz computer
with 512 megabytes of RAM. Constraint satisfaction problems were solved using
the `PrologIV` software. Mathematical programming models were solved using the
`Cplex 7.1` optimization library, the algorithms were coded using `Visual C++ 6`.
The CPU times are shown in milliseconds.

We compared the practical efficiency of the proposed approaches for several
classes of problems: the problem studied in [ ], the problem of saturation of
a railway point presented in [ ], and two classical problems of Petri nets analysis. In this section, we present some preliminary results for the classic problems
illustrated in figure : the *token ring protocol* and the *dining philosophers*. Additional results are presented in [ ]. The first problem represents a communication
protocol in a closed ring where computers are passing from hand to hand a token which gives them the right to send data across the network. The second one
represents philosophers around a table, who spend time eating spaghettis and
thinking. To eat, each one needs two forks, but there is only one available for two
people. Each of the entities (computer or philosopher) is provided with a *control
place*, allowing us to quantify how many times it has been active. The presence
of this unbounded place makes the corresponding reachability graph unbounded
too.

The size of each Petri net depends on the number e of entities used, more precisely, $N(e) = 5e$, $M(e) = 6e$ for the philosophers PN, and $N(e) = 4e$, $M(e) = 5e$
for the token ring PN. In the examples below, we vary this parameter from 3
to 7. A second parameter a is used to define the benchmarks. It corresponds
to the number of tokens in the control place at the final marking, all the other
places keeping their initial values. Our choice of those classic problems is motivated by the knowledge of the optimal value $k_{min} = f(a, e)$. For the philosophers
PN, k_{min} is equal to $6.a$ if the number of philosophers is odd, and $6.a + 1$ otherwise. For the token ring PN, k_{min} is equal to $e.(3.a + 1)$.

We present in table   the results of a *fixed depth search* for examples of
increasing difficulty. They were obtained using the formulations described in
sections   and   . The words *ho* and *tms* stand for *heap overflow* and *too
many scols*, meaning a memory overflow from `PrologIV`.

Firstly we should remark that the respective efficiency of the two approaches
considered *depends on the family of problems*. No approach dominates the other
one, each has its own skill domain.

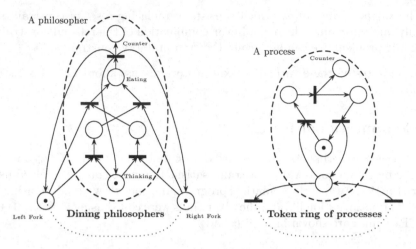

Fig. 3. The PN used for the numerical experiments

Concerning the dining philosophers class, mathematical programming dominates easily. The `Prolog` memory overflows for small values of a, while `Cplex` does not seem affected by the increasing difficulty. Even for values where `Prolog` does not explode, experimental results shows that it is more than 50 times slower than `Cplex`.

On the opposite, the token ring problems family is a class for which constraint programming technique seem well suited. Computed experiments show that `Prolog` gives the results approximately 30 times faster than `Cplex`. Unfortunately, memory issues eventually surface again.

A reasonable explanation for those particular behaviors remains is the particular structure of the reachability graphs corresponding to each family. Indeed, the dining philosophers example is characterized by the presence of deadlocks whereas on the other hand, the token ring PN is characterized by a weak behavioral parallelism. Each of those specificities benefits from a different approach.

These experiments allow us to compare the first two objective functions obj_1 and obj_2. One should remarks that obj_1 gives the better results: ILP using the first formulation terminate about 2.5 times faster. Again, a simple explanation can be advanced: since obj_1 vanishes identically, the optimization ends as soon as a solution has been found, without needing supplementary iterations.

The same case studies have been used to compare the performances of iterative searches for k_{min}. We present in table   the corresponding results. As above, constraint programming techniques were better when dealing with the token ring protocol. Thus, we present only the experiments regarding dining philosophers, for which we can compare the influence of relaxations. The amplitude used is 1, the objective function is obj_1.

Table 2. Fixed depth search

Families	e	a	k	$\mathrm{N}(e,a)$	$\mathrm{M}(e,a)$	#v	#c	Cplex$_{obj_1}$	Cplex$_{obj_2}$	Prolog
Dining Philosophers	3	1	7	15	18	105	250	40	41	551
	3	2	13	15	18	195	448	180	971	17154
	3	3	19	15	18	285	646	832	4046	164447
	3	4	25	15	18	375	844	82058	22413	ho
	5	1	7	25	30	175	416	130	260	1603
	5	2	13	25	30	325	746	570	501	85683
	5	3	19	25	30	475	1076	2664	20099	ho
	7	1	7	35	42	245	582	180	220	3094
	7	2	13	35	42	455	1044	1271	6890	ho
Token Ring	3	10	93	12	15	1116	2527	44083	80646	1152
	3	15	138	12	15	1656	3742	92543	263259	3264
	3	25	228	12	15	2736	6172	814792	1508649	18808
	3	40	363	12	15	4356	9817	$> 2.10^6$	$> 2.10^6$	tms
	5	5	80	20	25	1600	3626	150847	195621	6269
	5	10	155	20	25	3100	7001	$> 2.10^6$	$> 2.10^6$	84972
	5	15	230	20	25	4600	10376	$> 2.10^6$	$> 2.10^6$	tms
	7	1	28	28	35	784	1800	9934	21982	771
	7	5	112	28	35	3136	7092	934504	$> 2.10^6$	60036
	7	8	175	28	35	4900	11061	$> 2.10^6$	$> 2.10^6$	tms

#v ⇔ nb. of variables, #c ⇔ nb. of constraints

ho ⇔ heap overflow, tms ⇔ too many scols

Surprisingly, relaxations do not bring interesting improvements. Further attention helps to understand this phenomenon. Relaxed solutions occur very early in every experiment, typically for sequences of less than 5 steps. The profit that can be made by using continuous techniques instead of integer ones is minimal for those lengths. Sometimes, it has even the inverse effect since the use of relaxation simply adds one supplementary iteration.

6 Conclusion and Future Work

In this paper, we have presented two approaches for solving the Petri nets reachability problem using *Petri nets logical abstraction*. Our method is based on an *implicit exploration* of the PN reachability graph by constructing a single sequence of *partial steps*. This sequence represents *exactly* the behavior of the net. The first formulation reformulates the problem as a *constraint satisfaction program*. It has been used to *enumerate* all the reachability sequences of fixed length leading to a final marking. We have adapted it in order to use *mathematical programming*. This technique allowed us to search for the *shortest* reachability

Table 3. Finding k_{min} for the *dining philosophers*

e	a	k_{min}	Cplex$_{obj_1}$	Relaxed$_{obj_1}$	Prolog
3	1	7	120	121	1312
3	2	13	1021	981	41009
3	3	19	20950	19628	372155
3	4	25	1924358	1849950	*ho*
5	1	7	240	230	2443
5	2	13	1993	2043	174651
5	3	19	5849	5759	*ho*
7	1	7	400	390	4647
7	2	13	3255	3275	*ho*

ho ⇔ *heap overflow*

sequence leading to the final marking. These formulations have been validated and compared on large data sets, using `Prolog IV` and `Cplex solvers`.

Our results have shown that the two approaches are *complementary*. For some classes of problems (e.g. *token ring*), constraint programming presents a better efficiency because it is able to solve large problems (3000 variables, 7000 constraints) in a very short time. For other classes of problems (e.g. *dining philosophers*), mathematical programming is the only way to cope with the complexity of the system without producing memory overflows.

Preliminary techniques aiming at the improvement of the efficiency of *ILP-based explorations*, such as *relaxations*, have not revealed significant advantages for the considered examples. We propose to improve them in order to handle pathological cases like the *token ring* with performances as close as possible to those of constraint programming. For this purpose, we are currently following three promising directions:

- To test the use of binary variables coupled to an appropriate objective function like obj_3;
- To *refine our relaxations* in order to use the preceding results in the next stage, just like CSP incremental features;
- To decompose the problem exploiting the state equation. This last technique could also help us to develop *enumeration techniques*.

References

[1] A. Benasser. *L'accessibilité dans les réseaux de Petri : une approche basée sur la programmation par contraintes*. PhD thesis, Université des sciences et technologies de Lille, 2000.   ,   ,   ,   ,   ,

[2] A. Benasser and P. Yim. Railway traffic planning with petri nets and constraint programming. *JESA*, 33(8-9):959–975, 1999.   ,

[3] G. Berthelot. Transformations and decompositions of nets. In Brauer, W., Reisig, W., and Rozenberg, G., editors, *Advances in Petri Nets 1986 Part I, Proceedings of an Advanced Course*, volume 254, pages 359–376, 1986.

[4] T. Bourdeaud'huy, S. Hanafi, and P. Yim. Résolution du problème d'accessibilité dans les réseaux de Petri par l'abstraction logique et la programmation mathématique. Technical report, L. A. G. I. S., Ecole Centrale de Lille, 2004.

[5] C. Briand. Solving the car-sequencing problem using petri nets. In *International Conference on Industrial Engineering and Production Management*, volume 1, pages 543–551, 1999.

[6] Jean-CLaude Fernandez, Claude Jard, Thierry Jéron, and Laurent Mounier. "on the fly" verification of finite transition systems. *Formal Methods in System Design*, 1992.

[7] J. Gunnarsson. Symbolic tools for verification of large scale DEDS. In *Proc. IEEE Int. Conf. on Systems, Man, and Cybernetics (SMC'98), 11-14 October 1998, San Diego, CA*, pages 722–727, 1998.

[8] P. Huber, A. M. Jensen, L. O. Jepsen, and K. Jensen. Towards reachability trees for high-level petri nets. *Lecture Notes in Computer Science: Advances in Petri Nets 1984*, 188:215–233, 1985.

[9] J. Jaffar, Michaylov, P. Stuckey, and R. Yap. The clp(r) language and system. *ACM Transactions on Programming Languages and Systems*, 14(3):339–395, 1992.

[10] R. M. Keller. Formal verification of parallel programs. *Comm. of the ACM*, 19(7):371–384, 1976.

[11] S. R. Kosaraju. Decidability and reachability in vector addition systems. In *Proc. of the 14th Annual ACM Symp. on Theory of Computing*, pages 267–281, 1982.

[12] K. Lautenbach. Linear algebraic techniques for place/transition nets. In *Advances in Petri Nets 1986, Part I, Proceedings of an Advanced Course*, volume 254, pages 142–167, 1987.

[13] D. Y. Lee and F. DiCesare. Scheduling flexible manufacturing systems using petri nets and heuristic search. *IEEE Transactions on Robotics and Automation*, 10(2):123–132, 1994.

[14] M. Lindqvist. Parameterized reachability trees for predicate/transition nets. *Lecture Notes in Computer Science; Advances in Petri Nets 1993*, 674:301–324, 1993.

[15] Richard Lipton. The reachability problem requires exponential space. Technical report, Computer Science Dept., Yale University, 1976.

[16] T. Murata. Petri nets : properties, analysis ans applications. In *proceedings of the IEEE*, volume 77, pages 541–580, 1989.

[17] R. G. Parker and R. L. Rardin. *Discrete Optimization*. Academic Press, 1988.

[18] Antti Valmari. Stubborn sets for reduced state space generation. *Lecture Notes in Computer Science; Advances in Petri Nets 1990*, 483:491–515, 1991.

[19] F. Vernadat, P. Azéma, and P. Michel. Covering steps graphs. In *17 th Int. Conf on Application and Theory of Petri Nets 96*, 1996.

Generating Benders Cuts for a General Class of Integer Programming Problems

Yingyi Chu and Quanshi Xia

IC-Parc
Imperial College London, London SW7 2AZ, UK
{yyc,q.xia}@imperial.ac.uk

Abstract. This paper proposes a method of generating valid integer Benders cuts for a general class of *integer* programming problems. A *generic* valid Benders cut in disjunctive form is presented first, as a basis for the subsequent derivations of simple valid cuts. Under a qualification condition, a *simple valid* Benders cut in linear form can be identified. A cut generation problem is formulated to elicit it. The simple valid Benders cut is further generalized to a *minimally relaxed* Benders cut, based on which a complete Benders decomposition algorithm is given, and its finite convergency to optimality is proved. The proposed algorithm provides a way of applying the Benders decomposition strategy to solve *integer programs*. The computational results show that using the Benders algorithm for integer programs to exploit the problem structures can reduce the solving time more and more as the problem size increases.

1 Introduction

Benders decomposition is a strategy for solving large-scale optimization problems [ ]. The variables of the problem are partitioned into two sets: *master problem variables* and *subproblem variables*. The Benders algorithm iteratively solves a *master problem*, which assigns tentative values for the master problem variables, and a *subproblem*, obtained by fixing the master problem variables to the tentative values. In every iteration, the subproblem solution provides certain information on the assignment of master problem variables. Such information is expressed as a *Benders cut*, cutting off some assignments that are not acceptable. The Benders cut is then added to the master problem, narrowing down the search space of master problem variables and eventually leading to optimality. On one hand, Benders method is employed to exploit the problem structure: the problem is decomposed into a series of independent smaller subproblems, reducing the complexity of solving it [ ]. On the other hand, Benders method opens a dimension for 'hybrid algorithms' [, ,] where the master problem and the subproblems can be solved with different methods.

The generation of Benders cuts is the core of Benders decomposition algorithm. Indeed, *valid* Benders cuts guarantee the convergence of the iterations to the optimal solution of the original problem, and also the cuts determine how fast the algorithm converges.

J.-C. Régin and M. Rueher (Eds.): CPAIOR 2004, LNCS 3011, pp. 127–   , 2004.
© Springer-Verlag Berlin Heidelberg 2004

The classic Benders decomposition algorithm [ ] was proposed for linear programming problems, the cut generation of which is based on the strong duality property of linear programming [ ]. Geoffrion has extended it to a larger class of mathematical programming problems [ ].

For integer programming, however, it is difficult to generate valid integer Benders cut, due to the duality gap of integer programming in subproblems. One possible way is to use the *no-good* cut to exclude only the current tentative assignment of master problem variables that is unacceptable. Such no-good Benders cuts will result in an enumerative search and thus a slow convergence. For some specific problems, better Benders cuts can be obtained [ ]. For example, in the machine scheduling application, the cut that limits the incompatible jobs in the same machine is generally stronger. For more general integer programming, logic-based Benders decomposition [ ] was proposed to generate valid integer Benders cuts, but these cuts contain a large number of disjunctions [ ], the linearization of which leads to huge cuts with many auxiliary variables, complicating the master problem significantly.

This paper proposes a new method of generating valid Benders cuts for a class of integer programs, in which the objective function only contains the master problem variables.

As a foundation of our derivation, a *generic* valid integer Benders cut is firstly presented. However it is difficult to use due to its exponential size and nonlinearity. Instead, we can pick only one linear inequality from the disjunction of the generic cut, while preserving the validity. A qualification condition is then given, with which such a simple valid cut can be identified. A cut generation problem is formulated to determine the simple linear valid cut.

However, such an integer Benders cut is not always available. The *minimally relaxed* cut is then proposed as a generalization of it, obtainable in all cases. The simple integer Benders cut is just a special case of the minimally relaxed cut with 'zero' relaxation. Based on this, a complete Benders decomposition algorithm is presented, and its finite convergency to optimality is proved.

The paper is organized as follows. Section   introduces the integer programs under consideration and the principle of the Benders decomposition algorithm. Section   derives the integer Benders cut. Section   generalizes to the minimally relaxed integer Benders cut. Section   presents the complete Benders decomposition algorithm that could be used in practice. Section   gives computational results. Section   concludes the paper. The appendix gives the proofs of all the lemmas.

2   Preliminaries

2.1   Integer Programs

The programs we consider in the paper are written as the following form (Such programs arise from our study on a path generation application in network traffic engineering [ ]):

$$P: \quad \max_{y,x} \ c^T y$$
$$s.t. \quad \begin{cases} Dy \leq d, \\ Ay + Bx \leq b, \\ y \in \{0,1\}^m, x \in \{0,1\}^n \ . \end{cases}$$

The problem variables are partitioned into two vectors y (master problem variables) and x (subproblem variables), and the objective function only contains the master problem variables.

By the use of Benders decomposition, the problem P is decomposed into the master problem (MP) that solves only the y variables and the subproblem $(SP'(\bar{y}))$ that solves only the x variables, by fixing the y variables to the master problem solution, denoted by $\bar{y}$.

$$MP: \quad \max_{y} \ c^T y \qquad\qquad SP'(\bar{y}): \quad \max_{x} \ 0$$
$$s.t. \quad \begin{cases} Dy \leq d, \\ \text{Benders cuts,} \\ y \in \{0,1\}^m, \end{cases} \qquad s.t. \quad \begin{cases} Bx \leq b - A\bar{y}, \\ x \in \{0,1\}^n, \end{cases}$$

where the Benders cuts in MP are gradually added during the iterations. Note that the subproblem is a feasibility problem with a dummy objective.

2.2   Principle of Benders Decomposition Algorithm

The Benders decomposition algorithm iteratively solves the master problem and the subproblem. In each iteration k, the master problem $(MP^{(k)})$ sets a tentative value for the master problem variables $(\bar{y}^{(k)})$, with which the subproblem $(SP'(\bar{y}^{(k)}))$ is formed and solved. Using the subproblem solution, a valid Benders cut over the y variables is constructed and added to the master problem in the next iteration $(MP^{(k+1)})$. The Benders cut added in every iteration cuts off some infeasible assignments, thus the search space for the y variables is gradually narrowed down as the algorithm proceeds, leading to optimality.

Algorithm 1. *Benders Decomposition Algorithm*

1. *Initialization. Construct the initial master problem $MP^{(0)}$ without any Benders cut. Set $k = 0$.*
2. *Iteration.*
   (a) *Solve $MP^{(k)}$. If it is feasible, obtain the optimal solution $\bar{y}^{(k)}$ and the optimal objective $\phi_{MP}^{(k)}$. Otherwise, the original problem is infeasible; set $\phi_{MP}^{(k)} = -\infty$ and go to exit.*
   (b) *Construct the subproblem $SP'(\bar{y}^{(k)})$. If the subproblem is feasible, then optimality is found; obtain the optimal solution $\bar{x}^{(k)}$, and go to exit.*
   (c) *Cut Generation Procedure. Generate a valid integer Benders cut and add it to the master problem to construct $MP^{(k+1)}$. Set $k = k + 1$ and go back to step 2.*

3. *Exit. Return the current $\phi_{MP}^{(k)}$ as the optimal objective. If $\phi_{MP}^{(k)} = -\infty$ then the problem P is infeasible. If not, return the current $(\bar{\boldsymbol{y}}^{(k)}, \bar{\boldsymbol{x}}^{(k)})$ as the optimal solution of problem P.*

The Cut Generation Procedure *2(c)* is not specified. This is the key step for the algorithm, which must be selected carefully so that the algorithm eventually converges to the optimality of the original problem in finite iterations.

2.3   Always Feasible Subproblem

The subproblem $SP'(\bar{\boldsymbol{y}})$ can be reformulated to an equivalent one that is always feasible:

$$SP(\bar{\boldsymbol{y}}): \quad \min_{\boldsymbol{x},\boldsymbol{r}} \ \mathbf{1}^T \boldsymbol{r}$$
$$\text{s.t.} \quad \begin{cases} \boldsymbol{Bx} - \boldsymbol{r} \leq \boldsymbol{b} - \boldsymbol{A}\bar{\boldsymbol{y}}, \\ \boldsymbol{x} \in \{0,1\}^n, \boldsymbol{r} \geq \mathbf{0}, \end{cases}$$

which simply introduces slack variables $\boldsymbol{r}$ to the constraints of $SP'(\bar{\boldsymbol{y}})$. Obviously, $SP'(\bar{\boldsymbol{y}})$ is feasible *iff* $SP(\bar{\boldsymbol{y}})$ has 0 optimal value. In the Algorithm , once the objective of $SP(\bar{\boldsymbol{y}})$ equals 0 during iteration, the algorithm terminates (at step *2(b)*), and the optimal solution is found.

Dual values are very useful in the cut generation for linear programming. For integer programming, however, we need to introduce the *fixed subproblems* and their duals. A fixed subproblem is constructed from *any* feasible assignment $\tilde{\boldsymbol{x}}$ of subproblem $SP(\bar{\boldsymbol{y}})$. It just constrains the $\boldsymbol{x}$ variables to be equal to a *given* feasible $\tilde{\boldsymbol{x}}$.

$$SP_f(\bar{\boldsymbol{y}}, \tilde{\boldsymbol{x}}): \quad \min_{\boldsymbol{x},\boldsymbol{r}} \ \mathbf{1}^T \boldsymbol{r}$$
$$\text{s.t.} \quad \begin{cases} \boldsymbol{Bx} - \boldsymbol{r} \leq \boldsymbol{b} - \boldsymbol{A}\bar{\boldsymbol{y}}, \\ \boldsymbol{x} = \tilde{\boldsymbol{x}}, \\ \boldsymbol{x}\text{:free}, \boldsymbol{r} \geq \mathbf{0} \ . \end{cases} \tag{1}$$

The dual of fixed subproblem $SP_f(\bar{\boldsymbol{y}}, \tilde{\boldsymbol{x}})$ is:

$$DSP_f(\bar{\boldsymbol{y}}, \tilde{\boldsymbol{x}}): \quad \max_{\boldsymbol{u},\boldsymbol{v}} \ (\boldsymbol{A}\bar{\boldsymbol{y}} - \boldsymbol{b})^T \boldsymbol{u} + \tilde{\boldsymbol{x}}^T \boldsymbol{v}$$
$$\text{s.t.} \quad \begin{cases} -\boldsymbol{B}^T \boldsymbol{u} + \boldsymbol{v} = \mathbf{0}, \\ \boldsymbol{u} \leq \mathbf{1}, \\ \boldsymbol{v}\text{:free}, \boldsymbol{u} \geq \mathbf{0} \ . \end{cases} \tag{2}$$

The optimal solution of $DSP_f(\bar{\boldsymbol{y}}, \tilde{\boldsymbol{x}})$ depends on the value of $\tilde{\boldsymbol{x}}$ (while the feasible region of it does not). Let $(\tilde{\boldsymbol{u}}, \tilde{\boldsymbol{v}})$ denote the corresponding optimal solution of $DSP_f(\bar{\boldsymbol{y}}, \tilde{\boldsymbol{x}})$. Since any value $\tilde{\boldsymbol{x}} \in \{0,1\}^n$ is feasible for $SP(\bar{\boldsymbol{y}})$ (2^n possible combinations), there are $N = 2^n$ possible fixed subproblems, each with its dual.

Given $\tilde{\boldsymbol{x}}$, the fixed subproblem $SP_f(\bar{\boldsymbol{y}}, \tilde{\boldsymbol{x}})$ is itself a linear program. Therefore, strong duality holds for $SP_f(\bar{\boldsymbol{y}}, \tilde{\boldsymbol{x}})$ and $DSP_f(\bar{\boldsymbol{y}}, \tilde{\boldsymbol{x}})$. Furthermore, if $\tilde{\boldsymbol{x}}^*$ is an optimal solution of $SP(\bar{\boldsymbol{y}})$, then $SP(\bar{\boldsymbol{y}})$, $SP_f(\bar{\boldsymbol{y}}, \tilde{\boldsymbol{x}}^*)$ and $DSP_f(\bar{\boldsymbol{y}}, \tilde{\boldsymbol{x}}^*)$ have the same optimal value.

Relations between the optimal primal and dual solutions of $SP_f(\bar{\boldsymbol{y}}, \tilde{\boldsymbol{x}})$ and $DSP_f(\bar{\boldsymbol{y}}, \tilde{\boldsymbol{x}})$ can be established via the *complementary condition*, that is, the Karush-Kuhn-Tucker (KKT) [ ] constraints:

$$
\begin{cases}
\boldsymbol{u}_i(-\boldsymbol{Bx} + \boldsymbol{r} - \boldsymbol{A\bar{y}} + \boldsymbol{b})_i = 0 & \forall i, \\
\boldsymbol{r}_i(\boldsymbol{u} - \boldsymbol{1})_i = 0 & \forall i,
\end{cases}
\tag{3}
$$

together with the primal and dual constraints, ( ) and ( ).

Consider the complementary condition constraints. First, with the equation $\boldsymbol{x} = \tilde{\boldsymbol{x}}$ of ( ) and the equation $-\boldsymbol{B}^T\boldsymbol{u} + \boldsymbol{v} = 0$ of ( ), variables $\boldsymbol{x}$ and $\boldsymbol{v}$ can be replaced by $\tilde{\boldsymbol{x}}$ and $\boldsymbol{B}^T\boldsymbol{u}$. Secondly, the equation $\boldsymbol{r}_i(\boldsymbol{u}-\boldsymbol{1})_i = 0$ of ( ) means that $\boldsymbol{r}_i = \boldsymbol{r}_i\boldsymbol{u}_i$. Putting this into the equation $\boldsymbol{u}_i(-\boldsymbol{Bx} + \boldsymbol{r} - \boldsymbol{A\bar{y}} + \boldsymbol{b})_i = 0$ of ( ), we get $\boldsymbol{u}_i(-\boldsymbol{B\tilde{x}})_i + \boldsymbol{r}_i = \boldsymbol{u}_i(\boldsymbol{A\bar{y}} - \boldsymbol{b})_i$. The complementary condition constraints can then be simplified to:

$$
\begin{cases}
\boldsymbol{B\tilde{x}} - \boldsymbol{r} \le \boldsymbol{b} - \boldsymbol{A\bar{y}}, \\
\boldsymbol{u}_i(-\boldsymbol{B\tilde{x}})_i + \boldsymbol{r}_i = \boldsymbol{u}_i(\boldsymbol{A\bar{y}} - \boldsymbol{b})_i & \forall i, \\
\boldsymbol{r}_i(\boldsymbol{u} - \boldsymbol{1})_i = 0 & \forall i, \\
\boldsymbol{r} \ge 0, 0 \le \boldsymbol{u} \le 1 .
\end{cases}
$$

Finally, we can show the *redundancy* of $\boldsymbol{r}_i(\boldsymbol{u} - \boldsymbol{1})_i = 0$ in the following lemma (Note that the proofs of all the lemmas are given in the appendix):

Lemma 1. *The constraint $\boldsymbol{r}_i(\boldsymbol{u} - \boldsymbol{1})_i = 0$ is redundant in the presence of the constraints:*

$$
\begin{cases}
\boldsymbol{B\tilde{x}} - \boldsymbol{r} \le \boldsymbol{b} - \boldsymbol{A\bar{y}}, \\
\boldsymbol{u}_i(-\boldsymbol{B\tilde{x}})_i + \boldsymbol{r}_i = \boldsymbol{u}_i(\boldsymbol{A\bar{y}} - \boldsymbol{b})_i & \forall i, \\
\boldsymbol{r} \ge 0, 0 \le \boldsymbol{u} \le 1 .
\end{cases}
\tag{4}
$$

Thus, the complementary condition constraints are finally simplified to ( ).

3   Integer Benders Cut Generation

3.1   Generic Valid Integer Benders Cut

In general, the Benders cut is a logic expression over the $\boldsymbol{y}$ variables, generated using the information from the subproblem solution. The valid Benders cut must guarantee that Algorithm   finitely converges to the optimal solution. We define the *valid* Benders cut for integer programming.

Definition 1. *In a certain iteration of the Benders algorithm, a valid Benders cut is a logic expression over the master problem variables $\boldsymbol{y}$ that satisfies:*

Condition 1. *if the current master problem solution $\bar{\boldsymbol{y}}$ is infeasible, then the cut must exclude at least $\bar{\boldsymbol{y}}$;*

Condition 2. *any feasible assignment of $\boldsymbol{y}$ variables must satisfy the cut.*

Condition 1 guarantees finite convergence since y has a finite domain. *Condition 2* guarantees optimality since the cut never cuts off feasible solutions.

Lemma 2. *If a valid Benders cut is generated in every iteration (at step 2(c)), then Algorithm   finitely converges to the optimality of the original program P.*

Using the solutions of all N fixed subproblems (denoted by $SP_f(\bar{y}, \tilde{x}^i)$, $\forall i = 1, \cdots, N$), a *generic* integer Benders cut can be obtained as the disjunction of N linear inequalities:

$$
\begin{array}{c}
(Ay - b)^T \tilde{u}^1 + (\tilde{x}^1)^T \tilde{v}^1 \leq 0 \\
\vee \quad (Ay - b)^T \tilde{u}^2 + (\tilde{x}^2)^T \tilde{v}^2 \leq 0 \\
\cdots \qquad \cdots \\
\vee \quad (Ay - b)^T \tilde{u}^N + (\tilde{x}^N)^T \tilde{v}^N \leq 0,
\end{array}
\tag{5}
$$

where $\{\tilde{x}^1, \cdots, \tilde{x}^N\}$ are the list of all the possible $\tilde{x}$ values (i.e. an enumeration of $\{0, 1\}^n$). For each $\tilde{x}^i$, $(\tilde{u}^i, \tilde{v}^i)$ are the corresponding optimal dual solutions from $DSP_f(\bar{y}, \tilde{x}^i)$.

Similar to the Benders cut for linear programming, each linear inequality in the disjunction follows the expression of the objective function of $DSP_f(\bar{y}, \tilde{x})$. However, for integer programming, where a duality gap exists, we use a large number of dual fixed subproblem solutions, instead of a single dual subproblem solution for linear programming where no duality gap exists.

Lemma 3. *The generic integer Benders cut ( ) is a valid cut.*

The generic cut ( ) is valid, but it has a nonlinear (disjunctive) form and intrinsically contains all possible $\tilde{x}$ combinations. Although it has theoretical value, it is difficult to use it directly in practical algorithms.

3.2   Integer Benders Cut

Under certain conditions, one of the linear inequalities from the disjunction ( ) can still be a valid cut. In such cases the valid integer Benders cut becomes a simple linear inequality and the nonlinear disjunction disappears. We give the following sufficient condition under which such a simple valid cut can be identified:

Theorem 1. *If there exists a solution $\tilde{x} \in \{\tilde{x}^1, \cdots, \tilde{x}^N\}$ such that $\tilde{x}$ and the corresponding dual $\tilde{v}$ satisfy*

$$
\tilde{x}^T \tilde{v} \leq x^T \tilde{v} \quad \forall \ x \in \{\tilde{x}^1, \cdots, \tilde{x}^N\},
\tag{6}
$$

then the linear inequality

$$
(Ay - b)^T \tilde{u} + \tilde{x}^T \tilde{v} \leq 0
\tag{7}
$$

is a valid integer Benders cut.

Proof. (*for the valid cut condition 1*) If $\bar{y}$ is infeasible for the subproblem, then $SP(\bar{y})$ has positive objective value. Thus, any possible $SP_f(\bar{y}, x)$ ($x \in \{\tilde{x}^1, \cdots, \tilde{x}^N\}$) has a positive objective value, and so do all $DSP_f(\bar{y}, x)$. Therefore, all linear inequalities in ( ) are violated by $\bar{y}$. In particular, $(Ay - b)^T \tilde{u} + \tilde{x}^T \tilde{v} \leq 0$ is violated by $\bar{y}$, that is, the cut ( ) excludes the infeasible $\bar{y}$.

(*for the valid cut condition 2*) Let $\hat{y}$ be any feasible solution. There must exist a corresponding $(\hat{x}, \hat{u}, \hat{v})$ such that $SP_f(\hat{y}, \hat{x})$ and $DSP_f(\hat{y}, \hat{x})$ has 0 objective value as $(A\hat{y} - b)^T \hat{u} + \hat{x}^T \hat{v} = 0$. Since the feasible region of all $DSP_f(y, x)$ are identical and independent of the values of y and x, the values of $(\tilde{u}, \tilde{v})$ which are the *optimal* solution for $DSP_f(\bar{y}, \tilde{x})$ are also a *feasible* solution for $DSP_f(\hat{y}, \hat{x})$. Therefore,

$$(A\hat{y} - b)^T \tilde{u} + \hat{x}^T \tilde{v} \leq (A\hat{y} - b)^T \hat{u} + \hat{x}^T \hat{v} = 0 \ .$$

From the condition ( ), we have $\tilde{x}^T \tilde{v} \leq \hat{x}^T \tilde{v}$. Therefore,

$$(A\hat{y} - b)^T \tilde{u} + \tilde{x}^T \tilde{v} \leq (A\hat{y} - b)^T \tilde{u} + \hat{x}^T \tilde{v} \leq 0,$$

which means that the feasible $\hat{y}$ is not cut off by ( ). $\square$

If one can find an $\tilde{x}$ such that the condition ( ) holds, then the single linear inequality from the disjunction ( ) that corresponds to $\tilde{x}$ is a valid integer Benders cut by itself. However, the condition ( ) involves not only the selected $\tilde{x}$, but also all other possible assignments of x, making it difficult to express ( ) as a simple constraint. But the sufficient condition ( ) can be converted to an equivalent *sign condition*.

Lemma 4. *Inequalities ( ) are satisfied iff the following holds:*

$$\begin{aligned} \tilde{x}_i = 1 \Longrightarrow \tilde{v}_i \leq 0, \\ \tilde{x}_i = 0 \Longrightarrow \tilde{v}_i \geq 0, \end{aligned} \qquad \forall i = 1, \cdots, n \ . \tag{8}$$

The above sign condition can be enforced as the constraints:

$$\begin{cases} \tilde{x}_i \tilde{v}_i \leq 0 & \forall i, \\ (1 - \tilde{x})_i \tilde{v}_i \geq 0 & \forall i \ . \end{cases} \tag{9}$$

Unlike the condition ( ), the sign condition ( ) only involves the selected $\tilde{x}$ itself and the corresponding $\tilde{v}$.

3.3   Integer Benders Cut Generation

The integer Benders cut generation problem is to find a $\tilde{x}$ such that the sign condition ( ) is satisfied. We formulate a Cut Generation Program (*CGP*) to elicit it.

The sign condition relates $\tilde{x}$, an assignment that determines $SP_f(\bar{y}, \tilde{x})$, and $\tilde{v}$, the optimal dual solution from $DSP_f(\bar{y}, \tilde{x})$. The constraints between them

are established by ( ). Therefore, the program CGP is composed of constraints
( ), the sign condition ( ) and a dummy objective function.

$$CGP(\bar{y}): \quad \min_{\tilde{x}, r, u} \; 0$$

$$s.t. \quad \begin{cases} B\tilde{x} - r \leq b - A\bar{y}, \\ u_i(-B\tilde{x})_i + r_i = u_i(A\bar{y} - b)_i & \forall i, \\ \tilde{x}_i(B^T u)_i \leq 0 & \forall i, \\ (1 - \tilde{x})_i(B^T u)_i \geq 0 & \forall i, \\ \tilde{x} \in \{0, 1\}^n, r \geq 0, 0 \leq u \leq 1 \; . \end{cases} \quad (10)$$

Note that in CGP $\tilde{x}$ (together with r, u) is a *variable*. The CGP solves for
a value for $\tilde{x}$, which, together with the dual values, satisfies the sign condition.
If such a solution is found, a corresponding Benders cut is immediately obtained
as ( ).

Because $\tilde{x}_i \in \{0, 1\}$, all the bilinear terms $\tilde{x}_i u_j$ in the CGP can be linearized
by introducing the variables $w_{ij} = \tilde{x}_i u_j$ as:

$$\begin{cases} w_{ij} \leq \tilde{x}_i, \; w_{ij} \leq u_j, \\ w_{ij} \geq \tilde{x}_i + u_j - 1, \; w_{ij} \geq 0, \end{cases} \quad \forall \, i, j \; .$$

Thus, CGP can be in practice solved with MIP solvers such as XPRESS [ ].

Note that the CGP is not necessarily feasible due to the enforcement of the
additional sign condition constraints ( ), and hence the integer Benders cut ( )
is not always available in each iteration. Therefore, we need to generalize the cut
in order to give a complete Benders decomposition Algorithm .

4   Relaxed Integer Benders Cut

4.1   Relaxation

When the sign condition ( ) does not hold, one cannot directly use an inequality
from ( ) as the valid cut. However, we can still select one inequality but relax it
to some extent so that the sign condition is satisfied. This provides a generalized
way of constructing a valid Benders cut.

In fact, *any* inequality from the disjunction ( ):

$$(Ay - b)^T \tilde{u} + \tilde{x}^T \tilde{v} \leq 0 \quad (11)$$

can be relaxed by *inverting* the $\tilde{x}$ values for those elements that violate the sign
condition ( ) as follows:

$$\tilde{x}'_i = \begin{cases} \tilde{x}_i & \text{if (6) is satisfied for the ith element,} \\ 1 - \tilde{x}_i & \text{otherwise.} \end{cases}$$

In such way $\tilde{x}'$ and $\tilde{v}$ satisfy the sign condition, and the *relaxed cut* is given by:

$$(Ay - b)^T \tilde{u} + (\tilde{x}')^T \tilde{v} \leq 0 \; . \quad (12)$$

Lemma 5. *The relaxed cut ( ) satisfies the valid cut condition 2, and the relaxation gap from ( ) to ( ) is* $\sum_{i=1}^{n}(\tilde{\boldsymbol{x}} - \tilde{\boldsymbol{x}}')_i\tilde{\boldsymbol{v}}_i \geq 0$.

Note that ( ) does not necessarily satisfy the valid cut condition 1, that is, it may not cut off the infeasible $\bar{\boldsymbol{y}}$ in the current iteration due to the relaxation. In this case, however, it can be easily remedied by adding a no-good cut that excludes only one point (the infeasible $\bar{\boldsymbol{y}}$):

$$\sum_{j=1}^{m}\bar{\boldsymbol{y}}_j(1-\boldsymbol{y}_j) + \sum_{j=1}^{m}(1-\bar{\boldsymbol{y}}_j)\boldsymbol{y}_j \geq 1 \ . \tag{13}$$

4.2   Relaxed Cut Generation

Since *any* inequality from the disjunction ( ) can produce a relaxed cut, one can even avoid solving the CGP during iterations. Only the subproblem $SP(\bar{\boldsymbol{y}})$ is solved to find a solution $\tilde{\boldsymbol{x}}$, and $DSP_f(\bar{\boldsymbol{y}}, \tilde{\boldsymbol{x}})$ is solved to find the duals $\tilde{\boldsymbol{u}}$ and $\tilde{\boldsymbol{v}}$. Then a relaxed cut ( ), derived from this $\tilde{\boldsymbol{x}}$, can be generated. To ensure that the valid cut condition 1 is met, the value of $\bar{\boldsymbol{y}}$ is checked against the relaxed cut ( ). If it does violate ( ), then ( ) itself is a valid Benders cut that satisfies valid cut condition 1 and 2. If not, the conjunction of ( ) and ( ) constitutes a valid Benders cut.

The advantage of such a way of cut generation is its simplicity, since no CGP is involved. The disadvantage is that the selection of the inequality to be relaxed is rather arbitrary, and the generated cut can be loose. In particular, cut ( ), which is a tight cut that needs no relaxation, may exist but not be found.

Therefore it is desirable to find a *minimally relaxed cut*, that is, its corresponding relaxation gap (as is given in Lemma) is made as small as possible, and thus the cut is as tight as possible. This is indeed a generalization of the valid Benders cut ( ), which is just the special case when the minimum relaxation needed is zero.

The minimally relaxed cut can be generated by solving a Relaxed Cut Generation Program CGP_r, constructed by introducing slack variables $(\boldsymbol{p}, \boldsymbol{q})$ to the sign condition constraints of CGP.

$$CGP_r(\bar{\boldsymbol{y}}): \quad \min_{\tilde{\boldsymbol{x}},r,u,p,q} \quad \mathbf{1}^T\boldsymbol{p} + \mathbf{1}^T\boldsymbol{q}$$

$$s.t. \begin{cases} \boldsymbol{B}\tilde{\boldsymbol{x}} - \boldsymbol{r} \leq \boldsymbol{b} - \boldsymbol{A}\bar{\boldsymbol{y}}, \\ \boldsymbol{u}_i(-\boldsymbol{B}\tilde{\boldsymbol{x}})_i + \boldsymbol{r}_i = \boldsymbol{u}_i(\boldsymbol{A}\bar{\boldsymbol{y}} - \boldsymbol{b})_i & \forall i, \\ \tilde{\boldsymbol{x}}_i(\boldsymbol{B}^T\boldsymbol{u})_i - \boldsymbol{p}_i \leq 0 & \forall i, \\ (1 - \tilde{\boldsymbol{x}})_i(\boldsymbol{B}^T\boldsymbol{u})_i + \boldsymbol{q}_i \geq 0 & \forall i, \\ \tilde{\boldsymbol{x}} \in \{0,1\}^n, \boldsymbol{r} \geq \mathbf{0}, \mathbf{0} \leq \boldsymbol{u} \leq 1, \\ \boldsymbol{p}, \boldsymbol{q} \geq 0 \ . \end{cases} \tag{14}$$

As the program CGP, after simple linearization this program is solvable in practice with MIP solvers.

Lemma 6. *If the optimal solution of CGP_r is $(\tilde{x}, \tilde{u}, \tilde{v})$, and the optimal objective value is ϕ_{CGP_r}, then:*

$$\phi_{CGP_r} = \sum_{i=1}^{n} (\tilde{x} - \tilde{x}')_i \tilde{v}_i \ .$$

Since the right hand side of the above equation is just the relaxation gap and it is minimized, the derived cut ( ) is a minimally relaxed cut. In particular, if the optimal objective value of CGP_r is 0, then all the sign condition constraints are satisfied, and no relaxation is necessary. In this case the minimally relaxed cut is reduced to the basic valid Benders cut ( ).

In practice, the $CGP_r(\bar{y})$ is solved in every iteration (provided the algorithm does not terminate from step *2(a)* or *2(b)* before the cut generation). Its optimal solution gives a minimally relaxed cut as ( ). According to Lemma , cut ( ) satisfies the valid cut condition 2. If the optimal value is greater than 0, then the current (infeasible) assignment of master problem variables $\bar{y}$ is checked against the cut. If the cut is violated, then ( ) by itself satisfies both the valid cut conditions. If not, the conjunction of ( ) and the no-good cut ( ) constitutes a valid Benders cut.

5   Complete Algorithm

Based on the proposed integer Benders cut, the unspecified cut generation step *2(c)* in Algorithm   can now be given as:

Procedure 1. *Cut Generation Procedure (step 2(c) of Algorithm)*
Construct the cut generation program $CGP_r(\bar{y}^{(k)})$. Solve it to obtain the optimal solution $(\tilde{x}^{(k)}, \tilde{u}^{(k)}, \tilde{v}^{(k)})$ and its optimal objective value $\phi_{CGP_r}^{(k)}$. Generate the minimally relaxed cut:

$$(Ay - b)^T \tilde{u}^{(k)} + (\tilde{x}')^{(k)T} \tilde{v}^{(k)} \leq 0 \ .$$

There are three cases:

A. *if $\phi_{CGP_r}^{(k)} = 0$, then $\tilde{x}'^{(k)} = \tilde{x}^{(k)}$, the above cut is reduced to ( ), which is the valid Benders cut.*
B. *if $\phi_{CGP_r}^{(k)} > 0$ and the current $\bar{y}^{(k)}$ violates the above cut, then this cut is the valid Benders cut by itself.*
C. *if $\phi_{CGP_r}^{(k)} > 0$ and the current $\bar{y}^{(k)}$ satisfies the above cut, then this cut, in conjunction with the no-good cut ( ), is the valid Benders cut.*

Add the generated Benders cut to the master problem to construct $MP^{(k+1)}$. Set $k = k + 1$ and go back to step 2 of Algorithm .

Replacing step *2(c)* of Algorithm   with the above procedure, we have a complete Benders decomposition algorithm.

Theorem 2. *The Benders Decomposition Algorithm , with its step 2(c) instantiated by the Cut Generation Procedure , terminates in finite steps and returns the optimal solution of the original program P.*

The proof is trivial according to Lemma , since in all the three cases the cut being generated satisfies the valid cut condition 1 and 2.

6   Computational Experiments

This section presents computational results of using Benders decomposition with the proposed integer cuts in integer programming problems. The algorithm is implemented using the ECLiPSe [ ] platform. The test problems have bordered block structure in their coefficient matrices, so that the Benders algorithm can decompose the subproblem. The coefficients are generated randomly, and 20 cases are computed for each problem size configuration. The minimally relaxed cut derived from the CGP_r ( ) of Sect.   is used in the tests.

Table   summarizes the computational results for different problem sizes. The number of constraints is fixed to 300 and the number of blocks in the subproblem matrix is fixed to 10. Thus, the subproblem is decomposed into 10 smaller independent problems, each of which can generate a Benders cut in every iteration. We vary the number of master problem variables (MPV) and that of subproblem variables (SPV). For each problem size configuration, the average and maximum number of iterations (#Iter: avr, max) of the 20 test instances, and the average number of no-good cuts that have to be added (#NG) are recorded. Also the average and maximum solving time (Sol.Time: avr, max) of the 20 test instances, and the average percentages of solving time spent in the solution of master problem, subproblem and relaxed cut generation program (MP%, SP%, CGP%), are recorded. All the solving times are in seconds. For

Table 1. Computational Results using Minimally Relaxed Cut

SPV	MPV	#Iter		#NG	Sol.Time		MP%	SP%	CGP%	MIP.Time	#WIN
		avr	max	(avr)	avr	max	(avr)	(avr)	(avr)	(avr)	
300	100	10.40	14	0	84.50	132.06	5.1	3.6	91.3	785.07	7/20
300	150	11.55	16	0	104.04	187.39	16.2	3.4	80.4	109.84	6/20
300	200	12.40	20	0	136.11	260.56	27.1	2.9	70.0	176.20	13/20
300	250	13.30	25	0	195.67	562.64	46.9	2.1	51.0	318.35	12/20
400	100	12.10	18	0	152.03	245.67	3.4	3.9	92.7	1343.05	13/20
400	150	15.20	28	0.05	229.08	566.93	12.2	3.7	84.1	1697.62	17/20
400	200	14.50	21	0	215.85	434.22	19.4	3.5	77.1	889.04	19/20
400	250	18.05	30	0	371.47	851.96	35.1	2.8	62.1	3655.31	20/20
500	100	15.05	23	0.05	302.98	546.46	2.6	4.0	93.4	6482.65	20/20
500	150	18.20	36	0.10	409.43	873.56	7.9	3.8	88.3	8673.01	20/20
500	200	19.15	39	0.05	483.66	1441.65	15.3	3.4	81.3	8595.30	20/20
500	250	21.40	43	0.10	643.67	1929.80	30.7	3.0	66.3	10059.28	20/20

comparison purpose, every problem is also directly solved with MIP solver. The last two columns summarize the average MIP solving time (MIP.Time), and in how many cases (out of the total 20 cases) the Benders algorithm outperforms the directly solving (#WIN). The external solver used in both the decomposition algorithm and the direct solving is XPRESS 14.21 [ ].

Table   shows that as the problem size increases, the number of iterations and the solving time both increase. But throughout the test instances, no-good cuts being added are rare, which means that the generated cuts are usually tight enough to exclude the infeasible assignment in each iteration. It is also notable that a significant portion of the total solving time is spent in solving the relaxed cut generation program. However, in spite of the time spent in the cut generation, the Benders algorithm still wins over directly solving the problem in more cases when the problem size becomes larger. This shows the benefits of using Benders decomposition for integer programs to exploit the problem structures, that is, a problem is decomposed into a master problem and a series of smaller independent subproblems, reducing the complexity of solving it.

We observed that the decomposition algorithm is especially better for the hard instances. For those problems that take long time by direct solving, the Benders decomposition with integer cuts usually achieves high speedup in terms of solving time. Table   shows the comparison. Five hardest instances (in terms of direct MIP solving time) for each fixed subproblem size are recorded.

We also observed that, for all the test instances that take more than 200 seconds by directly solving, the decomposition algorithm invariably consumes less solving time than the direct solving.

Table 2. Solving Time Comparison for Hard Instances

SPV	MPV	#Iter	Sol.Time	MIP.Time
300	100	12	94.62	14181.90
300	250	17	270.74	1403.68
300	250	25	562.64	901.83
300	250	14	201.25	833.31
300	200	15	171.69	624.48
400	250	28	697.45	>20000.00
400	150	17	299.12	14577.89
400	100	17	235.38	11086.83
400	250	12	195.52	10335.30
400	250	30	851.96	7061.50
500	250	24	803.22	>20000.00
500	100	18	372.73	>20000.00
500	150	14	297.14	>20000.00
500	200	30	834.20	>20000.00
500	250	28	924.56	>20000.00

7 Conclusions

This paper studied the generation of valid Benders cuts for a general class of integer programming problems. The valid Benders cuts in the form of linear inequalities were derived, based on which a complete Benders algorithm was presented. The (relaxed) cut generation program was proposed to determine the valid cuts in practice. In theoretical aspect, the paper extended the application scope of Benders decomposition method to integer programming problems. In computational experiments, the results showed the benefits of using Benders algorithm with the proposed cut for integer programs.

The master problem discussed in the paper need not be restricted to linear integer programs. In fact, it can be any formulation and can be solved with any proper algorithm (such as Constraint Programming). More specifically, the linear objective function in problem P (i.e. $c^T y$) can be replaced with a general function $f(y)$. The first constraint in P (i.e. $Dy \leq d$) can be replaced with a general constraint $C(y)$ (even need not be arithmetic). Since the generalized objective and constraint are only handled in the master problem, they do not affect the theory and method proposed in the paper. Furthermore, the second constraint of P can be generalized to $h(y) + Bx \leq b$, that is, the part that relates to the master problem variables can be any function on y (i.e. $h(y)$), in place of the linear one, Ay. Accordingly, all the occurrences of Ay in the derivations are changed to $h(y)$, and the derivations remain valid. As the master problem is generalized as above, different modelling and solution methods could be combined via the method of Benders decomposition to cooperatively solve a given optimization problem.

References

[1] Flippo, O. E., Rinnoy Can, A. H. G.: Decomposition in General Mathematical Programming. Math. Programming. **60** (1993) 361–382

[2] Lasdon, L. S.: Optimization Theory for Large Systems. MacMillan, New York. (1970)   ,   ,

[3] Schimpf, J., Wallace, M.: Finding the Right Hybrid Algorithm: A Combinatorial Meta-Problem. Annals of Mathematics and Artificial Intelligence. **34** (2002) 259–269

[4] Eremin, A., Wallace, M.: Hybrid Benders Decomposition Algorithms in Constraint Logic Programming. In T. Walsh, editor, Principles and Practice of Constraint Programming - CP 2001, Springer. (2001) 1–15

[5] Benders, J. F.: Partitioning Procedures for Solving Mixed-Variables Programming Problems. Numerische Mathematik. **4** (1962) 238–252

[6] Geoffrion, A. M.: Generalised Benders Decomposition. Journal of Optimization Theory and Application. **10** (1972) 237–260

[7] Jain, V., Grossmann, I. E.: Algorithms for Hybrid MILP/CP Models for a Class of Optimisation Problems. INFORMS Journal on Computing. **13** (2001) 258–276

[8] Hooker, J. N., Ottosson, G.: Logic-Based Benders Decomposition. Math. Programming. **96** (2003) 33–60

[9] Eremin, A.: Using Dual Values to Integrate Row and Column Generation into Constraint Logic Programming. PhD Thesis. Imperial College London. (2003)

[10] Xia, Q., Simonis, H., Chu, Y.: Generating Primary/Secondary Path by Benders Decomposition Technique. IC-Parc Internal report, Imperial College London. (2003)

[11] Imperial College London: ECLiPSe 5.6 User's Manual. (2003)

[12] Dash Inc: Dash XPRESS 14.21 User's Manual. (2003)   ,

Appendix: Proofs of Lemmas

Lemma :

Proof. It suffices to show that any feasible solution of ( ) automatically satisfies $r_i(u-1)_i = 0$. Suppose $\hat{x}, \hat{r}, \hat{u}$ constitute a feasible solution of ( ).

The first constraint of ( ) implies

$$\hat{r}_i \geq (A\bar{y} + B\hat{x} - b)_i . \tag{15}$$

The second constraint of ( ) implies

$$\hat{r}_i = \hat{u}_i(A\bar{y} + B\hat{x} - b)_i . \tag{16}$$

Consider two cases on the non-negative value of $\hat{r}_i$.

Case 1: $\hat{r}_i = 0$. Then the constraint $r_i(u-1)_i = 0$ is trivially satisfied.

Case 2: $\hat{r}_i > 0$. Then we have $\hat{u}_i = 1$ (otherwise, $0 \leq \hat{u}_i < 1$. Then from ( ) $0 < \hat{r}_i = \hat{u}_i(A\bar{y} + B\hat{x} - b)_i < (A\bar{y} + B\hat{x} - b)_i$, which contradicts ( )). Since $\hat{u}_i = 1$, the constraint $r_i(u-1)_i = 0$ is again satisfied.   □

Lemma :

Proof. In every iteration, if $\bar{y}^{(k)}$ is feasible then the algorithm terminates from step *2(b)*. Otherwise a valid cut is added. Due to the valid cut condition 1, the feasible space of $MP^{(k+1)}$ must be smaller than that of $MP^{(k)}$, at least reduced by one point. Since the feasible space of master problem is finite domain, the algorithm terminates finitely. Due to the valid cut condition 2, the feasible space of $MP^{(k)}$ is always a relaxation of that of the original program P. If the algorithm terminates from *2(a)*, then $MP^{(k)}$ is infeasible, and so is the original program P. If the algorithm terminates from step *2(b)*, the current optimal solution of $MP^{(k)}$ is proved to be *feasible* for the subproblem and thus feasible for P. Since $MP^{(k)}$ is a relaxation of P, this solution is optimal for P.   □

Lemma :

Proof. For the valid cut condition 1, if $\bar{y}$ is infeasible for the subproblem, then $SP(\bar{y})$ has positive objective value. Thus any possible $SP_f(\bar{y}, x)$ has positive objective value, and so do all $DSP_f(\bar{y}, x)$. Therefore all inequalities in cut ( ) are violated, that is, cut ( ) excludes $\bar{y}$. For the valid cut condition 2, consider any feasible $\hat{y}$. There must exist one value $\hat{x}$ such that $SP_f(\hat{y}, \hat{x})$ has 0 objective value, and so do its dual $DSP_f(\hat{y}, \hat{x})$, that is, $(A\hat{y} - b)^T\hat{u} + \hat{x}^T\hat{v} = 0$ is satisfied, which means the disjunctive cut ( ) does not cut off the feasible $\hat{y}$.   □

Lemma :

Proof. For necessity, we show that if ( ) is violated, then ( ) must be violated. Suppose the ith element $\tilde{x}_i = 1$ and $\tilde{v}_i > 0$ (The case where the second condition of ( ) is violated can be proved similarly). Then $\tilde{x}_i \tilde{v}_i > 0$. Consider another assignment x, with the ith element $x_i = 0$ and with all other elements the same as $\tilde{x}$. It is easy to see that $\tilde{x}^T \tilde{v} > x \tilde{v}$, which means that ( ) is violated.

For sufficiency, we suppose ( ) is satisfied for every element. Then the inequality $\tilde{x}_i \tilde{v}_i \leq b \tilde{v}_i$ holds no matter b is 0 or 1. Therefore, for every element of any $x \in \{0,1\}^n$, we have $\tilde{x}_i \tilde{v}_i \leq x_i \tilde{v}_i$, which implies that ( ) holds.   □

Lemma :

Proof. We first prove that the relaxed cut satisfies the valid cut condition 2. Following the same reasoning as the proof of Theorem , we have:

$$(A\hat{y} - b)^T \tilde{u} + \hat{x}^T \tilde{v} \leq (A\hat{y} - b)^T \hat{u} + \hat{x}^T \hat{v} = 0 .$$

According to the construction of relaxed cut, we have $\tilde{x}'_i \tilde{v}_i \leq b \tilde{v}_i$ for any binary value $b \in \{0, 1\}$. In particular, $\tilde{x}'_i \tilde{v}_i \leq \hat{x}_i \tilde{v}_i, \forall i$. Therefore,

$$(A\hat{y} - b)^T \tilde{u} + \tilde{x}'^T \tilde{v} \leq (A\hat{y} - b)^T \tilde{u} + \hat{x}^T \tilde{v} \leq 0,$$

which means that the feasible $\hat{y}$ is not cut off by ( ).

The relaxation gap is directly obtained by subtracting the left hand side of ( ) from that of ( ). Because $\tilde{x}'_i \tilde{v}_i \leq \hat{x}_i \tilde{v}_i$, the relaxation gap $\sum_{i=1}^n (\tilde{x} - \tilde{x}')_i \tilde{v}_i \geq 0$.   □

Lemma :

Proof. Consider the program CGP_r. Since $v = B^T u$, the constraint $\tilde{x}_i (B^T u)_i - p_i \leq 0$ in ( ) becomes $\tilde{x}_i v_i \leq p_i$, and $(1 - \tilde{x})_i (B^T u)_i + q_i \geq 0$ becomes $-(1 - \tilde{x})_i v_i \leq q_i$.

For each element $\tilde{x}_i$ and its corresponding $\tilde{v}_i$, there are three cases.

Case 1: $\tilde{x}_i$ and $\tilde{v}_i$ satisfy the sign condition. Then the optimal slack variable p_i^* and q_i^* are 0, and $\tilde{x}_i = \tilde{x}'_i$. Therefore, $p_i^* + q_i^* = (\tilde{x} - \tilde{x}')_i \tilde{v}_i = 0$.

Case 2: $\tilde{x}_i$ and $\tilde{v}_i$ violates the sign condition as $\tilde{x}_i \tilde{v}_i > 0$, which implies that $\tilde{x}_i = 1$ and $\tilde{v}_i > 0$. The optimal slack variable $p_i^* = \tilde{v}_i > 0$, and q_i^* equals 0. Since the sign condition is violated, the value of $\tilde{x}_i$ will be changed from 1 to 0 ($\tilde{x}'_i = 0$) to construct the relaxed cut, and thus $(\tilde{x} - \tilde{x}')_i \tilde{v}_i$ equals $\tilde{v}_i$. Therefore, $p_i^* + q_i^* = (\tilde{x} - \tilde{x}')_i \tilde{v}_i > 0$.

Case 3: $\tilde{x}_i$ and $\tilde{v}_i$ violates the sign condition as $(1 - \tilde{x}_i) \tilde{v}_i < 0$, which implies that $\tilde{x}_i = 0$ and $\tilde{v}_i < 0$. The optimal slack variable $q_i^* = -\tilde{v}_i > 0$, and p_i^* equals 0. Since the sign condition is violated, the value of $\tilde{x}_i$ will be changed from 0 to 1 ($\tilde{x}'_i = 1$) to construct the relaxed cut, and thus $(\tilde{x} - \tilde{x}')_i \tilde{v}_i$ equals $-\tilde{v}_i$. Therefore, $p_i^* + q_i^* = (\tilde{x} - \tilde{x}')_i \tilde{v}_i > 0$.

In all cases the equation $p_i^* + q_i^* = (\tilde{x} - \tilde{x}')_i \tilde{v}_i \geq 0$ holds. Therefore,

$$\phi_{CGP_r} = \sum_{i=1}^n (p_i^* + q_i^*) = \sum_{i=1}^n (\tilde{x} - \tilde{x}')_i \tilde{v}_i \geq 0 .$$

□

A Constraint Programming Model
for Tail Assignment

Mattias Grönkvist

Carmen Systems AB
Odinsgatan 9, S-411 03 Göteborg, Sweden
mattias@carmensystems.com

Abstract. We describe a Constraint Programming model for the Tail Assignment problem in airline planning. Previous solution methods for this problem aim at optimality rather than obtaining a solution quickly, which is often a drawback in practice, where quickly obtaining solutions can be very important. We have developed constraints that use strong reachability propagation and tunneling to a column generation pricing problem to form a complete and flexible constraint model for Tail Assignment which is able to quickly find solutions. Results on real-world instances from a medium size airline are presented.

1 Introduction

Tail Assignment is a problem from the airline planning field. The problem consists in creating individual aircraft routes for a set of flights and ground activities, subject to a number of operational rules. The problem lies in the boundary region between planning and operations control, and it is therefore crucial to quickly be able to provide working solutions. In a longer perspective, e.g. when integrating the aircraft routes with crew planning, optimization also becomes important. For example, minimizing the number of aircraft swaps by flying crews can greatly improve the robustness of the complete solution in light of disruptions. Here, we will only be interested in finding a solution, ignoring the cost aspect altogether.

We have previously shown [ ] how constraint programming can be used to improve the performance of a mathematical programming model for the Tail Assignment problem. One problem with the mathematical programming model, which is based on column generation, is that it is not directly suited to finding solutions quickly. It rather puts optimization in focus, and is therefore very suitable indeed for the longer-term planning, but less suitable when a quick solution is desired. It is possible to obtain initial solutions more quickly with this model by integrating it with constraint programming, and forcing the mathematical model to work very aggressively.

However, it would be desirable to have a method that is more aimed at finding solutions quickly. Such a method could be used alone, but also to provide the initial solution to the column generator for further improvement, and thus make it possible to first provide an initial solutions very quickly, and then more

J.-C. Régin and M. Rueher (Eds.): CPAIOR 2004, LNCS 3011, pp. 142–   , 2004.
© Springer-Verlag Berlin Heidelberg 2004

optimized solutions after some more time. Finding a solution to Tail Assignment is a very hard problem, and practice has shown that greedy construction methods have problems finding solutions. We have therefore focused on constraint programming as a means to find initial solutions quickly. We have already had a partial constraint model before [ ], to use for integration with the column generator. But this model is not able to handle some of the more complicated constraints, and can therefore not be used to find proper solutions. Attempts have been made to complete the model [ ], but unfortunately the results were not that successful.

In this article we describe the development of new constraints and ordering heuristics to build a complete constraint model for Tail Assignment. Section we start by introducing Tail Assignment, and in Section   we describe the basic constraint model, and some simple extensions of it. Sections ,   and   describe the new constraints and ordering heuristics introduced to complete and solve the model, and Section   show computational results on real-world Tail Assignment instances. In Section   we conclude.

2 Tail Assignment

Tail Assignment is the problem of deciding which individual aircraft (identified by its *tail* number) should cover which flight. Each aircraft is thus assigned a *route* consisting of a sequence of flights, and possibly other activities such as maintenance, to perform. Tail Assignment deals with individual constraints, flights which are fixed in time, as well as individual rules for each tail. The planning period is typically one month. The purpose is to really create a solution that is possible to operate, satisfying all rules and regulations. The most basic rules are rules which only depend on two flights, so-called connection-based rules. For example, there must be a certain minimum buffer time between a landing and the next take-off.

Another important set of constraints are the *flight restriction* rules, which forbid certain aircraft to operate certain flights. There can be many reasons for the restriction – there can be a *curfew* for the arrival airport and some aircraft, because the aircraft violates noise or environmental restrictions. But there can also be more down-to-earth reasons, like the aircraft not having the required in-flight entertainment system or extra fuel tanks required for a long flight. Either way, the result is that an aircraft is restricted from operating a flight.

Finally, there are the maintenance rules. Aviation authorities require that all aircraft undergo various types of maintenance activities regularly. There are many maintenance types, depending om aircraft type, registration country, and airline. Typically, the rules specify that aircraft must undergo maintenance every X hours, or every Y landings. Airlines often also require that their aircraft return to a maintenance base frequently, even if no maintenance is done, to increase robustness in case disruptions occur. These rules typically specify that aircraft must come back to a maintenance base every Z days.

The normal representation of the Tail Assignment problem is in terms of a *flight network*. In the flight network, each node represents a flight, or some other activity such as a preassigned maintenance activity for specific aircraft, and each arc represents a connection between two flights or activities. For example, if operating flight f followed by flight f' is allowed according to connection rules, the connection from f to f' is considered *legal*, and the flight network will contain an arc between nodes f and f'. Since we are solving a dated problem, where flights are fixed in time, there are *carry-in* activities in the beginning of the period representing the last flights operated by each aircraft in the previous planning period, and the network is acyclic. The goal is now to find paths (routes) through the network for all aircraft, starting at the carry-in activities, such that all flight nodes are covered exactly once, and all rules are satisfied.

Variations of the Tail Assignment problem exist, for example the Aircraft Routing [ ] and Aircraft Rotation [,] Problems. [ ] contains a more thorough review of the literature about Tail Assignment and similar problems. Problems similar to Tail Assignment have been subject to constraint programming research, for example the Crew Scheduling Problem [,], and the Vehicle Routing Problem [,].

3   The Basic Model

This model was first described by Kilborn in [ ]. The model has three sets of variables. Let us call the set of flights F and the set of aircraft A. First, there are `successor` variables for all flights $f \in F$, containing the possible successors (legal connections) of the flight f, which we denote by S_f. Since the rules related to flight-to-flight connections are already modeled in the `successor` variables, no such constraints need to be explicitly added to the model. Then there are also `vehicle` variables for all flights, initially containing the aircraft in A that are allowed to operate the flight, A_f. These variables model preassigned activities as well as curfew rules.

Only two constraints are present. Firstly, since all flights must have unique successors to form disjoint routes through the network, all `successor` variables must take unique values. Therefore an `all_different` [ ] constraint over all `successor`s is added. Flight nodes which can represent route endings reconnect back to the carry-in activities. Secondly, to maintain consistency between the `successor` and `vehicle` variables, a special tunneling constraint is added. Observe that once completely instantiated, the `successor` and `vehicle` variables both describe the solution completely. The `successor` variables obviously give a direct route for an aircraft, and the `vehicle` variables specify which flights are assigned to a certain aircraft. Since the flights are fixed in time, simply sorting all flights assigned to an aircraft gives the route for this aircraft. The tunneling constraint is implemented as a single constraint that is propagated each time a variable is fixed, and takes appropriate action to keep the variables consistent:

```
vehicle[f] == a ⇒ POST  element(successor[f],vehicle,a)
successor[f] == f' ⇒ POST  vehicle[f] = vehicle[f']
```

The constraints are thus not posted all at once, but rather on demand whenever a `vehicle` or `successor` variable is fixed. It could have been possible to propagate each time a variable is *changed*, rather than fixed, but for simplicity and effectiveness we only propagate when variables are fixed. Observe that while the expressions above state the necessary actions to keep the variable consistent, other things can be enforced as well, to accelerate the propagation. For example, it is possible to directly remove j from the domains of overlapping flights when `vehicle[i] == j`. The constraints will make these assignments impossible, but from a performance point of view it is often beneficial to remove them directly. The `element(b,A,c)` constraint forces the `b`th value in vector `A` to take value `c`, i.e. `A[b] = c`.

3.1   Ordering Heuristics

To make this model behave properly is to crucial to define good variable and value ordering heuristics. Only `successor` variables are instantiated. The reason is that these variables are propagated much more than the others, thanks to the strong consistency algorithm for `all_different` due to Régin [ ]. The `successor` variables are fixed in order of increasing domain size, i.e. using the well-known *first-fail* ordering, with the exception of preassigned flights, which are fixed first. Values are chosen in increasing connection time order, except for long connections. If a connection is long, for example over night, we try to connect to the next preassigned activity first, rather than the next possible flight.

The result is a model that captures all Tail Assignment constraints except the maintenance rules. The model works very well for problems without many curfew rules, showing almost linear time behavior for problems of increasing size. For more results we refer the reader to [ ]. Unfortunately the model does not work well at all in the presence of many curfew rules. The reason is simply that the propagation for these constraints, which are embedded in the `vehicle` variables, is very weak. Simply put, we cannot get the partial routes we create to fit together, because the aircraft are incompatible with them.

3.2   Improvements of the Basic Model

It is fairly straightforward to improve the basic model above with a few more variables and constraints. Observe that the improvements in this section are only added to improve propagation, and as an effect the solution speed, of the basic model. Improvements to make the model more complete will be the topic of Sections   and .

Firstly, we can observe that since flights are fixed in time, flights which overlap in time can never be operated by the same aircraft. When we fix the `vehicle` of some flight we can therefore remove the fixed vehicle from the domains of all overlapping flights, as mentioned above. Also, we can add `all_different` constraints over all flights which pass a certain point in time. Observe that adding an `all_different` for *all* flights overlapping another flight is not a good idea, as an aircraft operating a flight overlapping only the beginning of the flight might be able to also operate a flight that overlaps only the end. The problem with adding all_different constraints is to decide where to add them. Since each all_different is only valid for one specific time, there is an infinite number of such constraints that could be added. Our strategy is to initially add one all_different for the start time of each flight which has some restriction, i.e. which cannot be operated by all aircraft:

$$\text{size}(\text{vehicle}[f]) \, ! = \text{size}(A) \Rightarrow \text{POST} \quad \text{all\_diff}(\text{successor}[G(f)])$$

$$G(f) \quad : \quad \text{Flights overlapping the start time of flight } f$$

Secondly, we add predecessor variables, representing the possible predecessor flights P_f of a flight. Considering predecessors explicitly has the benefit of finding flights that only has one predecessor, but whose predecessor has multiple successors. To keep the successor and predecessor variables consistent, we add an inverse constraint. The inverse(f,f') constraint, which has been implemented as a single constraint, simply ensures that f exists as a predecessor to f' if and only if f' exists as a successor to f:

> inverse(f,f'):
>
> successor[f] contains f' $\Longleftrightarrow$ predecessor[f'] contains f

We also include treatment for predecessor variables in the tunneling constraint, to keep it consistent with vehicle variables. These simple improvements have very positive effects on the model, decreasing the number of backtracks used and increasing the propagation.

4   Handling the Flight Restriction Rules

As we have already discussed, the flight restriction rules are in fact modeled already in the basic model, but when several restrictions are present, the performance of the search deteriorates. Since the propagation for the vehicle variables is poor, we end up generating partial routes, which do not fit together because of the restrictions. The effect is excessive backtracking, so-called *thrashing*. It is quite obvious that one way to deal with this problem is improved propagation for the vehicle variables. The introduction of all_different constraints over vehicle variables overlapping certain times in the previous section is an attempt at improving this, but it is not enough.

Instead, the key is to switch from the local 'successor-view' of the problem to a more route-oriented view. As our ultimate goal is to create routes for all aircraft, it seems like a good idea to introduce this point of view into the model. Since we want all flights to be operated by exactly one aircraft, we need to make sure at all times that each flight is contained in at least one route which satisfies the flight restrictions, for one of the aircraft. We do this by keeping track of the number of successor and predecessor flights that can be reached by each aircraft. That a flight can be reached by aircraft a means that a route from the carry-in of a to the flight exists. A flight can be reached in the forward direction, via

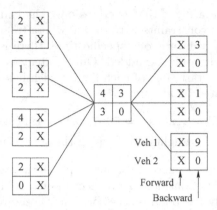

Fig. 1. An example of forward and backward labels with only two vehicles. Labels depending on flights not in the figure have been replaced by X

successors, and in the backward direction, via predecessors. The number of reachable neighbors is relatively cheap to update, and gives the information that we need. If no successor of a flight f can be reached by aircraft a, flight f itself can obviously not be reached, and thus not operated, by aircraft a. Similarly, if no predecessor can be reached by a, neither can f. But if at least one predecessor and one successor can be reached by a, and $a \in A_f$, then f can be operated by a.

For each flight, we maintain two *labels* for each aircraft. One label, called the *forward label*, counts how many predecessors have a non-zero forward label for the aircraft, i.e. how many predecessors can be reached, in the forward direction, by the aircraft. The other (the *backward label*) counts how many successor flights have a non-zero backward label for the vehicle. Let us call the forward label of flight f for aircraft a fl_f^a and the backward label bl_f^a. Let us further denote the set of carry-in flight by F_c, and let c_f^a be 1 if f is the carry-in flight of aircraft a and 0 otherwise. Now, the labels have the following relationship:

$$fl_f^a = c_f^a \quad \forall f \in F_c, \forall a \in A \tag{1}$$

$$bl_f^a = c_f^a \quad \forall f \in F_c, \forall a \in A \tag{2}$$

$$fl_f^a = \sum_{f' \in P_f \& fl_{f'}^a > 0} 1 \quad \forall f \notin F_c, \forall a \in A \tag{3}$$

$$bl_f^a = \sum_{f' \in S_f \& bl_{f'}^a > 0} 1 \quad \forall f \notin F_c, \forall a \in A \tag{4}$$

In the example in Figure , all of the predecessors have non-zero forward labels for vehicle 1, and the forward label is thus 4. Since none of the successors have non-zero backward labels for vehicle 2, the backward label is 0, and so on. As a consequence, the flight cannot be covered by vehicle 2.

4.1   Maintaining the Labels During Search

To initially set the forward and backward labels to their correct values is simple: Just set the forward labels by counting labels for predecessors, starting with the carry-ins and ending with the flight with the latest departure time. The backward labels are set in the opposite direction. The carry-in activities always have one forward and one backward label, and aircraft restricted from flying a flight of course always get forward and backward labels 0. More complicated is maintaining the correct labels once successor, predecessor and vehicle domains shrink, and in case backtracking occurs, increases in size.

Let us first assume that f' is removed from successor(f). In case f' is a carry-in activity, nothing happens, as we want to maintain one single label on carry-ins. If not, apply the following algorithm (in C-style pseudocode):

```
successor_removal(f, f')
begin
  for each aircraft a
    if (forward_label[f,a] == 0 OR
        forward_label[f',a] == 0)
      do nothing
    else
      --forward_label[f',a]
      if (forward_label[f',a] == 0)
        forward_remove_aircraft(a, f')
    end if
  end
end

forward_remove_aircraft(a, f')
begin
  q = empty queue
  q.push(f')
  while (q not empty)
    i = q.pop()
    for (all successors j of i)
      if (forward_label[j,a] != 0 AND
          j is not carry-in flight)
        --forward_label[j,a]
        if (forward_label[j,a] == 0)
          remove a~from vehicle(j)
          q.push(j)
        end if
      end if
    end
  end
end
```

What happens is that we first decrease the forward label on flight f' by 1 if both f and f' have non-zero forward labels, as one of the routes to flight f' is now removed. If this makes the forward label go to 0, this will potentially affect all successors of f', and we must call forward_remove_aircraft(). This function iteratively decreases forward labels as long as some label becomes 0 when decreased. The use of a queue makes sure that we treat the labels in the proper order, i.e. we treat all predecessors of f before we treat f. Upon termination of successor_removal, all forward labels are correct with respect to the variables. Backward labels are treated analogously.

Now instead imagine that aircraft a is removed from flight f. The algorithm for this case is shown below, and should be self-explanatory in light of the discussion above.

```
aircraft_removal(a, f)
begin
  if (forward_label[f,a] > 0)
    forward_label[f,a] = 0
    forward_remove_aircraft(a, f)
  end if
  // And the same for the backward label
end
```

Value insertions are treated much the same way as the removals, except that instead of updating forward/backward when the label becomes 0, we update when the label is increased from 0 to 1. We call this propagation algorithm the *reachability algorithm*.

Looking at the complexity of the propagation algorithm, the worst case is when we are forced to update all labels for all flights each time a value is removed or inserted. If we have p aircraft, n flights and each flight has m successors in average, the worst-case complexity of each successor removal is thus $\mathcal{O}(pnm)$, which is not that good. However, in practice we will not experience this behavior at all. The reason is that in the initial stage of the search, most flights can be reached via several routes, so most labels will have high values. Labels will thus seldom become 0, which means we will only have to update one or a few flights at each removal. As the problem gets more fixed, more labels take values close to 0. But on the other hand, each flight does not have as many successors at this stage, which means we will still not have to update that many labels. So in practice, this propagation algorithm works well, as Section   will show.

It should be observed that this propagation algorithm could probably have been stated in terms of simpler constraints rather than as a 'global' propagation algorithm. The propagation algorithm does not provide extra propagation because of the fact that it is implemented as a global algorithm. The decision to model it this way is rather motivated by performance. Also, the algorithm has been implemented as a constraint in the sense that it reacts to domain changes as a constraint, but it really only consists of the propagation algorithm.

5   Handling the Maintenance Rules

As discussed in Section , maintenance constraints are expressed as 'each aircraft must return to a maintenance base every X flying hours/Y landings/Z days for maintenance taking T hours'. Observe that several such rules can apply simultaneously, as there are several types of maintenance, which might take different time to perform, and might not all be possible to perform at the same base. Here, we will only require that the aircraft are provided maintenance *opportunities* with the proper intervals, and we will assume that whenever there is an opportunity the aircraft will be maintained. In practice an aircraft returning to base e.g. 3 nights in a row might not be (fully) maintained every night. But as long as there are enough opportunities, it will be possible to maintain all aircraft enough.

Before describing the constraints to handle the maintenance rules, let us briefly review the column generation approach to Tail Assignment [ ], as we will make use of it later. Column generation is a well-known mathematical programming technique [, , ,], used e.g. when the number of columns (variables) for a linear program are too many to enumerate. Instead, columns are generated dynamically, using dual cost information, until no more improving column exist. In the case of the Tail Assignment model, the columns are routes for individual aircraft. We have thus previously developed a module that given a set of dual costs can find a set of minimum cost routes, satisfying all flight restriction and maintenance rules. In this module, which is called the column generation *pricer*, the maintenance rules are modeled as *resource constraints*, making the pricing problem a resource-constrained shortest path problem. The implementation follows roughly that described in [ ].

Now, since we already have an implementation of the maintenance rules in the pricer module, and would like to avoid implementing the same thing twice, it seems reasonable to try to re-use the pricer for the constraint model as well. The pricer implementation is highly tuned and general, as it is user-customizable via the domain-specific Rave language [ ], making it possible to model any kind of rule that can be modeled as a resource constraint, not only maintenance rules. Completely re-implementing this functionality in the CP model would be cumbersome indeed. However, as the pricer is labeling-based [ ], it can easily allows us to check feasibility by checking whether

1. Each flight is reachable, i.e. there is at least one route arriving at each flight
2. There exists at least one legal route per aircraft

Unfortunately, this check is *incomplete*, in the sense that a partially fixed network can pass the feasibility check even if it is infeasible, e.g. if a flight is the only possible successor of several flights. However, for a fully fixed network it is complete, so we will never allow solutions violating the maintenance constraints.

---

[1] The least cost column is found, along with a predefined number of other low-cost columns, but not necessarily the least cost ones.

By creating a constraint that tunnels the changes in the variable domains to the internal structures used in the pricer, and does the above feasibility check, we have created a constraint that makes sure that a solution must be maintenance feasible. The constraint (which we call the *pricing constraint*) is not very strong in terms of propagation, but fortunately experience tells us that the maintenance rules are seldom very tight, so this is not be a huge problem in practice. The only propagation done by the constraint is that in case no route for a certain aircraft exists to a flight, this aircraft is removed from the `vehicle` domain.

One major concern with this constraint is that fact that it is fairly expensive to check. One of the drawbacks of using code 'unrelated' to the rest of our CP model is that it is difficult to customize it perfectly to our needs. We can customize it to a large extent by deciding on the number of labels, number of generated routes etc, but doing the check still takes too long for it to be done at every domain change. Instead, we have to settle for checking/propagating the constraint with a certain interval during the search, and at the very end. However, as the next section will show, this fits rather well into the full picture.

6 Ordering Heuristics

We have now described all necessary constraints to form the Tail Assignment model. However, as Section   will show, this is not enough unless well-working variable and value ordering heuristics are used. Unfortunately, the old variable ordering heuristic of instantiating preassigned activities first, and then fix according to increasing `successor` domain size, does not work well at all. Instead, we must again resort to routes. Instead of fixing single variables based on their local properties, e.g. domain size, we create an entire route, which we know satisfies all flight restrictions and maintenance rules, and let the route set the order in which the `successor` variables are instantiated. Once we have fixed all `successors` in a route, we create another route, and so on, until all aircraft have routes.

This goes hand-in-hand with the feasibility check performed in the pricing constraint, as this provides access to the necessary routes. When checking feasibility using the pricer, we do a labeling run which results in routes for all aircraft, if such exist. This suggests the following variable ordering strategy:

- Whenever all `successors` in a previous route have been fixed
  - Check feasibility of the pricing constraint
  - If infeasible, backtrack
  - If feasible, take one of the generated routes as the next route to fix

One question is which route to choose to fix. Our strategy is to fix the aircraft in order of decreasing number of flight restrictions. That is, to avoid them getting trapped when most of the network is fixed, we start by fixing routes for the aircraft which have the largest number of flight restrictions, i.e. the aircraft a for which $\sum_f f_a$, where f_a is 1 if flight f can be operated by aircraft a and 0 otherwise, is minimal. Also, to avoid the last few aircraft we fix from getting stuck because there are no maintenance opportunities left, we want to use as few maintenance opportunities as possible each time we fix a route. This can be at least approximately achieved by setting the dual costs properly for the pricing constraint. Remember that the pricer finds a set of least-cost routes with respect to the provided duals . Since long, typically overnight, connections are used to perform maintenance, we want to penalize the use of such connections. This is done by putting a large negative dual cost on using long connections. We therefore set the dual cost of all flights to the negation of the sum of the connection times to successors.

Finally, to speed up the search, in case we need to backtrack, we backtrack the entire route we are currently fixing instead of backtracking a single step. This makes the whole search incomplete, but is only done to avoid excessive backtracking, and excessive checking of the pricing constraint. If we find a conflict half-way through fixing a route, backtracking by single steps until we find a feasible node could potentially force us to check the pricing constraint far too many times to be efficient. So instead we backtrack all the way to the beginning of the route, where we know the pricing constraint has been checked. To avoid the same route from being re-generated, we add a penalty to the 'dual cost' of all flights we attempt to fix.

7   Computational Results

We have implemented the constraints and ordering heuristics as described above, and in this section we will present computational results for a set of real-world test instances. The instances come from different planning months at the same medium size airline. The aircraft included are a mix of M82, M83 and M88 aircraft. The underlying CSP solver is one designed in-house.

Table   presents the instances, and shows the running times of a fixing heuristic method [ ] based on column generation, which was the best known method for quickly producing solutions prior to this work, and the constraint model. It should be observed that the performance of the fixing heuristic also relies heavily on constraint propagation, as it uses the basic CP model to check that fixes proposed by the fixing heuristic do not lead to conflicts. The pure column generation fixing heuristic is not included in the comparison, as it has the property that it can leave flights unassigned. This is even more likely to occur if the algorithm is tuned so as to compete with the running times of the approaches presented here. Comparing running times of a heuristic that can leave flight unassigned with heuristics that cannot do this simply does not make sense, and hence this comparison was skipped.

The running times in the table, which are all rounded to even integer seconds, include problem setup and some preprocessing and bound-calculation steps as presented in [ ]. The number of flights reported is the number of activities given to the constraint model. Due to the preprocessing, each activity might consist

---

[2] In fact with respect to the reduced cost. But if real costs are set to 0, the optimization will only be over the duals.

Table 1. Comparison of solution times using a column generation-based fixing technique, and using the constraint model

| | | | | Fix. Heur. | CP |
Instance	Flights	Aircraft	Rules	Time (s)	Time (s)
Instance 1	2388	33	1	140	70
Instance 2	2965	31	1	247	98
Instance 3	2757	31	1	184	82
Instance 4	2379	31	2	232	99
Instance 5	2839	31	1	175	78
Instance 6	2652	31	2	209	99
Instance 7	2474	31	1	156	68
Instance 8	2858	31	1	1468	106
Instance 9	2604	31	1	218	66
Instance 10	2040	30	3	321	112
Instance 11	2232	30	3	235	112
Instance 12	2226	30	3	249	134
Instance 13	2163	30	3	184	100

of several atomic flights (*legs*). The actual number of flight legs is therefore roughly twice the number reported in table . The 'Rules' column gives the number of maintenance rules that are present for each instance. For all instances, a rule specifying that the aircraft should return to base with regular intervals in present. For some instances, so-called 'A-check' and lubrication maintenance rules are also present. It is clear that the constraint model produces solutions significantly faster than the old method. Roughly speaking, the constraint model is about twice as fast.

Table shows the benefit of the reachability propagation. The table shows the behavior of the basic model compared to the basic model with reachability propagation. When the reachability propagation is included, the variable order is changed by simply taking the first non-fixed flight from the most restricted aircraft first. We thus do not generate and fix entire routes, but still use the reachability labels to help guide the search. Entries marked with a star means that no solution was found within the predefined limits, which were set to maximum 1800 seconds or as many backtracks as there are flights in the problem. The measured times are in these cases the times to reach this many backtracks. The reachability propagation clearly helps, as without it only two instances can be solved within the limits, while when it is added, all but two instances are solved.

Table shows the importance of the ordering heuristics. The first columns show the full model using the old ordering heuristics, i.e. first-fail and short-connections-first, while the last columns show the full model with the route-fixing ordering. When using the old heuristics, we have applied the pricing constraint

154   Mattias Grönkvist

Table 2. Comparison of the basic model with and without reachability propagation

| | Basic model | | With reach. | |
Instance	BTs	Time(s)	BTs	Time(s)
Instance 1	2965*	122	18	63
Instance 2	2388*	73	2	91
Instance 3	0	57	1	74
Instance 4	2757*	74	2379*	450
Instance 5	2379*	74	44	75
Instance 6	2839*	76	29	71
Instance 7	2474*	786	5	58
Instance 8	2604*	72	34	67
Instance 9	2858*	467	2604*	1283
Instance 10	2040*	63	3	54
Instance 11	2232*	102	17	68
Instance 12	26	70	2	68
Instance 13	2163*	69	591	67

propagation every #flights/#aircraft search nodes, to approximately apply is once for every fixed aircraft. This was to get a fair comparison with the route-fixing odering, which applies the propagation exactly once per route. It is clear from the table that the old ordering heuristics do not work well at all, even in the presence of reachability propagation and the pricing constraint. No solution is found, and for all problems the excessive thrashing gives rise to very long running times.

8   Conclusions

We have presented a full constraint model for the Tail Assignment problem. The model includes a reachability algorithm that provides efficient propagation required to capture the flight restriction rules. The model also uses the pricing module from a column generation approach to Tail Assignment [ ] to model maintenance rules, which would otherwise require substantial effort to re-implement. It has previously been shown how constraint programming techniques can improve the performance of column generation for this problem [ ], and this model, by re-using a column generation pricing problem to check feasibility and improve ordering heuristics, shows that the reverse is also possible. Further integration between the approaches will likely benefit a lot from the results presented here, and will be researched further. We also conjecture that adding future constraints to a Tail Assignment model will often be easier for the constraint model than for the full column generation model, making it an excellent tool for experimentation.

Table 3. Comparison of constraint models using old ordering heuristics, no reachability propagation, and the full model

Instance	Full model, old ordering BTs	Full model, old ordering Time(s)	Full model, new ordering BTs	Full model, new ordering Time(s)
Instance 1	2388*	1182	177	70
Instance 2	1391*	1878	0	98
Instance 3	2757*	1238	0	82
Instance 4	2379*	1353	146	99
Instance 5	2839*	1191	116	78
Instance 6	2652*	1755	0	99
Instance 7	2474*	1458	0	68
Instance 8	1268*	1867	0	106
Instance 9	2604*	1227	0	66
Instance 10	2040*	1071	0	112
Instance 11	2226*	1172	0	112
Instance 12	2163*	752	0	134
Instance 13	2232*	1009	0	100

One potential drawback of the maintenance rule handling is that the propagation is not that strong. However, since these constraints are seldom extremely tight in practice, the limited propagation is enough to achieve reasonable performance. And the fact that the handling of these rules has not been re-implemented in the constraint model is immensely important, since the described implementation is part of a production system.

It is possible that the reachability algorithm, as well as the idea of re-using a column generartion pricing problem, could be of use also for other types of applications, similar to Tail Assignment. Finally, it should be mentioned again that this model is not designed to work alone to solve the Tail Assignment problem. It is primarily designed to quickly produce a feasible initial solution that can then be improved, given a cost function, using a combination of column generation and constraint propagation.

References

[1] L. W. Clarke, E. L. Johnson, G. L. Nemhauser, and Z. Zhu. The Aircraft Rotation Problem. *Annals of Operations Research*, 69:33–46, 1997.
[2] B. de Backer, V. Furnon, P. Kilby, P. Prosser, and P. Shaw. Solving Vehicle Routing Problems using Constraint Programming and Metaheuristics. *Journal of Heuristics*, 1(16), 1997.
[3] G. Desaulniers, J. Desrosiers, Y. Dumas, M. M. Solomon, and F. Soumis. Daily Aircraft Routing and Scheduling. *Management Science*, 43(6):841–855, July 1997.

[4] M. Desrochers and F. Soumis. A generalized permanent labelling algorithm for the shortest path problem with time windows. *INFOR*, 26(3):191–212, 1988.

[5] M. Elf, M. Jünger, and V. Kaibel. Rotation Planning for the Continental Service of a European Airline. In W. Jager and H.-J. Krebs, editors, *Mathematics – Key Technologies for the Future. Joint Projects between Universities and Industry*, pages 675–689. Springer Verlag, 2003.

[6] G. Erling and D. Rosin. Tail Assignment with Maintenance Restrictions - A Constraint Programming Approach. Master's thesis, Chalmers University of Technology, Gothenburg, Sweden, 2002.

[7] T. Fahle, U. Junker, S. Karisch, N. Kohl, M. Sellmann, and B. Vaaben. Constraint Programming Based Column Generation for Crew Assignment. *Journal of Heuristics*, 8(1):59–81, 2002.   ,

[8] M. Gamache, F. Soumis, G. Marquis, and J. Desrosiers. A Column Generation Approach for Large-Scale Aircrew Rostering Problems. *Operations Research*, 47(2):247–263, April-March 1999.

[9] R. Gopalan and K. T. Talluri. The Aircraft Maintenance Routing Problem. *Operations Research*, 46(2):260–271, March-April 1998.

[10] M. Grönkvist. Tail Assignment – A Combined Column Generation and Constraint Programming Approach. Lic. Thesis, Chalmers University of Technology, Gothenburg, Sweden, 2003.   ,   ,

[11] M. Grönkvist. Using Constraint Propagation to Accelerate Column Generation in Aircraft Scheduling. In *Proceedings of CPAIOR'03*, May 2003.   ,   ,

[12] C. Halatsis, P. Stamatopoulos, I. Karali, T. Bitsikas, G. Fessakis, A. Schizas, S. Sfakianakis, C. Fouskakis, T. Koukoumpetsos, and D. Papageorgiou. Crew Scheduling Based on Constraint Programming: The PARACHUTE Experience. In *In Proceedings of the 3rd Hellenic-European Conference on Mathematics and Informatics HERMIS '96*, pages 424–431, 1996.

[13] C. Hjorring and J. Hansen. Column generation with a rule modelling language for airline crew pairing. In *Proceedings of the 34th Annual Conference of the Operational Research Society of New Zealand*, pages 133–142, Hamilton, New Zealand, December 1999.

[14] E. Kilborn. Aircraft Scheduling and Operation – a Constraint Programming Approach. Master's thesis, Chalmers University of Technology, Gothenburg, Sweden, 2000.   ,   ,

[15] J.-C. Régin. A filtering algorithm for constraints of difference in CSPs. In *Proceedings of AAAI-94*, pages 362–367, 1994.   ,

[16] L.-M. Rousseau, M. Gendreau, and G. Pesant. Using Constraint-Based Operators to Solve the Vehicle Routing Problem with Time Windows. *Journal of Heuristics*, 8(1):43–58, 2002.

Super Solutions in Constraint Programming

Emmanuel Hebrard, Brahim Hnich, and Toby Walsh*

Cork Constraint Computation Centre, University College Cork
{e.hebrard,brahim,tw}@4c.ucc.ie

Abstract. To improve solution robustness, we introduce the concept of super solutions to constraint programming. An (a, b)-super solution is one in which if a variables lose their values, the solution can be repaired by assigning these variables with a new values and at most b other variables. Super solutions are a generalization of supermodels in propositional satisfiability. We focus in this paper on (1,0)-super solutions, where if one variable loses its value, we can find another solution by re-assigning this variable with a new value. To find super solutions, we explore methods based both on reformulation and on search. Our reformulation methods transform the constraint satisfaction problem so that the only solutions are super solutions. Our search methods are based on a notion of super consistency. Experiments show that super MAC, a novel search-based method shows considerable promise. When super solutions do not exist, we show how to find the most robust solution. Finally, we extend our approach from robust solutions of constraint satisfaction problems to constraint optimization problems.

1 Introduction

Where changes to a solution introduce additional expenses or reorganization, solution robustness is a valuable property. A robust solution is not sensitive to small changes. For example, a robust schedule will not collapse immediately when one job takes slightly longer to execute than planned. The schedule should change locally and in small proportions, and the overall makespan should change little if at all. To improve solution robustness, we introduce the concept of super solutions to constraint programming (CP). An (a, b)-super solution is one in which if the values assigned to a variables are no longer available, the solution can be repaired by assigning these variables with a new values and at most b other variables. An (a, b)-super solution is a generalization of both fault tolerant solutions in CP [ ] and supermodels in propositional satisfiability (SAT) [ ]. We show that finding (a, b)-super solutions for any fixed a is NP-Complete in general. Super solutions are computed offline and do not require knowledge about the likely changes. A super solution guarantees the existence of a small set of repairs when the future changes in a small way.

In this paper, we focus on the algorithmic aspects of finding (1,0)-super solutions, which are the same as fault tolerant solutions [ ]. A (1,0)-super solution

---

* All authors are supported by the Science Foundation Ireland and an ILOG grant.

J.-C. Régin and M. Rueher (Eds.): CPAIOR 2004, LNCS 3011, pp. 157–   , 2004.
© Springer-Verlag Berlin Heidelberg 2004

is a solution where if one variable loses its value, we can find another solution by re-assigning this variable with a new value, and no other variables are required for the other variables. We explore methods based both on reformulation and on search to find (1,0)-super solutions. Our reformulation methods transform the constraint satisfaction problem so that the only solutions are super solutions. We review two reformulation techniques presented in [ ], and introduce a new one, which we call the cross-domain reformulation. Our search methods are based on notions of super consistency. We propose two new search algorithms that extend the maintaining arc consistency algorithm (MAC [,]). We empirically compare the different methods and observe that one of them, super MAC shows considerable promise. When super solutions do not exist, we show how to find the most robust solution closest to a super solution. We propose a super Branch & Bound algorithm that finds the most robust solution, i.e., a solution with the maximum number of repairable variables. Finally, we extend our approach from robust solutions of constraint satisfaction problems to constraint optimization problems. We show how an optimization problem becomes a multi-criterion optimization problem, where we optimize the number of repairable variables and the objective function.

2   Super Solutions

Supermodels were introduced in [ ] as a way to measure solution robustness. An (a, b)-supermodel of a SAT problem is a model (a satisfying assignment) with the additional property that if we modify the values taken by the variables in a set of size at most a (breakage variables), another model can be obtained by flipping the values of the variables in a disjoint set of size at most b (repair variables).

There are a number of ways we could generalize the definition of supermodels from SAT to CP as variables now can have more than two values. A break could be either "losing" the current assignment for a variable and then freely choosing an alternative value, or replacing the current assignment with some other value. Since the latter is stronger and potentially less useful, we propose the following definition.

Definition 1. *A solution to a CSP is (a, b)-super solution iff the loss of the values of at most a variables can be repaired by assigning other values to these variables, and modifying the assignment of at most b other variables.*

Example 1. Let us consider the following CSP: $X, Y, Z \in \{1, 2, 3\}$ $X \le Y \wedge Y \le Z$. The solutions to this CSP are shown in Figure , along with the subsets of the solutions that are $(1, 1)$-super solutions and $(1, 0)$-super solutions.

The solution $\langle 1, 1, 1 \rangle$ is not a $(1, 0)$-super solution. If X loses the value 1, we cannot find a repair value for X that is consistent with Y and Z since neither $\langle 2, 1, 1 \rangle$ nor $\langle 3, 1, 1 \rangle$ are solutions. Also, solution $\langle 1, 1, 1 \rangle$ is not a $(1, 1)$-super solution since when X loses the value 1, we cannot repair it by changing the value assigned to at most one other variable, i.e., there exists no repair solution

solutions	$(1,1)$-super solutions	$(1,0)$-super solutions
$\langle 1,1,1 \rangle, \langle 1,1,2 \rangle$	$\langle 1,1,2 \rangle, \langle 1,1,3 \rangle$	$\langle 1,2,3 \rangle$
$\langle 1,1,3 \rangle, \langle 1,2,2 \rangle$	$\langle 1,2,2 \rangle, \langle 1,2,3 \rangle$	$\langle 1,2,2 \rangle$
$\langle 1,2,3 \rangle, \langle 1,3,3 \rangle$	$\langle 1,3,3 \rangle, \langle 2,2,2 \rangle$	$\langle 2,2,3 \rangle$
$\langle 2,2,2 \rangle, \langle 2,2,3 \rangle$	$\langle 2,2,3 \rangle, \langle 2,3,3 \rangle$	
$\langle 2,3,3 \rangle, \langle 3,3,3 \rangle$		

Fig. 1.   solutions, $(1,1)$-super solutions, and $(1,0)$-super solutions for the CSP: $X \leq Y \leq Z$

when X breaks since none of $\langle 2,1,1 \rangle$, $\langle 3,1,1 \rangle$, $\langle 2,2,1 \rangle$, $\langle 2,3,1 \rangle$, $\langle 2,1,2 \rangle$, and $\langle 2,1,3 \rangle$ is a solution. On the other hand, $\langle 1,2,3 \rangle$ is a $(1,0)$-super solution since when X breaks we have the repair solution $\langle 2,2,3 \rangle$, when Y breaks we have the repair solution $\langle 1,1,3 \rangle$, and when Z breaks we have the repair solution $\langle 1,2,2 \rangle$. We therefore have a theoretical basis to prefer the solution $\langle 1,2,3 \rangle$ to $\langle 1,1,1 \rangle$, as the former is more robust.

A number of properties follow immediately from the definition. For example, a (c,d)-super solution is a (a,b)-super solution if ($a \leq c$ or $d \leq b$) and $c+d \leq a+b$. Deciding if a SAT problem has an (a,b)-supermodel is NP-complete [ ]. It is not difficult to show that deciding if a CSP has an (a,b)-super solution is also NP-complete, even when restricted to binary constraints.

Theorem 1. *Deciding if a CSP has an (a,b)-super solution is NP-complete for any fixed a.*

Proof. To see it is in NP, we need a polynomial witness that can be checked in polynomial time. This is simply an assignment which satisfies the constraints, and, for each of the $O(n^a)$ (which is polynomial for fixed a) possible breaks, the $a + b$ repair values.

To show completeness, we show how to map a binary CSP onto a new binary problem in which the original has a solution iff the new problem has an (a,b)-super solution. Our reduction constructs a CSP which, if it has any solution, has an (a,b)-super solution for any $a + b \leq n$. The problem will even have a $(n,0)$-super solution. It is possible to show that if we have a $(n,0)$-super solution then we also have an (a,b)-super solution for any $a + b \leq n$. However, we will argue *directly* that the original CSP has a solution iff the constructed problem has an (a,b)-super solution.

We duplicate the domains of each of the variables, and extend the constraints so that the behave equivalently on the new values. For example, suppose we have a constraint $C(X,Y)$ which is only satisfied by $C(m,n)$. Then we extend the constraint so that is satisfied by just $C(m,n)$, $C(m',n)$, $C(m,n')$ and $C(m',n')$ where m' and n' are the duplicated values for m and n. Clearly, this binary CSP has a solution iff the original problem also has. In addition, any break of a variables can be repaired by replacing the a corresponding values with their primed values (or unpriming them if they are already primed) as well as any b other values.   □

A necessary but not sufficient condition to find supermodels in SAT or super solutions in CSPs is the absence of backbone variables. A *backbone variable* is a variable that takes the same value in all solutions. As a backbone variable has no alternative, a SAT or CSP problem with a backbone variable cannot have any (a, b)-supermodels or (a, b)-super solutions.

Another important factor that influences the existence of super solutions is the way the problem is modeled. For instance, the direct encoding into SAT (i.e., one Boolean variable for each pair variable-value in the CSP [ ]) of the problem in Example 1 has no $(1,0)$-supermodels, even though the original CSP had a $(1,0)$-super solution. Moreover, the meaning of a super solution depends on the model. For example, if a variable is a job and a value is a machine, the loss of a value may mean that the machine has now broken. On the other hand, if a variable is a machine and the value is a job, the loss of a value may mean that the job is now not ready to start. The CP framework gives us more freedom than SAT to choose what variables and values stand for, and therefore to the meaning of a super solution. For the rest of the paper, we just focus on $(1,0)$-super solutions and refer to them as super solutions when it is convenient.

3   Finding $(1, 0)$-Super Solutions via Reformulation

Fault tolerant solutions [ ] are the same as $(1,0)$-super solutions. The first reformulation approach in [ ] allows only fault tolerant solutions, but not *all* of them (see [ ] for a counter example). The second approach in [ ] duplicates the variables. The duplicate variables have the same domain as the original variables, and are linked by the same constraints. A not equals constraint is also posted between each original variable and its duplicate. The assignment to the original variables is a super solution, where the repair for each variable is given by its duplicate. We refer to the reformulation of a CSP P with this encoding as $P+P$.

We now present a third and new reformulation approach. Let $S = \langle v_1, v_2 \rangle$ be part of a $(1, 0)$-super solution on *two* variables X and Y. If v_1 is lost, then there must be a value $r_1 \in \mathcal{D}(X)$ that can repair v_1, that is $\langle r_1, v_2 \rangle$ is a compatible tuple. Symmetrically, there must exists r_2 such that $\langle v_1, r_2 \rangle$ is allowed. Now consider the following subproblem involving two variables:

v_1——v_2 means (v_1, v_2) is allowed.

Since it satisfies the criteria above, $S = \langle v_1, v_2 \rangle$ is a super solution whilst any other tuple is not. One may try to prune the values $r_1, r_2, i_1,$ and i_2 as they do not participate in any super solution. However r_1 and r_2 are *essential* for *providing* support to v_1 and v_2. On the other hand, i_1 and i_2 are simply not supported and can thus be pruned. So, we cannot simply reason about extending partial instantiations of values, unless we keep the information about the values that can be used as repair. So, let us instead think of the domain of the variables as pairs of values $\langle v, r \rangle$, the first element corresponding to the *super value* (which is part of a super solution), the second corresponding to the *repair value* (which can repair the former). Our cross-domain reformulation exploits this. We reformulate a CSP $P = \{\mathcal{X}, \mathcal{D}, \mathcal{C}\}$ such that any domain becomes its own cross-product (less the doubletons),i.e. $\mathcal{D}(\mathcal{X})$ becomes $\mathcal{D}(\mathcal{X}) \times \mathcal{D}(\mathcal{X}) - \{\langle v, v \rangle | v \in \mathcal{D}(\mathcal{X})\}$. The constraints are built as follows. Two pairs $\langle v_1, r_1 \rangle$ and $\langle v_2, r_2 \rangle$ are compatible iff

- v_1 and v_2 are compatible (the solution must be consistent at the first place);
- v_1 and r_2 are compatible (in case of a break involving v_2, r_2 can be a repair);
- v_2 and r_1 are compatible (in case of a break involving v_1, r_1 can be a repair).

The new domain $\mathcal{CD}(\mathcal{X})$ and $\mathcal{CD}(\mathcal{Y})$ of variable X and Y are:

$$\{\langle v_1, r_1 \rangle, \langle v_1, i_1 \rangle, \langle r_1, v_1 \rangle, \langle r_1, i_1 \rangle, \langle i_1, v_1 \rangle, \langle i_1, r_1 \rangle\}$$

$$\{\langle v_2, r_2 \rangle, \langle v_2, i_2 \rangle, \langle r_2, v_2 \rangle, \langle r_2, i_2 \rangle, \langle i_2, v_2 \rangle, \langle i_2, r_2 \rangle\}$$

The only one allowed tuple is $S = \langle \langle v_1, r_1 \rangle, \langle v_2, r_2 \rangle \rangle$. We refer to the cross-domain reformulation of a problem P as $P \times P$.

4 Finding $(1, 0)$-Super Solutions via Search

We first introduce the notion of super consistency for binary constraints, and then use it to build some new search algorithms.

4.1 Super Consistency

Backtrack-based search algorithms like MAC use local consistency to detect unsatisfiable subproblems. Local consistency can also be used to develop efficient algorithms for finding super solutions. We shall introduce three ways of incorporating arc consistency (AC) into a search algorithm for seeking super solutions.

AC+ is a naive approach that augments the traditional AC by a further condition, achieving a very low level of filtering.

AC($P \times P$) maintains AC on the cross-domain reformulation of P. This method allows us to infer *all* that can be inferred locally, just as AC does in a regular CSP [ ]. However, this comes at a high polynomial cost.

Super AC gives less inference than AC($P \times P$), but is a good tradeoff between the amount of pruning and complexity.

Informally, the consistent closure of a CSP contains only partial solutions for a given level of locality. However, the situation with super solutions is more complex because values that do not get used in any local super solution can still be essential as a *repair* and thus cannot be simply pruned.

AC+. If S is a super solution, then for every variable, at least *two* values are consistent with all the others values of S. Consequently, being arc consistent *and having non-singleton domains* is a necessary condition for the existence of super solution. AC+ is therefore defined as follows: for a CSP $P = \{\mathcal{X}, \mathcal{D}, \mathcal{C}\}$: AC+$(P) \Leftrightarrow$ AC$(P) \wedge \forall D \in \mathcal{D}, |D| > 1$. Whilst AC+ is usually too weak to give good results, it is the basis for an algorithm for the associated optimization problem (discussed in section   ).

AC$(P{\times}P)$. AC on $P{\times}P$ is the tightest local domain filtering possible. Note that $P{\times}P$ also has the same constraint graph topology as the original problem P. As a corollary, if the constraint graph of P is a tree, we can use AC on $P{\times}P$ to find $(1, 0)$-super solutions in polynomial time.

Super AC. AC on $P{\times}P$ allows us to infer all that can be inferred locally. In other words, we will prune any value in a cross-domain that is not locally consistent. However, this comes at high cost. Maintaining AC will be $O(d^4)$ where d is the initial domain size. We therefore propose an alternative that does less inference, but at just $O(d^2)$ cost.

The main reason for the high cost is the size of the cross-domains. A cross-domain is quadratic in the size of the original domain since it explicitly represents the repair value for each super value. Here we will simulate much of the inference performed by super consistency, but will only look at one value at a time, and not pairs. We will divide the domain of the variable into two separate sets of domains:

– The "super domain" (SD) where only super values are represented;
– The "repair domain" (RD) where repair values are stored.

We propose the following definition for super AC:

– A value v is in the super domain of X iff for any other variable Y, there exists v' in super domain of Y and r in repair domain of Y such that $\langle v, v' \rangle$ and $\langle v, r \rangle$ are allowed and $v' \neq r$.
– A value v is in repair domain of X iff for any other variable Y, there exists v' in super domain of Y such that $\langle v, v' \rangle$ is allowed.

The definition of super AC translates in a straightforward way into a filtering algorithm. The values are marked as either *super* or *repair*, and when looking for support of a super value, an additional and different support marked either as *super* or *repair* is required. The complexity of checking the consistency of an arc increases only by a factor of 2 and thus remains in $O(d^2)$.

Theoretical Properties. We now show that maintaining AC on $P{\times}P$ achieves more pruning than maintaining super AC on P, which achieve more pruning than maintaining AC on $P{+}P$ or AC+ on P (which are equivalent). For the theorem and the proof below, we use the notation (x)(P) to denote that the problem P is "consistent" for the filtering (x).

Fig. 2. The first graph shows the microstructure of a simple CSP, two variables and three values each, *allowed* combinations are linked. P is AC+ since the network is arc consistent and every domain contains 3 values. However, P is not super AC since the grayed values (in the second graph) are not in super domains, they have only one support. In the second step, the whitened variables (in the third graph) are also removed from both repair and super domains since they do not have a support in a super domain

Theorem 2 (level of filtering). *For any problem P, $AC(P \times P) \Rightarrow super\ AC(P)$ $\Rightarrow AC(P+P) \Leftrightarrow AC+(P)$.*

Proof. (1) **AC($P+P$) $\Rightarrow$ AC+(P):** Suppose that P is not AC+, then in the arc consistent closure of P, there exists at least one domain D_i such that $|D_i| \le 1$. $P+P$ contains P. In its arc consistent closure, we have $|D_i| \le 1$ as well. X_i is linked to a duplicate of itself in which the domain D_i' is then equal to D_i and therefore singleton (with the same value) or empty. However, recall that we force $X_i \ne X_i'$, thus $P+P$ is not AC.

(2) **AC+(P) $\Rightarrow$ AC($P+P$):** Suppose we have AC(P) and any domain D in P is such that $|D| > 1$, now consider $P+P$. The original constraints are AC since P is AC. The duplicated constraints are AC since they are identical to the original ones. The not equals constraints between original and duplicated variables are AC since any variable has at least 2 values.

(3) **super AC(P) $\Rightarrow$ AC+(P):** Suppose that P is not AC+, then there exists two variables X, Y such that any value of X has at most one support on Y, therefore the corresponding super domain is wiped-out, and P is not super AC.

(4) **super AC(P) $\nLeftarrow$ AC+(P):** See counter-example in Figure .

(5) **AC($P \times P$) $\Rightarrow$ super AC(P):** Suppose that AC($P \times P$), then for any two variables X, Y there exist two pairs $\langle v1, r1 \rangle \in D(X) \times D(X), \langle v2, r2 \rangle \in D(Y) \times D(Y)$, such that $\langle v1, r2 \rangle, \langle r1, v2 \rangle$ and $\langle v1, v2 \rangle$ are allowed tuples. Therefore $v1$ belongs to the super domain of X and $v1$ and $r1$ belong to the repair domain of X. Thus, the super domain of X is not empty and the repair domain of X is not singleton. Therefore, P is super AC.

(6) **AC($P \times P$) $\nLeftarrow$ super AC(P):** See counter-example in Figure . $\square$

4.2   Super Search Algorithms

We now present two new search algorithms: MAC+ and super MAC.

MAC+. This algorithm establishes AC+ at each node. That is, it maintains AC and backtracks if a domain wipes out *or* becomes singleton. In the MAC algorithm, we only prune future variables, since the values assigned to past variables are guaranteed to have a support in each future variable. Here, this

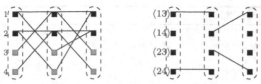

Fig. 3. The first graph shows the microstructure of a simple CSP, three variables and four values each, *allowed* combinations are linked. P is super AC since the black values have each one "black" support, and another "gray" or "black" for every neighbor, and all gray values have one "black" support. The super domains thus contain "black" values (size 2), and repair domains contain "black" and "gray" values (size 4). The second graph shows $P \times P$, which is not AC

also holds, but the condition on the size of the domains may be violated for an assigned variable because of an assignment in the future. Therefore, AC is first established on the whole network, and not only on the future variables. Second, variables are not assigned in a regular way (e.g. by reducing their domains to the chosen value) but one value is marked as super value, that is added to the current partial solution, and unassigned values are kept in the domain as potential repairs. The algorithm can be informally described as follows:

- Choose a variable X.
- Mark a value $v \in D(X)$ as assigned, but keep the unassigned values.
- For all $Y \neq X$, backtrack if Y has less than two supports for v.
- Revise the constraints as the MAC algorithm, and backtrack if the size of any domain falls bellow 2.

Super MAC. We give the pseudo code of super MAC in Figure . The algorithm is very similar to the MAC algorithm. Most of the differences are grouped in the procedure `revise-Dom`. The super domains ($\mathcal{SD}$) and repair domains ($\mathcal{RD}$) are both equal to the original domains for the first call. The values are pruned by maintaining super AC (`revise-dom`, loop 1). The algorithm backtracks if a super domain wipes out or a repair domain becomes singleton (line 2). Note that, as for MAC+, super AC is also maintained on the domains of the assigned variables (`super AC`, loop 1).

We have established an ordering relation on the different filterings. However, for the two algorithms above, *assigning* a value to a variable in the current solution does not give the same subproblem as in a regular algorithm. For a regular backtrack algorithm, the domains of assigned variables are reduced to the chosen value, whilst unassigned values are still in their domain for the algorithms above. We have proved that a problem P is AC+ iff $P+P$ is AC. However, consider the subproblem P' induced by the *assignment* of X by MAC+. P' may have more than one value in the domain of X, whereas the corresponding assignment in $P+P$ leaves only one value in the domain of X (see Figure). Therefore the ordering on the consistencies does not lift immediately to an ordering on the number of backtracks of the algorithms themselves. However, MAC($P \times P$) always backtracks when one of the other algorithms does, whilst MAC+ never

---

Algorithm 1: super MAC

    Data   : CSP: $P = \{\mathcal{X}, \mathcal{SD}, \mathcal{RD}, \mathcal{C}\}$, solution: $S = \emptyset$, variables: $\mathcal{V} = \mathcal{X}$

    Result   : Boolean // $\exists S$ a $(1,0)$-super solution

    if $\mathcal{V} = \emptyset$ then return True;

    choose $X_i \in \mathcal{V}$;

    foreach $v_i \in SD_i$ do

    | save $\mathcal{SD}$ and $\mathcal{RD}$;

    | $SD_i \leftarrow \{v_i\}$;

    | if super AC$(P, \{X_i\})$ then

    | | if super MAC$(P, S \cup \{v_i\}, \mathcal{V} - \{X_i\})$ then return True;

    | restore $\mathcal{SD}$ and $\mathcal{RD}$;

    return False;

Algorithm 2: super AC

    Data   : CSP: $P = \{\mathcal{X}, \mathcal{SD}, \mathcal{RD}, \mathcal{C}\}$, Stack: $\{X_i\}$

    Result   : Boolean // P is super arc consistent

    while *Stack is not empty* do

    | pop X_i from Stack;

1 | foreach $C_{ij} \in \mathcal{C}$ do

    | | switch revise-Dom(SD_j, RD_j, SD_i, RD_i) do

    | | | case *not-cons*

    | | | | return False;

    | | | case *pruned*

    | | | | push X_j on Stack;

    return True;

Procedure revise-Dom(SD_j, RD_j, SD_i, RD_i) : $\{$pruned,not-cons,nop$\}$

1   foreach $v_j \in SD_j$ do

    | if $\nexists v_i \in SD_i, v_i' \in RD_i$ such that $\langle v_i, v_j \rangle \in C_{ij} \wedge \langle v_i', v_j \rangle \in C_{ij} \wedge v_i \neq v_i'$ then

    | | $SD_j \leftarrow SD_j - \{v_j\}$;

    foreach $v_j \in RD_j$ do

    | if $\nexists v_i \in SD_i$ such that $\langle v_i, v_j \rangle \in C_{ij}$ then

    | | $RD_j \leftarrow RD_j - \{v_j\}$;

    if *at least one value has been pruned* then return pruned;

2   if $|SD_j| = 0 \vee |RD_j| < 2$ then return not-cons;

    return nop;

Fig. 4. super MAC algorithm

backtracks unless all the other algorithms do. Therefore any solution found by $MAC(P \times P)$ will eventually be found by the others, and MAC+ will only find solutions found by one of the other algorithms. We prove that MAC+ is correct and $MAC(P \times P)$ is complete. Hence all four algorithms are correct and complete.

Theorem 3. *For any given CSP P, the sets of solutions of MAC+(P), of super MAC(P), of MAC(P×P), and of MAC(P+P) are identical and equal to the super solutions of P.*

Proof. MAC+ is correct: suppose that S is not a super solution, then there exists a variable X assigned to v in S, such that $\forall w \in D(X), v \neq w, w$ cannot replace v in S. Therefore when all the variables are assigned, and so there remain in the domains only the values that are AC, $D(X) = \{v\}$ and thus S is not returned by MAC+.

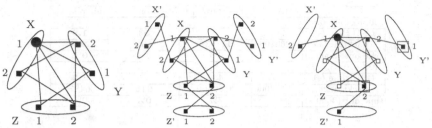

Fig. 5. Left: A CSP P, P is still AC+ after assigning X to 1. Middle: $P+P$, each variable has a duplicate which must be different, the constraints linking those variables are not represented here, the constraints on X' are exactly the same as the ones on X. Right: When the same assignment, $X = 1$ is done in $P+P$, we have the following propagation $X' \neq 1 \rightarrow Y \neq 1 \wedge Z \neq 1 \rightarrow Y' \neq 2 \wedge Z' \neq 2$. Now consider $(Y' : 1)$ and $(Z : 2)$. They are not allowed, and the network is no longer AC

$\mathrm{MAC}(P \times P)$ is complete: let S be a super solution, for any variables X, Y, let $v1$ be the value assigned to X in S, and $r1$ one of its possible repairs. Similarly $v2$ is the value assigned to Y and $r2$ its repair. It is easy to see that the pairs $\langle v1, r1 \rangle$ and $\langle v2, r2 \rangle$ are super arc-consistent, i.e, $\langle v1, v2 \rangle$, $\langle v1, r2 \rangle$ and $\langle r1, v2 \rangle$ are allowed tuples. $\qquad \square$

5   Extensions

Finding the Most Robust Solutions. Often super solutions do not exist. First, from a theoretical perspective, the existence of a backbone variable guarantees that super solutions cannot exist. Second, from an experimental perspective (see next section), it is quite rare to have super solutions in which *all* variables can be repaired. To cure both problems, we propose finding the "most robust" solution that is as close as possible to a super solution.

For a given solution S, a variable is said to be *repairable* iff there exists at least a value v in its domain different from the one assigned in S, and v is compatible with all other values in S. The most robust solution is a solution where the number of repairable variables is maximal. Such a robust solution is guaranteed to exist if the problem is satisfiable. In the worst case, none of the variables are repairable. We hope, of course, to find some of the variables are repairable. For example, our experiments show that satisfiable instances at the phase transition and beyond have a core of roughly $n/5$ repairable variables.

To find the most robust solutions, we propose a branch and bound algorithm. The algorithm implemented is very similar to MAC+ (see   ), where AC is established on the non-assigned as well as on the assigned variables. The current lower bound computed by the algorithm is the number of singleton domains. The initial upper bound is n. Indeed, each singleton domain corresponds to an *un-repairable* variable, since no other value is consistent with the rest of the solution. The rest of the algorithm is a typical branch and bound procedure. The first solution (or the proof of unsatisfiability) needs exactly the same time as

	MAC+	MAC on $P+P$	MAC on $P \times P$	super MAC
$\langle 50, 15, 0.08, 0.5 \rangle$				
CPU time (s)	788	43	53	**1.8**
backtracks	152601000	111836	**192**	2047
time out (3000 s)	12%	**0%**	**0%**	**0%**
$\langle 100, 6, 0.05, 0.27 \rangle *$				
CPU time (s)	2257	430	3.5	**1.2**
backtracks	173134000	3786860	**619**	6487
time out (3000 s)	66%	7%	**0%**	**0%**

Fig. 6. Results at the phase transition. (∗ only 50 instances of this class were given to MAC+)

the underlying MAC algorithm. Afterward, it will continue branching and discovering more robust solutions. It can therefore be considered as an *incremental anytime algorithm*. We refer to this algorithm as super Branch & Bound.

Optimization Problems. For optimization problems, the optimal solution may not be a super solution. We can look for either the most repairable optimal solution or the super solution with the best value for the objective function. More generally, an optimization problem then becomes a *multi-criterion* optimization problem, where we are optimizing the number of repairable variables and the objective function.

6   Experimental Results

We use both random binary CSPs and job shop scheduling problems. Random CSP instances are generated using the 4 parameters $\langle n, m, p_1, p_2 \rangle$ of Bessière's generator [ ], where n is the number of variables, m is the domain size, p_1 is the constraint density, and p_2 is the constraint tightness. The job shop scheduling problem consist of n jobs and m machines. Each job is a sequence of activities where each activity has a duration and a machine. The problem is satisfiable iff it is possible to schedule the activities such that their order is respected and no machine is required by two activities that overlap, within a given makespan mk. Instances were generated with the generator of Watson *et al.* [ ]. We define an instance with five parameters $\langle j, m, d_{min}, d_{max}, mk \rangle$ where j is the number of jobs, m the number of machine, d_{min} the minimum duration of an activity, d_{max} the maximum duration and mk the makespan. The actual duration of any activity is a random number between d_{min} and d_{max}.

Comparison. We compared the different solution methods using two samples of 100 random instances of the classes $\langle 50, 15, 0.08, 0.5 \rangle$ and $\langle 100, 6, 0.05, 0.27 \rangle$ at the phase transition. We observe, in Figure , that MAC on $P \times P$ prunes most, but is not practical when the domain size is large. As the problem size increases, super MAC outperforms all other algorithms in terms of runtimes.

Constrainedness and Hardness. We locate the phase transition of finding super solutions both experimentally and by an approximation based on the $\kappa appa$ framework [ ].

Empirical Approach. We fixed $n = 40$ and $m = 10$ and we varied p_1 from 0.1 to 0.9 by steps of 0.02 and p_2 from 0.1 to 0.32 by steps of 0.012. For every combination of density/tightness, a sample of 100 instances were generated and solved by MAC and super MAC, with *dom/deg* as a variable ordering heuristic for MAC and $(super\_dom)/deg$ for super MAC. The number of visited nodes are plotted in Figure   (a) for MAC and in Figure   (b) for super MAC. As expected, the phase transition for super MAC happens earlier. Also, the phase transition peak is much higher (two orders of magnitude) for super CSP than for CSP.

Probabilistic Approximation Approach. For a CSP $P = \{\mathcal{X}, \mathcal{D}, \mathcal{C}\}$ the expected number of solutions $\langle Sol \rangle$ is:

$$\langle Sol \rangle = (\prod_{x \in \mathcal{X}} m_x) \times (\prod_{c \in \mathcal{C}} (1 - p_c))$$

Where m_x is the domain size of x and p_c the tightness of the constraint c. A phase transition typically occurs around $\langle Sol \rangle = 1$ [ ]. In our case, domain sizes and constraint tightness are uniform, therefore the formula can be simplified as follows:

$$1 = m^n \times (1 - p_2)^{(\frac{n(n-1)}{2} p_1)}$$

We assume that the $P \times P$ reformulation behaves like a random CSP. Note that $P \times P$ has one solution iff P has a single super solution. We can derive the values m' and $(1 - p_2')$ of the CSP $P \times P$:

$$m' =_{def} (m^2 - m)$$

Moreover, we can see $(1 - p_2)$ as the probability that a given tuple of values on a pair of constrained variables satisfies the constraint. For a given *pair* $\langle \langle v_1, v_2 \rangle, \langle w_1, w_2 \rangle \rangle$, this pair satisfies the reformulated constraint iff $\langle v_1, w_1 \rangle \in c$ and $\langle v_1, w_2 \rangle \in c$ and $\langle v_2, w_1 \rangle \in c$, which has a probability of $(1 - p_2)^3$. Hence, we have:

$$(1 - p_2') = (1 - p_2)^3$$

The formula thus becomes:

$$1 = (m^2 - m)^n \times (1 - p_2)^{3(\frac{n(n-1)}{2} p_1)}$$

In Figure   (c), we plotted those equations along with the values gathered from the empirical study. To do so, we considered, for every p_1, the minimal value p_2 such that the sample $\langle n, m, p_1, p_2 \rangle$ has more than half of its instances unsatisfiable. We observe that our approximations are very close to the empirical findings.

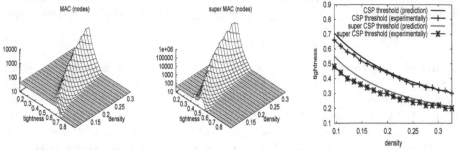

(a) Number of nodes visited (b) Number of nodes visited (c) Phase Transition Expec-
by MAC   by super MAC   tation

Fig. 7. Respective hardness to find a solution or a super solution

Job Shop Scheduling. We formulate the jobshop scheduling problem as a CSP,
with one variable for each activity, and a domain size equal to the makespan mk
minus its duration. We wish to minimize the makespan. We do so by iteratively
increasing the makespan (mk) and solving the resulting decision problem. When
a solution is found, we stop. We solved a sample of 50 problem instances for
$j, m \in \{3, 4, 5\}$, $d_{min} = 2$ and $d_{max} \in \{10, 20, 30\}$. Each sample was solved with
MAC, super MAC, and with super Branch & Bound. MAC and super Branch
& Bound stopped at the same value of mk for which the problem is satisfiable.
Whilst super MAC continued until the problem has a $(1, 0)$-super solution and
then we optimize the makespan.

- Makespan: In Figure   (a) we plot the makespan of the optimal solution
  found by MAC, the makespan of the optimal $(1,0)$-super solution returned
  by super MAC, and the worst possible makespan in case of a break to the
  optimal $(1,0)$-super solution. We observe that we have to sacrifice optimality
  to achieve robustness. Nevertheless, the increase in the makespan appears to
  be independent of the problem size and is almost constant.
- Search Effort: In Figure   (b), we plot the time needed by MAC to find the
  optimal solution, by super MAC to find the optimal $(1,0)$-super solution, and
  by super Branch & Bound to find the most robust solution with the optimal
  makespan as found by MAC. As expected, more search is needed to find
  more robust solutions. Super MAC is on average orders of magnitude worse
  than MAC. Super Branch & Bound requires little effort for small instances,
  but much more effort when the problem size increases.
- Repairability: In Figure   (c), we compare the percentage of repairable variable for the optimal solution found by MAC and the most robust optimal
  solution returned by super Branch & Bound. The optimal solutions returned

---

[1] In Figure   , for every sample $m \times m$ the first three histograms stand for $d_{max} = 10$,
the three following for $d_{max} = 20$, etc.

by MAC have on average 33% of repairable variables, whilst the most repairable optimal solutions found by super Branch & Bound have 58% on average.

Minimal Core of Repairable Variables. We generated two sets of 50,000 random CSPs with 100 variables, 10 values per variable, 250 constraints forbidding respectively 56 and 57 tuples. Those problems are close to the phase transition, which is situated around 54 or 55 disallowed tuples, and at 58, no satisfiable instances were found among the 50,000 generated. The satisfiable instances (a total of 2407 for the first set, and 155 for the second) have in average 22% and 19% of repairable variables, respectively. The worst cases being 9% and 11%, respectively.

7   Related Work

Supermodels [ ] and fault tolerant solutions [ ] have been discussed earlier.

The notions of neighborhood interchangeability [ ] and substitutability are closely related to our work, but whereas, for a given problem, interchangeability is a property of the values and works for all solutions, repairability is a property of the values in a given solution.

Uncertainty and robustness have been incorporated into constraint solving in many different ways. Some have considered robustness as a property of the algorithm, whilst others as a property of the solution (see, for example, dynamic CSPs [ ] [ ] [ ], partial CSPs [ ], dynamic and partial CSPs [ ], stochastic CSPs [ ], and branching CSPs [ ]). In dynamic CSPs, for instance, we can reuse previous work in finding solutions, though there is nothing special or necessarily robust about the solutions returned. In branching and stochastic CSPs, on the other hand, we find solutions which are robust to the possible changes. However both these frameworks assume significant information about the likely changes.

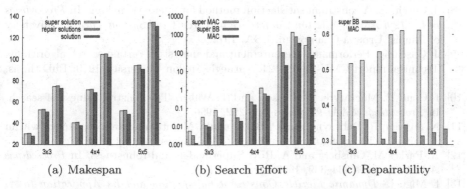

(a) Makespan   (b) Search Effort   (c) Repairability

Fig. 8. super solutions for the jobshop scheduling problem

8   Conclusion

To improve solution robustness in CP, we introduced the notion of super solutions. We explored reformulation and search methods to finding (1,0)-super solutions. We introduced the notion of super consistency, and develop a search algorithm, super MAC based upon it. Super MAC outperformed the other methods studied here. We also proposed super Branch & Bound, an optimization algorithm which finds the most robust solution that is as close as possible to a (1,0)-super solution. Finally, we extended our approach to deal with optimization problems as well.

The problem of seeking super solutions becomes harder when multiples repairs are allowed, i.e, for (1, b)-super solutions. We aim to generalize the idea of super consistency to (1, b)-super solutions. In a similar direction, we would like to explore tractable classes of (1, b)-super CSPs. Furthermore, as with dynamic CSPs, we wish to consider the loss of n-ary no-goods and not just unary no-goods.

References

[1] R. Dechter D. Frost, C Bessière and J. C. Régin. Random uniform CSP generator. url: http://www.ics.uci.edu/~dfrost/csp/generator.html, 1996.

[2] A. Dechter and R. Dechter. Belief maintenance in dynamic constraint networks. In *Proceedings AAAI-88*, pages 37–42, 1988.

[3] E. Hebrard B. Hnich and T. Walsh. Super CSPs. Technical Report APES-66-2003, APES Research Group, 2003.

[4] D. W. Fowler and K. N. Brown. Branching constraint satisfaction problems for solutions robust under likely changes. In *Proceedings CP-00*, pages 500–504, 2000.

[5] E. C. Freuder. A sufficient condition for backtrack-bounded search. *Journal of the ACM*, 32:755–761, 1985.

[6] E. C. Freuder. Partial Constraint Satisfaction. In *Proceedings IJCAI-89*, pages 278–283, 1989.

[7] E. C. Freuder. Eliminating Interchangeable Values in Constraint Satisfaction Problems. In *Proceedings AAAI-91*, pages 227–233, 1991.

[8] J. Gaschnig. A constraint satisfaction method for inference making. In *Proceedings of the 12th Annual Allerton Conference on Circuit and System Theory*. University of Illinois, Urbana-Champaign, USA, 1974.

[9] J. Gaschnig. Performance measurement and analysis of certain search algorithms. Technical report CMU-CS-79-124, Carnegie-Mellon University, 1979. PhD thesis.

[10] I. Gent, E. MacIntyre, P. Prosser, and T. Walsh. The constrainedness of search. In *Proceedings AAAI-96*, pages 246–252, 1996.

[11] N. Jussien, R. Debruyne, and P. Boizumault. Maintaining arc-consistency within dynamic backtracking. In *Proceedings CP-00*, pages 249–261, 2000.

[12] A. Parkes M. Ginsberg and A. Roy. Supermodels and robustness. In *Proceedings AAAI-98*, pages 334–339, 1998.

[13] I. Miguel. *Dynamic Flexible Constraint Satisfaction and Its Application to AI Planning*. PhD thesis, University of Edinburgh, 2001.

[14] T. Schiex and G. Verfaillie. Nogood recording for static and dynamic constraint satisfaction problem. *IJAIT*, 3(2):187–207, 1994.
[15] T. Walsh. SAT v CSP. In *Proceedings CP-2000*, pages 441–456, 2000.
[16] T. Walsh. Stochastic constraint programming. In *Proceedings ECAI-02*, 2002.
[17] J. P. Watson, L. Barbulescu, A. E. Howe, and L. D. Whitley. Algorithms performance and problem structure for flow-shop scheduling. In *Proceedings of the Sixteenth National Conference on Artificial Intelligence (AAAI-99)*, pages 688–695, 1999.
[18] R. Weigel and C. Bliek. On reformulation of constraint satisfaction problems. In *Proceedings ECAI-98*, pages 254–258, 1998.   ,   ,   ,

Local Probing Applied to Network Routing

Olli Kamarainen[1] and Hani El Sakkout[2]

[1] IC-Parc
Imperial College London, London SW7 2AZ, UK
ok1@icparc.ic.ac.uk
[2] Parc Technologies Ltd
Tower Building, 11 York Road, London SE1 7NX, UK
hani@parc-technologies.com

Abstract. Local probing is a framework that integrates (a) local search into (b) backtrack search enhanced with local consistency techniques, by means of probe backtrack search hybridization. Previously, local probing was shown effective at solving generic resource constrained scheduling problems. In this paper, local probing is used to solve a network routing application, where the goal is to route traffic demands over a communication network. The aim of this paper is (1) to demonstrate the wider applicability of local probing, and (2) to explore the impact of certain local probing configuration decisions in more detail. This is accomplished by means of an experimental evaluation on realistic networking scenarios that vary greatly in their characteristics. This paper yields a better understanding of local probing as well as a versatile local probing algorithm for network routing.

1 Introduction

1.1 Local Probing

Due to its systematic nature and support of constraint propagation, *backtrack search enhanced with local consistency techniques* (BT+CS) is effective at solving tightly-constrained problems with complex constraints. On the other hand, the quality of *local search*'s (LS) total assignment is more easily measurable, by comparison with the quality of conventional BT+CS's partial assignments. Also, the absence of systematicity allows LS's assignments to be modified in any order, and so early search moves do not necessarily skew search to focus only on particular sub-spaces. This leads to LS's superiority at optimizing loosely constrained problems. However, while BT+CS algorithms are usually **sat-complete** (i.e. in a finite number of steps, it either generates a solution, or proves that no solutions exist) and can be made **opt-complete** (i.e. in a finite number of steps, it either generates an *optimal* solution, or proves that no solutions exist), LS algorithms are incomplete in both senses.

   Local probing [ ] is a sat-complete *probe backtrack search* framework (PBT, [ ]) that executes a slave LS procedure at the nodes of the master BT+CS search tree. Local probing is designed to tackle practical constraint satisfaction problems (CSPs) and constraint satisfaction and optimization problems (CSOPs) that are difficult solve by pure BT+CS or pure LS algorithms. Its strength is derived from the combination of LS's non-systematic search characteristic with BT+CS's search systematicity. This enables local probing to satisfy complex constraints and prove infeasibility, while achieving good optimization performance, as well as (in some configurations) prove optimality.

In local probing, the constraints of the CSP or the CSOP to be solved must be divided into two sets, namely 'easy' constraints and 'hard' constraints. **The slave LS procedure** — the LS prober — solves sub-problems containing only the 'easy' constraints, at the nodes of a master BT+CS search tree. **The master BT+CS procedure** eliminates possible violations of 'hard' constraints by incrementally posting and backtracking additional 'easy' constraints.

The LS prober algorithm will return a solution to the 'easy' sub-problem. If the LS prober's solution happens to be feasible with 'hard' constraints also, a feasible solution to the entire problem is found. If the LS solution for the 'easy' sub-problem violates any 'hard' constraints, one of them must be selected for repair: A new 'easy' constraint, *forcing the LS prober to avoid returning solutions that violate the constraint in the same way*, is posted to the 'easy' sub-problem. If this leads to a failure (i.e. the new 'easy' sub-problem is unsatisfiable), the negation of the selected 'easy' constraint is imposed instead. Then, the new 'easy' sub-problem is solved with LS, and in the case of further 'hard' constraint violations, other 'easy' constraints are recursively imposed. This continues until a solution is found, or until all possibilities of posting 'easy' constraints are explored (i.e. a proof that no solutions exist is obtained). In the case of a CSOP, the search can be continued again by using a cost bound like in other branch and bound (B&B) methods.

A sat-complete local probing algorithm must satisfy the following conditions:

1. (a) Any 'hard' constraint must be expressible as a disjunction of *sets of 'easy' constraints* that apply to its variables $\langle ES_1, ES_2, \cdots, ES_k \rangle$, such that every solution that satisfies one of the 'easy' constraint sets is guaranteed to also satisfy the 'hard' constraint.
   (b) No solution exists that satisfies the 'hard' constraint, but does not satisfy any of the 'easy' constraint sets in this disjunction.
   For efficiency only, the next condition is also useful:
   (c) No solution to an 'easy' constraint set ES_i in the disjunction is also a solution to another 'easy' constraint set ES_j in the disjunction.
2. If a 'hard' constraint is violated, it must be detected and scheduled for repair by the master BT+CS procedure within a finite number of steps.

---

[1] This decomposition should not be confused with hard/soft constraints.

[2] We are theoretically guaranteed to find at least one partition satisfying 1 for any finite CS(O)P (define 'easy' constraints to be variable assignments or their negations).

3. The master BT+CS procedure must be capable of (eventually) unfolding all possible sets of 'easy' constraints whose satisfiability would guarantee that the 'hard' constraint is also satisfied.
4. It must systematically post 'easy' constraints (and on backtracking, their negations) to the LS prober until either the 'hard' constraint is satisfied, or it is proved impossible to satisfy.
5. The neighbourhood operator of the LS prober must guarantee to satisfy the posted 'easy' constraints if this is possible (and indicate infeasibility if it is not), by being sat-complete w.r.t. the 'easy' constraints.

Additionally, *opt-completeness* can be achieved if (a) the cost bound constraint generated by a B&B process at the master BT+CS level can be captured as an 'easy' constraint that is satisfied by the LS prober; or (b) the cost bound constraint can be dealt with at the master BT+CS level as a 'hard' constraint, ensuring the cost bound will be satisfied by the search if that is possible (which, in fact, is the case in the local probing algorithm detailed in this paper).

Related Work. Local probing belongs to the LS/BT+CS hybridization class where LS is performed in (all or some) search nodes of a BT+CS tree. In most of these hybrids, the task of the LS procedure is to support somehow the master BT+CS. They can be classified further as follows: **A.** Improving BT+CS partial assignments using LS, e.g. [,]. **B.** Enhancing pruning and filtering by LS at BT+CS tree nodes, e.g. [,]. **C.** Selecting variables by LS in a master BT+CS, e.g. [,]. **D.** Directing LS by a master BT+CS, e.g. [ ] . Local probing has aspects of B, C and D, although D is the key classification, since it is LS that creates assignments, and the task of BT+CS is to modify the sub-problem that LS is solving, directing LS to search regions where good solutions can be found.

In addition to LS/BT+CS hybrids, local probing belongs to the family of PBT hybrid algorithms [, , ,] (related ideas also in [ ]). Typically, they use LP or MIP, instead of LS, as the prober method to be hybridized with BT+CS.

1.2   Objectives of the Paper

In earlier work [ ], we demonstrated that local probing can be effective at solving a generic scheduling problem. We showed how the local probing hybridization framework can successfully marry the optimization strengths of LS and the constraint satisfaction capabilities of BT+CS, and we learned how it is possible to construct an effective sat-complete local probing hybrid that performs well

---

[3] The local probing algorithm presented in [ ] was not opt-complete.

[4] Other LS/BT+CS hybridizations either (a) perform BT+CS and LS serially as "loosely connected", relatively independent algorithms (dozens of published hybrids); or (b) use BT+CS in the neighbourhood operator of LS (including operators of GAs), e.g. [,]. An important sub-class of (b) is "shuffling" hybrids, e.g. [,].

[5] The results presented in [ ] are particularly promising to local probing research.

when compared to alternative algorithms. However, several questions remained unanswered.

Firstly, how readily applicable is local probing in different application domains? In particular, is it possible to build efficient sat-complete local probing hybrids for other applications?

Secondly, problem constrainedness can vary greatly. Can local probing be easily configured to trade-off satisfaction performance against optimization performance by changing the balance of effort between BT+CS and LS? Also, for problem instances with similar levels of constrainedness, it is likely that certain local probing configurations are more effective than others. However, could a single configuration remain competitive over problem instances that vary greatly in constrainedness, or are adaptive configuration mechanisms necessary?

The main objective of this paper is to address these questions by means of a detailed investigation of local probing performance on the selected network routing problem. In the process of doing so, we will establish a new and versatile family of algorithms based on local probing for solving this commercially important problem.

Next, we introduce the application domain. Section   details the algorithm to be used in the investigation study of Section . Section   concludes the paper.

1.3   Application Domain

The network routing problem solved here (NR) involves constraints including capacity, propagation delay, and required demands. In the NR, we have:

- a network containing a set of nodes N and a set of directed links E between them — each link (i,j) from a node i to a node j has a limited bandwidth capacity c_{ij} and a propagation delay d_{ij} that is experienced by traffic passing through it;
- a set of demands K, where each $k \in K$ is defined by:
  - a source node s_k, i.e. where the data is introduced into the network;
  - a destination node t_k, i.e. where the data is required to arrive;
  - a bandwidth q_k, i.e. the bandwidth that must be reserved from every link the demand is routed over; and
  - a maximum propagation delay d_k, i.e. the maximum source-to-destination delay that is allowed for the demand k;
- a subset of the demands $RK \subseteq K$ specifying which demands *must* have paths in any solution.

The aim is to assign paths over the network to demands such that:

1. the total unplaced bandwidth, i.e. the sum of the bandwidth requirements of the demands *not assigned to a path*, is minimized;
2. the following constraints are satisfied:
   - If a demand k belongs to the set of *required demands* RK, it must have a path.

- Every link in the network (i, j) has sufficient bandwidth to carry the demands that traverse it.
- For any path P_k assigned to a demand $k \in K$, the sum of the propagation delays of the links in the path does not exceed the maximum propagation delay d_k for k.

Note that, when the set of required demands is exactly the set of all the problem demands, i.e. $RK = K$, the problem is a pure CSP without any optimization dimension: each demand must have a feasible path. On the other hand, if RK is empty, it is always possible to find a trivial solution, because *even an empty set of paths is a solution*.

For each demand k, we use a boolean y_k to denote whether k is routed ($y_k = 1$) or not ($y_k = 0$). Another set of booleans x_{ijk} indicate whether a demand k is routed through a link (i, j) ($x_{ijk} = 1$), or not ($x_{ijk} = 0$). Next, we formally state the problem.

$$\min \sum_{k \in K} (1 - y_k) q_k, \tag{1}$$

s.t.

$$\forall k \in RK: \quad y_k = 1 \tag{2}$$
$$\forall k \in K \setminus RK: \quad y_k \in \{0, 1\} \tag{3}$$
$$\forall (i, j) \in E, \forall k \in K: \quad x_{ijk} \in \{0, 1\} \tag{4}$$

$\forall k \in K:$

$$y_k = 1 \Leftrightarrow \tag{5}$$
$$P_k = \langle (n_{k_1}, n_{k_2}), (n_{k_2}, n_{k_3}), \ldots, (n_{k_{l-1}}, n_{k_l}) \rangle,$$
$$\text{where}:$$
$$n_{k_1} = s_k$$
$$n_{k_l} = t_k$$
$$\forall (n_{k_p}, n_{k_q}) \text{ in } P_k: \quad (n_{k_p}, n_{k_q}) \in E$$
$$\text{alldifferent}\{n_{k_1}, n_{k_2}, \ldots, n_{k_l}\}$$
$$y_k = 0 \Leftrightarrow P_k = \langle \ \rangle \tag{6}$$

$\forall (i, j) \in E, \forall k \in K:$

$$x_{ijk} = 1 \quad \Leftrightarrow (i, j) \text{ in } P_k, \tag{7}$$
$$x_{ijk} = 0 \quad \Leftrightarrow (i, j) \text{ not in } P_k, \tag{8}$$
$$\forall (i, j) \in E: \quad \sum_{k \in K} q_k x_{ijk} \leq c_{ij} \tag{9}$$
$$\forall k \in K: \quad \sum_{(i,j) \in E} d_{ij} x_{ijk} \leq d_k \tag{10}$$

The objective function ( ) to be minimized is the sum of the bandwidths of the non-routed demands (i.e. those demands k with $y_k = 0$). The constraints of

( ) force demands in the set of required demands RK to have a path, and the remaining demands can either have a path or not ( ). The path constraints ( ) and ( ) force, for any placed demand k, its path P_k to be a connected loop-free sequence of links from the source node s_k to the destination node t_k, and for any non-routed demand, its path to be empty. The constraints in ( ) and ( ) tie the link-demand booleans x_{ijk} with the paths. The capacity constraints in ( ) ensure that for each link (i, j) the bandwidth reserved for demands passing through it does not exceed its capacity c_{ij}. The delay constraints ( ) restrict for each demand k the total delay of the links in its path to be no more than its maximum propagation delay d_k.

2   Algorithm Description

The problem decomposition and the master BT+CS of local probing are introduced first. Then the LS prober and its neighbourhood operator are described.

2.1   Problem Constraint Decomposition into 'Easy' and 'Hard'

The NR is a typical large-scale combinatorial optimization problem (LSCO) in that it can be divided into different sub-problems. The constraints of the problem are divided into the 'easy' and 'hard' constraints here as follows:

- 'Hard' constraints (not guaranteed to be satisfied by the prober):
    - Capacity constraints: the sum of the bandwidth requirements of the demands routed via a link must not exceed the *bandwidth of the link* ( )
- 'Easy' constraints (guaranteed to be satisfied by the prober):
    - The sum of the propagation delays of the links in a path is less than or equal to the *maximum propagation delay* of the demand ( )
    - 'Easy' constraints that the master BT+CS procedure can impose during search (but might also exist in the original problem):
        * If the *demand k is dropped* $(y_k = 0)$, it cannot have a path ( ).
        * If the *demand k is required* $(y_k = 1)$, it must have a valid path ( ).
        * If a *link is forbidden* $(x_{ijk} = 0)$ from a demand, its path is not allowed to contain the forbidden link ( ).
        * If a *link is forced* $(x_{ijk} = 1)$ for a demand, its path must contain the forced link ( ).

2.2   Master BT+CS Level

The last four of the constraints above are used by the master BT+CS procedure to repair violations on 'hard' constraints, i.e. link capacity constraints, in the following way. Assume that the prober returns a probe (a set of paths for all non-dropped demands), and the master BT+CS procedure detects that the bandwidth usage on link (i, j) exceeds its capacity, and then selects this link to be repaired (see Fig.). 1. The algorithm picks one demand k traversing (i, j),

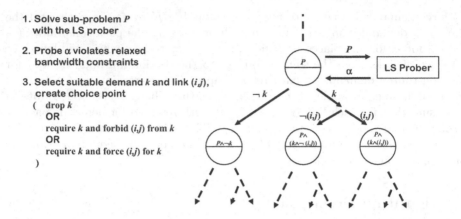

Fig. 1. Illustration of local probing for NR

and *drops* it, i.e. posts a constraint $y_k = 0$. This first search branch does not subsequently allow a path for the demand k. **2.** If dropping k does not lead to a solution in the subsequent nodes, a constraint that requires that demand k has a path, i.e. $y_k = 1$, and a constraint that forbids the link (i, j) from the demand k, i.e. $x_{ijk} = 0$, are posted to the sub-problem, and local probing continues having these constraints present in all the search nodes within this second branch. **3.** If forbidding the link (i, j) from the demand k does not lead to a feasible solution, the third choice is to require k and force it to use the link (i, j), i.e. $x_{ijk} = 1$, at the child nodes in the third branch. If this decision leads to failure as well, the search backtracks to higher choice points in the tree. These three choices:

1. drop demand k,
2. require demand k but forbid it from using link (i, j), and
3. require demand k and force it to use link (i, j),

fully partition the search space of the sub-problem at the master BT+CS tree node. This decomposition allows us to build a local probing algorithm that satisfies the conditions for sat-completeness and also for opt-completeness. When a solution is found, a B&B cost bound CB, enforcing $\sum_{k \in K}(1 - y_k)q_k < CB$, is imposed, and local probing is continued from the root of the search tree. Due to the 'easy' constraint decisions $y_k = 0$ and $y_k = 1$ (drop/require demand) and the form of the objective function, we can immediately prune branches that are infeasible w.r.t. the cost bound. Because the algorithm can deal with the cost bound constraint as a 'hard' constraint, opt-completeness is achieved.

---

[6] The PBT decomposition above for network routing was initially suggested by Liatsos et al, and it is used in a MIP-based PBT application. A version (that does not include the first branch) of their algorithm is published in [ ], which tackles a related problem to NR (having *all* demands required and a different objective function).

Several heuristics can be used in selecting the link (i, j) to be repaired among the violated links at a BT+CS search node, as well as in selecting the demand k to be dropped/required. In this implementation, the procedure selects (1) the link (i, j) that is most violated, i.e. its bandwidth availability is exceeded most; and (2) the demand k (a) routed via (i, j) and (b) having the largest bandwidth requirement from the set of demands passing over (i, j) that are *not forced to use the link* (i, j) (i.e. the set of demands that are not treated by an imposed constraint $x_{ijk} = 1$ at the master BT+CS tree nodes). These heuristics were chosen because they tend to lead the search quickly towards a feasible solution or a failure, thus reducing the size of the master BT+CS tree. Also various consistency checks are made before the local probing search is allowed to continue in the branches (reasoning with the capacity constraints and the imposed master level decisions for dropping/requiring demands and forbidding/forcing links).

2.3 Local Search Prober

The LS prober returns a probe, which is a set of paths to all non-dropped demands. As an LS strategy, simulated annealing (SA, [ ]) is used because it is easy to implement and can avoid local minima. At each search step, the neighbourhood operator suggests for the sub-problem a candidate neighbour that satisfies all the 'easy' constraints (recall Section   ).

2.4 Neighbourhood Operator of LS Prober

The neighbourhood operator procedure applies a "shuffle" approach. First, a subset of demands to be re-routed are selected, and heuristically ordered for re-routing. Then, before evaluation, the selected demands are routed separately by utilizing Dijkstra's shortest path first algorithm. The neighbourhood operator gives us the neighbour candidate in *two sets of paths*: (1) the first set S_1 is a set of paths that *together satisfy all the constraints* - including capacity constraints, and (2) the second set S_2 is a set of paths that satisfy all the 'easy' constraints but *were left outside of S_1* by the LS heuristics because they caused capacity constraints to be violated. The procedure includes the following four steps:

1. Select Demands to be Re-routed. From the current assignment, the paths of (1) all the demands that had paths in S_2, *and* (2) a *percentage fraction P* of the demands that had paths in S_1 are chosen for re-routing. In the latter case, paths are selected in three stages, depending on how many demands the percentage fraction allows us to select: for a randomly selected bandwidth-violated link, randomly select paths from S_1 from the following sets until P paths have been selected (start with the set D_1, and move to the next set when it is empty): **1.** Set D_1: Paths traversing the selected link. **2.** Set D_2: Paths traversing any

---

[7] Note that demands can be be dropped only by the master BT+CS search decisions.
[8] Another "shuffle"-based LS algorithm for network routing is presented in [   ].
[9] The LS *initialization* includes slightly modified Steps 1 and 2, and Step 3 as it is.

of the links of the paths in D_1. This tends to free bandwidth in the vicinity of the problematic link. **3.** Set D_3: All the other paths in S_1. The *demands to be re-routed* are the demands of the selected paths.

2. Order Demands to be Re-routed. The selected demands to be re-routed are ordered for routing in three groups (the order within each group is randomized): (1) *required* demands; (2) *non-required* demands that were in S_2 in the current assignment; and (3) *non-required* demands that were in S_1 in the current assignment. This robust heuristic is used, since we prefer finding feasible solutions (routing required demands first) to optimization (minimizing unplaced total bandwidth).

3. Re-route Demands. Routing is carried out in two phases. Although capacity constraints are relaxed in the LS sub-problem, the local probing performance is dependent on the LS prober producing good quality probes. Therefore, the first routing phase tries to route demands such that the *links that do not have enough bandwidth available are not seen* by the single-demand routing procedure used. If a path is found, the bandwidth availabilities of the links are updated immediately, and *the path is placed in S_1*. If a path is not found, the routing is postponed to the second phase.

In the second phase, all the selected demands, still without a path, are routed again in the same order as in the first routing phase, but now *without taking the capacity constraints into account*, i.e. also bandwidth-infeasible links are seen by the single-demand routing procedure used. This phase is guaranteed, for each demand, to find a feasible path w.r.t. all constraints except the capacity constraints. *The set of paths generated in the second phase is S_2.*

4. Evaluate. The value of the neighbour candidate is the sum of (1) the unplaced bandwidth, which is the sum of the bandwidth requirements of the demands that *have paths in S_2*, and (2) the penalty component, which is the *number of required demands in S_2* multiplied by a large constant. According to this value and the SA strategy, the assignment may or may not be accepted as the new LS assignment. (When the termination condition is met, the LS prober returns the best set of paths found throughout this prober search (the best $S_1 \bigcup S_2$).)

Single-Demand Routing Component of Neighbourhood Operator. At Step 3 of the neighbourhood operator, a single demand is routed by using Dijkstra's *shortest path first* algorithm (SPF) that finds the shortest path between two nodes in a graph. The "length" of a path is the sum of the metric costs of the links in the path. The shortest path is the path where the sum of the edge costs is the lowest. Here, SPF is run over a network that *excludes* (1) the links that are forbidden for the demand to be routed, and (in the case the path

---

[10] This causes capacity constraint violations.

must made bandwidth-feasible) (2) the links that do not have sufficient capacity. The default metric used for the cost of a link $(i, j) \in E$ is the *propagation delay* d_{ij} of the link. Since this may leave some areas of the sub-problem search space unexplored during the LS prober, the propagation delays of the links are occasionally replaced by *random metrics* in the first routing phase of Step 3 of the neighbourhood operator (a resulting bandwidth-feasible path is immediately re-calculated with the default metric if it is infeasible with the delay constraints).

Routing demands with forced links. On its own, SPF cannot guarantee to satisfy forced link constraints. If SPF is run to get a path, and this path does not contain all the forced links, then an additional procedure is applied as follows. First, the forced links that are already in the path are identified. Then, the ordering they are in the path is used to define a "partial forced-link sequence" Seq. The remaining forced links are incrementally inserted in randomly selected positions in Seq such that the insertion is backtracked on failure (all possible positions in Seq are explored in the worst case). After each insertion of a forced link, SFP is applied to connect the disconnected path segments. If the completed path violates the delay constraint, backtracking to the previous insertion node occurs. The first delay-feasible path containing all the forced links is returned. If all extensions to Seq lead to failure, the same extension process is repeated for other possible permutations of Seq (in random order). The complexity of this procedure is exponential, even though the resulting path is not guaranteed to be the shortest possible. However, the average complexity is low due to the high probability of obtaining a valid path before the space is exhaustively searched.

3   Local Probing Configuration Investigation

The previous section showed a way of constructing a sat- and opt-complete local probing algorithm for the NR problem. In this section, we seek answers to the following questions: **Q1:** Can local probing be easily configured to trade-off satisfaction performance against optimization performance? **Q2:** What is the best computational balance between the BT+CS and LS components in terms of satisfaction and optimization behaviour? For problem instances with similar level of constrainedness, it is likely that certain local probing configurations are more effective than others. **Q3:** Could a single configuration remain competitive over problem instances that vary greatly in constrainedness, or are adaptive configuration mechanisms necessary?

3.1   Algorithm Parameters

Controlling the Computational Balance between the BT+CS and LS Components. Over all the generated instances, we compared 12 local probing algorithms with different prober termination conditions that vary the balance

---
[11] Although they would eventually be explored due to the master BT+CS decisions.

of search effort between BT+CS and LS. The maximum number of neighbour candidate evaluations in the LS prober was set to {1, 2, 3, 6, 12, 25, 50, 100, 200, 400, 800, 1600}. We denote the local probing algorithms with these termination conditions by Probe(SA1), Probe(SA2), Probe(SA3), ... , Probe(SA1600), respectively. This range of termination conditions covers a whole range of behaviours from low (small amount of effort spent optimizing the probe by LS) to high (large amount of effort spent optimizing the probe by LS). Within the timeout selected, these configurations lead to thousands LS steps (see the averages in Tables   and   ). *Note that there is a key difference between* Probe(SA1) *and all the other configurations.* In Probe(SA1) the prober restores probe consistency w.r.t. the 'easy' constraints, but does nothing else. However, all other variants additionally perform a "shuffle" on *heuristically selected demand subsets.* At high temperatures, the neighbours are always accepted. As temperature decreases SA is more selective. **Other parameters:** In the experiments, we used a timeout of 1000 seconds. The initial temperature for the SA prober is 1000000000, and after each neighbour evaluation, the temperature is reduced by multiplying it by the cooling factor 0.9.   The multiplier for the penalty term for penalizing infeasible paths for required demands is the sum of the bandwidth requirements of all the demands in the instance. The probability of using randomized metrics in the first routing phase of the LS neighbourhood operator is 0.0001, and the percentage affecting the neighbourhood size is 0.1%. The algorithms were implemented on ECL^{i}PS^e 5.5, and the test were run on Pentium IV 2GHz PCs.

3.2   Problem Instances

The experiments are carried out on two different real-world network topologies, named **Network 1** and **Network 2**, with artificially generated demand data that is based on realistic demand profiles. The Network 1 topology contains 38 nodes and 86 bi-directional links, whereas the Network 2 topology is larger: 208 nodes and 338 bi-directional links. The demands are generated randomly by respecting a realistic *bandwidth profile*; features of the network topology; and the network load factor. For each network, we apply three network load factors (affecting the average bandwidth requirement), and generate 20 different sets of demands for each factor, creating 60 demand sets. For the Network 1 instances, a demand exists between each pair of nodes, ending up with 1406 demands. The used network load factors were {0.6, 1.0, 1.4}. For the Network 2 instances, the demands are generated for fewer pairs of nodes, resulting in 182 demands. The network load factors used were {0.3, 0.4, 0.5}. The bandwidth profile used for generating relative bandwidth requirements of demands was different for the networks. For Network 2, it resulted in more combinatorial problems: most demands have a significant impact on the result, unlike in Network 1 whose results are

---

[12] A neighbour candidate is accepted if it is better than the current assignment or $p < exp^{(C_{curr} - C_{new})/T}$ where: p is a random number between 0 and 1; C_{new} is the cost of the neighbour candidate; C_{curr} is the cost of the current assignment; and T is the temperature.

Table 1. Overall satisfaction results, **Network 1**

Configuration	Av. no. of LS steps	Solution found (#)	Proved infeasible (#)	Timeout w/o solution (#)	Rank
Probe(SA1)	965.70	499	110	51	7
Probe(SA2)	1845.48	**501**	113	46	2
Probe(SA3)	2623.46	500	**116**	**44**	1
Probe(SA6)	4403.40	**501**	113	46	2
Probe(SA12)	7013.62	500	114	46	2
Probe(SA25)	10219.10	500	112	48	5
Probe(SA50)	13621.08	500	108	52	8
Probe(SA100)	16569.83	**501**	109	50	6
Probe(SA200)	18950.45	500	103	57	9
Probe(SA400)	21057.47	497	93	70	10
Probe(SA800)	22456.14	491	82	87	11
Probe(SA1600)	24451.50	480	69	111	12

Table 2. Overall satisfaction results, **Network 2**

Configuration	Av. no. of LS steps	Solution found (#)	Proved infeasible (#)	Timeout w/o solution (#)	Rank
Probe(SA1)	2176.39	479	31	150	6
Probe(SA2)	2604.19	**499**	37	124	2
Probe(SA3)	2819.54	495	**42**	**123**	1
Probe(SA6)	3183.90	494	41	125	3
Probe(SA12)	3691.76	495	36	129	4
Probe(SA25)	4378.13	489	32	139	5
Probe(SA50)	5074.26	479	27	154	7
Probe(SA100)	5525.13	457	23	180	8
Probe(SA200)	5878.84	443	22	195	9
Probe(SA400)	6512.35	424	11	225	10
Probe(SA800)	6883.67	400	7	253	11
Probe(SA1600)	6997.31	385	2	273	12

dominated by a small set of demands with very large bandwidth requirements. Finally, for each demand instance we vary the proportion of required demands. The number of required demands is the number of all the demands multiplied by a *constrainedness* factor in the set $\{0, 0.1, 0.2, \ldots 1\}$. With these 11 factors for each of the 60 instances, we end up with 660 instances for each network, and a total of 1320 problem instances.

3.3 Constraint Satisfaction Results

Overall Constraint Satisfaction Performance. First, we look at the overall constraint satisfaction characteristics over all instances. We are interested in **1.** how often a configuration found a solution; **2.** how often a configuration proved infeasibility; and **3.** how often an instance *remained unsolved*, i.e. how often the algorithm was terminated because of a *timeout before any solutions were found* (i.e. neither a solution nor a proof of infeasibility). The third measure is the most revealing, since it represents what is left after the first two. We therefore rank the algorithm configurations in terms of the number of timeouts. These are shown over the Network 1 and Network 2 instances in Tables   and .

1. Solutions Found. In terms of solutions found for Network 1 (Table   ), there is only a small variation between the best and worst configurations (21

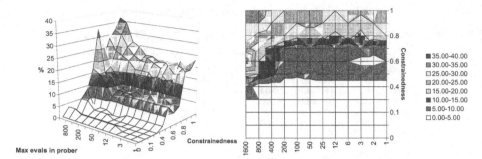

Fig. 2. Timeouts without a solution (percentage of unsolved instances), **Network 1**

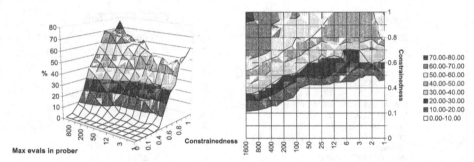

Fig. 3. Timeouts without a solution (percentage of unsolved instances), **Network 2**

instances). For Network 2 (Table   ), the variation is greater (114 instances). **2. Infeasibility proofs.** For the Network 1 instances, among the local probing algorithms, the best configuration for proving infeasibility is Probe(SA3) (116 instances). The Network 2 results indicate the same: Probe(SA3) is the best with 42 infeasibility proofs. For the Network 2 instances, Probe(SA1600) is able to prove infeasibility only in 2 cases. **3. Timeouts without a solution.** As explained above, this is the best measure for ranking constraint satisfaction performance, since it is an accurate measure of algorithm *constraint satisfaction* failure. In Network 1 cases, the best algorithm is Probe(SA3) leaving 44 instances unsolved. Probe(SA1) is seventh best, the worst configuration is Probe(SA1600). Probe(SA3) is the best configuration for constraint satisfaction also in Network 2 cases, with 123 instances unsolved. In general, the main reason for Probe(SA3)'s success in both networks, is its superiority on proving infeasibility rather than on finding solutions when there are fewer differences between configurations.

Robustness of Constraint Satisfaction Performance as Constrainedness Varies. First, we want to know the the percentage of unsolved instances, when the **proportion of required demands** vary. These are illustrated in

Table 3. Timeouts without solution over network load (number of unsolved instances)

Configuration	Network 1 load 0.6	Network 1 load 1.0	Network 1 load 1.4	Network 2 load 0.3	Network 2 load 0.4	Network 2 load 0.5
Probe(SA1)	0	12	39	7	49	94
Probe(SA2)	0	12	**34**	2	32	90
Probe(SA3)	0	10	**34**	1	36	**86**
Probe(SA6)	0	10	36	1	**30**	94
Probe(SA12)	0	10	36	0	32	97
Probe(SA25)	0	10	38	1	33	105
Probe(SA50)	0	8	44	0	48	106
Probe(SA100)	0	9	41	1	57	122
Probe(SA200)	1	11	45	2	64	129
Probe(SA400)	4	12	54	5	80	140
Probe(SA800)	8	17	62	9	100	144
Probe(SA1600)	9	24	78	20	107	146

Figs.   and   .   The results indicate that, from constraint satisfaction perspective, hybridization is useful since performance degrades at the extremes Probe(SA1) and Probe(SA1600), but that only a limited investment in LS is sufficient to obtain the best performance from a constraint satisfaction perspective (the same is not true for optimization). The figures indicate that the best performing configuration, Probe(SA3), remains effective as problem constrainedness varies.

The **network load factor** is another way of varying problem constrainedness. Table   shows the numbers of unsolved instances (out of 220) for each network load factor. In the Network 1 instances, all local probing configurations up to Probe(SA100) can find a solution or prove infeasibility for all instances that have a network load of 0.6. The number of unsolved instances increases with instances of larger network loads. The bigger the network load, the more variation there is between the configurations, and fewer LS steps per prober are required to obtain the best constraint satisfaction performance. This is true also for Network 2 instances. However, the best constraint satisfaction configuration, Probe(SA3), appears to be reasonably robust also in terms of network load.

3.4   Optimization Results

The solution quality graphs in Figs.   and   show the average unplaced bandwidth   for each constrainedness (proportion of required demands) subset over *the instances where all algorithm configurations found a solution*, to enable a fair comparison for optimization performance.   In total, there were 477 such instances for Network 1, and 372 for Network 2, each out of 660.

As expected, Probe(SA1) performs worst on optimization due to the loss of LS's optimization characteristic. In the Network 1 instances, the more LS neighbour candidate evaluations performed during a prober, the better the results.

---

[13] The rightmost graphs represent top-down views of the surfaces in the left graphs.

[14] The quality is scaled to the interval [0,1] such that 0 represents the best quality (all configurations routed all the demands), and 1 represents the worst quality.

[15] As constrainedness increases, more demands are required, and the costs of solutions fall because less bandwidth can be left unplaced.

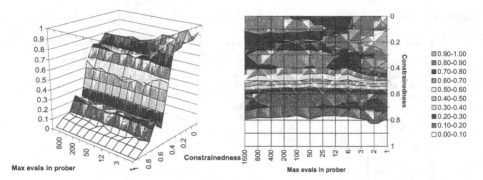

Fig. 4. Optimization, restricted subset of **Network 1**

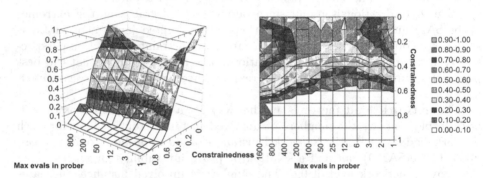

Fig. 5. Optimization, restricted subset of **Network 2**

However, in the Network 2 instances, the results start to get worse again when the prober evaluation limit increases beyond 12. This is due to the fact that, in terms of the demand bandwidth distribution profile, the Network 2 instances are more combinatorial (more demands significantly impact solution quality), and therefore the differences between local minima are greater. The Network 2 instances indicate that *local probing's ability to force local search away from local minima leads to significant improvements* in the optimization characteristics. As the interaction of LS with BT+CS decreases, by increasing the number of the LS prober steps beyond 12, optimization performance degrades.

4   Conclusion

The main objective of this paper was to address questions on the applicability and configuration of local probing, by means of a detailed investigation of local probing performance on a new application, the NR. In the process of doing so, we established a new and versatile family of algorithms based on local probing for solving this commercially important problem.

This paper presented a *sat-* and *opt-complete* local probing hybrid for solving the NR. The local probing configurations were evaluated in a detailed experimental investigation. The hybridization balance between BT+CS and LS components was investigated by controlling the configuration of the LS prober termination condition. This successfully traded-off satisfaction performance against optimization performance **(Q1)**. Relatively low LS settings (Probe(SA3) for satisfaction & Probe(SA12) for optimization) performed best overall — several thousand LS steps were sufficient per run **(Q2)**. In terms of either satisfaction only or optimization only, each configuration's relative performance remained robust over different constrainedness levels **(Q3)**. However, introducing adaptive behaviour for local probing may prove useful, because the (more BT+CS oriented) configuration that was best for pure satisfaction performance was different from the (more LS oriented) configuration that was best for pure optimization performance. Future research will investigate alternative local probing neighbourhood operators and compare performance with alternative strategies for the NR problem.

Acknowledgements

We would like to thank Vassilis Liatsos, Neil Yorke-Smith and Wided Ouaja for their helpful input to this work.

References

[1] F. Ajili and H. El Sakkout. A probe based algorithm for piecewise linear optimization in scheduling. *Annals of Operations Research*, 118:35–48, 2003.

[2] C. Beck and P. Refalo. A hybrid approach to scheduling with earliness and tardiness costs. *Annals of Operations Research*, 118:49–71, 2003.

[3] T. Benoist and E. Bourreau. Improving global constraints support by local search. In *Proc. of CoSolv'03*, 2003.

[4] Y. Caseau and F. Laburthe. Disjunctive scheduling with task intervals. Technical Report LIENS-95-25, École Normale Supérieure, Paris, France, 1995.

[5] Y. Caseau and F. Laburthe. Heuristics for large constrained vehicle routing problems. *Journal of Heuristics*, 5(3):281–303, 1999.

[6] H. El Sakkout and M. Wallace. Probe backtrack search for minimal perturbation in dynamic scheduling. *Constraints*, 5(4):359–388, 2000.

[7] F. Focacci and P. Shaw. Pruning sub-optimal search branches using local search. In *Proc. of CP-AI-OR'02*, 2002.

[8] J.N. Hooker, H.-J. Kim, and G. Ottosson. A declarative modeling framework that integrates solution methods. *Annals of Operations Research*, 104:141–161, 2001.

[9] N. Jussien and O. Lhomme. Local search with constraint propagation and conflict-based heuristics. *Artificial Intelligence*, 139:21–45, 2002.

[10] O. Kamarainen and H. El Sakkout. Local probing applied to scheduling. In *Proc. of CP'02*, pages 155–171, 2002.

---

[16] Q1, Q2 and Q3 were described in the beginning of Section .

[11] S. Kirkpatrick, C. Gelatt Jr., and M. Vecchi. Optimization by simulated annealing. *Science*, 220:671–680, 1983.

[12] V. Liatsos, S. Novello, and H. El Sakkout. A probe backtrack search algorithm for network routing. In *Proc. of CoSolv'03*, 2003.   ,

[13] S. Loudni, P. David, and P. Boizumault. On-line resources allocation for ATM networks with rerouting. In *Proc. of CP-AI-OR'03*, 2003.

[14] B. Mazure, L. Saïs, and É. Grégoire. Boosting complete techniques thanks to local search. *Annals of Mathematics and Artificial Intelligence*, 22:319–331, 1998.

[15] S. Prestwich. A hybrid search architecture applied to hard random 3-sat and low-autocorrelation binary sequences. In *Proc. of CP'00*, pages 337–352, 2000.

[16] L. Rousseau, M. Gendreau, and G. Pesant. Using constraint-based operators to solve the vehicle routing problem with time windows. *J. Heuristics*, 8:45–58, 2002.

[17] M. Sellmann and W. Harvey. Heuristic constraint propagation: Using local search for incomplete pruning and domain filtering of redundant constraints for the social golfer problem. In *Proc. of CP-AI-OR'02*, 2002.

[18] P. Shaw. Using constraint programming and local search methods to solve vehicle routing problems. In *Proc. of CP'98*, pages 417–431, 1998.

[19] M. Wallace and J. Schimpf. Finding the right hybrid algorithm - a combinatorial meta-problem. *Annals of Mathematics and Artificial Intelligence*, 34:259–269, 2002.

Dynamic Heaviest Paths in DAGs
with Arbitrary Edge Weights

Irit Katriel*

Max-Planck-Institut für Informatik
Saarbrücken, Germany
irit@mpi-sb.mpg.de

Abstract. We deal with the problem of maintaining the heaviest paths in a DAG under edge insertion and deletion. Michel and Van Hentenryck [ ] designed algorithms for this problem which work on DAGs with strictly positive edge weights. They handle edges of zero or negative weight by replacing each of them by (potentially many) edges with positive weights. In this paper we show an alternative solution, which has the same complexity and handles arbitrary edge weights without graph transformations. For the case in which all edge weights are integers, we show a second algorithm which is asymptotically faster.

1 Introduction

The *Heaviest Paths* (HP) problem is as follows. Given a Directed Acyclic Graph (DAG) $G = (V, E)$ with a weight $w(e)$ for each edge e, compute for each node $v \in V$ the weight of the heaviest path from the source of G to v, where the weight of a path is the sum of the weights of its edges. The *Dynamic Heaviest Paths* (DHP) problem is to efficiently update this information when a small change is performed on G. Here "efficient" means that the running time is proportional to the size of the portion of the graph that is affected by the change [ ].

Formally, for each $v \in V$ denote by $\hbar(v)$ the weight of the heaviest path in G from the source of G to v. Let G' be the graph obtained by performing an operation on G (such as adding or deleting an edge) and let $\hbar'(v)$ be the weight of the heaviest path to v in G'. We define $\delta = \{v | v \in V \wedge \hbar(v) \neq \hbar'(v)\}$ to be the set of nodes that were affected by the operation. That is, the nodes for whom the weight of the heaviest path changed. We define $|\delta|$ to be the number of nodes in δ and $\|\delta\|$ to be $|\delta|$ plus the number of edges that are adjacent to at least one node in δ. Assuming that any algorithm would have to do something for each node in δ and to examine each edge adjacent to a node $v \in \delta$, we use $|\delta|$ and $\|\delta\|$ as the parameters for the complexity of an update.

Michel and Van Hentenryck [ ] recently studied the DHP problem. Their research was motivated by scheduling applications, where one wishes to represent the current solution by a DAG and to perform tasks such as to evaluate the

---

* Partially supported by DFG grant SA 933/1-1.

makespan and update it upon a small change to the DAG. They present algorithms that solve the DHP problem on DAGs with strictly positive edge weights.
Their algorithms run in time $O(\|\delta\| + |\delta| \log |\delta|)$ for an edge insertion and $O(\|\delta\|)$
for an edge deletion. They show that it is possible to replace edges with zero or
negative weights by edges with positive weights and then apply their algorithm.
However, the number of edges added may be large. They conclude by asking
whether there is an efficient algorithm that can handle arbitrary edge weights
without graph transformations.

We answer their question by showing such an algorithm. In fact, their algorithm for updating the graph upon the deletion of an edge works also in the
presence of non-positive edge weights. So what we need to show is an algorithm
for edge insertion that can handle arbitrary weights. Our solution has the same
asymptotic complexity as theirs and is not more complicated to implement.

In addition, we discuss the case in which the edge weights are integers and
show an algorithm that runs in time $O(\|\delta\| + |\delta| \log \log \min\{|\delta|, \Delta_{\max}+1\})$ where
$\Delta_{\max} = \max_{v \in V} \{\hbar'(v) - \hbar(v)\}$ is the maximum change to the heaviest path value
of a node due to the edge insertion.

2 Fibonacci Heaps

Our algorithms use *Fibonacci Heaps* [ ]. A min-sorted (max-sorted) Fibonacci
Heap is a data structure that supports the following operations:

- *Insert*(i, p) inserts the item i with priority p.
- *UpdatePriority*(i, p): If i is not in the heap, performs *Insert*(i, p). If i is in the
  heap and has priority p', its priority is updated to $\min\{p, p'\}$ $(\max\{p, p'\})$.
- *ExtractMin* (*ExtractMax*) returns the item with minimal (maximal) priority
  and removes it from the heap.

Any sequence of n_i *Insert* operations, n_u *UpdatePriority* operations and n_e
ExtractMin(*ExtractMax*) operations on an initially empty Fibonacci Heap can
be performed in time $O(n_i + n_u + n_e \log n)$, where n is the maximal number of
items in the heap at any point in time. Certainly, $n \leq n_i + n_u$.

3 Intuition

Figure   shows the impact of adding the dashed edge to the DAG. The nodes in
δ are the ones inside the rectangle and the changes to their $\hbar$ values are shown.
Clearly, these changes propagate along paths in the graph. That is, if we add
an edge (x, y) then $\hbar(v)$ for a node $v \neq y$ can only change if $\hbar(u)$ changed
for a predecessor u of v. Furthermore, the changes propagate in a monotonous
manner. That is, the change to $\hbar(v)$ is not larger than the maximum of the
changes at v's predecessor.

The algorithm in [ ] would proceed from y in topological order and update
the $\hbar$ value for each node only after the updated $\hbar$ values for all of its predecessors

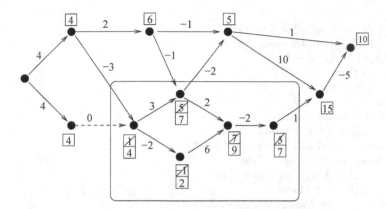

Fig. 1. The impact of inserting the dashed edge to the DAG

are known. This is possible when edge weights are strictly positive because the previous $\hbar$ values provide us with the relative topological order of the nodes. When zero or negative edge weights exist in the graph, it may be the case that $\hbar(v) \leq \hbar(u)$ while u precedes v in the topological order (see [ ], Section 6 for an example). In other words, we do not know the topological order of the nodes so we need a different method to traverse them, which would still guarantee that we do not traverse the same subgraph more than once.

The rule we use is as follows. We begin by updating $\hbar(y) \leftarrow \max\{\hbar(y), \hbar(x) + w(x, y)\}$. This is the final value for $\hbar(y)$ because it can only change due to the edge (x, y), so we label y as *processed*. Then we go over the edges outgoing from y and for each such edge $e = (y, v)$, we insert the node v into a set which we call the *frontier* and which contains the nodes which are not processed but have a processed predecessor. In addition, we compute the change $\Delta^{(y,v)}$ that would occur to $\hbar(v)$ by advancing along the edge e: $\Delta^{(y,v)} = \hbar(y) + w(y, v) - \hbar(v)$. We select v for which $\Delta^{(y,v)}$ is maximal, update $\hbar(v) \leftarrow \hbar(v) + \Delta^{(y,v)}$, remove v from the frontier and mark it as processed. As we will show later, we have found the final value for $\hbar(v)$. We then continue in the same manner: for each successor u of v, we insert u into the frontier if it is not already there and compute $\Delta^{(v,u)}$. Since u may have already been in the frontier, we set $\Delta(u) = \max\{\Delta^{(u',u)}|u' \text{ is processed}\}$. We then select the node v' from the frontier with maximal $\Delta(v')$ value and advance on an edge (w, v') such that $\Delta(v') = \Delta^{(w,v')}$. We repeat this until there is no node v in the frontier with $\Delta(v) > 0$.

The only point to add is that when we process a node v and compute the value Δ^e for an edge $e = (v, u)$, u might already belong to the frontier. In that case, we do not want to add it again but rather to update $\Delta(u)$ if necessary. For this reason, we use a max-sorted Fibonacci Heap for the nodes in the frontier, where the priority of a node u is $\Delta(u)$.

Function $Insert(G, (x, y), \hbar, w(x, y))$
    $Fh \Leftarrow []$ (* Empty Fibonacci Heap. *)
    $\Delta = \hbar(x) + w(x, y) - \hbar(y)$
    if $\Delta > 0$ **then**
        $Fh.Insert(y, \Delta)$
    endif
    while $Fh.NotEmpty$ **do**
        $(u, \Delta) \Leftarrow Fh.ExtractMax$
        $\hbar(u) \Leftarrow \hbar(u) + \Delta$ (* Update $\hbar(u)$ *)
        foreach $e = (u, v) \in E$ **do**
            $\Delta \Leftarrow \hbar(u) + w(u, v) - \hbar(v)$
            if $\Delta > 0$ **then**
                $Fh.UpdatePriority(v, \Delta)$
            endif
        endfor
    end while
end

Fig. 2. Algorithm for updating $\hbar$ values upon an edge insertion

4   The Algorithm

Figure   shows the algorithm for updating the heaviest path values of the affected nodes upon insertion of an edge to the DAG. It receives a DAG $G = (V, E)$, the function $\hbar$ for G, an edge (x, y) which is to be inserted to G and the weight $w(x, y)$ of this edge. It updates the function $\hbar$ for the graph $G' = (V, E \cup \{(x, y)\})$.

Initially, it inserts the node y into the Fibonacci Heap Fh with priority equal to $\Delta_y = \hbar(x) + w(x, y) - \hbar(y)$. It then enters the while loop, which continues as long as Fh is not empty. It extracts from Fh a node u with maximal priority and updates $\hbar(u)$ for this node. Then it traverses the edges outgoing from u and updates the Δ values for the target of each of these edges. Recall that the operation $Fh.UpdatePriority(v, \Delta)$ inserts v to Fh if it is not already there, and otherwise sets its priority to the maximum among Δ and its previous priority.

4.1   Example

Before turning to the correctness proof, we show how this algorithm operates on the example shown in Figure . Figure   reproduces the part of the graph that the algorithm explores, with labels on the nodes.

Initially, y is inserted to Fh with priority $\Delta = \hbar(x) + w(x, y) - \hbar(y) = 4 + 0 - 1 = 3$. Then the algorithm enters the while loop and performs five iterations.

Iteration 1: $(u, \Delta) \leftarrow (y, 3)$ and $\hbar(y) \leftarrow \hbar(y) + \Delta = 1 + 3 = 4$. Next, the successors of y are checked. a is inserted to Fh with priority $\Delta_a = \hbar(y) + w(y, a) - \hbar(a) = 4 + 3 - 5 = 2$ and b is inserted with priority $\Delta_b = \hbar(y) + w(y, b) - \hbar(b) = 4 - 2 + 1 = 3$.

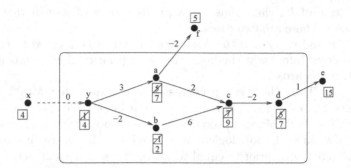

Fig. 3. The portion of the graph of Figure   that the algorithm explores

Iteration 2: $(u, \Delta) \leftarrow (b, 3)$, $\hbar(b) \leftarrow \hbar(b) + \Delta = -1 + 3 = 2$. c is inserted to Fh with priority $\Delta_c = \hbar(b) + w(b, c) - \hbar(c) = 2 + 6 - 7 = 1$.

Iteration 3: $(u, \Delta) \leftarrow (a, 2)$, $\hbar(a) \leftarrow \hbar(a) + \Delta = 5 + 2 = 7$. The priority of c is updated from 1 to $\hbar(a) + w(a, c) - \hbar(c) = 7 + 2 - 7 = 2$ and $\Delta_f = \hbar(a) + w(a, f) - \hbar(f) = 7 - 2 - 5 = 0$ so f is not inserted to Fh.

Iteration 4: $(u, \Delta) \leftarrow (c, 2)$, $\hbar(c) \leftarrow \hbar(c) + \Delta = 7 + 2 = 9$. d is inserted to Fh with priority $\Delta_d = \hbar(c) + w(c, d) - \hbar(d) = 9 - 2 - 5 = 2$.

Iteration 5: $(u, \Delta) \leftarrow (d, 2)$, $\hbar(d) \leftarrow \hbar(d) + \Delta = 5 + 2 = 7$. $\Delta_e = \hbar(d) + w(d, e) - \hbar(e) = 7 + 1 - 15 = -7$ so e is not inserted to Fh.

Now, Fh is empty so the algorithm terminates.

4.2   Proof of Correctness

For a node $v \in V$, let $\hbar(v)$ be the weight of the heaviest path to v in the input DAG G and let $\hbar'(v)$ be the weight of the heaviest path to v in the DAG G' that we get by adding the edge (x, y) to G. Let $\bar{\Delta}(v) = \hbar'(v) - \hbar(v)$ be the change that occurs in the heaviest path value for v due to the insertion of this edge. To show that the algorithm computes the correct value for $\hbar'(v)$ for all $v \in V$, we will show that (1) If $\bar{\Delta}(v) > 0$ then v is inserted to Fh and when it is extracted from Fh, its priority is equal to $\bar{\Delta}(v)$. (2) If $\bar{\Delta}(v) \leq 0$ then v is not inserted to Fh.

It is easy to see that if (1) and (2) hold then the algorithm computes the correct value of $\hbar'(v)$ for all nodes.

Lemma 1. *For all $v \in V$, if $\bar{\Delta}(v) > 0$ then v is inserted into Fh and when it is extracted from Fh, its priority is equal to $\bar{\Delta}(v)$.*

Proof. Assume the converse and let v be minimal w.r.t. the topological order such that $\bar{\Delta}(v) > 0$ and the claim does not hold for v. By construction, the priority of a node in Fh can never be higher than its $\bar{\Delta}$ value. So we need to show that v is inserted into the queue and that its priority when it is extracted is not less than $\bar{\Delta}(v)$.

By definition of $\hbar'$, there must be a predecessor u of v such that $\hbar'(v) = \hbar'(u) + w(u, v)$. There are two cases.

Case 1: $u = x$ and $v = y$ and (u, v) is the edge that was inserted to the graph. Then v is inserted into Fh at the beginning with priority $\bar{\Delta}(v)$ and is extracted immediately afterwards.

Case 2: $v \neq y$. Then there is a predecessor u of v such that $\hbar'(v) = \hbar'(u) + w(u, v)$. Since the edge (u, v) was in the graph before the insertion, $\hbar'(u) > \hbar(u)$, hence $\bar{\Delta}(u) > 0$. Since u precedes v in the topological order, and by the minimality of v w.r.t. the topological order, we know that u was inserted to Fh and was extracted with priority equal to $\bar{\Delta}(u)$. If v was not extracted from Fh before u, then after u was extracted, v was either inserted with priority $\bar{\Delta}(v)$ (if it was not already in the queue) or its priority was updated to $\bar{\Delta}(v)$ (if it was already in Fh). Hence, v was inserted to Fh and when it was extracted, its priority was $\bar{\Delta}(v)$.

Assume that v was extracted before u. Let $P = <z_1, \ldots, z_n>$ be a heaviest path in G' from y to v through u. That is, $z_1 = y$, $z_{n-1} = u$ and $z_n = v$ and for each $1 < i \leq n$, $\hbar'(z_i) = \hbar'(z_{i-1}) + w(z_{i-1}, z_i)$. Clearly, for all $1 \leq i < n$, $\bar{\Delta}(z_i) \geq \bar{\Delta}(z_{i+1}) > 0$. Let i be maximal such that z_i was extracted from Fh before v. Since z_i precedes v in the topological order and by the minimality of v w.r.t. the topological order, we know that z_i was extracted when its priority was $\bar{\Delta}(z_i)$. Hence, after it was extracted, z_{i+1}'s priority became $\hbar(z_i) + \bar{\Delta}(z_i) + w(z_i, z_{i+1}) - \hbar(z_{i+1}) = \hbar'(z_i) + w(z_i, z_{i+1}) - \hbar(z_{i+1}) = \hbar'(z_{i+1}) - \hbar(z_{i+1}) = \bar{\Delta}(z_{i+1})$. Since v was extracted before z_{i+1} but when z_{i+1} was in Fh with priority $\bar{\Delta}(z_{i+1})$, the priority of v when it was extracted was at least $\bar{\Delta}(z_{i+1}) \geq \bar{\Delta}(v)$. □

Lemma 2. *For all $v \in V$, if $\bar{\Delta}(v) = 0$ then v is not inserted to Fh.*

Proof. If $\bar{\Delta}(v) = 0$, this means that for every predecessor u of v, $\hbar'(u) + w(u, v) \leq \hbar(v)$. If none of the predecessors of v were inserted to Fh then v was never a candidate for insertion. If there are predecessors of v which were inserted to Fh, then whenever one of them was extracted and the edge leading from it to v was examined, $\Delta \leq 0$ so v was not inserted. □

Corollary 1. *The algorithm in Figure   correctly updates $\hbar(v)$ for all $v \in V$.*

4.3   Complexity Analysis

By Lemma , when a node v is extracted from Fh, its priority is equal to $\bar{\Delta}(v)$ so $\hbar(v)$ is updated to its final value. This implies that v will never be inserted to Fh again; whenever another of its predecessors will be extracted from Fh, Δ will not be positive. In addition, each edge outgoing from a node in δ is examined once to determine whether its target should be inserted into Fh (or its priority updated if it is already there). We get that throughout the algorithm, Fh needs to support a total of $|\delta|$ insertions, $|\delta|$ extractions and at most $\|\delta\|$ *UpdatePriority* operations. The total running time is therefore $O(\|\delta\| + |\delta| \log |\delta|)$.

5   Integer Edge Weights

In this section we show an alternative algorithm which works for general inputs, but its advantages come to play when all edge weights are integers.

Again, for every node v let $\hbar(v)$ be the heaviest path value of v before the change to the graph and $\hbar'(v)$ the value after the edge insertion. Let $\bar{\Delta}(v) = \hbar'(v) - \hbar(v)$ be the change that occurred at v.

For every edge $e = (u, v)$, let $\Delta\bar{\Delta}(u, v) = \hbar(u) + w(u, v) - \hbar(v)$. Note that $\Delta\bar{\Delta}(u, v) \leq 0$. Intuitively, $\Delta\bar{\Delta}(u, v)$ measures by how much the change in the $\bar{\Delta}$ values decreases as the update propagates along the edge e. Formally,

Lemma 3. *For each node* v, $\bar{\Delta}(v) = \max_{u \in \mathrm{pred}(v)}\{\bar{\Delta}(u) + \Delta\bar{\Delta}(u, v)\}$.

Proof. By definition, $\hbar'(v) = \max_{u \in \mathrm{pred}(v)}\{\hbar'(u) + w(u, v)\}$. So $\hbar'(v) - \hbar(v) = \max_{u \in \mathrm{pred}(v)}\{\hbar'(u) - \hbar(u) + \hbar(u) + w(u, v) - \hbar(v)\} = \max_{u \in \mathrm{pred}(v)}\{\bar{\Delta}(u) + \Delta\bar{\Delta}(u, v)\}$.   □

We define a new weight function w_ϵ over the edges of the DAG where $w_\epsilon(u, v) = -\Delta\bar{\Delta}(u, v)$. This function enables us to prove Lemma   which characterizes the set δ. In the following a *path* is a directed path in the DAG and for a path $P = <v_1, \ldots, v_n>$, $w_\epsilon(P) = \sum_{i=1}^{n-1} w_\epsilon(v_i, v_{i+1})$ is the weight of the path with respect to w_ϵ. For a pair of nodes u, v, $w_\epsilon^P(u, v)$ is the minimal $w_\epsilon(P)$ over all paths P from u to v.

Lemma 4. *Assume that an edge* (x, y) *was inserted into* G. *A node* v *is in* δ *iff* $w_\epsilon^P(y, v) < \bar{\Delta}(y)$. *That is, there is a path* P *from* y *to* v *such that* $w_\epsilon(P) < \bar{\Delta}(y)$.

Proof. Assume that there is such a path $P = <y = v_1, \ldots, v_n = v>$. By definition, for each $1 \leq i < n$ we have $\bar{\Delta}(v_{i+1}) \geq \bar{\Delta}(v_i) - w_\epsilon(v_i, v_{i+1})$. So $\bar{\Delta}(v_n) \geq \bar{\Delta}(y) - w_\epsilon(P) > 0$, which implies $v_n \in \delta$.

For the other direction, we show by induction that if $v \in \delta$, which means that $\bar{\Delta}(v) > 0$, then there is a path P from y to v with $w_\epsilon(P) = \bar{\Delta}(y) - \bar{\Delta}(v)$. By definition we know that there is a predecessor u of v such that $\bar{\Delta}(v) = \bar{\Delta}(u) - w_\epsilon(u, v)$. If $u = y$ then since $\bar{\Delta}(v) > 0$ we have that the edge (u, v) is a path from y to v such that $w_\epsilon(u, v) = \bar{\Delta}(u) - \bar{\Delta}(v) = \bar{\Delta}(y) - \bar{\Delta}(v)$.

Assume that $u \neq y$. Then by the induction hypothesis, there is a path P_u from y to u such that $w_\epsilon(P_u) = \bar{\Delta}(y) - \bar{\Delta}(u)$. Let P_v be the path from y to v that we get by appending the edge (u, v) to P_u. Then $w_\epsilon(P_v) = w_\epsilon(P_u) + w_\epsilon(u, v) = \bar{\Delta}(y) - \bar{\Delta}(u) + w_\epsilon(u, v) = \bar{\Delta}(y) - \bar{\Delta}(u) + \bar{\Delta}(u) - \bar{\Delta}(v) = \bar{\Delta}(y) - \bar{\Delta}(v)$.   □

We now know how to identify the nodes of δ. We begin at y and compute shortest paths w.r.t. the weight function w_ϵ to nodes that are reached from y, but only as long as the length of the path is less than $\bar{\Delta}(y)$. The following lemma states that once we have found the length of the shortest path from y to v, we also know $\bar{\Delta}(v)$ and can update $\hbar(v)$.

Lemma 5. *Assume that an edge* (x, y) *was inserted into* G. *Let* v *be a node in* δ *and let* $w = w_\epsilon^P(y, v)$. *Then* $\bar{\Delta}(v) = \bar{\Delta}(y) - w$.

Proof. We have shown in the proof of Lemma   that if $v \in \delta$ then there is a path P from y to v with $w_\epsilon(P) = \bar{\Delta}(y) - \bar{\Delta}(v)$. Assume that it is not minimal. That is, there is a path $Q = < y = v_1, \ldots, v_n = v >$ with $w_\epsilon(Q) < \bar{\Delta}(y) - \bar{\Delta}(v)$. For all $1 \le j \le n$, let $Q_j = < v_1, \ldots, v_j >$ and let i be minimal such that $w_\epsilon(Q_i) < \bar{\Delta}(y) - \bar{\Delta}(v_i)$. Note that $i > 1$ because $w_\epsilon(Q_1 = < v_1 >) = \bar{\Delta}(y) - \bar{\Delta}(v_1) = 0$. $w_\epsilon(Q_{i-1}) = \bar{\Delta}(y) - \bar{\Delta}(v_{i-1})$, so $w_\epsilon(v_{i-1}, v_i) = w_\epsilon(Q_i) - w_\epsilon(Q_{i-1}) < \bar{\Delta}(v_{i-1}) - \bar{\Delta}(v_i)$. By substituting $\hbar(v_i) - \hbar(v_{i-1}) - w(v_i, v_{i-1})$ for $w_\epsilon(v_{i-1}, v_i)$ and $\hbar'(v_{i-1}) - \hbar(v_{i-1}) - \hbar'(v_i) + \hbar(v_i)$ for $\bar{\Delta}(v_{i-1}) - \bar{\Delta}(v_i)$ we get that $\hbar'(v_{i-1}) + w(v_{i-1}, v_i) < \hbar'(v_i)$, contradicting the definition of $\hbar'$.

This leads to the algorithm in Figure   as an alternative to the one in Figure . In this version, the nodes of the frontier are inserted into a min-sorted Fibonacci Heap where the priority of a node at any point in time is the length of the minimal path leading to it which was discovered so far.

We now prove correctness of this algorithm.

Lemma 6. *For all $v \in V$, if $w_\epsilon^P(y, v) < \bar{\Delta}(y)$ then v is inserted into Fh and when it is extracted from Fh, its priority is equal to $w_\epsilon^P(y, v)$.*

Proof. Assume the converse and let v be minimal w.r.t. the topological order such that $w_\epsilon^P(y, v) < \bar{\Delta}(y)$ and the claim does not hold for v. By construction, the priority of a node in Fh can never be lower than $w_\epsilon^P(y, v)$. So we need to show that v is inserted into Fh and that its priority when it is extracted is not higher than $w_\epsilon^P(y, v)$.

Since y was extracted with priority $0 = w_\epsilon^P(y, y)$, $v \neq y$. By definition of w_ϵ, there must be a predecessor u of v such that $w_\epsilon^P(y, v) = w_\epsilon^P(y, u) + w_\epsilon(u, v)$.

```
Function Insert(G, (x, y), ℏ, w(x, y))
    Fh ⇐ [] (* Empty Fibonacci Heap. *)
    Δ̄_y = ℏ(x) + w(x, y) − ℏ(y)
    if Δ̄_y > 0 then
        Fh.Insert(y, 0) (* The minimum path from y to y has length 0. *)
    endif
    while Fh.NotEmpty do
        (u, P) ⇐ Fh.ExtractMin
        ℏ(u) ⇐ ℏ(u) + Δ̄_y − P (* Update ℏ(u) *)
        foreach e = (u, v) ∈ E do
            P' ⇐ P + w_ε(u, v)
            if P' < Δ̄_y then
                Fh.UpdatePriority(v, P')
            endif
        endfor
    end while
end
```

Fig. 4. Alternative algorithm for updating $\hbar$ values upon an edge insertion

Since u precedes v in the topological order, and by the minimality of v w.r.t. the topological order, we know that u was inserted to Fh and was extracted with priority equal to $w_\epsilon^P(y, u)$. If v was not extracted from Fh before u, then after u was extracted, v was either inserted with priority $w_\epsilon^P(y, v)$ (if it was not already in Fh) or its priority was updated to $w_\epsilon^P(y, v)$ (if it was already in Fh). Hence, v was inserted to Fh and when it was extracted, its priority was $w_\epsilon^P(y, v)$.

Assume that v was extracted before u. Let $P = < z_1, \ldots, z_n >$ be a path from y to v through u such that $w_\epsilon(P) = w_\epsilon^P(y, v)$. That is, $z_1 = y$, $z_{n-1} = u$ and $z_n = v$ and for each $1 < i \leq n$, $w_\epsilon^P(y, z_i) = w_\epsilon^P(y, z_{i-1}) + w_\epsilon(z_{i-1}, z_i)$. Clearly, for all $1 \leq i < n$, $w_\epsilon^P(y, z_i) \leq w_\epsilon^P(y, z_{i+1}) \leq w_\epsilon^P(y, v) < \bar{\Delta}(y)$. Let i be maximal such that z_i was extracted from Fh before v. Since z_i precedes v in the topological order and by minimality of v w.r.t. the topological order, we know that z_i was extracted when its priority was $w_\epsilon^P(y, z_i)$. Hence, after it was extracted, z_{i+1}'s priority became $w_\epsilon^P(y, z_i) + w_\epsilon(z_i, z_{i+1}) = w_\epsilon^P(y, z_{i+1})$. Since v was extracted before z_{i+1} but when z_{i+1} was in Fh with priority $w_\epsilon^P(y, z_{i+1})$, the priority of v when it was extracted was at most $w_\epsilon^P(y, z_{i+1}) \leq w_\epsilon^P(y, v)$. $\square$

Lemma 7. *For all $v \in V$, if $w_\epsilon^P(y, v) \geq \bar{\Delta}(y)$ then v is not inserted to Fh.*

Proof. If $w_\epsilon^P(y, v) \geq \bar{\Delta}(y)$, this means that for every predecessor u of v, $w_\epsilon^P(y, u) + w_\epsilon(u, v) \geq \bar{\Delta}(y)$. If none of the predecessors of v were inserted to Fh then v was never a candidate for insertion. If there are predecessors of v which were inserted to Fh, then by Lemma   whenever one of them (say u) was extracted from Fh, its priority was $w_\epsilon^P(y, u)$ so v was not inserted into Fh. $\square$

Lemma 8. *The algorithm in Figure   correctly updates $\hbar(v)$ for all $v \in V$.*

Proof. If $w_\epsilon^P(y, v) \geq \bar{\Delta}(y)$ then by Lemma   , $v \notin \delta$, so $\hbar'(v) = \hbar(v)$, and by Lemma   , v is never inserted to Fh so $\hbar(v)$ is never updated.

If $w_\epsilon^P(y, v) < \bar{\Delta}(y)$ then by Lemma   , v is inserted into Fh and when it is extracted its priority is $w_\epsilon^P(y, v)$. So the algorithm sets $\hbar'(v)$ to $\hbar(v) + \bar{\Delta}_y - w_\epsilon^P(y, v)$, which by Lemma   is equal to $\hbar(v) + \bar{\Delta}_v$. $\square$

5.1   Example

To illustrate how the algorithm works, we include here a trace of its execution on the example in Figure   . Initially, $\bar{\Delta}(y) = 4 + 0 - 1 = 3$ is computed and y is inserted to Fh with priority 0. Then the algorithm enters the while loop and performs five iterations.

Iteration 1: $(u, P) \leftarrow (y, 0)$ and $\hbar(y) \leftarrow \hbar(y) + \bar{\Delta}(y) - P = 1 + 3 - 0 = 4$. Next, the successors of y are checked. a is inserted to Fh with priority $P'_a = P + w_\epsilon(y, a) = 0 + 1 = 1$ and b is inserted with priority $P'_b = P + w_\epsilon(y, b) = 0 + 0 = 0$.

Iteration 2: $(u, P) \leftarrow (b, 0)$, $\hbar(b) \leftarrow \hbar(b) + \bar{\Delta}(y) - P = -1 + 3 - 0 = 2$. c is inserted to Fh with priority $P'_c = P + w_\epsilon(b, c) = 0 + 2 = 2$.

Iteration 3: $(u, P) \leftarrow (a, 1)$, $\hbar(a) \leftarrow \hbar(a) + \bar{\Delta}(y) - P = 5 + 3 - 1 = 7$. The priority of c is updated from 2 to $P + w_\epsilon(a, c) = 1 + 0 = 1$. $P'_f = P + w_\epsilon(a, f) = 1 + 2 = 3$ so f is not inserted to Fh.

Iteration 4: $(u, P) \leftarrow (c, 1)$, $\hbar(c) \leftarrow \hbar(c) + \bar{\Delta}(y) - P = 7 + 3 - 1 = 9$. d is inserted to Fh with priority $P'_d = P + w_\epsilon(c, d) = 1 + 0 = 1$.
Iteration 5: $(u, P) \leftarrow (d, 1)$, $\hbar(d) \leftarrow \hbar(d) + \bar{\Delta}(y) - P = 5 + 3 - 1 = 7$. $P'_e = P + w_\epsilon(d, e) = 1 + 9 = 10$ so e is not inserted to Fh.
Now, Fh is empty so the algorithm terminates.

5.2   Complexity Analysis

When all edge weights are integers, we have that all w_ϵ values are non-negative integers and all priorities in the queue are in the interval $[0, \bar{\Delta}(y)]$. This means that we can use $\bar{\Delta}(y) + 1$ buckets for the nodes and place the buckets in the queue instead of the individual nodes. Since $\bar{\Delta}_y = \Delta_{\max}$ we get that if we use a Fibonacci Heap, the algorithm runs in time $O(\|\delta\| + |\delta| \log \min\{|\delta|, \Delta_{\max} + 1\})$.

If we use Thorup's integer priority queue [ ] we can achieve an asymptotic running time of $O(\|\delta\| + |\delta| \log \log \min\{|\delta|, \Delta_{\max} + 1\})$. However, this priority queue is more complicated to implement.

References

[1] M. L. Fredman and R. E. Tarjan. Fibonacci heaps and their uses in improved network optimization algorithms. *J. Assoc. Comput. Mach.*, 34:596–615, 1987.

[2] Laurent Michel and Pascal Van Hentenryck. Maintaining longest paths incrementally. In *Proceedings of the 9th International Conference on Principles and Practice of Constraint Programming (CP 2003)*, volume 2833 of *LNCS*, pages 540–554, 2003.   ,   ,

[3] G. Ramalingam and Thomas Reps. On the computational complexity of dynamic graph problems. *Theor. Comput. Sci.*, 158(1-2):233–277, 1996.

[4] Mikkel Thorup. Integer priority queues with decrease key in constant time and the single source shortest paths problem. In *Proc. 35th ACM Symp. on Theory of Computing (STOC)*, pages 149–158, 2003.

Filtering Methods
for Symmetric Cardinality Constraint

Waldemar Kocjan[1] and Per Kreuger[2]

[1] Mälardalen University, Västerås, Sweden
waldemar.kocjan@mdh.se
[2] Swedish Institute Of Computer Science, Kista, Sweden
piak@sics.se

Abstract. The symmetric cardinality constraint is described in terms of a set of variables $X = \{x_1, \ldots, x_k\}$, which take their values as subsets of values $V = \{v_1, \ldots, v_n\}$. It constraints the cardinality of the set assigned to each variable to be in an interval $[l_{x_i}, c_{x_i}]$ and at the same time it restricts the number of occurrences of each value $v_j \in V$ in the sets assigned to variables in X to be in an other interval $[l_{v_j}, c_{v_j}]$. In this paper we introduce the symmetric cardinality constraint and define set constraint satisfaction problem as a framework for dealing with this type of constraints. Moreover, we present effective filtering methods for the symmetric cardinality constraint.

1 Introduction

The symmetric cardinality constraint is specified in terms of a set of variables $X = \{x_1, \ldots, x_k\}$, which take their values as subsets of $V = \{v_1, \ldots, v_n\}$. The cardinality of the set assigned to each variable is constrained by the interval $[l_{x_i}, c_{x_i}]$, where l_{x_i} and c_{x_i} are non–negative integers. In addition, it constraints the number of occurrences of each value $v_j \in V$ in the sets assigned to variables in X to be an interval $[l_{v_j}, c_{v_j}]$. Both l_{v_j} and c_{v_j} are non–negative integers.

The symmetric cardinality constraint problems arise in many real–life problems. For example, consider an instance of a project management problem. The main task of the problem is to assign personnel with possibly multiple specialized competences to a set of tasks, each requiring a certain number of people (possibly none) of each competence. In this instance we consider a project consisting of 7 activities numbered from 1 to 7, each demanding a number of persons to be accomplished. An activity which demands minimum 0 persons is optional.

There are 6 members of personnel which can be assigned to this project. Each of those persons, here referred to as x with a respective index, is qualified to perform respective activities as shown in Fig. . A non-zero value of a lower bound indicates that members of the staff represented by the variable must be assigned to some activities in the project.

The goal is to produce an assignment which satisfy the following constraints:

- every member of staff must be assigned to a minimum and maximum number of activities in the project,

J.-C. Régin and M. Rueher (Eds.): CPAIOR 2004, LNCS 3011, pp. 200–   , 2004.
© Springer-Verlag Berlin Heidelberg 2004

	1	2	3	4	5	6	7
x_1	1	1	0	0	0	0	0
x_2	0	1	1	0	0	0	0
x_3	1	0	1	0	0	0	0
x_4	0	1	0	1	0	0	0
x_5	0	0	1	1	1	1	0
x_6	0	0	0	0	0	1	1

person	activities
x_1	2..3
x_2	0..2
x_3	0..1
x_4	1..1
x_5	0..3
x_6	0..2

task	persons
1	1..3
2	1..2
3	0..1
4	1..2
5	1..3
6	0..2
7	1..1

Fig. 1. Project assignment specification. In the left table, a 1 indicates that a person represented by x_i is qualified to perform corresponding activity in the project. The table in the center and to the right specifies the number of activities each person can perform and the number of persons required by each activity

- every activity must be performed by a minimum and maximum number of persons,
- each person can be assign only to an activity he/she is qualified to perform and, by symmetry, each activity must be performed by qualified personnel.

In this paper we show how such problem can be modeled as a constraint satisfaction problem. First, in Section   we give some preliminaries on graphs and flows. Then, in Section   we define set constraint satisfaction problem and give a formal definition of the symmetric cardinality constraint. The following section, , gives a method for checking consistency of symmetric cardinality constraint. Finally, we describe a filtering method for symmetric cardinality constraint.

2   Preliminaries

2.1   Graph

The following definitions are mainly due to [ ].

A *directed graph* $G = (X, U)$ consists of a set of nodes (vertices) X and arcs (edges) U, where every pair (u, v) is an ordered pair of distinct nodes. An *oriented graph* is a directed graph having no symmetric pair of arcs.

A *directed network* is a directed graph whose nodes and/or arcs have associated numerical values. In this paper we do not make distinction between terms "network" and "directed network".

An arc (u, v) connects node u with node v, i.e. in directed graph it is an arc oriented from node u to node v. A *path* in a graph G from v_1 to v_k is a sequence of nodes $[v_1, \ldots, v_k]$ such that each (v_i, v_{i+1}) is an arc for $i \in [1, \ldots, k-1]$. The path is *simple* if all its nodes are distinct. A path is a *cycle* if $k > 1$ and $v_1 = v_k$.

A subgraph of a directed graph G, which contains at least one directed path from every node to every other node is called a *strongly connected component of* G.

2.2   Flows

Let N be a directed network in which each arc e is associated with two non–negative integers $l(e)$ and $c(e)$ representing lower bound and a capacity of flow on e. A flow $f(e)$ on arc e represents the amount of commodity that the arc accommodates. More formally:

Definition 1. *A flow in a network N is a function that assigns to each arc e of the network a value $f(e)$ in such way that*

1. *$l(e) \leq f(e) \leq c(e)$, where $l(e)$ is a lower bound of the flow in the arc and $c(e)$ is a capacity of e*
2. *for each node p in the network N it is true that $\sum_n f(e(n, p)) = \sum_r f(e(p, r))$ where $e(x, y)$ is an arc from node x to y*

The second property is known as a conservation law and states that the amount of flow of some commodity incoming to each node in N is equal the amount of that commodity leaving each node.

Three problems from the flow theory are referred to in this paper:

– *the feasible flow problem* which resolves if there exists a flow in N which satisfies lower bound and capacity constraints for all arcs in N.
– *the problem of maximum flow from m to n* which consists of finding a feasible flow in N such that the value $f(m, n)$ is maximum
– *the problem of minimum flow from m to n* which consists of finding a feasible flow in N such that the value $f(m, n)$ is minimum.

It is a well known fact that if the lower bounds and capacities of a flow problem are integral and there exists a feasible flow for the network, then the maximum and minimum flows between any two nodes flows are also integral on all arcs in the network. Hence, if there exists a feasible flow in a network there also exists an integral feasible flow. In this paper when we refer to a feasible flow we always mean an *integral* feasible flow.

We refer in this paper also to the *residual network*, which is a network representing the utilization and remaining capacity in the network with respect to a flow f.

Definition 2. *Given a flow f from s to t in the network N, the residual network for f, denoted by $R(f)$, consists of the same set of nodes as N. The arc set of $R(f)$ is defined as follows. For all arcs (m, n) in N*

– *if $f(m, n) < c(m, n)$ then (m, n) is an arc of $R(f)$ with residual capacity $res(m, n) = c(m, n) - f(m, n)$,*
– *if $f(m, n) > l(m, n)$ then (n, m) is an arc of $R(f)$ with residual capacity $res(n, m) = f(m, n) - l(m, n)$,*

3   Set Constraint Satisfaction Problem

We define a set constraint satisfaction problem as follows.

Definition 3. *A set constraint satisfaction problem (sCSP) is a triple (X, D, Cs) where*

1. *$X = \{x_1, \ldots, x_n\}$ is a finite set of variables.*
2. *$D = \{D_1, \ldots, D_n\}$ is a set of finite sets of elements such that for each i, x_i takes as value a subset of D_i.*
3. *Cs is a set of constraints on the values particular subsets of the variables in X can simultaneously take. Each constraint $C \in Cs$ constrains the values of a subset $X(C) = \{x_{C_1}, \ldots, x_{C_k}\}$ of the variables in X and may be thought of as a subset $T(C)$ of the Cartesian product $= C_{C_1} \times \ldots, \times C_{C_k}$ where each $C_{C_i} = \{C \mid C \subseteq D_{C_i}\}$.*

Let

$$D(C) = \bigcup_{i \in \{i \mid x_i \in X(C)\}} D_i$$

be the set of values that can be taken by any variable in $X(C)$. Furthermore, for a given assignment P, let $P(x_i)$ denote the value assigned to the variable x_i by P and $\#(x_i, P)$, the cardinality $|P(x_i)|$ of the set $P(x_i)$ and for any constraint C and element $v_j \in D(C)$, $\#(v_j, C, P)$ denote the number of occurrences of v_j in the values assigned by P to the variables in $X(C)$, i.e:

$$\sum_{x_i \in X(C)} \begin{cases} 1 \text{ if } v_i \in P(x_i) \\ 0 \text{ otherwise} \end{cases}$$

Definition 4. *A sCSP $\langle X, D, Cs \rangle$ is consistent if and only if there exists an assignment P with the following properties:*

1. *For each variable $x_i \in X$ with domain D_i, the value $P(x_i)$ assigned to x_i by P must be a subset of D_i.*
2. *For each constraint $C \in Cs$ and each variable in $X(C) = \{x_{C_1}, \ldots, x_{C_k}\}$ the tuple $\langle P(x_{C_1}), \ldots, P(x_{C_k}) \rangle \in T(C)$.*

Moreover, a value $v \in D(x_{C_i})$ for x_{C_i} is consistent with C iff $\exists P(P(X(C)) \in T(C))$ such that v is an element in the value $P(x_i)$.

An n–ary constraint can be seen in terms of its *value graph* ([ ]), i.e the bipartite graph $G(C) = (X(C), D(C), E)$, where for all $x \in X(C), v \in D(C), (x_i, v) \in E$ iff $v \in D_i$. This graph establishes an immediate correspondence between any assignment P and a special set of edges in a value graph.

We formulate this notion in the following proposition.

Proposition 1. *For any $C \in Cs$ every $P(X(C))$ corresponds to a subset of edges in $G(C)$ and the number of edges connecting $x_i \in X(C)$ with any $v_j \in D(C)$ is equal to the cardinality of the subset $P(x_i)$.*

4   Consistency of the Symmetric Cardinality Constraint

We define the symmetric cardinality constraint as follows.

Definition 5. *A symmetric cardinality constraint is a constraint C over a set of variables $X(C)$ which associates with each variable $x_i \in X(C)$ two non-negative integers l_{x_i} and c_{x_i}, and with each value $v_j \in D(C)$ two other non-negative integers l_{v_j} and c_{v_j} such that a restriction of an assignment P to the variables in $X(C)$ is an element in $T(C)$ iff*
$\forall i \, (l_{x_i} \le \#(x_i, P) \le c_{x_i})$ *and* $\forall j \, (l_{v_j} \le \#(v_j, C, P) \le c_{v_j})$.

From the symmetric cardinality constraint we propose to build a particular oriented graph, which we denote $N(C)$. This extends the value network of the global cardinality constraint as described in [ ] to handle sets of nonnegative cardinality assigned to the variables. Then, we will show an equivalence between the existence of a feasible flow in such graph and the consistency of the symmetric cardinality constraint.

Let C be a symmetric cardinality constraint, the *value network* $N(C)$ of C is an oriented graph with a capacity and a lower bound on each arc. The value network $N(C)$ is obtained from the value graph $G(C)$ by

- orienting each edge of $G(C)$ from values to variables. Since each value can occur in a subset assigned to a variable at least 0 and at most 1 time for each arc $(v, x) : l(v, x) = 0, c(v, x) = 1$
- adding a source node s and connecting it which each value. For every arc $(s, v_i) : l(s, v_i) = l_{v_i}, c(s, v_i) = c_{v_i}$
- adding a sink node t and an arc from each variable to t. For each such arc $(x_i, t) : l(x_i, t) = l_{x_i}, c(x_i, t) = c_{x_i}$
- adding an arc (t, s) with $l(t, s) = 0$ and $c(t, s) = \infty$

Proposition 2. *Let C be a symmetric cardinality constraint and $N(C)$ be the value network of C. The following properties are equivalent:*

- *C is consistent;*
- *there exists a flow from s to t which satisfies lower bounds and capacities of the arcs in $N(C)$.*

Proof. Suppose that C is consistent then $T(C) \ne \emptyset$. Consider $P \in T(C)$. We can build a function f in $N(C)$ as follows:

1. $\forall x_i \in X(C), f(x_{c_i}, t) = \#(x_{c_i}, P)$
2. $\forall x_i \in X(C), f(v, x_i) = 1$ if v appears in the subset $P(x_i)$, otherwise $f(v, x_i) = 0$
3. $\forall v_j \in D(C), f(s, v_j) = \#(v_j, C, P)$

---

[1] Actually, the orientation of the graph has no importance. Here, we have chosen the same orientation as in [ ].

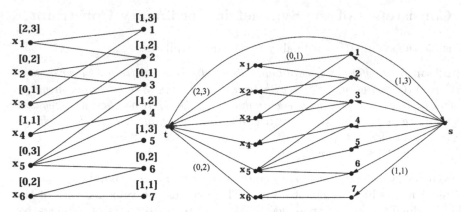

Fig. 2. Value graph for assignment problem from Figure   and its value network

Since C is consistent then $\forall x_i \in X(C), 0 \le l_{x_i} \le \#(x_i, P) \le c_{x_i}$ and $\forall v_j \in D(C), 0 \le l_{v_j} \le \#(v_j, C, P) \le c_{v_j}$, which satisfies the lower bound and capacity constraint for the flow $f(v_i, x_i)$. Furthermore from ( )- ( ) follows that v_j must appear $\#(v_j, C, P)$ times in $P(X(C))$, which means that $\sum_{i=1}^{k} f(v_j, x_i) = f(s, v_j)$ which satisfies the conservation law for all $v_j \in D(C)$. From this and ( ) follows that for each $x_i \in X(C)$ the number of arcs with flow value 1 entering x_i is equal to $\#(x_i, P)$. According to ( ): $\forall x_i \in X(CS), f(x_i, t) = \#(x_i, P)$ which satisfies the conservation law for all $x_i \in X$.

On the other hand, suppose there exists a feasible flow f from s to t. Since $f(v_j, x_i) = 1$ or $f(v_j, x_i) = 0$ and by the conservation law $f(s, v_j) = \sum_{i=1}^{k} f(v_j, x_i)$ then for each v_j the number of edges with the flow of value 1 leaving v_j is equal $f(s, v_j)$. Consequently, the number of $x_i \in X(C)$ connected with each v_j by an arc with a flow equal to 1 is equal to $f(s, v_j)$. Furthermore, due to the conservation law, the number of arcs for which $f(v_j, x_i) = 1$ entering each x_i is equal the value of $f(x_i, t)$. Thus the set of arcs such that $f(v_j, x_i) = 1$ corresponds to a set of edges in the value graph. By the capacity constraint $\forall v_i \in D(C) : l_{v_i} \le f(s, v_i) \le c_{v_i}$ and by the conservation law $f(s, v_i) = \sum_x f(v_i, x) \le c_i$ therefore $l_i \le \#(v_i, C, P) \le c_i$. Similarly, $\forall x_i \in X : l_{c_{x_i}} \le \sum_{i=1}^{k} f(v_j, x_i) \le c_{x_i}$ and by the conservation law $f(x_i, t) = \sum f(v_j, x_i) \le c_{x_i}$, therefore $l_{x_i} \le \#(x_i, P) \le c_{x_i}$, thus C is consistent. □

This proposition gives a way of checking the consistency of a symmetric cardinality constraint by computing a feasible flow in $N(C)$. Different algorithms for computing feasible flows are given in the literature on flow theory, e.g. in [ ].

In the next section we will show how to implement filtering of symmetric cardinality constraint by considering certain properties common to all feasible flows in the value graph.

5   Filtering Algorithm for Symmetric Cardinality Constraint

Let f be a feasible flow in a network N, $R(f)$ be a residual graph for f. If there is a simple path p from n to m in $R(f) - \{(n,m)\}$, then we can obtain a new feasible flow f' in N such that $f'(m,n) > f(m,n)$ (see [ ]). We call such a path p an augmenting path. Similarly, if there exists a simple path p from m to n in $R(f) - \{(m,n)\}$, then we can obtain a new feasible flow f' such that $f'(m,n) < f(m,n)$). We refer to such simple path p as a reducing path.

Moreover, the maximum and minimum flow are defined as follows ([ ]).

Definition 6. *A flow f from m to n is* maximum *if and only if there is no augmenting path from m to n for f.*

A flow f from m to n is minimum *if and only if there is no reducing path from m to n for f.*

The following Theorem   gives a way of determining if an arbitrary arc in N is contained in any feasible flow in the network. The theorem is similar to Theorem 4 from [ ], but here the computation is performed on the residual graph of f which includes both (t,s) and, in the case when $f(t,s) > 0$, also (s,t).

Theorem 1. *Let N be a network for which each arc is associated with two non-negative integers, f be an arbitrary feasible flow in N, $R(f)$ be the residual graph for f and (m,n) be an arbitrary arc in N. For all feasible flows f' in N, $f'(m,n) = f(m,n)$ if and only if neither (m,n) nor (n,m) are contained in any simple cycle in $R(f)$ involving at least three nodes.*

Proof. If (m,n) is not contained in a simple cycle in $R(f)$ involving at least three vertices it means that there is no augmenting path from n to m for f. By the definition , the flow f is the maximum flow from m to n.

If (n,m) is not contained in a simple cycle in $R(f)$ which involves at least three nodes then there is no reducing path from n to m in N, so by the definition f is the minimum flow from m to n.

Similarly, if (t,s) is not contained in a simple cycle in $R(f)$ with at least 3 nodes, then there is no augmenting path from s to t and by Definition f is a maximum flow in N. Moreover, if (s,t) is not contained in a simple cycle in $R(f)$ involving at least three nodes then there is no reducing path for f in $R(f)$ and by Definition f is a minimum flow in N.   □

Let $C \in Cs$ be a symmetric cardinality constraint and f be an arbitrary feasible flow in $N(C)$. By Proposition   a value v of a variable x is not consistent with C if and only if there exists no feasible flow in $N(C)$ which contains (v,x). So, by the Theorem   if $f(v,x) = 0$ and (v,x) is not contained in a simple cycle in $R(f)$ involving at least three nodes then the value v of a variable x is not consistent with C. Furthermore since two nodes m and n can be contained in the such a cycle only if they belong to the same strongly connected component in $R(f)$, we can determine if a value for a variable is consistent with a symmetric cardinality constraint using an algorithm that search for strongly connected components in a graph, e.g. the algorithm described in [ ].

This leads to the following corollary:

Corollary 1. *The consistency of any value v of variable x with symmetric cardinality constraint C can be computed by finding an arbitrary feasible flow f from s to t in $N(C)$ and by computing strongly connected components in $R(f)$.*

6   Notes on Complexity

The complexity of the proposed filtering algorithm is dominated by the computation of a feasible flow in the value network of the symmetric cardinality constraint. Methods for finding feasible flows have in the worst case the same complexity as that of finding a maximum flow, which is $O(n^2m)$ (see e.g. [,]), where n is the number of nodes (i.e. number of variables, $|X|$, plus the number values, $|D(C)|$) and m is the number of arcs which, for a bipartite graph of the type used in the value graph, is bounded by $(\frac{n}{2})^2 + n$.

The complexity of finding strongly connected components in the residual graph using the method proposed in [ ] is $O(m + n)$. This gives worst case complexity for filtering of the symmetric cardinality constraint of $O((|X| + |D(C)|)^4)$, which is the same as for the global cardinality constraint introduced in [ ].

7   Conclusion

In this paper we have introduced the symmetric cardinality constraint derived from the global cardinality constraint. Moreover, we have formalized set constraint satisfaction problem and defined symmetric cardinality constraint in the context of this problem. We have also presented efficient methods for filtering domains of symmetric cardinality constraint.

Acknowledgement

The work on this paper was supported by Swedish Institute of Computer Science and Mälardalen University. The authors wish to thank Nicolas Beldiceanu, Irit Katriel, Markus Bohlin and the anonymous reviewers for their helpful comments on this paper.

References

[1] Ahuja, R. K., Magnanti, T. L., Orlin, J. B.: Network flows. Theory, algorithms and applications. Prentice–Hall Inc. (1993)   ,   ,
[2] Laurière, J.-L.: A language and a program for stating and solving combinatorial problems. Artificial Intelligence, **10** (1978) 29–127

[3] Régin, J.-Ch.: Generalized Arc Consistency for Global Cardinality Constraint.Proc. of the Fourteenth National Conference on Artificial Intelligence (AAAI-96) (1996)   ,   ,

[4] Lawler, E.: Combinatorial Optimization: Network and Matroids, Holt, Rinehart and Winston (1976)

[5] Ahuja, R. K., Kodialam, A., Mishra, A. K., Orlin, J. B.: Computational Investigations of Maximum Flow Algorithms. European Journal of Operational Research, **97** (1997) 509–542

[6] Tarjan, R. E.: Depth–First Search and Linear Graph Algorithms. SIAM J. Computing, **1** (1972) 146–160

Arc-Consistency Filtering Algorithms
for Logical Combinations of Constraints

Olivier Lhomme

ILOG
1681, route des Dolines, F-06560 Valbonne
olhomme@ilog.fr

Abstract. The logical connectives between constraints, sometimes called meta-constraints, although extremely useful for modelling problems, have either poor filtering algorithms or are not very efficient. We propose in this paper new filtering algorithms that achieve arc-consistency over those logical connectives. The principle is to export supports from a constraint.

1 Introduction

A conventional wisdom is that Constraint Programming is strong in making different algorithms work together: a subproblem can be solved by a specific technique (*e.g.* an OR algorithm) encapsulated in a constraint, and the propagation through the domains of variables is responsible for communicating with the other parts of the problem. Nevertheless, in practice, this is not so true. The first constraint model of experienced people does not work so often, and interaction between constraints in general does not work as well as it should. Improving interactions needs a lot of know-how and algorithmic skills from the developper. The final working model typically adds redundant constraints and/or specifically developped constraints that improve the interaction of some of the constraints in the first model.

In this paper, we focus on improving the interaction of constraints in a logical combination of constraints. Such logical combinations occur frequently in the first constraint models of unexperienced constraint programmers. Unfortunately, the logical connectives between constraints, although extremely useful for modelling problems, have either poor filtering algorithms or are not very efficient.

We propose new filtering algorithms for logical combinations of constraints. We mainly focus on constraints given in extension. Constraints given in extension occur frequently in some kinds of applications. For example, solving configuration problems needs to deal with potentialy huge tables of data, which express different kinds of compatibility or incompatibility relations, between variables that can be classical integer variables, set variables or multiset variables (see [ ]).

Consider a simple example: Let w, x, y, z, t be five integer variables. Let C be the constraint: t must have the value a, excepted for some given combinations of

J.-C. Régin and M. Rueher (Eds.): CPAIOR 2004, LNCS 3011, pp. 209–   , 2004.
© Springer-Verlag Berlin Heidelberg 2004

values for w, x, y, z and, in this case, t has a value that can be computed from the values of w, x, y, z. Thus, assuming there is a table with the given combinations for w, x, y, z and the corresponding value for t, the constraint can be expressed as: (1) or (2), where (1) is "(w, x, y, z, t) is in the table" and (2) is "(w, x, y, z) is not in the subtable that corresponds to the first four columns of the table, and $t = a$".

In general, it is not realistic to give this constraint fully in extension. For example, assume the number of the given combinations of values for w, x, y, z is 10000, and that each variable can take a value in [1..100]. These 10000 given combinations are a compact form for representing the constraint, but if the constraint is given fully in extension, this compactness is lost: for each possible combination of values for w, x, y, z (given or not), there is a value for t, and thus the cardinality of the allowed set of combinations of values for w, x, y, z, t is $100^4 = 10^8$.

It is possible to design a specific constraint for this problem but this will need time and constraint programming expertise. The approach we propose in this paper allows a user to express the constraint exactly as it is defined above (i.e. "(1) or (2)"), while achieving the same pruning than a specific constraint, and with the same efficiency.

Our approach is based on the concept of *support*, which is a crucial concept for constraints given in extension. The common idea of the filtering algorithms we propose is *exporting supports from constraints*. Then, supports are combined to build supports for the combination of constraint. For the disjunction, this idea leads to a filtering algorithm that is more efficient than constructive disjunction while keeping the same domain reduction (an earlier work appeared in [ ]). For the conjunction of constraints, the idea of exporting the supports is combined with an additional improvement that accelerates drastically the search for a support in a table.

The paper is organized as follows. In Section 2 we give background on constraint networks. We then present algorithms to achieve arc consistency on a disjunction of constraints (Section 3), negation of a constraint (Section 4) and conjunction of constraints (Section 5). Section 6 addresses arbitrary logical combinations of constraints and Section 7 gives some perspectives.

Although the implication constraint occurs frequently for certain kinds of application (like configuration, or time-tabling problems), it does not deserve a specific study in this paper as the classical transformation in terms of disjunction and negation can be applied without loss of property.

2   Background

Constraint Network. A (*constraint network*) $\mathcal{N} = (X, \mathcal{D}, \mathcal{C})$ is defined as a set of n *variables* $X = \{x_1, \ldots, x_n\}$, a set of *domains* $\mathcal{D} = \{D(x_1), \ldots, D(x_n)\}$ where $D(x_i)$ is the finite set of possible *values* for variable x_i, and a set $\mathcal{C}$ of *constraints* between variables. Let $<_d$ be a total ordering on $D(x_i), \forall x_i \in X$.

---

[1] (if a then b) is equivalent to (not a or b)

Constraint. A constraint C on the ordered set of variables $X(C) = (x_{i_1}, \ldots, x_{i_r})$ is a subset of the Cartesian product $D(x_{i_1}) \times \cdots \times D(x_{i_r})$ that specifies the *allowed* combinations of values for the variables $x_{i_1}, \ldots, x_{i_r}$.

Tuple. An instantiation of the variables in $X(C)$ is called a *tuple* on $X(C)$. Two tuples τ and τ' on $X(C)$ can be ordered by the natural lexicographic order $\prec_{lo}$ in which $\tau \prec_{lo} \tau'$ iff $\exists k / \tau[1..k-1] = \tau'[1..k-1]$ and $\tau[k] <_d \tau'[k]$. ($\tau[1..k]$ being the prefix of size k of τ, and $\tau[k]$ the kth value of τ.)

The value of variable x in a tuple τ is denoted by $\tau[x]$. By extension, if V is a subset of $X(C)$, $\tau[V]$ denotes the values of the variables of V in the tuple τ.

A tuple τ on $X(C)$ is *valid* iff $\forall x \in X(C), \tau[x] \in D(x)$.

Support and Conflict. Let $\mathcal{N} = (X, \mathcal{D}, \mathcal{C})$ be a constraint network, and let C be a constraint in $\mathcal{C}$. A value a for a variable x is often denoted by (x, a).

Let $a \in D(x)$ be a value, let τ be a valid tuple on $X(C)$, such that $a = \tau[x]$. τ is called a *support* for (x, a) on C iff it is allowed by C. τ is called a *conflict* for (x, a) on C iff it is not allowed by C.

Arc Consistency. Let $\mathcal{N} = (X, \mathcal{D}, \mathcal{C})$ be a constraint network, C a constraint in $\mathcal{C}$. A value $a \in D(x)$ is *consistent with* C iff $x \notin X(C)$, or $\exists \tau$ such that τ is a support for (x, a) on C. C is *arc consistent* iff $\forall x \in X(C), D(x) \neq \emptyset$ and $\forall a \in D(x)$, a is consistent with C.

3   Disjunction of Constraints

This section first reviews two existing approaches of dealing with the disjunction of constraints, and then presents a new filtering algorithm.

3.1   Existing Approaches

The standard implementation for the constraint $or(c_1, c_2)$ consists in applying the following propagation rules:

- if c_1 is false on the current domains then $add(c_2)$ to the solver.
- if c_2 is false on the current domains then $add(c_1)$ to the solver.

A side-effect of such a propagation rule is that, as long as both constraints can be true, the or constraint will not be able to prune anything.

Example 1. Let x be a variable and its domain be $D(x) = \{1, 2, 3\}$. Let us consider the constraint $or(x = 1, x = 2)$. It is easy to see that the value 3 is not possible for x. Nevertheless, the above propagation rule will not be able to prune the value 3 since the two constraints $x = 1$ and $x = 2$ can be true.

Indeed, using the standard propagation rule does not ensure arc-consistency for the variables in the disjunction. Instead it ensures arc-consistency for an equivalent constraint system in which some boolean variables appear. The auxiliary variables are constrained to be equal to the truth value of the constraints in the disjunction. The equivalent system is the following: $b_1 = truthValue(c_1)$, $b_2 = truthValue(c_2)$, $booleanOr(b_1, b_2)$.

In this equivalent constraint system, $(x, 3)$ is supported by $(b_1, 0)$ in the constraint $b_1 = truthValue(c_1)$, and $(b_1, 0)$ is itself supported by $(b_2, 1)$ in the constraint $booleanOr(b_1, b_2)$. Thus the value $(x, 3)$ is arc-consistent in this equivalent constraint system. In some sense, the standard propagation rule loses the globality of the **or** constraint.

Another approach to implement the **or** constraint is the use of *constructive disjunction* [ ]. The idea is to propagate independently each term of the disjunction. The domains of the variables are the union of their domains in the different branches of the disjunction. In our example, we have:

$constraint(x = 1) \longrightarrow D_1(x) = \{1\}$

and:

$constraint(x = 2) \longrightarrow D_2(x) = \{2\}$.

Then the union of $D_1(x)$ and $D_2(x)$ can be computed: $\{1\} \cup \{2\} = \{1, 2\}$. Thus $D(x)$ can be reduced to $D_1(x) \cup D_2(x) = \{1, 2\}$.

Whereas the standard propagation rule for the **or** constraint in general leads to a simple generate-and-test behavior, with no pruning at all, the constructive disjunction performs a very good pruning of the domains. In fact it is easy to show the following proposition:

Proposition 1. *Let C_1 and C_k be k constraints whose filtering algorithms achieve arc-consistency. Then constructive disjunction applied on $or(C_1, ..., C_k)$ achieves arc-consistency.*

Proof. Consider a variable x whose domain after reduction is $D(x)$. The domain $D(x)$ is, by definition, the union of the domains of x in the different branches $D(x) = D_1(x) \cup ... \cup D_k(x)$. Thus if value a is in $D(x)$ it is also in one domain $D_i(x)$ at least. In the branch i, we know that C_i is arc-consistent. Thus (x, a) is arc-consistent in the branch i. For a disjunction, to be consistent, it suffices that one branch is consistent, thus (x, a) is arc-consistent for $or(C_1, ..., C_k)$.□

Constructive disjunction thus performs an optimal reduction of the domains. Nevertheless, constructive disjunction is not so often used in constraint programming applications. This may be explained by the following drawbacks of constructive disjunction:

- Constructive disjunction needs a local backtrack between the different constraints in the disjunction, and this is not very efficient.
- Maintaining arc-consistency during search with a constructive disjunction in an incremental way is not simple to do.

– Another problem is the lack of generality of constructive disjunction: it cannot handle complex logical combinations of constraints like $or(c_1, or(c_2, and(c_3, c_4)))$.

Indeed, the main reason of the limited use of constructive disjunction in constraint applications seems to be that it is much more expensive in term of CPU time than the standard filtering for disjunction, and, even if its pruning may be much stronger, constructive disjunction does not always pay off [ ].

3.2   A New Filtering Algorithm for the Disjunction of Constraints

In this section, we introduce a filtering algorithm for **or** constraint that performs the same pruning as constructive disjunction, and thus achieves arc-consistency, but which is much more efficient. Furthermore, it does not have the drawbacks of constructive disjunction which are listed in the previous section.

Constructive disjunction needs arc consistency on each constraint in the disjunction to be achieved. In this section, we show that it is not necessary in general to performs so much work. The intuition of the ideas presented in this section can be given as follows:

– A variable x that is shared by all the constraints in the disjunction needs only one support per value to be found. When a support for the value (x, a) is found on the constraint C_1, we know that this constraint can be true. Thus, the disjunction can be true for the value a. Thus, it is useless to find the supports for the value (x, a) on the other constraints.
– When any support for a constraint C_1 in the disjunction can be found, then this constraint can be true. Thus, all the other constraints of the disjunction can be false. Therefore, the variables of the disjunction that are not in the constraint C_1 cannot be pruned at all. Thus, it is useless to find supports for their values.

Let us consider the following example:

Example 2. Let x, y, z be three variables, let their respective domains be $D(x) = [1, 1000], D(y) = [1, 2], D(z) = [1, 1000]$, and let C be the constraint $or(C_1, C_2)$, where C_1 corresponds to the constraint $x = y$ given in extension, and C_2 corresponds to the constraint $y = z$ given in extension. The tuple $\{(x, 1), (y, 1)\}$ is a support for the constraint C_1. That is to say, when $x = 1$ and $y = 1$, the constraint C_1 is true, and thus C is true, even if the constraint C_2 is false. In other words, for any value v of z in $D(z)$, the tuple $\{(x, 1), (y, 1), (z, v)\}$ is a support on the constraint C.

Symmetrically, $\{(y, 2), (z, 2)\}$ is a support on the constraint C_2, and thus $\{(x, v), (y, 2), (z, 2)\}$ is a support on C for every value $v \in D(x)$.

Indeed, the two above support checks are sufficient to prove arc-consistency of the constraint C.

If one wants to use constructive disjunction to prove arc-consistency of C, it will first compute arc-consistency for constraint C_1, and then for constraint C_2, doing at least 1000 constraints checks for each constraint.

Example   should have given to the reader the intuition of the new filtering algorithm for the **or** constraint. Proposition   formalizes this intuition.

Proposition 2. *Let C be the constraint $or(C_1, C_2)$. Let V_1 be the variables of C_1 that are not in C_2, let V_2 be the variables of C_2 that are not in C_1, and let V_3 be the variables that are shared by C_1 and C_2. That is: $V_1 = X(C_1) - X(C_2), V_2 = X(C_2) - X(C_1), V_3 = X(C_1) \cap X(C_2)$.*

- *Let τ_1 be a support on C_1. Let τ be a tuple on C such that:*
  1. $\tau[X(C_1)] = \tau_1[X(C_1)]$
  2. for all variable z in V_2, $\tau[z] \in D(z)$.
  Then τ is a support on C.
- *Let τ_2 be a support on C_2. Let τ be a tuple on C such that:*
  1. $\tau[X(C_2)] = \tau_2[X(C_2)]$
  2. for all variable x in V_1, $\tau[x] \in D(x)$.
  Then τ is a support on C.

Proof. we give the proof for the first assertion, the other one is symmetric. The values of variables in $X(C_1)$ are in the domains since τ_1 is a support and thus is valid. Then, in τ, the values of variables $X(C_1)$ are in the domains. As, by construction the values of variables V_2 in τ are also in the domains, we know that τ is valid. As τ makes the constraint C_1 true, the constraint $or(C_1, C_2)$ is true, and thus τ is an allowed tuple, hence it is a support. $\square$

Then, a support τ on C can be simply derived from a support τ_1 on C_1 (resp. τ_2 on C_2). It suffices to complete τ_1 (resp. τ_2) by adding values for the variables in V_2 (resp. V_1): any value in their domains can be taken.

Corollary 1. *For the variables that are in V_3, it suffices to find one support per value, either in C_1 or in C_2.*

Indeed, corollary   uses for the meta-constraint level the principle of lazy support that was introduced in AC-6 [ ]. Whereas AC-4 [ ] needs to compute all supports for a given value, AC-6 introduces the idea that only one support is sufficient. As AC-4, constructive disjunction computes several supports for a value: one for each constraint in the disjunction. Corollary   tells us that only one support is sufficient.

Corollary 2. *When a support τ_1 on C_1 is found, we can derive directly a support on C for all the values (z, b) where z is a variable in V_2 and $b \in D(z)$. Thus, τ_1 plays the role of a generic support for the variables in V_2, and there is no reason to waste time in searching individual supports in C_2 for the variables in V_2. (A symmetric argument holds for a support in C_2 and variables in V_1.)*

The filtering algorithm in Figure   simply applies those principles. Once we know there exists a support of (x, a) on C, it is clearly a waste of time to try to find another one.

```
procedure GlobalOrFiltering(C₁, C₂)
    τ₁ = getAnySupport(C₁)
    if τ₁ = nil
        replace this constraint by C₂
        return
    τ₂ = getAnySupport(C₂)
    if τ₂ = nil
        replace this constraint by C₁
        return
    For each variable x in V₃
        For each value a in D(x)
            τ = getSupport(C₁, (x, a))
            if τ = nil
                τ = getSupport(C₂, (x, a));
                if τ = nil
                    remove a from D(x)
```

Fig. 1. A global filtering algorithm for or(C_1, C_2)

Description of the Algorithm. The algorithm first computes any support for C_1, thanks to the method getAnySupport(C_1) which is supposed to exist. In other terms, a support is *exported* from C_1. If such a support can be found, then it is a generic support for every variable in V_2. If none exists, then C_1 is false; thus C_2 must be true, it is added to the constraint solver (and the constraint or(C_1, C_2) can be removed from the constraint solver).

A similar generic support is sought for C_2.

Then the algorithm has to find a support for each value of each variable in V_3. It suffices to find one support of this pair variable/value in C_1 or in C_2. If none exists, the value can be removed from the domain of the variable. The method getSupport($C, (x, a)$) is supposed to exist for each constraint C_1 and C_2.

The following proposition is a direct consequence of the proposition :

Proposition 3. *The filtering algorithm globalOr achieves arc-consistency for the constraint* or(C_1, C_2).

Following the GAC-schema [ ], two important characteristics for arc-consistency algorithms are:

- *incrementality* of arc-consistency maintenance during search;
- taking into account *multidirectionality* of supports (i.e., a support for a given pair variable/value is also a support for every pairs variable/value that compose the support)

It is easy to modify the algorithm of Figure   to take into account incrementality: this can be done by storing supports, reconsidering only supports that are made invalid, and starting the search for a new support from the old one thanks to a method `getNextSupport()`. Taking into account multidirectionnality is also easy to do. But, in this paper, for clarity we have decided to present the algorithms without incrementality and multidirectionality. The reader is referred to [ ] where incrementality and multidirectionality are well described. However, it should be noted that incrementality will be still more efficient on this algorithm than with constructive disjunction since incrementality is directly related to the number of supports that are maintained, which are much less numerous as the next paragraph will show.

Complexity of the Algorithm. Let us assume a general model with p constraints in disjunction $or(C_1, C_2, ..., C_p)$. There is no difference in complexity if we write a specific code for the p-ary disjonction constraint, or if we write $or(C_1, or(C_2, or(C_3, or(...))))$ since the globalOr implements the protocol (see Section 6). Assume that:

- there are $k > 0$ variables that are shared in the p constraints.
- all the variables have d values in their domains,
- all the constraints have the same arity r,
- we do not take into account multidirectionnality of support in our analysis since its impact on the complexity is quite complex to measure and depends on the tuples of the constraints and on the supports that are considered first.

Constructive disjunction will find $p * r * d$ supports. The globalOr filtering algorithm will find $k * d$ supports. As $k \leq r$, globalOr save at least a factor of p in CPU time. The gain may be much larger, for example, if there are two variables that are shared by 100 constraints whose arity is 10, and with domains of size 20: there are 40 supports to find in one case and 20000 in the other case.

4   Negation of a Constraint

In a given solver, the negation of a primitive constraint is sometimes a primitive constraint. For example, the negation of $x < y$ is $x \geq y$, both these constraints are generally primitive constraints of a given solver. Nevertheless, there generally exist in a solver primitive constraints whose negation is not a primitive constraint. Furthermore, the negation of a combination of constraints like $or(and(C_1, C_2), or(C_3, C_4))$ is generally not a primitive constraint.

The usual filtering algorithm for the non primitive constraint $not(C_1)$ is to check whether C_1 is *entailed*, i.e., whether it is always true for the given domains. If this is the case, the constraint $not(C_1)$ is false and thus the filtering algorithm return a failure. That means that the usual filtering algorithm *never* reduces the domains of the variables, except to raise a failure.

We propose a new filtering algorithm for the negation of a constraint that is able to achieves arc-consistency and thus to prune the search tree.

The principle is that a support for a value (x, a) on the constraint $not(C_1)$ is a conflict for (x, a) on the constraint C_1.

```
procedure not-Filtering(C)
    For each variable x in X(C)
        For each value a in D(x)
            τ = getConflict(C, (x, a))
            if τ = nil
                remove a from D(x)
```

Fig. 2. A global filtering algorithm for not(C)

Let us assume there exists a function getConflict(C, (x, a)) that returns a conflict. The filtering algorithm we propose for the *not* constraint is the following:

Indeed, this algorithm is not so new. It corresponds, in another form, to the GAC-Schema for a constraint expressed by forbidden tuples [ ].

5   Conjunction of Constraints

5.1   Filtering Algorithm

Now, let us consider the constraint $C = and(C_1, C_2)$. Achieving arc-consistency over C generally reduces more the domains than achieving arc-consistency for the constraint system $\{C_1, C_2\}$, but is more costly.

Let V_1 denote the variables that are in C_1 but not in C_2, V_2 the variables that are in C_2 but not in C_1, and V_3 the variables that are shared by C_1 and C_2.

Each support for the constraint C can be obtained by joining two compatible supports for C_1 and C_2, *i.e.* two supports that share the same values for the shared variables. This is formalized in the following proposition:

Proposition 4. *If τ_1 is a support for C_1 and τ_2 is a support for C_2 such that $\tau_1[V_3] = \tau_2[V_3]$, then τ such that $\tau[V_1] = \tau_1[V_1]$, $\tau[V_2] = \tau_2[V_2]$, $\tau[V_3] = \tau_1[V_3] = \tau_2[V_3]$ is a support for C.*

Conversely, if τ is a support for C, then $\tau[V_1 \cup V_3]$ is a support for C_1, $\tau[V_2 \cup V_3]$ is a support for C_2.

Thus to find a support for C, one can try to find two *compatible* supports in C_1 and C_2. This may be costly. However note this is not always a difficult task:

Corollary 3. *if $|V_3| = 0$ or $|V_3| = 1$, arc-consistency over the constraint system $\{C_1, C_2\}$ is equivalent to arc-consistency over and(C_1, C_2).*

A possible algorithm that achieves arc-consistency on the conjunction of two constraints is given in Figure : it focuses on the difficult and more interesting case, and assumes there are at least two variables that are shared between C_1 and C_2, i.e., $|V_3| \geq 2$.

The algorithm achieves arc-consistency for the three sets of variables V_1, V_2, V_3. It first achieves arc-consistency for each constraint independently (note this first step is not necessary, this is only an optimization).

```
procedure and-Filtering(C₁, C₂)
    AC-filtering(C₁)
    AC-filtering(C₂)
    For each variable x in V₁ ∪ V₂ ∪ V₃
        For each value a in D(x)
            found = false
            τ₁ = getSupport(C₁, (x, a))
            while τ₁ ≠ nil and not found
                let τ₁[V₃] be the values of the variables V₃ in τ₁
                τ₂ = getCommonSupport(C₂, V₃, τ₁[V₃])
                if τ₂ ≠ nil
                    found = true
                    % to get a support for (x, a), merge τ₁ and τ₂
                else
                    τ₁ = getNextSupport(C₁, (x, a), τ₁)
            if not found
                remove a from D(x)
```

Fig. 3. A global filtering algorithm for and(C_1, C_2)

Then, the algorithm loops over the pairs variable/value to be supported. For each pair (x, a), it tries to find a support τ_1 on C_1, and then tries to find a support on C_2, τ_2, which is compatible with τ_1. That is, the variables in V_3 have to take the same values in τ_1 and in τ_2. This is done thanks to a function getCommonSupport(), which is supposed to exist. Function getCommonSupport() seeks a support on C_2 which is common for different pairs variable/value. If such a common support does not exist on C_2, another support τ_1 for (x, a) on C_1 is generated by the function getNextSupport(), which is also assumed to exist.

5.2 Efficient Search of Compatible Supports

The key function in the algorithm of Figure is the function getCommonSupport(), which takes as parameters a constraint C and a set of pairs variable/value $\{(x_1, a_1), ..., (x_p, a_p)\}$. It returns a support τ on the constraint C such that $\tau[\{x_1, ..., x_p\}] = \{(x_1, a_1), ..., (x_p, a_p)\}$.

This is in general a quite expensive function. In this section we propose an efficient method when the constraint is given in extension as a set of tuples.

A good data structure for representing allowed tuples in a constraint is the following:

```
class tuple
  value: array of int
  nextValue: array of pointer
```

The values are stored in the field `value`, and `nextValue[i]` contains a pointer to the next tuple where the i-th variable is assigned to `values[i]`. This data structure allows efficient implementation of `getNextSupport()` to be done.

Nevertheless, for the function `getCommonSupport()`, even with only two common values as parameters, say (x, a) and (y, b), this data structure is not sufficient to reach a good efficiency. We could iterate over the list of supports of (x, a) and check each support to see whether it is also a support of (y, b). The problem is that this algorithm may lead to a complete exploration of the list of supports of (x, a), as shown in the following example.

Example 3. Consider the constraint over w, x, y, z_1, z_2 that ensures that

- $w = a \rightarrow (x = a, y = a)$
- $w = b \rightarrow (x = b, y = b)$
- $w = c \rightarrow (x = a, y = b)$

Assume the allowed values for z_1 and z_2 are the integers in [1, 100]. As the constraint is given in extension, values for variables z_1 and z_2 also appear in the tuples. The allowed tuples are:

w	x	y	z_1	z_2
a	a	a	1	1
a	a	a	1	2
...	...	...	...	...
a	a	a	1	100
a	a	a	2	1
...	...	...	...	...
a	a	a	100	100
b	b	b	1	1
...	...	...	...	...
b	b	b	1	100
...	...	...	...	...
b	b	b	100	100
c	a	b	1	1
...	...	...	...	...

Assume we want to find a support where $(x = a)$ and $(y = b)$. Thus, we iterate over the list of supports of (x, a) and check each support to see whether it is also a support of (y, b). This needs $100*100=10000$ support checks in this case. The same problem occurs if we want to iterate over the list of supports of (y, b) and check each support to see whether it is also a support of (x, a).

The example should have convinced the reader that iterating over the list of supports of a given value and checking the other value may have poor performance. Thus, we need to improve the search for a common support. First of all, assume a total order on the tuples of the constraint, for example the lexicographic order $\prec_{lo}$, and assume the lists of supports of a given value are ordered.

If we know that there is no common support smaller than a support τ_1 of (x, a), then seeking for a common support smaller than τ_1 in the list of supports of (y, b) is useless. (The symmetric argument also holds.)

Thus, a good idea would be to perform an interleaved exploration of both the list of supports of (x, a) and the list of supports of (y, b): we select at each step the tuple τ which is the greatest among the current tuple of the list of supports of (x, a) and the current tuple of the list of supports of (y, b). Then τ is taken as the starting point for the next iteration: i.e. the next tuple from the list of supports of (x, a) and the next tuple from the list of supports of (y, b) should both be greater or equal to τ. In the above example, this method needs only 3 supports checks instead of 10000.

Now, assume τ the greatest tuple comes from the list of supports of (x, a), and does not have value (y, b), but, say, value (y, c). It is easy to find the next support of (x, a) from τ since it suffices to follow the pointer nextValue[x]. But, the pointer nextValue[y] points to the next tuple of (y, c). So, how can we find the next support of (y, b) from τ? Indeed, this algorithm needs to have in each tuple not only a pointer to the next tuple of every pair variable/value in the current tuple, but also the next tuple of every *possible* pair variable/value. The field nextValue thus becomes a bi-dimensional array, indexed by variables and values. That is to say, we will pay this algorithmic improvement with an increase in space consumption for storing the tuples. The increase in space is a factor of d, where d is the size of the domains. Generally, the tuple set of a constraint is large, d is also large, and in that case, this approach is not possible.

We propose in this section a better approach. With a small and constant increase in the consumption space it will be able to use the starting point as above in time linear with d.

For doing this, we introduce a data structure, called hologram-tuples: it adds to the above tuple data structure two additional arrays, whose indexes range over the variables of the constraint, and an additional pointer:

```
class hologram-tuple
   value: array of int
   nextValue: array of pointer
   nextTuple: pointer
   redundantValue: array of int
   redundantNextValue: array of pointer
```

Assume all the tuples are stored in an increasing order in a global list, thanks to the pointer nextTuple. Assume also that nextValue[i], that links a sub-list of this global list, preserves that order.

Assume the possible values for the variable i in the set of tuples are $\{0, 1, 2, .., d_i\}$. The value redundantValue[i] is a value among $\{0, 1, 2, .., d_i\}$. It is precomputed in the following way:

- first tuple in the global list has value 0 in redundantValue[i];
- second tuple in the global list has value 1 in redundantValue[i];
- j-th tuple in the global list has value $(j-1)$ *modulo* d_i in redundantValue[i];

The field `redundantNextValue[i]` is a pointer to the next tuple τ in the global list where $\tau[i] = $ `redundantValue[i]`.

Then, we can state the following proposition which is the core of the method:

Proposition 5. *Let τ a tuple. Let $next(\tau, x, a)$ the smallest tuple that is greater than τ and which is a support of (x, a). With the hologram-tuples, we can find $next(\tau, x, a)$ in a time linear with d_i.*

Proof: It suffices to iterate over the global list starting from τ. Each explored tuple is checked to see if it contains (x, a). If this is the case, we have found $next(\tau, x, a)$. Otherwise, we check if the redundant value `redundantValue[x]` is equal to a. If this is the case, `redundantNextValue[x]` contains a pointer to $next(\tau, x, a)$. Otherwise, we continue iterating over the global list. As by construction `redundantValue[x]` $= a$ each $d_i + 1$ tuple, we know we explore at most d_i tuples. $\square$

The proof directly leads to an algorithm. Different uses more or less complex of the hologram-tuples data structure are possible to keep a good complexity in every case, but they are out of the scope of this paper. Nevertheless, in the worst case we can explore all the tuples, thus the upper bound of the algorithmic complexity is not improved, it remains linear in the number of tuples, and thus exponential in the arity of the constraint. But, the key point is that this data structure may save a number of iterations over the list of tuples that is exponential in the arity of the constraint.

6   Arbitrary Logical Combinations of Constraints

The previous three sections assume the existence of some functions over the constraints that can appear in the not, or, and constraints. Those functions (getSupport(), getNextSupport(), getCommonSupport(), getConflict,...) defines a protocol to export supports and conflicts from constraints. This protocol is quite simple to implement for constraints given in extension. For those constraints, we only detailed the most interesting function getCommonSupport(). The other functions already exist in the literature; for example, our getConflict is nothing else than the seekSupport of [ ] over a constraint whose forbidden tuples instead of allowed tuples are given in extension.

Thus, for now, we are able to use not, or, and constraints over constraints given in extension, but we can have only flat logical combinations. Now we address the question of an arbitrary logical combinations of constraints given in extension like $or(c_1, or(c_2, and(c_3, not(c_4))))$.

Indeed, it suffices that the not, or, and constraints themselves implement the protocol to export supports. This protocol is not difficult to implement over those constraints. Consider the function getSupport(C, (x, a)). It can be derived immediately from the filtering algorithms we gave, since a filtering algorithm is typically a loop over the values to get their supports.

As for the function getConflict, there is no more difficulty. GetConflict over a constraint $not(C)$ amounts to finding a support over C. GetConflict over a constraint $or(C_1, C_2)$ is equivalent to get a support over constraint $and(not(C_1), not(C_2))$. GetConflict over a constraint $and(C_1, C_2)$ is equivalent to get a support over constraint $or(not(C_1), not(C_2))$.

Thanks to this protocol, the **not**, **or**, **and** constraints can themselves be combined through logical connectives, and thus arbitrary logical combinations of constraints can benefit from the filtering algorithms we propose. Thus, by recurrence we can show the following proposition:

Proposition 6. *Arc-consistency over an arbitrary complex logical formula on constraints given in extension can be computed.*

Of course, the problem of solving an arbitrary complex logical formula on constraints given in extension is an NP-complete problem, and so arc-consistency filtering for such a formula is also NP-complete.

7   Related Work and Perspectives

As for related work on conjunctions of constraints, an interesting point is raised in [ ]: they remark that it is sometimes worthwhile to solve independently a subproblem, that is a given conjunction of constraints, in order to solve the whole problem more efficiently. Thus, they propose to estimate the benefits of developping a new constraint that corresponds to a subproblem by first solving the subproblem "on the fly". In our approach it will be more simple to try to solve some subproblems (just put a *and()* around the constraints of the subproblem). Furthermore, sometimes our approach will be as efficient as a specific constraint, thus the developement of a new constraint may be avoided.

Among the perspectives of this work we can cite:

- For constraints that are not given in extension but in intension like $x < y$, the protocol can still be used if a support can be provided. Nevertheless, it is much better to generalize this protocol to exploit for example the structure of monotonic or arithmetic constraints as in AC-5 [ ].
- An arbitrary formula is an NP-complete problem, and several equivalent logical formulas are possible for a given constraint. An interesting open problem is to find a formula that gives the best efficiency of the filtering algorithms. This problem is very important in databases and is known as query optimization.
- The hologram-tuples data structure is applicable in other contexts. We already mentioned databases algorithms, where the join operator on relations is quite close to the **and** constraint. To stay in CP, the AC filtering algorithms can be improved with the `getCommonSupport()` method and the hologram-tuples: when seeking for a support for (x, a), we can take into account the values that remain in the domains of the other variables. For each other variable y, whose values in the domain are noted v_i, we know that a support for (x, a) should be also a support for at least one value v_i of y. Let $minSup(y)$ the min over the v_i of the smallest supports common with (x, a) and (y, v_i). Then, all the tuples that are less than the max of $minSup(y)$ over all variables y can be skipped.

8   Conclusion

The usual logical connectives between constraints have poor filtering algorithms. Some simply do not filter at all the domains (as the standard filtering algorithms for not, or and imply). Other are costly in CPU time, difficult to implement and cannot be combined (as constructive disjunction).

In this paper, we propose new filtering algorithms over those logical connectives. More precisely, the contributions are the following:

- We propose filtering algorithms that achieve arc-consistency for the or, and and not constraints.
- The filtering algorithm we propose for the or constraint is always better than constructive disjunction in efficiency for the same pruning. It improves constructive disjunction in the same way AC-6 [ ] improved AC-4 [ ]. Furthermore, the approach can be generalized to all meta constraints on cardinality like: atmost (or at least) p constraints are true among q.
- We introduce the hologram-tuples data structure for constraints defined in extension. It allows to drastically reduce the number of supports checks in the filtering algorithm for the and.
- Another contribution is the definition of a protocol to be implemented by constraints to export supports. Thanks to this protocol, the constraints combined through logical connectives can benefit from the filtering algorithms we propose for not, or, and. Thus, arc-consistency over an arbitrary complex logical formula can be computed.

We think that some applications that suffer from poor performance due to the lack of domain reduction of logical connectives may be improved by the use of the new filtering algorithms proposed in this paper. Furthermore, developpers generally know the poor domain reductions performed by standard filtering algorithms over logical connectives. Thus, they try to avoid them, by designing specific constraints or trying to design more complex models. We hope that, thanks to those new filtering algorithms, a first step is done that will make the use of logical connectives in constraint programming much more frequent and successful.

References

[1] ILOG: ILOG JConfigurator 2.0 User's Manual. ILOG S. A. (2003)
[2] Lhomme, O.: An efficient filtering algorithm for disjunction of constraints. In: Proc. of the the Conference on Principles and Practice of Constraint Programming. (2003) 904–908
[3] Van Hentenryck, P., Saraswat, V., Deville, Y.: Design, implementation, and evaluation of the constraint language cc(FD). In Podelski, A., ed.: Constraint Programming: Basics and Trends. LNCS 910. Springer (1995) (Châtillon-sur-Seine Spring School, France, May 1994)

[4] Würtz, J., Müller, T.: Constructive disjunction revisited. In Görz, G., Hölldobler, S., eds.: 20th German Annual Conference on Artificial Intelligence. Volume 1137., Dresden, Germany, Springer-Verlag (1996) 377–386

[5] Bessière, C.: Arc-consistency and arc-consistency again. Artificial Intelligence **65** (1994) 179–190   ,

[6] Mohr, R., Henderson, T. C.: Arc and path consistency revisited. Artificial Intelligence **28** (1986) 225–233   ,

[7] Bessière, C., Régin, J. C.: Arc consistency for general constraints networks: preliminary results. In: IJCAI'97, Nagoya (1997) 398–404   ,   ,   ,

[8] Bessière, C., Régin, J. C.: Enforcing arc consistency on global constraints by solving subproblems on the fly. In: CP'99, Alexandria, VA, USA (1999) 103–117

[9] Van Hentenryck, P., Deville, Y., Teng, C.: A generic arc-consistency algorithm and its specializations. Artificial Intelligence **57** (1992) 291–321

Combining Forces to Solve
the Car Sequencing Problem

Laurent Perron and Paul Shaw

ILOG SA
9 rue de Verdun, 94253 Gentilly cedex, France
{lperron,pshaw}@ilog.fr

Abstract. Car sequencing is a well-known difficult problem. It has resisted and still resists the best techniques launched against it. Instead of creating a sophisticated search technique specifically designed and tuned for this problem, we will combine different simple local search-like methods using a portfolio of algorithms framework. In practice, we will base our solver on a powerful LNS algorithm and we will use the other local search-like algorithms as a diversification schema for it. The result is an algorithm is competitive with the best known approaches.

1 Introduction

Car sequencing [, , , , ,] is a standard feasibility problem in the Constraint Programming community. It is known for its difficulty and there exists no definitive method to solve it. Some instances are part of the CSP Lib repository.

As with any difficult problem, we can find two approaches to solve it. The first one is based on complete search and has the ability to prove the existence or the non-existence of a solution. This approach uses the maximum amount of propagation [ ]. The second approach is based on local search methods and derivatives. These methods are, by nature, built to find feasible solutions and are not able to prove the non-existence of feasible solutions. Recent efforts have shown important improvements in this area [ ].

Structured Large Neighborhood Search methods have been proved to be successful on a variety of hard combinatorial problems, *e.g.*, vehicle routing [,], scheduling (job-shop: shuffle [ ], shifting bottleneck [ ], forget-and-extend [ ]; RCPSP: block neighborhood [ ]), network design [ ], frequency allocation [ ]. It was natural to try structured LNS on the car sequencing problem.

The first phase of our work was rewarded by mixed results. Even with the implementation of special improvements (or tricks), the Large Neighborhood Search Solver was not giving satisfactory results as the search was stuck in local minima. At this point, we saw two possible routes. The first one involved creating dedicated and complex neighborhoods. The second one relied on the portfolio of algorithms [ ] and uses specific search algorithms as diversification routines on top of the LNS solver. We chose the second route as it required much less effort and, to this effect, we designed two simple and specific local search improvement routines.

J.-C. Régin and M. Rueher (Eds.): CPAIOR 2004, LNCS 3011, pp. 225–   , 2004.

In the end, we ended up with a effective solver for the car sequencing problem, built with pieces of rather simple search algorithms.

The rest of the article is divided as follows: section   describes the car sequencing problem in more detail, section   describes the LNS solver and all the improvements we tried to add to it, the successful ones as well as the unsuccessful ones. We then present the block swapping method in section   and the sequence shifting method in section . These three methods are combined using an algorithm portfolio. We give experimental results in section .

2   The Car Sequencing Problem

The car sequencing problem is concerned with ordering the set of cars to be fed along a production line so that various options on the cars can be installed without over-burdening the production line. Options on the cars are things like air conditioning, sunroof, DVD player, and so on. Each option is installed by a different working bay. Each of these bays has a different capacity which is specified as the proportion of cars on which the option can be installed.

2.1   The Original Problem

The original car sequencing problem is stated as follows: We are given a set of options O and a set of configurations $K = \{k|k \subseteq O\}$. For each configuration $k \in K$, we associate a demand d_k which is the number of cars to be built with that configuration. We denote by n the total number of cars to be built: $n = \sum_{k \in K} d_k$. For each option $o \in O$ we define a *sequence length* l_o and a *capacity* c_o which state that no more than c_o cars in any sequence of l_o cars can have option o installed. The throughput of an option o is given as a fraction c_o/l_o. Given a sequence of configurations on the production line $s = < s_1, \ldots, s_n >$, we can state that:

$$\forall o \in O, \forall x \in \{1 \ldots n - l_o + 1\} \sum_{i=x}^{x+l_o-1} U_o(s_i) \le c_o$$

where $U_o(k) = 1$ if $o \in k$ and 0 otherwise.

This statement of the problem seeks to produce a sequence s of cars which violates none of the capacity restrictions of the installation bays; it is a decision problem.

2.2   From a Decision Problem to an Optimization Problem

In this paper, we wish to apply local search techniques to the car sequencing problem. As such, we need available a notion of quality of a sequence even if it does not satisfy all the capacity constraints. Approaches in the past have softened the capacity constraints and added a cost when the capacity of any installation bay is exceeded; for example, see [,]. Such an approach then seeks a solution of violation zero by a process of improvement (or reduction) of this violation.

However, when Constraint Programming is used, this violation-based representation can be undesirable as it results in little propagation until the violation of the partial sequence closely approaches or meets its upper bound. (This upper bound could be imposed by a local search, indicating that a solution better than a certain quality should be found.) Only at this point does any propagation into the sequence variables begin to occur.

We instead investigate an alternative model which keeps the capacity constraints as hard constraints but relaxes the problem by adding some additional cars of a single new configuration. This additional configuration is an 'empty' configuration: it requires no options and can be thought of as a stall in the production line. Such configurations can be inserted into a sequence when no 'real' configurations are possible there. This allows capacity restrictions to be respected by lengthening the sequence of real configurations. The idea, then, is to process all the original cars in the least possible number of slots. If all empty configurations come at the end of the sequence, then a solution to the original decision problem has been found. We introduce a cost variable c which is defined to be the number of slots required to install all cars minus n. So, when $c = 0$, no empty configurations are interspersed with real ones, and a solution to the original decision problem has been found.

3   Large Neighborhood Search

Large Neighborhood Search (LNS) is a technique which is based upon a combination of local and tree-based search methods. As such, it is an ideal candidate to be used when one wishes to perform local search in a Constraint Programming environment. The basic idea is that one iteratively relaxes a part of the problem, and then re-optimizes that part, hoping to find better solutions at each iteration. Constraint programming can be used to add bounds on the search variable to ensure that the new solution found is not worse than the current one.

Large Neighborhood Search has its roots in the shuffling procedures of [ ], but has become more popular of late thanks to its successes [, ,].

The main challenge in Large Neighborhood Search is knowing which part of the problem should be relaxed and re-optimized. A random choice rarely does as well as a more reasoned one, as was demonstrated in [ ], for example. As such, in this paper, much of the investigation will be based on exactly how LNS can be used successfully on this car sequencing model, and what choices of relaxation need to be made to assure this.

3.1   Basic Model

A LNS Solver is based on two components: the search part and the neighborhood builder. Here, the search part is a depth-first search assignment procedure that instantiates cars in one direction from the beginning to the end of the sequence. The values are chosen randomly with a probability of choosing a value v proportional to the number of unassigned cars for the same configuration v.

We have implemented two types of LNS neighborhoods. The solver chooses one of them randomly at each iteration of the improvement loop with the same probability. They both rely on a size parameter s. The first one is a purely random one that freezes all cars except for the $4s$ randomly chosen ones. The second one freezes all cars except cars in an interval of length s randomly placed on the sequence, and except cars in the tail. Cars in the tail are defined as all cars beyond the initial number of cars. (These cars are in fact pushed there by stalls.) We have observed that removing one neighborhood degrades performances.

The solver is then organized in the following way. Given an intermediate solution, we try to improve it with LNS. Thus, we choose one neighborhood randomly and freeze all variables that do not appear in the neighborhood. Then we choose a small search limit (here a fail limit), and we start a search on the unfrozen variables with the goal described above. At each iteration, we select the first acceptable solution (strictly better or equal in case of walking).

Finally, the goal is written in such a way that it will never insert a stall in the sequence, thus stalls only appears through propagation and not through goal programming.

3.2   Improvements and Non-improvements

In the first phase of our work, we tried to improve the performance of our solver by adding special techniques to the base LNS solver. We briefly describe them and indicate how successful these modifications were.

Walking. This improvement is important as we can see at the end of the experimental results section. Walking implies accepting equivalent intermediate solutions in a search iteration instead of requiring a strictly better one. Intuitively, it allows walking over plateaus.

Tail Optimization. We tried to periodically fully re-optimize the tail of the sequence. As this tail is generally small, this is not computationally expensive. Unfortunately, no improvements came from this method.

Swap-based Local Search. We also tried to apply a round of moves by swapping cars. Unfortunately, the space of feasible solutions is sparse and to find possible moves with this method, we had to relax the objective constraint. This allowed the local search to diverge towards high cost solutions and it never came back. We did not keep this method.

Circle Optimization. When there is only one stall remaining in the sequence, we can try to re-optimize a section around this stall. Thus, given a distance ρ, we can freeze all cars whose distance to the stall is greater than ρ. Then we can try to re-optimize the unfrozen part. Unfortunately, this never worked. Even with ρ around 15, meaning 31 cars to re-optimize, this method just wasted time and never found one feasible solution to the problem.

Growing the Neighborhood. We also tried to grow the size of the neighborhood slowly after repeated failures. Unfortunately, this did not allow the solver to escape the local minimum and just slowed down each iteration. Thus the system was slower and was not finding new solutions. We did not keep this change.

4   Block Swapping Methods

Following the long sequence of unsuccessful attempts at improving the LNS solver, we thought of inserting between each LNS iteration another improvement technique that would have the same characteristics in terms of running time, and that would perform a search that would be orthogonal. Thus we would not spend too much time on this method while it could act as a diversification schema for the LNS solver.

The first diversification schema we implemented was a method that cuts the sequence into blocks and tries to rearrange them.

The current solution is cut into blocks of random length (the minimum and the maximum of this length are parameters of this method). We also make sure there are no 'empty' configurations in the block and we remove the portions before the first and after the last empty configuration. Then we try to rearrange these blocks without changing the sequence inside a block.

This method is simple but describes moves that are not achievable with our LNS neighborhood and with the LNS approach in general, as it may changes many (and potentially all) variables if we insert a block near the beginning of the sequence.

There are many improvements that can be added to this simple block swapping schema.

4.1   Walking

Walking is always possible as a meta-heuristic on top of the block swapping routine. We will show the effect of block swapping on the performance of the complete solver in the experimental section.

4.2   Changing the Order of Instantiations of Blocks

As we have a limited number of fails, only a small number of swaps will be tried in a call to the block swapping improvement method. We have tried different block instantiation heuristics. The first one is simply to assign them from the beginning to the end of the sequence and to do a simple depth first search. This heuristic can be improved by using SBS [ ] instead of depth-first search.

We can also change the order of blocks when they are assigned, trying to prioritize the small blocks or to give a random order. These two orders did not improve the performance of the solver.

However, we kept the SBS modification as it did improve the results of our solver a little. This is compatible with the results we obtained in [ ].

4.3   Changing the Size of Blocks

We also tried to change the size of blocks dynamically during the search in case of repeated failure to improve the current solution. We found no interesting results when doing so and thus we did not keep any block size modification schema in our solver.

Yet, we did experiment with different maximum block sizes. The results are reported in the experimental section. The minimum block size is set to three.

5   Sequence Shifting Methods

The block swapping method did improve the performance of the solver a lot as can be seen in the experimental section phase. Unfortunately, there were still soluble problems that were not solved with the combination of LNS and block swapping. We then decided to implement another improvement method that would also add something to the block swapping and the LNS methods.

We decided to implement a basic improvement schema that tries to shift the sequence of cars towards the beginning, leaving the tail to be re-optimized.

Here is another search method that could be seen as a specialized kind of LNS. We shift the cars towards the start of the sequence by a given amount and re-optimize the unassigned tail using the default search goal.

As with the other two methods, we tried different modifications to this method.

5.1   Walking

At this point in our work, walking was a natural idea to try. Strangely, it did not improve our solver and even degraded its performance. Thus it was not kept.

5.2   Growing

The conclusion here is the same as with the block swapping phase. During all our experiments, we did not find any simple length changing schema that would improve the results of the solver. Thus, we did not keep any in our solver.

Yet, we did experiment with different fixed maximum lengths and these results are visible in the experimental results section. In those experiments, the minimum shift length is set to one.

6   Experimental Results

We will give two kinds of results in this section. The first part deals with comparative results between our solver and a re-implementation of the Comet article [ ]. The second part describes the effect of parameter tuning on the performance of our solver.

6.1   Experimental Context

In order to compare ourselves with the competition, we used what we thought was the best competitor, namely the Comet code from [ ]. Unfortunately, at the time of writing the article, we did not have access to the comet solver, thus we re-implemented the algorithm described in Pascal and Laurent's article from scratch (hereafter, we refer to this algorithm as RComet, for "re-implemented Comet"). While this was helpful, it is not completely satisfactory and we plan to do a comparison with the real thing (the real Comet code) as soon as possible.

Note that the Comet optimizer code uses a different optimization criterion from us: namely violations of the capacity constraints. To perform a more reasonable comparison, we decoded the order obtained by RComet to produce a number of empty configurations. This comparison is more meaningful than comparing violations to empty configurations, but it is far from perfect. The main problem is that RComet code has no incentive to minimize stalls, and it is possible to have two solutions a and b where a has lower violations but b has lower stalls. Thus, although we give figures in terms of stalls for the RComet code, comparisons only have real merit between run times for which both solvers found a solution of cost 0.

All tests were run on a Pentium-M 1.4 GHz with 1 GB of RAM. The code was compiled with Microsoft Visual Studio .NET 2003. Unless otherwise specified, the LNS approach uses block swapping with minimum size 3 and maximum size 10 and sequence shifts with minimum shift 1 and maximum shift 30. Furthermore, as we have tried avoid the addition of more control parameters, all three basic methods use the same fail limit, and are applied in sequence with the same number of calls.

All our experiments rely on randomness. To deal with unstable results, we propose to give results of typical runs. By typical, we mean that, given a seed s for the random generator, we run all instances with the same seed and then we do multiple full runs with different seeds. Then we select the results for the seed that produced the median of the results. We applied this methodology to both the RComet runs and runs from our solver.

All tests use a time limit of ten minutes. The tables report the time to get to the best solution in the median run.

6.2   Results on the CSP Lib Data Set

Table   presents the result we obtained on all the randomly generated instances of the CSP library. Unfortunately, except for one notable exception (90-05), all are easy and most can be solved in a few seconds. Thus, except for the 90-05 instance, we will not use these problems in the rest of the article.

Table   presents our results on the hard problems from the CSP Lib.

This result demonstrates the competitiveness of the LNS approach. In particular, our implementation find solutions where RComet does and, on average, finds solutions for problem 16-81, when RComet does not. On the other hand, when both solvers find a solution, the RComet is significantly faster than our solver.

Table 1. Results on random CSP Lib instances

Problem	Solving Time (s)
60-**	0-9
65-**	1-10
70-**	1-9
75-**	2-11
80-**	2-19
85-**	2-27
90-**	1-92

Table 2. Results on hard CSP Lib instances

Problem	our stalls	our time (s)	RComet stalls	RComet time (s)
10-93	3	228	6	50
16-81	0	296	1	250
19-71	2	51	2	55
21-90	2	38	2	15
26-82	0	93	0	29
36-92	2	241	2	65
4-72	0	412	0	95
41.66	0	23	0	12
6-76	6	12	6	9

6.3   Results on Randomly Generated Hard Instances

As we can see from the previous section, nearly all instances of the CSP Lib are easy and all except one are solved in less than one minute of CPU time. Thus we decided to generate hard instances. The random generator is described in the next section. The following section describes the results we obtained on these instances.

The Random Generator. The random generator we created takes three parameters: the number of cars n, the number of options o, and the number of configurations c. We first generate the o options, without replacement on the set of option throughputs: $\{p/q \mid p,q \in Z, 1 \le p \le 3, p < q \le p+2\}$.

We then generate the c configurations. The first o configurations are structured, the ith configuration involving only one option: option i. We generate the remaining $c - o$ configurations independently of each other, except for the stipulation that no duplicate nor empty configurations are created. Each of these configurations is generated by including each option independently with probability $1/3$. Finally, the set of cars to construct is generated by choosing a configuration randomly and uniformly for each car.

Problems are then rejected on two simple criteria: too easy, or trivially infeasible. A problem is considered too easy if more than half of its options are used at less than 91%, and trivially infeasible if at least one of its options is

Table 3. Results on hard random instances

Problem	our stalls	our time (s)	RComet stalls	RComet time (s)
00	0	24	0	11
01	8	41	10	60
02	5	43	5	180
03	9	25	10	30
04	0	85	1	90
05	2	421	7	50
06	0	19	0	13
07	3	119	4	20
08	0	232	0	90
09	1	152	1	580
10	5	80	7	100
11	0	17	1	130
12	14	14	14	20
13	0	34	0	13
14	3	397	5	190
15	4	132	6	340
16	6	50	7	140
17	7	47	8	412
18	0	47	0	21
19	0	160	1	520

used at over 100%. The use of an option o is defined as: $\mu_o(\sum_{k \in K|U_o(k)=1} d_k)/n$. where $\mu_o(t)$ is the minimum length sequence needed to accommodate option o occurring t times: $\mu_o(t) = l_o((t-1) \textbf{ div } c_o) + ((t-1) \textbf{ mod } c_o) + 1$.

This approach can take a long time to generate challenging instances due to the large number of instances that are rejected. However, it does have the advantage that there is little bias in the generation, especially when it comes to the distribution of configurations. The random generator and the generated instances are freely available from the authors. For this paper, we generated problems with $n = 100, o = 8, c = 20$.

The Results. Table   give results on the 20 instances we have created with the previous problem generator with $n = 100, o = 8, c = 20$.

On this problem suite, again our solver finds solutions to all five problems solved by RComet, plus three more. On the problems where both methods find solutions, RComet is again faster, but to a much lesser extent than on the CSP Lib problems.

The conclusion from our experiments is that our solver is more robust and provides solutions to harder problems when RComet cannot. However, for the easier of the problems, RComet's local search approach is more efficient.

6.4   Analysis of the Portfolio of Algorithms

We measure the effect of each method on the overall results. To achieve this, we test all combinations of algorithms on a selected subset of problems.

Focus on Hard Feasible Instances. In this section, we focus on 12 hard feasible instances. We will try to evaluate the effect of different parameters on our search performance. We will consider instances 16-81, 26-92, 4-72, 41-22, 90-05 of the CSP Lib and instances 00, 04, 06, 08, 11, 18, 19 of our generated problems. For these problems and each parameter set, we will give the number of feasible solutions found with a cost of 0 (feasible for the original problem) and the running time, setting the time of a run to 600 seconds if the optimal solution is not found.

Experimental Results. Table   shows the results of all combinations of the three methods (LNS, Block Swapping and Shift) on the twelve hard feasible instances. The following table shows, for a given time, the number of problems solved in a time below the given time.

The results are clear. None of the extra methods (Block Swapping and Sequence Shifting) are effective solving methods as they are not able to find one feasible solution on the twelve problems.

Furthermore, block swapping is much better at improving our LNS implementation than sequence shifting. Furthermore, only the combination of the three methods is able to find results for all twelve problems.

Finally, we can see that LNS + shift is slower than LNS alone. Thus sequence shifting can only be seen as an extra diversification routine and not as a stand-alone improvement routine at all.

6.5   Tuning the Search Parameters

We keep the same twelve hard feasible instances to investigate the effect of parameter tuning on the result of the search. In particular, we will check the effect of changing the maximum size of blocks used in block swapping, of changing the

Table 4. Removing methods from the portfolio of algorithms

Methods	50s	100s	150s	200s	250s	300s	350s	400s	450s	500s	550s	600s
Full	3	6	7	9	10	11	11	11	12	12	12	12
Block + Shift	0	0	0	0	0	0	0	0	0	0	0	0
LNS + Shift	0	0	0	1	1	2	4	4	4	4	4	4
LNS + Block	3	6	7	7	7	8	8	8	8	8	8	8
LNS	0	1	2	2	3	3	4	4	4	4	4	4
Block	0	0	0	0	0	0	0	0	0	0	0	0
Shift	0	0	0	0	0	0	0	0	0	0	0	0

Table 5. Modifying the maximum block size (MBS)

MBS	50s	100s	150s	200s	250s	300s	350s	400s	450s	500s	550s	600s
5	4	6	7	7	7	7	7	8	8	8	8	9
8	4	7	7	8	8	9	9	10	11	11	11	11
10	3	6	7	9	10	11	11	11	12	12	12	12
12	3	6	6	8	9	9	9	9	9	9	10	10
15	3	5	6	6	8	8	8	8	9	9	10	10
18	2	6	7	8	9	9	9	9	9	9	9	9
20	2	5	5	6	7	8	9	10	10	10	10	10
22	2	5	6	6	7	7	8	9	9	9	10	10

maximum length of a shift, and the effect of LNS walking, block walking or shift walking.

In order to conduct our experiments, we fix all parameters except one and see the effect of changing one parameter only.

Changing Block Swapping Maximum Size. By changing the maximum size of the block built in the block swapping phase, we can influence the complexity of the underlying search and we can influence the total number of blocks. The bigger the blocks, the fewer they are and the smaller the search will be.

With small blocks, we can better explore the effect of rearranging the end of the sequence using block swapping. Thus we find easy solutions more quickly. However, due to the depth-first search method employed, the search is concentrated at the end of the sequence. Thus if a non-optimal group of cars is located early in the sequence, it is likely it will not be touched by the block swapping routine as the maximum block size is small. Thus we will find easy solutions faster and may not find the hard solutions at all.

On the other hand, if the maximum block size is big, then we add diversification at the expense of intensification; more tries will be unsuccessful, thus slowing down the complete solver. For example, the number of solutions found around 200s goes down as the maximum size of the blocks increases. Furthermore, the number of solutions found after fifty seconds slowly decreases from four to two as the maximum block size grows from four to twenty-two.

Changing the Fail Limit. By changing the fail limit used in the small searches of each loop, we explore the trade off between searching more and searching more often. Our experience tell us that above a given threshold which gives poor results, the performance of the LNS system degrades rapidly as we allow more breadth in the inner search loops.

We see a peak at around eighty fails per inner search. There is also a definite trend indicating that having too small a search limit is better than having a too large one, which is consistent with our expectations. Also interesting is the fact that a larger fail limit results in more solutions earlier, but is overtaken by

Table 6. Modifying the fail limit (FL)

FL	50s	100s	150s	200s	250s	300s	350s	400s	450s	500s	550s	600s
20	3	4	5	7	7	8	9	10	10	10	10	10
30	1	5	6	7	9	9	9	9	9	9	9	9
40	2	5	8	8	9	9	9	9	10	10	10	10
50	4	5	6	7	7	8	9	9	9	9	9	9
60	4	6	7	8	8	9	9	9	9	9	9	9
70	3	6	7	7	8	8	8	8	8	9	9	10
80	3	6	7	9	10	11	11	11	12	12	12	12
90	3	4	5	6	7	7	8	8	9	10	10	10
100	3	6	7	7	9	9	9	9	9	9	9	9
110	1	6	7	7	8	8	8	8	8	8	8	8
120	3	5	6	7	7	8	8	9	9	9	9	9
130	3	6	8	9	9	9	9	9	9	9	9	9

smaller limit after around three minutes of search. Thus, it would appear that a larger fail limit can be efficient on easier problems, by tends to be a burden on more difficult ones.

Changing the Sequence Shifting. By modifying the maximum length of a shift, we can quite radically change the structure of the problem solved. Indeed, the space filled by the shift is solved rather naively by the default search goal. Thus a longer shift allows a fast recycling of a problematic zone by shifting it out of the sequence, at the expense of creating long badly optimized tails. A shorter shift sequence implies less freedom in escaping a local minima while keeping quite optimized tails.

Changing the maximum length of a shift is hard to interpret. There is a specific improvement around thirty where we can find solutions for all twelve problems. But we lack statistical results to offer a definitive conclusion.

Table 7. Modifying the shift length (MSL)

MSL	50s	100s	150s	200s	250s	300s	350s	400s	450s	500s	550s	600s
05	3	6	8	9	9	9	10	10	10	10	10	10
10	3	5	8	9	9	9	9	9	10	10	10	10
15	3	6	7	7	7	7	8	8	8	8	8	9
20	3	5	6	7	7	7	7	8	10	10	10	11
25	4	6	6	6	7	7	9	9	9	9	9	9
30	3	6	7	9	10	11	11	11	12	12	12	12
35	3	7	8	8	9	9	9	10	10	10	10	10
40	4	7	7	8	8	9	9	9	9	9	9	10
45	3	7	8	8	9	9	9	9	9	9	9	9
50	4	5	6	8	8	9	9	9	9	10	10	10

Table 8. Modifying the LNS walking parameter (LNSW)

LNSW	50s	100s	150s	200s	250s	300s	350s	400s	450s	500s	550s	600s
Allowed	3	6	7	9	10	11	11	11	12	12	12	12
Forbidden	0	0	0	0	0	0	0	0	0	0	0	0

Table 9. Modifying the block walking parameter (BW)

BW	50s	100s	150s	200s	250s	300s	350s	400s	450s	500s	550s	600s
Allowed	3	6	7	9	10	11	11	11	12	12	12	12
Forbidden	3	6	7	8	8	8	10	10	10	10	10	10

Table 10. Modifying the shift walking parameter (SW)

SW	50s	100s	150s	200s	250s	300s	350s	400s	450s	500s	550s	600s
Allowed	3	6	6	7	8	8	8	8	8	8	8	9
Forbidden	3	6	7	9	10	11	11	11	12	12	12	12

Forbidding LNS Walking. We can make the same tests with LNS and walking meta-heuristics. These results are shown in table .

LNS walking is a mandatory improvement to the LNS solver. Without it, our solver is not able to find any solution to the twelve problems. This is quite different from our experience with other problems as this parameter was not so effective. This is probably due to structure of the cost function: in car sequencing problems, there are large numbers of neighboring solutions with equal cost.

Forbidding Block Walking. As we have seen in the basic Large Neighborhood Search approach, walking in the LNS part is a key feature with regard to the robustness of the search. We can wonder if this is the case with block swapping, because if walking allows the solver to escape plateau areas, it also consumes time and walking too much may in the end consume too much time and degrade search performance. These results are shown in table .

Walking is a plus for the block swapping routine as it allows the solver to find more solutions more quickly. In particular, we are able to find all twelve solutions with it while we find only ten of them without it.

Allowing Shift Walking. We can make the same tests with shift and walking meta-heuristics. These results are shown in table .

Strangely enough, allowing shift walking does not improve the results of our solver and even degrades its performance. We do not have any clear explanation at this time. We just report the results.

7 Conclusion and Future Work

The main contribution of this article is the different usage of the portfolio of algorithms framework. Instead of combining multiple competing search techniques to achieve robustness, we combine a master search technique (LNS) with other algorithms whose purpose is not to solve the problem, but to diversify and create opportunities for the master search. This approach means that several simple searches can be used instead of one complicated search, leading to a more modular and easily understood system. We also investigated many possible improvements and modifications to the base LNS architecture and report on them, even if most of them were not successful.

Our method works well on the car sequencing problem, especially harder instances. We have achieved good results by being able to solve all twelve hard instances. In the future, we hope to compare against the Comet solver itself when it becomes available.

examples in order to have real statistics on the performance of our solver.

In the future, we would like to further study our model based on insertion of stalls vs. the more standard violation-based model. Good algorithms for one model may not carry over into the other, and the full consequences of the modeling differences warrant further analysis and understanding.

Finally, we would like to examine the effect of the introduction of problem-specific heuristics into the LNS search tree, such as those explored in [ ]. These heuristics are reported to work well, even using a randomized backtracking search, and such a comparison merits attention.

We would like to thank our referees for taking the the time to provide excellent feedback and critical comment.

References

[1] Parrello, B., Kabat, W., Wos, L.: Job-shop scheduling using automated reasoning: a case study of the car-sequencing problem. Journal of Automated Reasoning **2** (1986) 1–42

[2] Dincbas, M., Simonis, H., Hentenryck, P. V.: Solving the car-sequencing problem in constraint logic programming. In Kodratoff, Y., ed.: Proceedings ECAI-88. (1988) 290–295

[3] Hentenryck, P. V., Simonis, H., Dincbas, M.: Constraint satisfaction using constraint logic programming. Artificial Intelligence **58** (1992) 113–159

[4] Smith, B.: Succeed-first or fail-first: A case study in variable and value ordering heuristics. In: Proceedings of PACT'97. (1997) 321–330 (Presented at the ILOG Solver and ILOG Scheduler 2nd International Users' Conference, Paris, July 1996)

[5] Regin, J. C., Puget, J. F.: A filtering algorithm for global sequencing constraints. In Smolka, G., ed.: Principles and Practice of Constraint Programming - CP97, Springer-Verlag (1997) 32–46 LNCS 1330

[6] Gent, I.: Two results on car sequencing problems. Technical Report APES-02-1998, University of St. Andrews (1998)

[7] Michel, L., Hentenryck, P. V.: A constraint-based architecture for local search. In: Proceedings of the 17th ACM SIGPLAN conference on Object-oriented programming, systems, languages, and applications, ACM Press (2002) 83–100 ,

[8] Shaw, P.: Using constraint programming and local search methods to solve vehicle routing problems. In Maher, M., Puget, J. F., eds.: Proceeding of CP '98, Springer-Verlag (1998) 417–431 ,

[9] Bent, R., Hentenryck, P. V.: A two-stage hybrid local search for the vehicle routing problem with time windows. Technical Report CS-01-06, Brown University (2001)

[10] Applegate, D., Cook, W.: A computational study of the job-shop scheduling problem. ORSA Journal on Computing **3** (1991) 149–156 ,

[11] J. Adams, E. B., Zawack, D.: The shifting bottleneck procedure for job shop scheduling. Management Science **34** (1988) 391–401

[12] Caseau, Y., Laburthe, F.: Effective forget-and-extend heuristics for scheduling problems. In: Proceedings of the First International Workshop on Integration of AI and OR Techniques in Constraint Programming for Combinatorial Optimisation Problems (CP-AI-OR'99). (1999)

[13] Palpant, M., Artigues, C., Michelon, P.: Solving the resource-constrained project scheduling problem by integrating exact resolution and local search. In: 8th International Workshop on Project Management and Scheduling PMS 2002. (2002) 289–292

[14] Le Pape, C., Perron, L., Régin, J. C., Shaw, P.: Robust and parallel solving of a network design problem. In Hentenryck, P. V., ed.: Proceedings of CP 2002, Ithaca, NY, USA (2002) 633–648

[15] Palpant, M., Artigues, C., Michelon, P.: A heuristic for solving the frequency assignment problem. In: XI Latin-Iberian American Congress of Operations Research (CLAIO). (2002)

[16] Gomes, C. P., Selman, B.: Algorithm Portfolio Design: Theory vs. Practice. In: Proceedings of the Thirteenth Conference On Uncertainty in Artificial Intelligence (UAI-97), New Providence, Morgan Kaufmann (1997)

[17] Gottlieb, J., Puchta, M., Solnon, C.: A study of greedy, local search and ant colony optimization approaches for car sequencing problems. In: Applications of evolutionary computing (EvoCOP 2003), Springer Verlag (2003) 246–257 LNCS 2611 ,

[18] Chabrier, A., Danna, E., Le Pape, C., Perron, L.: Solving a network design problem. To appear in Annals of Operations Research, Special Issue following CP-AI-OR'2002 (2003)

[19] Perron, L.: Fast restart policies and large neighborhood search. In: Proceedings of CPAIOR 2003. (2003) ,

[20] Beck, J. C., Perron, L.: Discrepancy-Bounded Depth First Search. In: Proceedings of CP-AI-OR 00. (2000)

Travelling in the World of Local Searches in the Space of Partial Assignments*

Cédric Pralet and Gérard Verfaillie

LAAS-CNRS, Toulouse, France
{cpralet,gverfail}@laas.fr

Abstract. In this paper, we report the main results of a study which has been carried out about the multiple ways of parameterising a local search in the space of the partial assignments of a *Constraint Satisfaction Problem* (CSP), an algorithm which is directly inspired from the *decision repair* algorithm [ ]. After a presentation of the objectives of this study, we present the generic algorithm we started from, the various parameters that must be set to get an actual algorithm, and some potentially interesting algorithm instances. Then, we present experimental results on randomly generated, but not completely homogeneous, binary CSPs, which show that some specific parameter settings allow such *a priori* incomplete algorithms to solve almost all the consistent and inconsistent problem instances on the whole constrainedness spectrum. Finally, we conclude with the work that remains to do if we want to acquire a better knowledge of the best parameter settings for a local search in the space of partial assignments.

1   Local Search in the Space of Partial Assignments

1.1   Depth-First Tree Search in Constraint Satisfaction

The basic algorithm, designed to solve *Constraint Satisfaction Problems* (CSP [ ]), consists of a *depth first* search in the space of the partial assignments, organised into a tree the root of which is the empty assignment, leaves are complete assignments, and each node, except the root, results from the assignment of an unassigned variable in its father node. Variable and value heuristics are used to choose pertinently the next variable to assign and the value to assign to it [ ]. Constraint propagation algorithms are also used to enforce any local consistency property, like for example *arc consistency*, at each node of the tree [,]. These algorithms allow domains of the unassigned variables to be reduced and eventually wiped out. In the latter case, inconsistency of the associated subproblem is proven and another value for the last assigned variable is chosen. Producing and recording value removal explanations allow other forms of backtracking than this chronological one to be performed, like for example *conflict directed backjumping* [ ].

---

* The study reported in this paper has been initiated when both authors were working at ONERA, Toulouse, France.

J.-C. Régin and M. Rueher (Eds.): CPAIOR 2004, LNCS 3011, pp. 240–   , 2004.
© Springer-Verlag Berlin Heidelberg 2004

The main advantage of such a tree search is its completeness: if the problem instance is consistent, a solution is found; if not, inconsistency is established. Another advantage is its limited space complexity, which is only a linear function of the instance size. But its lack of scalability in terms of CPU-time prevents it from being used for solving large instances: at least for the hardest instances at the frontier between consistency and inconsistency, its time complexity is an exponential function of the instance size.

1.2   Local Search in Constraint Satisfaction

On the contrary, local search algorithms perform an unordered search in the space of the complete assignments: each move goes from a complete assignment to another one in its neighbourhood, generally randomly chosen with a heuristic-biased probability distribution (choice of the variable to unassign and of the new value to assign to it). When solving CSPs, each neighbour assignment is evaluated via constraint checking [ ]. Meta-heuristics, such as *simulated annealing*, *tabu search*, or others, allow various search schemes to be defined, including *restart* mechanisms [ ].

A first advantage of local search is its limited space complexity, which is, as with depth first tree search, only a linear function of the instance size. Its main advantage is however its far better scalability in terms of CPU-time. But it suffers from at least three shortcomings:

- its incompleteness: if the problem instance is consistent, there is no guarantee that a solution will be found; if not, there is no mechanism which allows inconsistency to be established;
- its non deterministic behaviour: when random choice mechanisms are used, running twice the same algorithm on the same instance may give different results;
- its difficulty handling constraints: on the one hand, because local search is basically an optimisation method, it has some difficulty handling correctly together constraints and criterion; on the other hand, because only complete assignments are considered, constraint propagation is not used and is replaced by a simple constraint checking.

1.3   Hybridisations between Tree Search, Local Search, and Constraint Propagation

This landscape, which has been recognised for a long time [ ], pushed researchers and practitioners to explore combinations of tree search, local search, and constraint propagation and to propose various hybridisation schemes.

Many of these schemes led to loose combinations: two kinds of search on the same problem in sequence or in parallel [,   ]; a kind of search on a master subproblem and another kind on the complementary slave subproblem [ ]. Stronger combinations have been however proposed with large neighbourhood local searches [,   ,   ]: when neighbourhoods become large, they can no longer be enumerated, as they are in basic local search; combinations of tree search and constraint propagation become thus good candidates for exploring them.

1.4   A Stronger Hybridisation Scheme between Local Search and Constraint Propagation

Under the name of *decision repair*, a stronger hybridisation scheme has been proposed in [ ]. It originated from the following observations: strong combination between search and deduction, such as exists between tree search and constraint propagation, is not possible with basic local search, because it manages only complete assignments; but, it is possible as soon as one considers a local search which manages partial assignments.

More precisely, *decision repair* starts from any assignment A (empty, partial, or complete). It applies any local consistency enforcing algorithm (constraint propagation) on the associated subproblem. In case of consistency, the neighbourhood is the set of assignments that result from the assignment of any unassigned variable in A. In case of inconsistency, it is on the contrary the set of assignments that result from the unassignment of any assigned variable in A. This procedure is iterated until a complete consistent assignment (a solution) has been found, inconsistency has been proven, or a time limit has expired.

1.5   Expected Gains and Costs

Expected gains and costs can be analysed from the local search and tree search starting points.

- From the local search starting point, constraint propagation from partial assignments may avoid either exploring uselessly locally inconsistent subproblems, or considering inconsistent values in locally consistent subproblems. But, because constraint propagation from a partial assignment is far more costly than constraint checking on a complete assignment, it is certain that the time taken by a local move will increase. In fact, we hope that, as it has been observed for the combination between tree search and constraint propagation, the decrease in number of moves will compensate for the increase in the cost of each move. To get this result, it is necessary to design constraint propagation algorithms which are both incremental and decremental, that is which support efficiently variable assignments as well as unassignments .

---

[1] With depth first tree search, we need only incremental algorithms which support variable assignments, because the search is performed in a tree. Any backtrack, even in case of backjumping, comes back to a previously visited partial assignment for which constraint propagation has been already performed and its results recorded for example on a stack. If we allow any assigned variable to be unassigned in case of backtrack, the search is performed no more in a tree, but in a neighbourhood graph. The partial assignment that results from a backtrack may have never been visited. Results of constraint propagation are not available and must be decrementally computed. This is what occurs with *dynamic backtracking* [,], which is a particular case of *decision repair*.

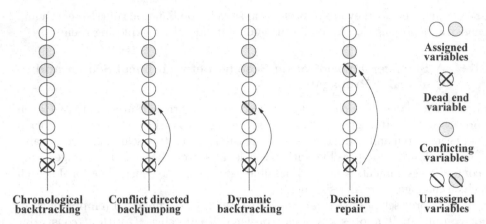

Fig. 1. Various forms of backtracking

– From the tree search starting point, the main novelty is the complete free-dom of choice of the variable to unassign. With *chronological backtracking*, the last assigned variable is systematically unassigned. With *conflict directed backjumping* [ ], all the assigned variables from the last one to the last one involved in the current conflict are unassigned. With *dynamic backtrack-ing* [ ], which is no longer a tree search, only the last one involved in the current conflict is unassigned. With *partial order backtracking* [,], still more freedom is available by following only a partial order between variables. In fact, *decision repair* is similar to *incomplete dynamic backtracking* [ ]. As does *incomplete dynamic backtracking*, it lays down completeness for effi-ciency and scalability. Potentially, this freedom allows the main shortcoming of depth first tree search (the possibility of getting stuck in an inconsistent subtree because of wrong assignments of the first variables) to be fixed (see Figure). But, as it has been already said, the need for incremental and decremental constraint propagation algorithms may lead to more costly as-signments and unassignments. We hope that the more intelligent search will compensate for that.

1.6   Study Objectives and Article Organisation

In [ ], N. Jussien and O. Lhomme proposed a specific instance of *decision repair*, called *tabu decision repair*, and evaluated it on open-shop scheduling problems. The objective of our study was slightly different: starting from a minimal *decision repair* scheme, we aimed at listing all the parameters that must be set to get an actual algorithm, and at exploring and evaluating *a priori* interesting algorithm instances. In other words, our main objective was not to exhibit an efficient instance, but to acquire knowledge about all the possible parameter settings and their relative efficiency, with the long-term objective of acquiring about local

Data   : a CSP P and a starting assignment A.
          n is the number of variables in P.
begin
  | $bool \leftarrow Initial\_Filter(P, A)$
  | **repeat**
  | | **if** $bool$ **then**
  | | | $v \leftarrow Extend\_Assignment(P, A)$
  | | | **if** $v = n + 1$ **then return** yes
  | | | **else** $bool \leftarrow Incremental\_Filter(P, A, v)$
  | | **else**
  | | | $v \leftarrow Repair\_Assignment(P, A)$
  | | | **if** $v = 0$ **then return** no
  | | | **else** $bool \leftarrow Decremental\_Filter(P, A, v)$
  | **until** $Stop()$
  | **return** ?
end

Fig. 2. A generic decision repair algorithm

search in the space of partial assignments of CSPs the same level of knowledge as has been acquired about depth-first tree search.

After a presentation of the generic algorithm we started from in Section , we present its degrees of freedom in Section , some possible algorithm instances in Section , the results of the experiments we carried out in Section , and directions for future work in Section .

In this paper, we only consider satisfaction problems (no optimisation criterion) involving unary and binary constraints, although *decision repair* can deal with constraints of any arity.

2 A Generic Algorithm

We started from a very generic version of *decision repair* (see Figure) derived from the first algorithm presented in [ ].

The starting assignment A may be empty, partial, or complete. Function *Initial_Filter* uses constraint propagation to enforce any local consistency property on the problem P restricted by A. It returns *True* if P is locally consistent and *False* if not. Function *Extend_Assignment* chooses a variable v which is not assigned in the current assignment A and extends A by choosing a value a for v. It returns $n + 1$ if there is no such variable. It is the case when A is complete. Conversely, function *Repair_Assignment* chooses a variable v which is assigned in the current assignment A and repairs A by unassigning v. It returns 0 if there is no such variable. It is the case when A is empty or when inconsistency explanations (sets of assigned variables the assignments of which imply inconsistency) are produced and recorded and the current inconsistency explanation is empty. Functions *Incremental_Filter* and *Decremental_Filter* use respectively

	Consistent instances	Inconsistent instances
Tree search	*Yes*	*No*
Local search	*Yes* or *?*	*?*
Decision repair	*Yes* or *?*	*No* or *?*

Fig. 3. Possible outputs on consistent and inconsistent problem instances

incremental and decremental algorithms to enforce local consistency, without computing everything again from scratch. As does *Initial_Filter*, they return *True* if the current subproblem is locally consistent and *False* if not. Function *Stop* implements any stopping criterion. The algorithm ends with a proof of consistency of P and an associated solution (answer *yes*), with a proof of inconsistency of P (answer *no*), or with nothing in case of activation of the stopping criterion (answer *?*). See Figure   for a comparison with the outputs of classical tree or local searches. Note the complete symmetry of the algorithm, with regard to extension and repair and to consistency and inconsistency.

To put things in a more concrete form, we show in Figures   and   possible traces of a *decision repair* algorithm on a consistent graph colouring problem P_c and on an inconsistent one P_i (see Figure   ).

In both cases, after each assignment of a variable v, the algorithm performs forward checking from v to the current domains of the unassigned variables. For each removed value, the singleton $\{v\}$ is recorded as a removal explanation. When the domain of a variable v' is wiped out, a variable v'' is chosen to be unassigned in the current inconsistency explanation, that is in the union of the value removal explanations in the domain of v'. After unassignment of v'', a value removal explanation is created for its previous value (the previous inconsistency explanation minus v''). Irrelevant value removal explanations, those that involve v'', are then removed. The associated values are restored. But forward checking must be performed again from the assigned variables to the current domains of the unassigned ones. For assignment, the variable of lowest index among the variables of smallest current domain is chosen. Values are tried in the order $\{R, G, B\}$ for P_c and $\{R, B\}$ for P_i. For unassignment, a variable is randomly chosen in the current inconsistency explanation.

Because *decision repair* in not a tree search but a local search in the space of partial assignments, its trace is only a sequence of states, each state being

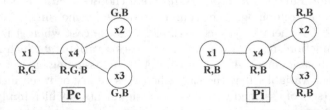

Fig. 4. A consistent graph colouring problem P_c and an inconsistent one P_i

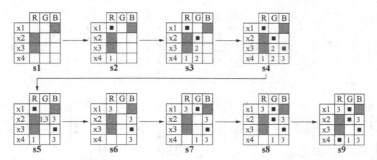

Fig. 5. A possible execution of *decision repair* on P_c

composed of a partial assignment and of a set of value removal explanations. In each state, initially forbidden values are pointed out in dark grey, current assignments by a small black square, and currently removed values by the indices of the variables that are involved in their removal explanation. Values the removal explanation of which is empty, those that are inconsistent whatever the assignment of the other variables is, are pointed out in light grey.

For example, on P_c (see Figure   ), in state s_4, the domain of variable x_4 is wiped out and all the other variables are involved in the inconsistency explanation. We assume that variable x_2 is chosen to be unassigned. This is what is done in state s_5. Value G is removed from the domain of x_2 with $\{x_1, x_3\}$ as an explanation. The value removal explanations in which x_2 was involved are forgotten and associated values restored: value G for x_3 and x_4. But forward checking the current domain of x_2 removes value B with $\{x_3\}$ as an explanation.

On P_i, in state s_6 (see Figure   ), the domain of variable x_4 is wiped out with $\{x_2\}$ as an explanation. Variable x_2 is unassigned and its previous value R is removed with an empty explanation. It is thus sure that value R for x_2 does not take part to any solution and can be removed from its domain. Similarly, when, in state s_{10}, the domain of variable x_3 is also wiped out with $\{x_2\}$ as an explanation, variable x_2 is unassigned and its previous value B is removed with an empty explanation. It is thus sure that value B for x_2 does not take part to any solution and can be removed from its domain. Because the domain of x_2 is now empty, inconsistency of P_i is proven.

3   Parameter Settings

We can now list all the parameters that take place in such an algorithm:

1. the *local consistency property which is enforced* via constraint checking or constraint propagation at each step of the algorithm (functions *Initial_Filter*, *Incremental_Filter*, and *Decremental_Filter*). If a partial assignment is considered as consistent when all the completely assigned constraints are satisfied, we get the local search counterpart of *backward checking*. If it is considered as consistent when no domain of any unassigned vari-

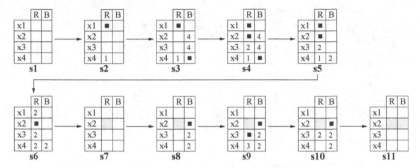

Fig. 6. A possible execution of *decision repair* on P_i

able is wiped out by forward checking, we get the local search counterpart of *forward checking* [ ]. If it is considered as consistent when no domain of any unassigned variable is wiped out by *arc consistency* enforcing, we get the local search counterpart of *MAC* [ ]. Other forms of local consistency may be obviously considered;

2. the *way local consistency is enforced* at the beginning of the search and at each step of the search after variable assignment or unassignment. Questions arise particularly about variable unassignments (function *Decremental_Filter*) and the way of avoiding propagating constraints again from scratch after each unassignment. If no information is recorded when propagating constraints, constraint graph information can be used to limit the work to do in case of unassignment [,]. If the variable assignment order is recorded, it can also be used to limit this work. In [ ], *conflict counts* associate with any value of any variable the number of current conflicting assignments and are used with *forward checking* to know at each step of the algorithm if a value of a variable belongs or not to its current domain (null or not null counter). Finally, if value removal justifications or explanations are recorded when propagating constraints, more value removals can be saved in case of unassignment [, , ,];

3. the *way these value removal explanations are handled* when they are produced and recorded. Can more than one removal explanation be recorded per value (functions *Incremental_Filter* and *Decremental_Filter*)? Must irrelevant removal explanations (inconsistent with the current assignment) be maintained (function *Decremental_Filter*)? When irrelevant removal explanations are maintained, how should the learning memory size be limited?

4. the *variables and values that are affected by constraint propagation* (functions *Initial_Filter*, *Incremental_Filter*, and *Decremental_Filter*). With depth first tree search, only the variables that are not assigned yet and the values in their domains that have not been removed yet are affected by constraint propagation. This is justified by the fact that the search goes always forward: more variables are assigned, more values are removed in the domains of the unassigned variables. The backward moves are not free and use

a stack to recover previous states. With *decision repair*, the search goes freely forward and backward: any variable can go at any step from the unassigned to the assigned state and inversely; the same way, any value of any variable can go at any step from the present to the removed state and inversely. In such conditions, it may be interesting both for the sake of efficiency and of information quality to perform constraint propagation on all the variables (unassigned and assigned) and all the values of the initial domains (present or removed). Such a systematic computing is obviously more costly, but can make the computing at each step easier and provide information for better heuristic choices (see below);

5. the *heuristics* that are used by function *Extend_Assignment* *to choose the variable to assign* in case of local consistency and *the value to assign to it*. All the studies that have been carried out about that in the context of depth first tree search [ ] are *a priori* reusable in this larger local search context;

6. the *heuristics* that are used by function *Repair_Assignment* *to choose the variable to unassign* in case of inconsistency. Because this choice is the main novelty of *decision repair* with regard to depth first tree search, completely new heuristics must be designed and experimented for that. Various criteria should be *a priori* taken into account to judge the interest in unassigning a variable v of current value $a(v)$, initial domain $d(v)$, and current domain $d'(v)$: (i) the fact that v is involved in the current inconsistency, whatever the way this inconsistency is computed (constraint graph information, value removal explanations ...), (ii) the number of constraints in which v is involved (its degree in the constraint graph), (iii) the number of values in the domains of the other variables that are inconsistent with $a(v)$ (in the consistency graph), (iv) the quality of $a(v)$ with regard to any static heuristic ordering of $d(v)$, (v) the size of $d(v)$ and $d'(v)$, (vi) the removal explanations already produced and recorded for the values in $d(v) - d'(v)$, (vii) the rank of v in the current assignment order, (viii) the value removal explanations that would be destroyed and forgotten in the domains of the other variables in case of unassignment of v .... Note that we can build from these criteria unassignment heuristics which aim at building a consistency proof (a solution) by removing bad choices (for example, to unassign a variable the value of which is inconsistent with the greatest number of values in the domains of the other variables) and other ones which aim at building an inconsistency proof (for example, to unassign a variable the domain size of which is the smallest, or a variable for which the number of destroyed value removal explanations in the domains of the other variables would be the smallest in case of unassignment, in order to destroy as less as possible the proof in progress);

7. the presence or absence of priority for assignment (resp. unassignment) to the variable that has been unassigned (resp. assigned) just before, when an assignment (resp. unassignment) immediately follows an unassignment (resp. assignment). See for example [ ]. Such priorities limit the choice of the variable to assign or to unassign in these situations, but favour the

production of inconsistency proofs. When both priorities are present, we say that the associated algorithm *perseveres*.

8. finally, the *stopping criterion* which is used by function *Stop* when no result *yes* or *no* has been already produced: CPU-time, number of assignment extensions or repairs ...

4   Some Algorithm Instances

This very generic algorithm scheme includes many known instances.

- *Depth first tree search* with *chronological backtracking* is an instance of *decision repair* where the last assigned variable is systematically chosen to be unassigned in case of inconsistency. In such conditions, the current state of the algorithm is a partial assignment equipped with an assignment order and the neighbourhood graph becomes a tree. A stack can be used to avoid performing constraint propagation again in case of unassignment. Completeness is guaranteed without the help of any inconsistency explanation.
- *Dynamic backtracking* [,   ] is another instance where value removal explanations are produced and recorded as long as they remain relevant (consistent with the current assignment) and where, in case of inconsistency, the last assigned variable involved in the current inconsistency explanation is systematically chosen to be unassigned. The neighbourhood graph is no more a tree. A stack cannot be used anymore. But completeness is guaranteed.
- *Partial order backtracking* [,   ] is an extension of *dynamic backtracking* where, in case of inconsistency, an assigned variable, involved in the current inconsistency explanation and following a partial order between variables which results from the previously recorded value removal explanations, is chosen to be unassigned. Completeness is still guaranteed.
- *Incomplete dynamic backtracking* [ ] is another instance where constraint propagation is performed via *forward checking* on all the variables (assigned or not) and all their values (removed or not), where *conflict counts* associate with any value of any variable the number of current conflicting assignments, and where the choice of the variable to unassign in case of inconsistency is free (either random or following a given heuristics). As a consequence of this complete freedom, completeness is no more guaranteed.
- Even *min conflicts* [ ] can be considered as an instance of *decision repair* where constraint checking is only performed on complete assignments (all the partial assignments are assumed to be consistent), where the explanation for the inconsistency of a complete assignment is the union of the variables of the unsatisfied constraints, where, in case of inconsistency, a variable is randomly chosen in the current inconsistency explanation to be unassigned, and where

---

[2] In fact, *incomplete dynamic backtracking* allows several variables to be unassigned at the same time. The number of unassigned variables at each backtrack step is a new algorithm parameter. This option could be introduced in the generic *decision repair* scheme.

a new value is randomly chosen among those that minimise the number of unsatisfied constraints if they were selected. In absence of inconsistency explanation recording, completeness is not guaranteed.

Beyond these known algorithms and without any exhaustivity intention, new potentially interesting algorithm instances could be considered round specific unassignment heuristics.

- A first obvious option consists in choosing the variable to unassign randomly in the current inconsistency explanation, as in *min conflicts*, whatever the way this explanation is built. Such a heuristic we refer to as *UHrand* could be profitably integrated into a global *randomisation* and *restart* scheme, inspired from [ ].
- A second option consists in choosing the variable to unassign randomly among the variables of the current inconsistency explanation the assignment of which is the most doubtful. If a static heuristic value is associated with each value of each variable, we can for example consider that the doubtful nature of an assignment is measured by the difference between the heuristic values of the first and of the second value. Such a heuristic we refer to as *UHmostdoubt* aims at undoing quickly doubtful choices in order to produce solutions.
- A third option consists in choosing the variable to unassign randomly among the variables of the current inconsistency explanation that are the least involved in value removal explanations in the domains of the other variables. Such a heuristic we refer to as *UHmindestroy* aims at keeping recorded as long as possible value removal explanations in order to build inconsistency proofs. It can be improved by giving higher weights to removal explanations that have been built by backtrack than to removal explanations that have been directly built by constraint propagation, because the first ones generally needed more work to be produced, and by giving higher weights to smaller removal explanations. It can be also improved by performing constraint propagation on all the variables and not only on the unassigned ones in order to get more precise heuristic information.

5   Experiments

We carried out experiments on many problems and instances, with many algorithmic variants, but we report in this paper only the ones that have been carried out on randomly generated, but not completely homogeneous, binary CSPs, with a limited number of algorithmic variants: the ones that appeared to be potentially more efficient than the classical tree or local search algorithms.

5.1   Problem Instances

We considered binary CSPs, randomly generated with the usual four parameters: number of variables n, domain size d (the same for all the variables), graph connectivity p_1, and constraint tightness p_2 (the same for all the constraints), but we broke their homogeneity by partitioning the set of variables into nc clusters of the same size and by introducing a graph connectivity p_1 inside each cluster (the same for all the clusters) and a lower one p_{1c} between clusters.

The experimental results that are shown in Figures   and   have been obtained with $n = 50$, $d = 10$, $nc = 5$, $p_1 = 30$, $p_{1c} = 20$, and p_2 varying between 30 and 40 around the complexity peak. 10 instances have been generated for each value of p_2. Note that, for $p_2 \leq 32$, all the generated instances are consistent, that, for $p_2 \geq 34$, all of them are inconsistent, and that, for $p_2 = 33$, only one generated instance out of 10 is inconsistent.

5.2   Algorithms

The algorithms we compared are *backtrack* (BT), *conflict directed backjumping* (CBJ), *dynamic backtracking* (DBT), *min conflicts* (MC), and four variants of *decision repair* (DR(rand), DR(mostdoubt), DR(mindestroy,uvar), and DR(mindestroy,avar)).

Except MC, all of these algorithms perform forward checking. Except MC and BT, all of them compute and record value removal explanations, and maintain only those that remain relevant. Forward checking is only performed on the current domains. Except for the fourth variant of DR (DR(mindestroy,avar)), it is only performed on the unassigned variables. Except MC, all of these algorithms persevere. Except for MC, the assignment heuristics consists in choosing an unassigned variable of smallest ratio between its current domain size and its degree in the constraint graph. Except for MC and the second variant of DR (DR(mostdoubt)), the value heuristic is random. The four variants of DR have in common the choice of the variable to unassign inside the current inconsistency explanation. However, they differ mainly in the way of making this choice.

- DR(rand) uses *UHrand* as an unassignment heuristic (see Section   ).
- DR(mostdoubt) uses *UHmostdoubt* as an unassignment heuristic (see Section   ) and *VHmineff* as a static value heuristic. This value heuristic results from the pre-computing for each value of each variable of the number of values it removes in the domains of the other variables. Values are ordered in each domain according to an increasing value of this number.
- DR(mindestroy,uvar) uses *UHmindestroy* as an unassignment heuristic (see Section   ).
- DR(mindestroy,avar) differs from DR(mindestroy,uvar) only in that forward checking is performed on all the variables.

5.3   Experimental Results

Because all the considered algorithms involve random choices, each of them has been run 20 times on each instance. Each run has been given a maximum CPU-time of 120 seconds. A CPU-time of 120 seconds is associated with any run which did not finish by the deadline. Figure   reports the median CPU-time as

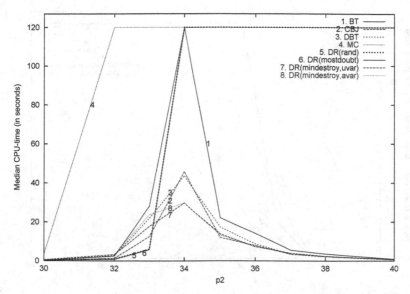

Fig. 7. Median CPU-time on randomly generated problem instances

a function of p_2. Figure   reports the number of runs which did not finish by the deadline, among the 200 $(20 \cdot 10)$ for each value of p_2. These results allow us to make the following observations.

- If MC may be efficient on consistent instances, it becomes quickly inefficient when approaching the consistency/inconsistency frontier. It is, as other classical local search algorithms, unable to solve inconsistent instances.
- BT, CBJ, and DBT present the same usual behaviour with a peak of complexity at the consistency/inconsistency frontier. Results of CBJ and DBT are similar and both clearly better than those of BT.
- DR(rand) and DR(mostdoubt) produce quasi identical results and present a behaviour which is similar to that of MC. Although they are the most efficient on consistent instances at the consistency/inconsistency frontier, they are, as MC and other classical local search algorithms, unable to solve inconsistent instances.
- Although they are basically local search algorithms in the space of partial assignments, and thus *a priori* incomplete, DR(mindestroy,uvar) and DR(mindestroy,avar) present the same behaviour as do complete algorithms such as BT, CBJ, and DBT. Moreover they are the most efficient on inconsistent instances at the consistency/inconsistency frontier and allow the greatest number of instances to be solved by the deadline. Note a small advantage for DR(mindestroy,uvar) when compared with DR(mindestroy,avar) in terms of CPU-time on consistent instances at the consistency/inconsistency frontier: performing forward checking on all the variables is too costly. On the contrary, note a small advantage for DR(mindestroy,avar) when com-

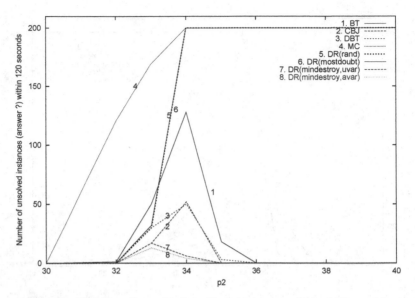

Fig. 8. Number of unsolved problem instances within 120 seconds

pared with DR(mindestroy,uvar) in terms of number of solved instances at this frontier: performing forward checking on all the variables provides the algorithm with a better unassignment heuristic.

6   Future Work

Beyond these first results, many questions remain unanswered and need further experimental and theoretical studies. Among them:

- Are there unassignment heuristics that can better solve consistent instances and other ones that can better solve inconsistent ones? If we know nothing about consistency or inconsistency, would it be profitable to run two algorithms concurrently, one with the objective of building a solution and the other one with the objective of building an inconsistency proof?
- What must be the unassignment heuristics to allow inconsistent instances to be solved? Can we define sufficient conditions on these heuristics to guarantee algorithm termination without any stopping criterion, and thus algorithm completeness? Can we, still better, define necessary conditions?
- What is the influence of the level of local consistency, checked on each partial assignment, on the efficiency of this kind of local search (forward checking, arc consistency ...), and, beyond that, the precise influence of all the parameters listed in section   ?

Acknowledgements

Many thanks to Narendra Jussien and Olivier Lhomme for the *decision repair* algorithm and particularly to Narendra for fruitful discussions about this algorithm.

References

[1] Jussien, N., Lhomme, O.: Local Search with Constraint Propagation and Conflict-based Heuristics. Artificial Intelligence **139** (2002)   ,   ,   ,

[2] Tsang, E.: Foundations of Constraint Satisfaction. Academic Press Ltd. (1993)

[3] Dechter, R., Pearl, J.: Network-based Heuristics for Constraint Satisfaction Problems. Artificial Intelligence **34** (1987)   ,

[4] Haralick, R., Elliot, G.: Increasing Tree Search Efficiency for Constraint Satisfaction Problems. Artificial Intelligence **14** (1980)   ,

[5] Sabin, D., Freuder, E.: Contradicting Conventional Wisdom in Constraint Satisfaction. In: Proc. of ECAI (1994)   ,

[6] Prosser, P.: Hybrid Algorithms for the Constraint Satisfaction Problems. Computational Intelligence **9** (1993)   ,

[7] Minton, S., Johnston, M., Philips, A., Laird, P.: Minimizing Conflicts: a Heuristic Repair Method for Constraint Satisfaction and Scheduling Problems. Artificial Intelligence **58** (1992)   ,

[8] Aarts, E., Lenstra, J., eds.: Local Search in Combinatorial Optimization. John Wiley & Sons (1997)

[9] Langley, P.: Systematic and Nonsystematic Search Strategies. In: Proc. of AAAI (1992)

[10] Kask, K., Dechter, R.: GSAT and Local Consistency. In: Proc. of IJCAI (1995)

[11] Schaerf, A.: Combining Local Search and Look-Ahead for Scheduling and Constraint Satisfaction Problems. In: Proc. of IJCAI (1997)

[12] Zhang, J., Zhang, H.: Combining Local Search and Backtracking Techniques for Constraint Satisfaction. In: Proc. of AAAI (1996)

[13] Pesant, G., Gendreau, M.: A View of Local Search in Constraint Programming. In: Proc. of CP (1996)

[14] Shaw, P.: Using Constraint Programming and Local Search Methods to Solve Vehicle Routing Problems. In: Proc. of CP (1998)

[15] Lobjois, L., Lemaître, M., Verfaillie, G.: Large Neighbourhood Search using Constraint Propagation and Greedy Reconstruction for Valued CSP Resolution In: Proc. of the ECAI Workshop on "Modelling and Solving with Constraints" (2000)

[16] Ginsberg, M.: Dynamic Backtracking. Journal of Artificial Intelligence Research **1** (1993)   ,   ,

[17] Jussien, N., Debruyne, R., Boizumault, P.: Maintaining Arc-Consistency within Dynamic Backtracking. In: Proc. of CP (2000)   ,   ,

[18] Ginsberg, M., McAllester, D.: GSAT and Dynamic Backtracking. In: Proc. of KR (1994)   ,

[19] Bliek, C.: Generalizing Partial Order and Dynamic Backtracking. In: Proc. of AAAI (1998)   ,

[20] Prestwich, S.: Combining the Scalability of Local Search with the Pruning Techniques of Systematic Search. Annals of Operations Research **115** (2002)   ,   ,

[21] Berlandier, P., Neveu, B.: Maintaining Arc Consistency through Constraint Retraction. In: Proc. of ICTAI (1994)
[22] Georget, Y., Codognet, P., Rossi, F.: Constraint Retraction in CLP(FD): Formal Framework and Performance Results. Constraints : An International Journal **4** (1999)
[23] Bessière, C.: Arc-Consistency in Dynamic Constraint Satisfaction Problems. In: Proc. of AAAI (1991)
[24] Debruyne, R.: Arc-Consistency in Dynamic CSPs is no more Prohibitive. In: Proc. of ICTAI (1996)
[25] Fages, F., Fowler, J., Sola, T.: Experiments in Reactive Constraint Logic Programming. Journal of Logic Programming **37** (1998)
[26] Sabin, D., Freuder, E.: Understanding and Improving the MAC Algorithm. In: Proc. of CP (1997)
[27] Gomes, C., Selman, B., Kautz, H.: Boosting Combinatorial Search Trough Randomization. In: Proc. of AAAI (1998)

A Global Constraint for Nesting Problems[*]

Cristina Ribeiro[1,2] and Maria Antónia Carravilla[1,2]

[1] FEUP — Faculdade de Engenharia da Universidade do Porto
[2] INESC — Porto
Rua Dr. Roberto Frias s/n, 4200-465 Porto, Portugal
{mcr,mac}@fe.up.pt

Abstract. Nesting problems are particularly hard combinatorial problems. They involve the positioning of a set of small arbitrarily-shaped pieces on a large stretch of material, without overlapping them. The problem constraints are bidimensional in nature and have to be imposed on each pair of pieces. This all-to-all pattern results in a quadratic number of constraints.

Constraint programming has been proven applicable to this category of problems, particularly in what concerns exploring them to optimality. But it is not easy to get effective propagation of the bidimensional constraints represented via finite-domain variables. It is also not easy to achieve incrementality in the search for an improved solution: an available bound on the solution is not effective until very late in the positioning process.

In the sequel of work on positioning non-convex polygonal pieces using a CLP model, this work is aimed at improving the expressiveness of constraints for this kind of problems and the effectiveness of their resolution using global constraints.

A global constraint "outside" for the non-overlapping constraints at the core of nesting problems has been developed using the constraint programming interface provided by Sicstus Prolog. The global constraint has been applied together with a specialized backtracking mechanism to the resolution of instances of the problem where optimization by Integer Programming techniques is not considered viable.

The use of a global constraint for nesting problems is also regarded as a first step in the direction of integrating Integer Programming techniques within a Constraint Programming model.

Keywords: Nesting, Constraint programming, Global constraints

1 Introduction

Nesting problems are particularly hard combinatorial optimisation problems, belonging to the more general class of cutting and packing problems where one or more pieces of material or space must be divided into smaller pieces.

---

[*] Partially supported by FCT, POSI and FEDER (POSI/SRI/40908/2001 (CPackMO))

J.-C. Régin and M. Rueher (Eds.): CPAIOR 2004, LNCS 3011, pp. 256–   , 2004.
© Springer-Verlag Berlin Heidelberg 2004

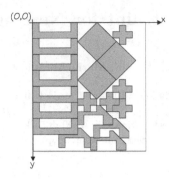

Fig. 1. A solution of a nesting problem

In nesting problems a given set of small pieces must be placed over a large stretch of material (the large plate) while trying to minimise the total length used (see Figure). A more detailed description of nesting problems and cutting and packing problems can be found in [,].

The input data in a nesting problem are the width and an upper bound on the length of the plate, together with the description of the small pieces (polygons). The output data are the positioning points (X_i, Y_i) of all the pieces. The objective is to determine all the positioning points, such that the pieces do not overlap and the length of the plate used to place them is minimised. In the variant of the problem that will be handled rotation of the pieces will not be considered.

Nesting problems have been tackled by several approaches, ranging from simple heuristics to local optimisation techniques and meta-heuristics [, , , , ,]. Another traditional way of tackling combinatorial optimization problems, building mixed integer programming models and solving them with appropriate software, allows only the resolution of nesting problems of very small size.

Constraint programming has been proven applicable to this category of problems, particularly in what concerns exploring them to optimality. A CLP approach has been developed by the authors since 1999. In [,] only convex shapes were considered. The focus of [ ], was on the development and application of some ideas for handling non-convex polygons.

Applying Constraint Programming to nesting has been a good demonstration of the strengths and weaknesses of the constraint models for combinatorial problems. Constraint programming provides the means to keep the formulation of a problem close to its natural expression, with variables that are meaningful in the problem domain and a wealth of built-in specialized algorithms for handling widely used combinatorial constraints. The general-purpose nature of constraint programs makes it easy to include any extra constraints that, although not strictly required to define the solution, may contribute to pruning the search space. This is the case of redundant constraints, symmetry constraints, and those that incorporate information on partial solutions which are known not to be viable.

On the other hand, the flexibility provided by the general-purpose nature of constraint programming may have a high efficiency cost. The problems may require the combination of built-in constraints in a manner that originates little or no propagation, reducing the constraint program to a generate-and-test one. On the other hand, as noted in [ ], constraint programs for optimisation problems frequently exhibit a loose coupling between the constraints and the objective function, therefore lacking an efficient form of optimality reasoning.

In the nesting problem the constraints are bidimensional in nature and have to be imposed on each pair of pieces. This all-to-all pattern results in a quadratic number of constraints. It is not easy to get effective propagation of the bidimensional constraints represented via finite-domain variables. It is also not easy to achieve incrementality in the search for an improved solution: an available bound on the solution is not effective until very late in the positioning process.

The right structure for the constrained variables in the nesting problem would be points in a 2-D space, but current constraint programming systems do not offer such structures. The expression of the constraints in integer finite-domain variables requires some geometrical manipulations of the representations for the pieces. On the optimisation side, point coordinates are also not the more natural support. Although the optimisation criterium is strictly unidimensional (minimum length of the plate), most of the indicators that can be used for evaluating the quality of a partial positioning (waste area, area required for the remaining pieces) are bidimensional in nature.

In the sequel of the work on positioning non-convex polygonal pieces using a CLP model, this work is aimed at improving the expressiveness of constraints for this kind of problems and the effectiveness of their resolution using global constraints. A global constraint "outside" for the non-overlapping constraints (the core of nesting problems) has been developed using the constraint programming interface provided by Sicstus Prolog [,]. The constraint has been applied together with a specialized backtracking mechanism to the resolution of instances of the problem where optimization by Integer Programming techniques is not considered viable.

It has been argued that global constraints are a suitable basis for the integration of Constraint Programming and Integer Programming [ ] and the development of a global constraint for nesting problems can be seen as a first step in this direction [,].

The paper is organized as follows. In the next section the basic concept to satisfy the geometric constraints between poligonal pieces, the nofit polygon, is introduced. This concept is used in the following section, where the evolution of the CLP approach to the nesting problem is described. The first approach was based on reification of constraints and a naive strategy to search the solution space. Section   explains the limitations of the reified constraints, the development of a global constraint for the nesting problem and the replacement of the naive search by selective backtracking. The final section points out relations with ongoing work and future developments.

2   The Nofit Polygon

In the nesting problem, the pieces to be positioned are non-convex polygons represented as lists of vertex coordinates. For each polygon an arbitrary vertex is chosen as a reference for positioning (the positioning point). For two polygons A and B the nofit polygon of B with respect to A, NFP_{AB}, identifies the region in the plane where the positioning point of B may not be located in order to avoid the overlap of A and B. In Figure   the construction of a nofit polygon is represented. In this construction A is fixed and B circulates around it; B's positioning point draws the NFP_{AB}.

The concept of nofit polygon was first introduced by Art [ ]; Mahadevan [ ] has presented a comprehensive description of an algorithm to build it.

The nofit polygon is a convenient geometric transformation for handling the constraints of the nesting problem. Instead of stating that each point in the interior of polygon A may not coincide with any point in the interior of polygon B, it is possible to reason on the basis of a single point for polygon B (its positioning point) using the nofit polygon instead of polygon A itself. Imposing the original constraints would amount to calculating pairwise intersections for the edges of A and B. Using the nofit polygon this reduces to calculating the relative positions of B's positioning point with respect to each of the edges of the nofit polygon.

Using the nofit polygon concept, the fact that polygon B does not overlap polygon A is equivalent to the positioning point of polygon B being over or outside the nofit polygon of B with respect to A (see Figure   ).

Handling the constraints based on the nofit polygon reduces the number of operations from quadratic on the number of edges of the polygons to linear on the number of edges of the nofit polygon, which is no larger than the sum of the number of edges of the two polygons. The nofit polygons can be obtained once for each set of polygons and therefore influence only the setup part of the running time. The complexity of computing nofit polygons is estimated as $O(m^2n)$, where m and n are the number of vertices of the polygons, $m \geq n$, [ ].

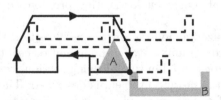

Fig. 2. Nofit polygon construction

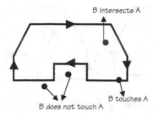

Fig. 3. NFP_{AB} and the positioning point of polygon B

3 Solving Nesting Problems with CLP and Reified Constraints

The prototype applications described here use CLP-FD of Sicstus Prolog [,], which handles constraints on finite domains. Although the nesting problems are not intrinsically finite domain problems, it is common to solve them with some predefined granularity. This is also convenient for benchmarking and comparison, as many state-of-the-art approaches to the solution of nesting problems adopt discrete models for the positioning space.

In the approach used, the input is the width of the big piece, an estimate of its total length and the piece specifications. The output is a list of 4-tuples, one for each piece, describing the type of the piece, its identifier and the X and Y coordinates of the positioning points. The type of the piece and the identifier are constants, and the X and Y coordinates are fd-variables.

For identical pieces (pieces with the same shape), symmetric positionings are avoided. This is accomplished by imposing additional constraints on their X and Y coordinates. Identical pieces are assigned in an 'a priori' arbitrary order. For identical pieces A and B such that piece A precedes piece B, the constraint $X_a \leq X_b$ is imposed. In case $X_a = X_b$, then the constraint $Y_a \leq Y_b$ is imposed.

3.1 Geometric Constraints Based on the Nofit Polygon

If polygon B does not overlap polygon A, the positioning point of B is over or outside the nofit polygon of B with respect to A. This can be turned into a relation between the positioning point of B and the edges of the nofit polygon.

If both pieces are convex, the nofit polygon is also convex. Considering that the contour of NFP_{AB} is travelled clockwise, then piece B does not intersect piece A if the positioning point of B is over or on the left-hand side of at least one of the edges of NFP_{AB}.

If the nofit polygon is non-convex, the same strategy can be applied, provided that the nofit polygon is decomposed into convex subpolygons [ ]. In this case, the above condition must be verified for each subpolygon, with just a slight

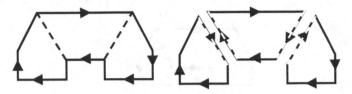

Fig. 4. Nofit polygon decomposition

adjustment: edges resulting from the decomposition ("false edges" of the nofit polygon) are not valid positioning points (see Figure).

3.2   CLP Implementation with Reified Constraints

The order in which the pieces are chosen is given by a list with a sequence of pairs of variables representing their positioning points. For each piece, it is sufficient to state the non-overlapping constraint with respect to the following pieces in the sequence. This is due to the fact that labelling is performed according to the same piece ordering. Taking into account the decompositions of the non-convex nofit polygons, stating the non-overlapping constraint of a pair of pieces A and B amounts to constraining the positioning point of B to be over or outside each of the sub-polygons of NFP_{AB}. As the NFPs are calculated with coordinates relative to the positioning point of piece A, as soon as piece A is positioned its NFPs are subject to the same translation.

The core constraint involves one positioning point and one polygon (possibly resulting from a decomposition). To express the fact that the positioning point is over or outside the polygon, we may say that it must be over or on the left-hand side of at least one of the edges of the polygon. The polygon is represented as a list of vertex coordinates. Between each pair of coordinates there is a flag to indicate whether the corresponding edge is a "true" edge (an edge of the original nofit polygon) or a "false" one (an extra edge issued by the decomposition).

```
% "true" edge
  edgeConstraints([X1, Y1, 0, X2, Y2|Vertex], Xa, Ya, Xb, Yb, [B|Bools]):-
     (X1-X2)*(Y1+Ya-Yb)-(Y1-Y2)*(X1+Xa-Xb) #=< 0 #<=>B,
     edgeConstraints([X2, Y2|Vertex], Xa, Ya, Xb, Yb, Bools).

% "false" edge followed by "true" edge
  edgeConstraints([X1, Y1, 1, X2, Y2, 0|Vertex], Xa, Ya, Xb, Yb, [B|Bools]):-
     (X1-X2)*(Y1+Ya-Yb)-(Y1-Y2)*(X1+Xa-Xb) #< 0 #<=>B,
     edgeConstraints([X2, Y2, 0|Vertex], Xa, Ya, Xb, Yb, Bools).

% "false" edge followed by "false" edge
  edgeConstraints([X1, Y1, 1, X2, Y2, 1|Vertex], Xa, Ya, Xb, Yb,
                  [BVertex, BEdge|Bools]):-
     (X1-X2)*(Y1+Ya-Yb)-(Y1-Y2)*(X1+Xa-Xb) #< 0 #<=>BEdge,
     (Xb #= Xa + X2 #/~Yb #= Ya + Y2) #<=>BVertex,
     edgeConstraints([X2, Y2, 1|Vertex], Xa, Ya, Xb, Yb, Bools).
```

An elementary constraint to be considered involves one positioning point and one edge. To express the fact that the positioning point is over or on the left-hand side of a particular edge, it suffices to impose a "less than" or "less than or equal" constraint using the vertex coordinates and the expression for the edge's slope. Note that the adopted coordinate system has its origin in the upper left corner of the plate (see Figure   ).

The core constraint for B's positioning point and polygon A is based on the reification of the elementary constraints of B with respect to A's edges, imposing that at least one of them holds.

Let X_a, Y_a, X_b, Y_b be the coordinates of the positioning points of pieces A and B, and assume that there is a list of sub-polygons generated by the decomposition of NFP_{AB}. For each sub-polygon the set of vertices is fetched and a list of $0, 1$ values resulting from the reified edge constraints is obtained. The Sicstus global constraint "sum" is used to constrain this list to contain at least one 1 value.

```
coreConstraints(Xa, Ya, Xb, Yb, [SubPolygon |SubPolygons]):-
    pol(SubPolygon, Verts),
    edgeConstraints(Verts, Xa, Ya, Xb, Yb, Bools),
    sum(Bools, #>=, 1),
    coreConstraints(Xa, Ya, Xb, Yb, SubPolygons).
```

3.3   Exploring the Search Space

Solutions to nesting problems are obtained and improved using a systematic search for positioning patterns that satisfy the non-overlapping constraints. The naive approach is to merely try the possible values for the x and y coordinates of the positioning point for each individual piece. In what follows, we assume that the necessary constraints are in force and describe the labelling process in its more basic form: general backtracking is used and no early pruning strategy exists.

The input for the labelling predicate is a "layout" list with unbound fd-variables. The output is the total length of the layout; the fd-variables representing the positioning points of the pieces become bound.

Positioning of the pieces starts using the order in which pieces are sequenced in the data input, searching for alternative layouts each time a complete one is generated.

The Labelling Strategy. The main part of the labelling module is a loop that binds the coordinates of each piece. The x-coordinate is chosen first, selecting a value from its domain, in ascending order. The same strategy is used for the y-coordinate. Choosing values for the x and y-coordinates of the positioning points explores the bidimensional domain for each positioning point. This can be viewed as an exploration of a search tree, where the path from the root to each leaf is a possible layout, with a length.

Optimization. For each complete layout, the total length is registered. A failure-driven loop is used to force backtracking and try to find better solutions with alternative piece positions.

The search space is not explored exhaustively. Instead, paths that do not lead to improved solutions are avoided. The knowledge of the length of the best path obtained so far is used to cut some branches of the search space. At any point in a labelling path, if the x-positioning of the current piece results in a total length greater or equal to the recorded best solution, a failure occurs and backtrack generates an alternate positioning for the previous piece.

Pruning should occur as a result of the constraints imposed on the variable domains. The propagation of constraints is expected to remove from the domain of one variable those values for the coordinates which are known to be no longer feasible due to the labelling of some other variable.

4   A Global Constraint and Improved Search Strategy

The use of built-in constraints and reification has proven not to produce effective propagation for this problem. As expected, the reified constraints detect unfeasible locations late in the process of positioning a piece. The generic backtracking strategy for obtaining a solution has also shown poor performance, exploring parts of the search tree where no solutions could be expected.

Constraint programming is a natural paradigm for building an operational environment where both the constraints required for defining solutions and the extra knowledge on the problem nature can be put to work in the search for a solution. Experimenting with the basic approach has been very useful for identifying features of the problem that helped solving it better.

Constraints were the first aspect that has received our attention. It is possible to express "non-overlappedness", in a bidimensional space, in a form that can be made operationally interesting, cutting off branches of the tree where overlapping occurs. This is done through the definition of a "global constraint" that is used in programs to solve nesting problems and works by reducing the domains of the Y coordinates. The "outside" global constraint is described next. This kind of domain reduction can be related to the sweepline approach of [ ]; it operates after some piece has been positioned and is specialized for non-convex polygons.

The second aspect that called for improvement was the search strategy. Due to the nature of the problem, it is possible to make a choice for the position of some piece that leads to an increased cost (because the piece becomes the rightmost one) and then position several other pieces without affecting the cost (they fit in the available space up to the rightmost X). If, at some point down the tree, we backtrack to the intermediate pieces (the ones that have been positioned without affecting maximum length), it is superfluous to explore alternative locations for them: at this point only changes in the position of the original piece can improve the current solution. This feature comes from the nature of the problem and the way in which it has been used is detailed in the next section.

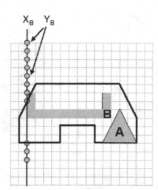

Fig. 5. Global Constraint outside

4.1   The Global Constraint "outside"

The global constraint "outside" expresses the non-overlapping of a pair of polyg-onal pieces. For pieces A and B, "outside" states that the positioning point of piece B must be outside the nofit polygon of B with respect to A, NFP_{AB}. The constraint has been implemented with the global constraint programming interface of Sicstus Prolog [ ].

The inputs of the global constraint are the fd-variables X_A, Y_A, X_B and Y_B, the coordinates of the positioning points of pieces A and B, and the correspond-ing nofit polygon NFP_{AB}, represented as a list of vertex coordinates.

The constraint is activated when piece A is already positioned (X_A and Y_A are ground) and a value has been chosen for X_B. The activation of the constraint leads to the reduction of the domain of Y_B by the points inside NFP_{AB} (see Figure).

To determine the points at coordinate X_B inside NFP_{AB} it is necessary to find the intersections of NFP_{AB} (positioned according to the location of A) with the vertical line $X = X_B$.

The problem of finding the intersections of a convex polygon with a vertical line is rather straightforward. There can be 0, 1, 2 or an infinite number of intersections, as follows:

– the vertical line does neither intersect nor touch the polygon;
– the vertical line is tangent to the polygon in one point;
– the vertical line intersects the polygon in two points;
– the vertical line is tangent to an edge of the polygon.

The third case is the only one that may lead to a reduction in the domain of Y_B.

If the polygon is non-convex, there are many more different situations to be analyzed. In our approach, the problem of finding the intersections of the polygon with a vertical line is dealt with in two steps.

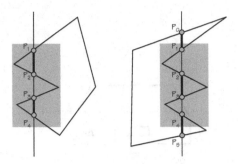

Fig. 6. Two interpretations of the subset of intersection points $\{P_1, P_2, P_3, P_4\}$

In the first step, the polygon is traversed edge by edge and the intersections are collected in a list, together with annotations on their nature. We have identified 12 different cases that have to be distinguished to be able to preserve the context and decide, after obtaining all the intersections, the correct ranges of values to be excluded.

In the second step, the list obtained in the first step is sorted and organized in intervals (pairs of Y-values) that will be cut from the domain of Y_B.

The need to collect more than the location of the intersection points is illustrated in Figure , where two situations that have an identical subset of intersection points give raise to different interpretations of the intervals to be excluded, based on the context in which they appear.

Figure represents two situations involving polygons with vertical edges, that we have named vertical edge—limit and vertical edge—crossing and the intersection of these polygons with a vertical line.

In both these cases, we have two successive vertices with an X-value equal to X_B. In the first one the Y values for the two vertices are kept in the intersection list, annotated as extremes of a vertical edge. In the second, they are kept with an annotation meaning that any of them might be an interval extreme, but not both.

Those two situations are detected by the following code.

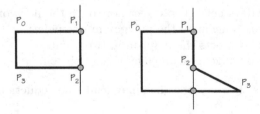

Fig. 7. Two types of vertical edges: vertical edge—limit and vertical edge—crossing

```
% Vertical edge - limit
 inside(X, [X0, _, X, Y1, X, Y2, X3, Y3|Rest], [Y1, v, Y2|ListYY]):-
   ((X0>X, X3>X);(X0<X, X3<X)), !,
   inside(X, [X, Y1, X, Y2, X3, Y3|Rest], ListYY).

% Vertical edge - crossing
 inside(X, [X0, _, X, Y1, X, Y2, X3, Y3|Rest], [Y1, a, Y2|ListYY]):-
   ((X0>X, X3<X);(X0<X, X3>X)), !,
   inside(X, [X, Y1, X, Y2, X3, Y3|Rest], ListYY).
```

A more involved case occurs when a piece fits tightly and vertically in a con-cave region of another piece. The corresponding nofit polygon has a spike. In the example in Figure   we can see the two pieces in gray, and the contour of the resulting nofit polygon. The nofit is non-convex and has a spike corresponding to the position where the cross-shaped piece exactly fits into the concave region of the U-shaped piece. The following code snippet handles the corresponding situation.

```
% Spike

 inside(X, [X0, Y0, X, Y1, X, Y2, X, Y3|Rest], [Y2, 1|ListYY]):-
   Y2<Y3, Y2<Y1, !,
   inside(X, [X0, Y0, X, Y1, X, Y3|Rest], ListYY).
```

We will not go into the details of the extra information associated to the intersection points to preserve their context of occurrence in the original polygon. When all the intersections have been determined, it is necessary to decide the right pairing of intersection points to define the intervals to be excluded. Figure shows that this decision can be easily made if we consider points in ascending or descending order. With the intervals obtained it is possible to remove from the domain of Y_B values that would cause the pieces to overlap, and therefore prune branches of the search tree where no solutions can be found.

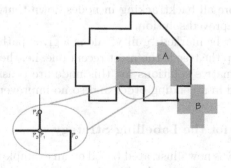

Fig. 8. Nofit polygon with a spike

4.2   Selective Backtracking

Generically, the labelling process is similar to the one used in the basic approach, differing only in the strategy used to prune the search space.

Avoiding Branches with no Improved Solutions. There are now two ways to prune the search space in the labelling process.

1. At any point in a labelling path, if the X-positioning of the current piece results in a total length greater or equal to the recorded best solution, a failure occurs and backtrack generates an alternate positioning for the previous piece.
2. Upon completion of a labelling path, i.e. when a solution has been found, backtracking is forced to the node in the tree that immediately precedes the piece responsible for the total length.

   To achieve this, a fact is added to the database for each X-positioning, keeping the identification number of each piece, a state value and the current X-value for the positioning point of the piece.

   The state of a piece is "guilty" in the recorded facts, if the current value of the X-coordinate of the piece results in at least one of its vertices becoming the rightmost vertex of the layout. The piece is marked "innocent" otherwise.

It is necessary to distinguish, when traversing the tree, between two situations:

- Going down the tree – labelling proceeds as described
- Going up the tree – failing the labelling in all the nodes up to the node in the tree that immediately precedes the piece responsible for the total length (first guilty piece in the backtracking path).

The rationale for this pruning procedure is as follows: If piece P is responsible for the total length in a layout (meaning that one of its vertices is one of the rightmost vertices in the layout), keeping this piece in place does never improve the solution. Therefore all backtracking in nodes down that path is guaranteed not to produce an improved solution.

Several pieces may be marked "guilty" along a given path, however the place to stop when going up the tree is the most recent one, i.e. the one located deeper in the tree. No alternative solutions for this node are considered, because the X-domain is explored in ascending order and so no improvement could result.

4.3   An Example for the Labelling Strategy

The labelling strategy is now illustrated based on an example where 4 non-convex pieces are positioned in a rectangular big piece. The pieces are positioned in the same order in which they are sequenced in the data input. Figure   shows the data input, with 4 pieces and their sequencing.

Fig. 9. Input: 4 pieces and their sequencing

Figure   is the labelling tree, where each node represents the positioning of a single piece, involving the choice of its X and Y coordinates. A depth-first left-to-right traversal of the tree shows the sequence of positionings that are actually generated. A path from the root to a leaf node represents a complete layout, with a total length which is recorded. Nodes at the same level in the tree represent positionings for the same piece at different phases in the positioning process.

There are branches in the tree that do not lead to a leaf; these are paths that have been abandoned as a result of pruning. A branch is abandoned if the best result that can be obtained from exploring it does not improve over an existing solution.

Some branches of the tree are not explored as a result of the constraints imposed on the variable domains. The propagation of constraints has the effect of removing from the domain of one variable those values for the coordinates which are known to be no longer feasible due to the labelling of some other variable in the course of positioning some piece. The positionings that would result from binding variables to values which are removed from the domain are not shown in the search tree because they are not actually explored in the labelling.

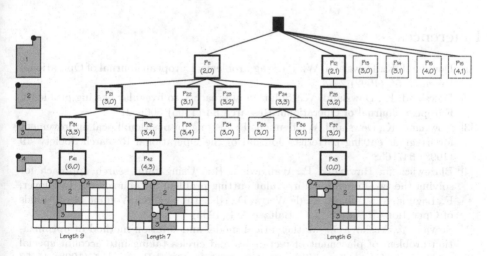

Fig. 10. The labelling tree

5   Conclusions and Ongoing Work

Nesting problems are a challenging example for the application of Constraint Programming to combinatorial problems. The geometrical bidimensional constraints are difficult to capture in a CP system using built-in constraints and the bounds on the solution do not effectively contribute to cut the search space, compromising an effective optimisation strategy.

We presented a global constraint for nesting problems that was built on the experience obtained with an early prototype where nesting problems were modelled in a finite-domain constraint logic programming system with built-in constraints and solved with a basic backtracking strategy. This global constraint achieves a better propagation action and, together with a selective backtracking strategy, provides a better solution for this kind of problems.

With global constraints it is possible to keep the model closer to the problem, simplifying both modelling and maintenance [ ]. Global constraints have been shown to be a convenient level of modelling for integrating constraint programming and integer programming techniques. We are in the course of further exploring global constraints and the possibilities of integrating MIP techniques for achieving a more optimisation-oriented strategy. For this kind of problems this will possibly involve the decomposition of the problem into subproblems of manageable size for MIP. Another aspect that has not yet been handled with the constraint approach is piece rotation, which is relevant in many problem instances.

This work has been developed within a group where approaches to the same problem using heuristics and local optimisation techniques are producing relevant results. Constraint programming will contribute to this line of research by providing an environment where feasibility of solutions is accounted for by the global constraint, allowing a better focus on the optimisation techniques.

References

[1] Dowsland, K., Dowsland, W.: Packing problems. European Journal of Operational Research **56** (1992) 2–14
[2] Dowsland, K., Dowsland, W.: Solution approaches to irregular nesting problems. European Journal of Operational Research **84** (1995) 506–521
[3] Dowsland, K., Dowsland, W., Bennell, J.: Jostling for position: Local improvement for irregular cutting patterns. Journal of the Operational Research Society **49** (1998) 647–658
[4] Błażewicz, J., Hawryluk, P., Walkowiak, R.: Using tabu search approach for solving the two-dimensional irregular cutting problem in tabu search. In Glover, F., Laguna, M., Taillard, E., de Werra, D., eds.: Tabu Search. Volume 41 of Annals of Operations Research. J. C. Baltzer AG (1993)
[5] Stoyan, Y., Yaskov, G.: Mathematical model and solution method of optimization problem of placement of rectangles and circles taking into account special constraints. International Transactions on Operational Research **5** (1998) 45–57

[6] Milenkovic, V., Daniels, K.: Translational polygon containment and minimal enclosure using mathematical programming. International Transactions in Operational Research **6** (1999) 525–554

[7] Bennell, J. A., Dowsland, K. A.: Hybridising tabu search with optimization techniques for irregular stock cutting. Management Science **47** (2001) 1160–1172

[8] Gomes, A. M., Oliveira, J. F.: A 2-exchange heuristic for nesting problems. European Jornal of Operational Research **141** (2002) 359–370

[9] Ribeiro, C., Carravilla, M. A., Oliveira, J. F.: Applying constraint logic programming to the resolution of nesting problems. In: Workshop on Integration of AI and OR techniques in Constraint Programming for Combinatorial Optimization Problems. (1999)

[10] Ribeiro, C., Carravilla, M. A., Oliveira, J. F.: Applying constraint logic programming to the resolution of nesting problems. Pesquisa Operacional **19** (1999) 239–247

[11] Carravilla, M. A., Ribeiro, C., Oliveira, J. F.: Solving nesting problems with non-convex polygons by constraint logic programming. International Transactions in Operational Research **10** (2003) 651–663

[12] Milano, M., Ottosson, G., Refalo, P., Thorsteinsson, E. S.: The Role of Integer Programming Techniques in Constraint-Programming's Global Constraints. INFORMS Journal on Computing **14** (2002) 387–402

[13] Swedish Institute of Computer Science: SICStus Prolog User's Manual. (1995)

[14] Carlsson, M., Ottosson, G., Carlson, B.: An Open-Ended Finite Domain Constraint Solver. In Glaser, H., Hartel, P., Kucken, H., eds.: Programming Languages: Implementations, Logics, and Programming. Volume 1292 of Lecture Notes in Computer Science., Southampton, Springer-Verlag (1997) 191–206

[15] Ottosson, G., Thorsteinsson, E. S., Hooker, J. N.: Mixed Global Constraints and Inference in Hybrid CLP-IP Solvers. Annals of Mathematics and Artificial Intelligence **34** (2002) 271–290

[16] Bockmayr, A., Kasper, T.: Branch-and-Infer: A Unifying Framework for Integer and Finite Domain Constraint Programming. INFORMS Journal on Computing **10** (1998) 287–300

[17] Art, R.: An Approach to the Two-Dimensional, Irregular Cutting Stock Problem. Technical Report 36.008, IBM Cambridge Centre (1966)

[18] Mahadevan, A.: Optimization in Computer-Aided Pattern Packing. PhD thesis, North Carolina State University (1984)

[19] Fernandéz, J., Cánovas, L., Pelegrín, B.: Algorithms for the decomposition of a polygon into convex polygons. European Journal of Operational Research **121** (2000) 330–342

[20] Beldiceanu, N., Carlsson, M.: Sweep as a Generic Pruning Technique Applied to the Non-Overlapping Rectangles Constraint. In Walsh, T., ed.: CP'2001, Int. Conf. on Principles and Practice of Constraint Programming. Volume 2239 of Lecture Notes in Computer Science., Pisa, Springer-Verlag (2001)

Models and Symmetry Breaking
for 'Peaceable Armies of Queens'

Barbara M. Smith[1], Karen E. Petrie[1], and Ian P. Gent[2]

[1] School of Computing & Engineering
University of Huddersfield, Huddersfield, West Yorkshire HD1 3DH, U.K.
{b.m.smith,k.e.petrie}@hud.ac.uk
[2] School of Computer Science
University of St. Andrews, St Andrews, Fife KY16 9SS, U.K.
ipg@dcs.st-and.ac.uk

Abstract. We discuss a difficult optimization problem on a chess-board, requiring equal numbers of black and white queens to be placed on the board so that the white queens cannot attack the black queens. We show how the symmetry of the problem can be straightforwardly eliminated using SBDS, allowing a set of non-isomorphic optimal solutions to be found. We present three different ways of modelling the problem in constraint programming, starting from a basic model. An improvement on this model reduces the number of constraints in the problem by introducing ancillary variables representing the lines on the board. The third model is based on the insight that only the white queens need be placed, so long as there are sufficient unattacked squares to accommodate the black queens. We also discuss variable ordering heuristics: we present a heuristic which finds optimal solutions very quickly but is poor at proving optimality, and the opposite heuristic for which the reverse is true. We suggest that in designing heuristics for optimization problems, the different requirements of the two tasks (finding an optimal solution and proving optimality) should be taken into account.

1 Introduction

Robert Bosch introduced the "Peaceably Coexisting Armies of Queens" problem in his column in Optima in 1999 [ ]. It is a variant of a class of problems requiring pieces to be placed on a chessboard, with requirements on the number of squares that they attack: Martin Gardner [ ] discusses more examples of this class. In the "Armies of Queens" problem, we are required to place two equal-sized armies of black and white queens on a chessboard so that the white queens do not attack the black queens (and necessarily v.v.) and to find the maximum size of two such armies. Bosch asked for an integer programming formulation of the problem and how many optimal solutions there would be for a standard 8×8 chessboard.

Here we discuss a range of possible models of the problem as a CSP, and show how Symmetry-Breaking During Search (SBDS) [ ] can be used to eliminate the symmetry in each model, and hence find all non-isomorphic optimal solutions.

J.-C. Régin and M. Rueher (Eds.): CPAIOR 2004, LNCS 3011, pp. 271–   , 2004.
© Springer-Verlag Berlin Heidelberg 2004

We have implemented some of the models in both ECL^{i}PS^e and ILOG Solver. Constraint programming tools can differ, for instance in the way that apparently equivalent constraints propagate: by comparing two different implementations of our models, we can draw conclusions that are independent of a particular constraint programming tool.

2   Basic Model

In a later issue of Optima, Bosch gives an IP formulation due to Frank Plastria. This has two binary variables for each square of the board:

$$b_{ij} = 1 \text{ if there is a black queen on square } (i, j)$$
$$= 0 \text{ otherwise}$$
$$w_{ij} = 1 \text{ if there is a white queen on square } (i, j)$$
$$= 0 \text{ otherwise}$$

For the general case of an $n \times n$ board:

$$\text{maximize} \sum_{i=1}^{n} \sum_{j=1}^{n} b_{ij}$$
$$\text{subject to} \sum_{i=1}^{n} \sum_{j=1}^{n} b_{ij} = \sum_{i=1}^{n} \sum_{j=1}^{n} w_{ij}$$
$$b_{i_1 j_1} + w_{i_2 j_2} \le 1 \text{ for all } ((i_1, j_1), (i_2, j_2)) \in M$$
$$b_{ij}, w_{ij} \in \{0, 1\} \text{ for all } 1 \le i, j \le n$$

where M is the the set of ordered pairs of squares that share a line (row, column or diagonal) of the board. Bosch reported that finding an optimal solution for an 8×8 board (with value 9) took just over 4 hours using CPLEX on a 200 MHz Pentium PC.

A straightforward model of the problem as a CSP is similar to this IP formulation. There is no difficulty in having variables with more than 2 values, so the number of variables can be reduced to n^2:

$$s_{ij} = 2 \text{ if there is a white queen on square } (i, j)$$
$$= 1 \text{ if there is a black queen on square } (i, j)$$
$$= 0 \text{ otherwise}$$

We can express the 'non-attacking' constraints as:

$$s_{i_1 j_1} = 1 \Rightarrow s_{i_2 j_2} \ne 2$$
$$\text{and } s_{i_1 j_1} = 2 \Rightarrow s_{i_2 j_2} \ne 1 \text{ for all } ((i_1, j_1), (i_2, j_2)) \in M$$

or more compactly as:

$$s_{i_1 j_1} + s_{i_2 j_2} \neq 3 \text{ for all } ((i_1, j_1), (i_2, j_2)) \in M$$

In both ECL^{i}PS^e and Solver, the single constraint gives the same number of backtracks as the two implication constraints, but is faster.

Constrained variables w, b count the number of white and black queens respectively (using the counting constraints provided in constraint programming tools such as ECL^{i}PS^e and Solver). We then have the constraint $w = b$, and the objective is to maximize w. This is achieved by adding a lower bound on w whenever a solution is found, so that future solutions must have a larger value of w; when there are no more solutions, the last one found has been proved optimal.

The model has n^2 search variables and approximately $4n^3$ binary constraints, as well as the counting constraints which have arity n^2.

3   Symmetry

The problem has the symmetry of the chessboard, as in the familiar n-queens problem; in addition, in any solution we can swap all the white queens for all the black queens, and we can combine these two kinds of symmetry. Hence the problem has 16 symmetries. It is well-known that symmetry in CSPs can result in redundant search, since subtrees may be explored which are symmetric to subtrees already explored. If only one solution is required, these difficulties do not always arise in practice. However, if a complete traversal of the search tree is required, either because there is no solution, or because all solutions are wanted, symmetry must lead to wasted search unless dealt with. This means that symmetry will cause difficulties in optimization problems, where proving optimality entails a complete search to prove that there is no better solution.

Symmetry Breaking During Search [ ] is ideal for problems such as this since it only requires a simple function to describe the effect of each symmetry (other than identity) on the assignment of a value to a variable. Hence, in this case, just 15 such functions are required. Briefly, on backtracking to a choice point in the search, represented by the two constraints $var = val$ and $var \neq val$, SBDS adds a constraint to the second branch for any symmetry which has not yet been broken along the path from the root of the search tree to this node. The constraint is the symmetric equivalent of $var \neq val$ and prevents exploration of partial solutions equivalent under this symmetry to those which have already been explored following the choice $var = val$. If the effect of each individual symmetry is described, SBDS will eliminate all symmetry: all solutions produced are non-isomorphic to each other, and the search never explores any part of the search tree which is symmetric to a subtree already explored.

The seven board symmetries for which symmetry functions are required can be labelled x, y, d1, d2, r90, r180 and r270 (reflection in the horizontal, vertical and both diagonal axes, and rotations through 90°, 180° and 270°, respectively).

An assignment $s_{ij} = v$ is passed to each function as a constraint, and the equivalent constraint under the relevant symmetry is returned. For instance, if the rows and columns of the board are numbered $1, .., n$, $\text{r90}(s_{ij} = v)$ is the constraint $s_{j,n+1-i} = v$. The symmetry which interchanges the black and white queens, bw, returns $s_{ij} = v'$, where $v' = 0$ if $v = 0$, and otherwise $v' = 3 - v$. We also need to describe the 7 symmetries which combine a board symmetry with interchanging black and white: for instance, the symmetry $\text{bw} \circ \text{r90}$ returns $s_{j,n+1-i} = v'$.

4   Basic Model: Results

The square variables are assigned in a predefined (static) order: top row, left to right, 2nd row, left to right, and so on. To ensure that good solutions are found early, values are assigned in descending order; otherwise, the first solution found has 0 assigned to every variable, corresponding to no queens of either colour, which is valid but far from optimal. The running times given relate to a 1.6GHz Pentium 4 PC for ECL^{i}PS^e and a 1.7GHz Celeron PC with 128MB RAM for Solver. The implementation of SBDS in ECL^{i}PS^e used in these experiments is due to Warwick Harvey.

Tables 1 and 2 show that using SBDS makes a huge difference to the time required to prove optimality, although not to the time to find the optimal solution. There is a more than 10-fold reduction in the number of fails, except for the smallest values of n, though the reduction in running time is less. It would be possible to achieve some of the speed-up without SBDS, by adding constraints to the model, for instance that the top half of the board contains more white queens than the bottom, but simple constraints of this kind cannot remove all the symmetry. Table 3 compares finding all solutions with and without symmetry breaking using SBDS. It proved impracticable to find all solutions for the 8×8 board without any symmetry breaking: there are evidently hundreds of possible solutions, although only 71 are distinct.

Table 1. Search effort and running time to find an optimal solution to the armies of queens problem, with no symmetry breaking. Value = optimal number of queens of each colour; F = number of fails (backtracks) to find the optimal solution; P = number of fails to prove optimality; sec. = running time in seconds

n	Value	ECL^{i}PS^e			ILOG Solver		
		F	P	sec.	F	P	sec.
2	0	7	7	<0.01	7	14	< 0.01
3	1	6	18	0.01	6	24	< 0.01
4	2	0	134	0.01	0	148	< 0.01
5	4	25	978	0.13	30	1,031	0.05
6	5	10	21,469	3.2	9	24,210	1.48
7	7	64	393,806	78	51	435,598	38.1
8	9	4339	10,846,300	3,500	5,270	12,002,608	1,660

Table 2. Search effort and running time to find an optimal solution to the armies of queens problem, with SBDS

n	Value	ECL^{i}PS^e			ILOG Solver		
		F	P	sec.	F	P	sec.
2	0	1	1	< 0.01	1	2	< 0.01
3	1	4	6	0.01	4	11	0.01
4	2	0	15	0.04	0	16	0.01
5	4	24	96	0.16	29	131	0.03
6	5	10	1,609	1.7	9	1,854	0.35
7	7	64	29,255	27	51	33,529	7.96
8	9	4,339	806,056	640	5,270	924,505	331

Table 3. Search effort and running time to find all optimal solutions to the armies of queens problem, with ECL^{i}PS^e

n	No symmetry breaking			With SBDS		
	Solutions	backtracks	sec.	Solutions	backtracks	sec.
2	1	0	< 0.01	1	0	< 0.01
3	16	15	0.02	1	2	0.01
4	112	219	0.05	10	18	0.06
5	18	1856	0.02	3	169	0.24
6	560	44400	5.8	35	3306	3.0
7	304	822108	130	19	59876	48
8	*not completed*			71	1604456	1130

5   Combining Squares and Lines

The basic model has a constraint between two variables if they represent squares which are on the same line (row, column or diagonal) of the board. We could consider an alternative model in which the lines are also represented by variables in the CSP. Any line must have either only white queens on it, or only black queens, or be empty, so we could create the line variables with three values corresponding to these possibilities. More compactly, we can have two possible values for each line variable: 0 means that there is no white queen on the line, and 1 means that there is no black queen on the line (unoccupied lines can have either value).

The advantage of adding the line variables is that we can reduce the number of constraints. Whenever a queen is placed on a square the values of the corresponding line variables are set accordingly. Thereafter, a queen of opposite colour cannot be placed on any of these lines, and we no longer need the constraints between square variables to enforce this.

Taking the rows as an example, we have n variables $r_1, ..., r_n$, and constraints:

$$s_{ij} = 1 \Rightarrow r_i = 0$$
$$\text{and } s_{ij} = 2 \Rightarrow r_i = 1 \text{ for all } 1 \leq i, j \leq n$$

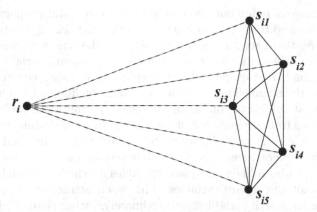

Fig. 1. Constraints in the two encodings of the 'armies of queens' problem

As before, we can reduce the pair of constraints to a single constraint:

$$s_{ij} + r_i \neq 2 \text{ for all } 1 \leq i, j \leq n$$

The combined model has more variables than the basic model (another $6n$, approximately), but we can still use just the n^2 square variables as the search variables. There are approximately $4n^2$ constraints, each between a line variable and a square variable, rather than $4n^3$ constraints between pairs of square variables as before.

Adding the line variables to the model makes no difference to the number of fails in ECL^{i}PS^e (there are some differences in Solver), but reduces the running time to solve the problems optimally by about one-third in Solver and about one-sixth in ECL^{i}PS^e, for $n = 8$.

Note that since the search variables, and hence the branching decisions made during search, are unchanged, SBDS is unaffected by the change in the model.

6   Combined Model: Discussion

Figure   compares the constraints required in the two models: it shows the constraints required to express that row i cannot have queens of different colours, in the case $n = 5$. The solid lines show the clique of constraints required between the variables corresponding to the squares on row i in the basic model. The dotted lines are the constraints that replace them in the combined model: we have replaced an n-clique of constraints by just n.

In addition to the constraints expressing that we cannot have queens of different colours on any line, we also need constraints on the number of queens of each colour. These would be difficult if not impossible to express solely in terms of the line variables. Hence, adding the line variables to the basic model is not exactly analogous to the idea of redundant modelling [ ]: in redundant modelling, two models are combined, either of which could be used independently. Moreover, the claimed advantage of redundant modelling is that constraint propagation within either model can feed through to the other, via the channelling constraints which link them. Here, there are no constraints between the line variables, and in the combined model, the only constraints between the square variables are those counting the number of queens. We are replacing all other constraints of the basic model by the channelling constraints linking the line and square variables. This is somewhat similar to combining models of permutation problems, where the benefit comes from propagation of the channelling constraints, which in that case can replace $\neq$ constraints between the variables of the original model [ ].

Although we cannot have a model with line variables alone, we could in theory have a model with both line and square variables in which we search on the line variables and not the square variables. This is an attractive idea, since there would only be $6n$ search variables, approximately, rather than n^2. However, in practice it leads to a number of difficulties. It would introduce new symmetries, since the values 0 and 1 are interchangeable if a line is unoccupied. We could avoid this by having three values rather than two for the line variables, but even then, a complete assignment to the line variables does not always uniquely determine the values of the square variables, so that not all non-isomorphic optimal solutions would be found.

7   Counting Unattacked Squares

In trying to solve the armies of queens problem by hand, it becomes apparent that we need only place the queens of one colour, say white, provided that we check as each queen is placed that the number of squares not so far attacked is at least equal to the number of white queens on the board. A black queen can be placed on any square which is not attacked by any white queen; hence if there are k white queens on the board and at least k unattacked squares, we can extend the current assignment to a complete solution with value k.

This leads to a new model of the problem. As in the basic model, there is a variable s_{ij} for each square on the board, but now with possible values 0 and 1, where 0 signifies that the square is either empty or occupied by a black queen and 1 that it contains a white queen. For each square, we also construct the set of squares, A_{ij}, that a queen placed on this square would attack. A set variable U represents the unattacked squares on the board. If $s_{ij} = 0$ but $(i, j) \notin U$, square (i, j) must be empty.

The constraints are:

$$s_{ij} = 1 \Leftrightarrow A_{ij} \cap U = \emptyset \quad 1 \leq i, j \leq n$$

The minimum of the number of white queens and the number of unattacked squares gives the size of the two equal-sized armies; this minimum is the objective to be maximised, as before.

The unattacked squares model does not affect the optimal value, but it does lead to a different way of counting distinct optimal solutions. The earlier models produce solutions with equal numbers of white and black queens, but now solutions with different numbers of the two colours can be produced. For example, it is possible to place 9 white queens and 10 black queens on an 8×8 board: such a solution leads to 10 distinct solutions with exactly 9 of each colour, all obtained by omitting one black queen. In the new model we count such a configuration as just one solution, but we note that an unequal number of queens appear. That is to say, we report only solutions in which no queen of either colour can be added to any square, even if this unbalances the numbers. All previous solutions can be obtained from the smaller number of solutions we now find. Unfortunately, the number of solutions in the previous model cannot be calculated trivially from the number in the new model, a point we will discuss further below.

Optimal solutions with different numbers of black and white queens led to other subtle problems that we had to address in implementation. We added constraints so that a square being unattacked by a white queen is equivalent to having a black queen on it, and v.v. Without these constraints, the model produces spurious solutions in which an unattacked square does not have a black queen on it. This is incorrect as we seek only solutions in which no queens of either colour can be added to the solution. The constraints are implemented by introducing a matrix of Boolean variables b_{ij} to indicate if a black queen is on square (i, j), and a set variable W to represent the squares occupied by white queens. We then add the constraints:

$$s_{ij} = 1 \Leftrightarrow (i, j) \in W \quad 1 \leq i, j \leq n$$

$$b_{ij} = 1 \Leftrightarrow (i, j) \in U \quad 1 \leq i, j \leq n$$

$$b_{ij} = 1 \Leftrightarrow A_{ij} \cap W = \emptyset \quad 1 \leq i, j \leq n$$

The first two constraints simply give the intended meanings to the sets W and U, while the last (together with the primary constraint given above) equates squares occupied by black queens with squares not attacked by white queens.

These additional constraints are expensive to reason with and rarely have an effect during search. To save runtime, we only add them when an assignment satisfying all the other constraints has been found, backtracking if this then causes a failure. For example, when $n = 8$, these constraints eliminated just one solution out of 46 candidates; thus, they are not important except to get an exact count of genuinely non-isomorphic solutions.

The model has $4n^2 + 1$ binary constraints, as opposed to $O(n^3)$ for the squares model and about $4n^2$ for the combined model with line and square variables. However, as just described, only $n^2 + 1$ of these constraints are active during most of the search. The search variables are the n^2 s_{ij} variables. This is the same number as in the previous models, but now each variable has only two possible values rather than three.

It is easy to implement SBDS in the unattacked squares model. The functions for the seven board symmetries are exactly as in the previous models. For symmetries that swap the black and white queens, we could obviously make $s_{ij} = 1$

equivalent to $b_{ij} = 1$. However, this would necessitate the inclusion of the inefficient constraints throughout search, so instead we make $s_{ij} = 1$ map to $(i, j) \in U$. That is, since black queens are on all unattacked squares, the symmetric version of a white queen existing on a square is that the same square is unattacked by any white queen. This combines with the symmetries of the board to give eight more functions.

Table 4 shows the results for this model using ILOG Solver. In reporting the number of solutions, we give the total number of distinct solutions found, but also note how many of these have unbalanced numbers of queens of each colour. As discussed above, in these cases the number of solutions in this table is different to that reported in Table 3. We can obtain a set of solutions with equal numbers of each by dropping any extra queens in all possible ways, but we cannot guarantee these are all symmetrically distinct without further checking. For example, for $n = 3$ there are two distinct ways to have two queens of one colour and one of the other, but when the extra queen is removed in both possible ways, the four resulting solutions are all symmetrically equivalent to the unique solution with one of each colour. In short, there is no trivial way to match the numbers of solutions in Table 3 and Table 4, although it would be possible to generate all solutions with equal numbers of queens from the set of solutions in the unattacked squares model and discard symmetric duplicates.

In comparison with the previous models, the number of fails is almost quartered for $n = 8$, and the running time also greatly improved compared to the model combining square and line variables (171 sec. to 60 sec.). The difference is still larger when $n = 9$: the combined model takes over 31 million fails and 5500 sec. to find and prove the optimal solution.

Table 4. Search effort and running time to solve the armies of queens problem and find all non-isomorphic optimal solutions, with the unattacked squares model, SBDS and lexicographic ordering, using ILOG Solver. The number of solutions in brackets is the number found that have more queens of one colour

n Value	Finding & proving optimal solution			Finding all solutions		
	F	P	sec.	Solutions	fails	sec.
2 0	0	1	< 0.01	1 (1)	1	< 0.01
3 1	0	4	0.01	2 (2)	3	< 0.01
4 2	0	12	< 0.01	7 (2)	8	0.01
5 4	26	112	0.02	3	101	0.03
6 5	8	1,062	0.27	21 (3)	1,462	0.42
7 7	39	12,318	3.21	19	17,587	4.85
8 9	1,150	185,504	60.4	45 (3)	260,237	81.3
9 12	591,486	3,101,026	1,040	18	3,283,004	1,320

Fig. 2. (Left) the optimal solution for an 8×8 board that is closest to having 10 queens of each colour. A white queen could be placed on the square marked $\times$ *or* a black queen on the square marked •, but not both, since they would then attack each other. (Right) a possible chess position, with the king and all nine possible queens of each colour

7.1   A Puzzle for the Standard Chessboard

In the solution set for the standard chessboard, we noticed how remarkably close to placing 10 queens of each colour we can get. Of the three solutions with 10 of one colour and 9 of the other, two differ in just one place, as illustrated in Figure   (left).

From this layout we can derive that shown on the right of Figure   , containing the king and all nine possible queens of each colour, i.e. the original queen and eight promoted pawns. Thus, the figure solves the following puzzle, and in fact gives the unique solution up to symmetry: *Put the king and the 9 queens of each colour on a chessboard in such a way that no queen is on the same row, column or diagonal as any piece of the opposite colour.*

8   Variable Ordering

The unattacked squares model was derived from trying to solve the problem by hand. This also led to an algorithm for constructing a solution and from that a variable ordering heuristic. The algorithm places a white queen on the square attacking fewest squares that are not already attacked; hence, it tries to keep the number of unattacked squares as large as possible. The algorithm terminates when no more white queens can be placed without reducing the number of unattacked squares below the number of white queens. Often, the solution found is optimal or near optimal. The first white queen placed is in a corner square, and the lexicographic ordering used so far assigns the variable representing the top left corner first. However, after assigning the first variable, the lexicographic ordering diverges from the algorithm. We have therefore experimented with a dynamic variable ordering heuristic that chooses next the variable

Table 5. Search effort and running time to solve the armies of queens problem and find all non-isomorphic optimal solutions, with the unattacked squares model, SBDS and the most-unattacked-squares variable ordering heuristic, using ILOG Solver

n	Value	Finding & proving optimal solution			Finding all solutions		
		F	P	sec.	Solutions	fails	sec.
2	0	0	1	< 0.01	1 (1)	1	< 0.01
3	1	3	10	0.01	2 (2)	6	< 0.01
4	2	2	18	< 0.01	7 (2)	11	0.02
5	4	4	46	0.03	3	64	0.03
6	5	47	598	0.13	21 (3)	681	0.22
7	7	504	3,387	1.06	19	5,401	1.56
8	9	2,890	40,751	15.1	45 (3)	55,723	24.8
9	12	40,195	320,589	141	18	482,537	217
10	14	14,752	4,581,194	3,030	149 (17)	7,498,801	5,180
11	17	208,573	61,076,593	43,400	168 (25)	95,112,446	67,600
12	21	131,988,279	717,853,580	681,000	*insufficient time to run*		

representing a square which is already attacked itself and where a white queen would attack fewest unattacked squares.

The fewest-unattacked-squares heuristic finds optimal solutions very quickly, but is worse than lexicographic ordering at proving optimality. For instance, when $n=8$, it finds an optimal solution immediately, with no backtracking, but then takes more than 320,000 fails and 78 sec. to prove optimality. It is also poor at enumerating solutions: for $n = 8$ and 9 it takes 118 sec. and 2,020 sec. respectively, again both worse than lexicographic ordering.

Since this heuristic is so poor at proving optimality, it seemed worthwhile to try exactly the opposite heuristic, i.e. choose the square where a white queen will attack *most* unattacked squares. Not surprisingly, this takes much longer to find the optimal solution (though not as long as lexicographic ordering for the larger values of n), but it is overall much faster than either the fewest-unattacked-squares heuristic or lexicographic ordering. The results are shown in Table 5. For 8×8 and 9×9, this heuristic runs more than 10 times faster than lexicographic ordering.

The 10×10, 11×11 and 12×12 problems can now be solved, although the latter takes more than a week of cpu time. In the first two cases, we have found all optimal solutions; there are solutions with an extra queen of one colour, and a solution with two extra queens in the 11×11 case. The fewest-unattacked-squares heuristic again finds optimal solutions very quickly for these two problems.

Since the most-unattacked-squares heuristic performs so much better than either of the other variable orderings considered, and yet is not especially good at finding optimal solutions, it is worth trying to explain why it does well. Figure shows a configuration with 6 white queens attacking all but 6 squares on a 11 × 11 board, and an optimal solution with 17 queens of each colour.

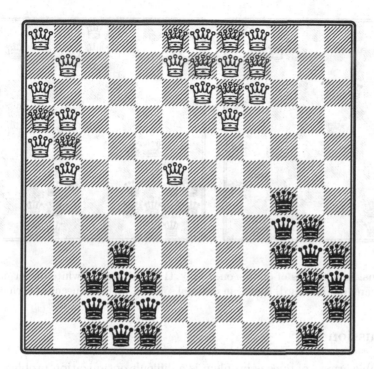

Fig. 3. An optimal solution for a 12 × 12 board, with 21 queens of each colour

The most-unattacked-squares heuristic is biased towards producing configurations with a few white queens attacking all but a few squares, as on the left of Figure . However, once an optimal solution has been found, such configurations become infeasible. In fact, it would no longer be possible to place 6 white queens as shown: a branch of the search tree leading to this configuration would be pruned as soon as fewer than 17 unattacked squares are left. Hence, we conjecture that the heuristic is successful because it can prune branches of the search tree when only a few variables have been assigned, i.e. it tends to find small nogoods. On the other hand, a heuristic which tries to place as many white queens as possible before leaving fewer than the optimal number of unattacked squares (as the fewest-unattacked-squares heuristic does) will tend to prune the search much lower down the tree.

Both heuristics could be used with the earlier models, but would be expensive to implement, since the information on unattacked squares is not readily available. Here, we compute $|A_{ij} \cap U|$ for each unassigned variable s_{ij}, and choose the variable for which this is smallest or largest, depending on the heuristic.

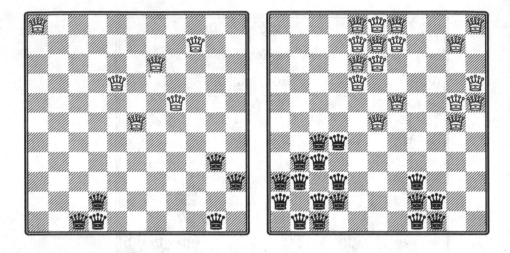

Fig. 4. Equal sized armies of queens on a 11 × 11 board. Left, 6 white queens attack all but 6 squares on the board. Right, an optimal solution with 17 queens of each colour

9 Discussion

The peaceable armies of queens problem is a difficult optimization problem that was hard to solve using an integer programming model. The constraint programming models considered here have all done reasonably well in solving the 8 × 8 problem; even so, problems larger than 10 × 10 are taking a very long time to solve, even for the best model we have found. Related problems have been investigated by Velucchi [ ]. His results give optimal solutions for the armies of queens problem up to 10 × 10 and for 12×12, but it is unclear if optimality was proved, and certainly the number of optimal solutions is not reported. This suggests that constraint programming is competitive with other methods that have been tried for this problem, and indeed is capable of obtaining new results in the field. However, the problem has no practical importance and it is the experience of trying to solve it that is useful, rather than the solutions themselves.

Starting from a basic constraint programming model with no symmetry breaking, we have shown that the time to solve the 8 × 8 problem can be reduced from 1,660 sec. to 15 sec. (using ILOG Solver), a more than 100-fold improvement. The results for 8×8 and 9×9 are summarized in Table . Note that even our initial technique is similar in speed to Plastria's IP solution, allowing for differences in machine, so our final technique is literally a hundred times better. With increasing size the improvement ratio gets better: for 9 × 9 we saw a 400-fold runtime improvement from the first model (65,300sec.) to the last (141sec.), using the same environment on the same machine.

A major part of the improvement is due to eliminating the symmetry using SBDS. Given an implementation of SBDS, it requires no ingenuity on the part of the user to write the 15 functions to describe the effects of the individual

Table 6. Performance of different models in solving the 8×8 and 9×9 armies of queens problems

8×8 Model	F	P	sec.
Basic model, no SBDS	5,270	12,002,608	1,660
Basic model, with SBDS	5,270	924,505	331
Combined model, with SBDS	5,270	924,505	171
Unattacked squares model, with SBDS	1,150	185,504	60
Unattacked squares model, with SBDS & most-unattacked-squares heuristic	2,890	40,751	15

9×9 Model	F	P	sec.
Basic model, no SBDS	8,049,706	273,820,671	65,300
Basic model, with SBDS	7,502,848	31,407,249	12,500
Combined model, with SBDS	7,502,848	31,407,249	6,620
Unattacked squares model, with SBDS	591,486	3,101,026	1,040
Unattacked squares model, with SBDS & most-unattacked-squares heuristic	40,195	320,589	141

symmetries of the problem. For the 8×8 problem, eliminating the symmetry reduces the time to solve the problem optimally from 1,660 sec. to 331 sec.; it also allows a set of non-isomorphic solutions to be found, whereas without symmetry breaking, it took too long to find all the possible solutions, which would in any case have been uninformative.

Further reductions in running time are due to remodelling the problem. We have described three different ways of modelling it, starting from a basic model not very different from an integer programming formulation.The combined model introduces ancillary variables (one for each row, column or diagonal) in order to reduce the number of constraints, from $4n^3$ to $4n^2$, approximately. This significantly reduces running time, although the search effort is largely unaffected.

The unattacked squares model has the same number of search variables as the other models, but with fewer possible values, so that the number of possible assignments is reduced. The model also has fewer constraints than the previous models, which probably contributes to the reduction in running time. However, the binary constraints are between an integer variable and a set variable, so that constraint propagation may be more expensive than with binary constraints involving two integer variables.

Devising new models does require ingenuity. The different models we have presented can be seen as viewing the problems at different levels. The basic model expresses that a single white queen and a single black queen are inconsistent if they are on the same row, column or diagonal. The combined model takes the perspective of a line (row, column or diagonal) of the board: any number of queens can be placed on a line provided that they are all the same colour. The unattacked squares model expresses that any number of white queens can be placed anywhere on the board, as long as there are at least as many unattacked squares as white queens. Hence, each model takes a broader view of the problem than the previous model. Moreover, whereas the first two models are only concerned with whether the white and black queens attack each other, the final model also has something of the optimization criterion built into it: not only must the white and black queens not attack each other, but there must be enough of each of them. Trying to view the problem from several different angles is likely to be a fruitful source of ideas for remodelling; we found that constructing solutions by hand facilitated this and gave useful insights into key features of the problem.

The final improvement in modelling the problem came from a variable ordering heuristic. We have presented two: one finds optimal or near-optimal solutions very quickly, but is poor at proving optimality. The other is its exact opposite and takes much longer to find an optimal solution, but then is much better at proving optimality. Although it is intuitively clear that finding optimal solutions and proving optimality are different in nature, it is surprising to see it demonstrated in such a clear-cut way. Again, the first heuristic was inspired by trying to construct solutions by hand. There may be other problems where a good heuristic for proving optimality is the exact opposite of a good heuristic for finding an optimal solution, and this will be investigated further. Variable ordering heuristics have hitherto mainly been investigated in the context of constraint satisfaction rather than optimization: our experience with this problem suggests that variable ordering heuristics for satisfaction problems and for optimization problems may need to be designed separately. For some optimization problems, it may be better to use two different heuristics, the first to find a good solution and the second to improve that solution if possible and to prove optimality.

Acknowledgements

We are extremely grateful to Warwick Harvey for his advice and help in using ECL^{i}PS^e, and for his implementation of SBDS. The authors are members of the APES and CP Pod research groups (http://www.dcs.st-and.ac.uk/~apes and http://www.dcs.st-and.ac.uk/~cppod) and would like to thank the other members, as well as Graeme Bell. This work is supported by EPSRC grants GR/R29666 and GR/R29673, as well as a Royal Society of Edinburgh SEELLD Support Fellowship to the third author.

References

[1] R. A. Bosch. Peaceably Coexisting Armies of Queens. *Optima (Newsletter of the Mathematical Programming Society)*, 62:6–9, 1999.

[2] B. M. W. Cheng, K. M. F. Choi, J. H. M. Lee, and J. C. K. Wu. Increasing constraint propagation by redundant modeling: an experience report. *Constraints*, 4:167–192, 1999.

[3] M. Gardner. Chess Queens and Maximum Unattacked Cells. *Math Horizon*, pages 12–16, November 1999.

[4] I. P. Gent and B. M. Smith. Symmetry Breaking During Search in Constraint Programming. In W. Horn, editor, *Proceedings ECAI'2000*, pages 599–603, 2000.

[5] B. M. Smith. Dual Models of Permutation Problems. In *Proceedings of CP'01: the 7th International Conference on Principles and Practice of Constraint Programming*, LNCS 2239, pages 615–619. Springer, 2001.

[6] M. Velucchi. For me, this is the best chess-puzzle: Non-dominating queens problem. http://anduin.eldar.org/~problemi/papers.html. Accessed January 2004.

A Global Constraint
for Graph Isomorphism Problems

Sébastien Sorlin and Christine Solnon

LIRIS, CNRS FRE 2672, bât. Nautibus
University of Lyon I
43 Bd du 11 novembre, 69622 Villeurbanne cedex
France
{sebastien.sorlin,christine.solnon}@liris.cnrs.fr

Abstract. The graph isomorphism problem consists in deciding if two
given graphs have an identical structure. This problem can be modeled as a constraint satisfaction problem in a very straightforward way,
so that one can use constraint programming to solve it. However, constraint programming is a generic tool that may be less efficient than
dedicated algorithms which can take advantage of the global semantic of
the original problem.

Hence, we introduce in this paper a new global constraint dedicated to
graph isomorphism problems, and we define an associated filtering algorithm that exploits all edges of the graphs in a global way to narrow
variable domains. We then show how this global constraint can be decomposed into a set of "distance" constraints which propagate more domain
reductions than "edge" constraints that are usually generated for this
problem.

1   Introduction

Graphs provide a rich mean for modeling structured objects and they are widely
used in real-life applications to represent, e.g., molecules, images, or networks. In
many of these applications, one has to compare graphs to decide if their structure
is identical. This problem is known as the Graph Isomorphism Problem (GIP).

More formally, a *graph* is defined by a pair (V, E) such that V is a finite
set of vertices and $E \subseteq V \times V$ is a set of edges. In this paper, we shall restrict
our attention to graphs without self-loops, *i.e.*, $\forall (u, v) \in E$, $u \neq v$. Two graphs
$G = (V, E)$ and $G' = (V', E')$ are *isomorphic* if there exists a bijective function
$f : V \rightarrow V'$ such that $(u, v) \in E$ if and only if $(f(u), f(v)) \in E'$. We shall say
that f is an *isomorphism function*. The GIP consists in deciding if two given
graphs are isomorphic.

There exists many dedicated algorithms for solving GIPs, such as [, ,].
These algorithms are often very efficient (eventhough their worst case complexities are exponential). However, such dedicated algorithms can hardly be used
to solve more general problems, such as isomorphism problems with additional
constraints, or larger problems that include GIPs.

J.-C. Régin and M. Rueher (Eds.): CPAIOR 2004, LNCS 3011, pp. 287–   , 2004.
© Springer-Verlag Berlin Heidelberg 2004

An attractive alternative to these dedicated algorithms is to use Constraint Programming (CP), which provides a generic framework for solving any kind of Constraint Satisfaction Problems (CSPs). Indeed, GIPs can be transformed into CSPs in a very straightforward way [   ], so that one can use generic constraint solvers to solve them. However, when transforming a GIP into a CSP, the global semantic of the problem is lost and replaced by a set of binary constraints. As a consequence, using CP to solve isomorphism problems may be less efficient than using dedicated algorithms which have a global view of the problem.

Outline of the Paper. The goal of this paper is to allow constraint solvers to handle GIPs in a global way so that they can solve them efficiently without loosing CP's flexibility. To this aim, we introduce a new global constraint for modeling GIPs, and we show how one can take benefit of this globality to solve more efficiently GIPs.

Section 2 gives some complexity results for GIPs and an overview of existing approaches for solving these problems. Section   presents some properties of the GIP which are used to define our filtering algorithm. In section   , we introduce a new global constraint for modeling GIPs on non directed graphs, and we define filtering technics for this global constraint. In section   , we discuss the extension of our work to directed graphs and to the subgraph isomorphism problem.

2   Solving Graph Isomorphism Problems

Complexity. The theoretical complexity of the GIP is not exactly stated: the problem is in NP but it is not know to be in P or to be NP-complete [   ] and its own complexity class, *isomorphism-complete*, has been defined. However, some topological restrictions on graphs (e.g., planar graphs [   ], trees [ ] or bounded valence graphs [   ]) make this problem solvable in a polynomial time.

Dedicated Algorithms. To solve a GIP, one has to find a one to one mapping between the vertices of the two graphs. The search space composed of all possible mappings can be explored in a "Branch and Cut" way: at each node of the search tree, some graph properties (such as edges distribution, vertices neighbourhood) can be used to prune the search space [,   ]. This kind of approach is rather efficient and can be used to solve GIPs up to 1000 vertices very quickly (less than 1 second).

originally used to detect graph automorphisms (*i.e.*, non trivial isomorphisms between a graph and itself). The idea is to compute for each vertex v_i a unique label that characterizes the relationships between v_i and the other vertices of the graph, so that two vertices are assigned with a same label if and only if they can be mapped by an isomorphism function. This approach is implemented in the system *nauty* which is, to our knowledge, the most efficient solver for the graph isomorphism problem. The time needed to solve a GIP with *nauty* is comparable to "Branch and Cut" methods but *nauty* is often the quickest for large graphs [ ].

Hence dedicated algorithms are very efficient to solve GIPs in practice, even-though their worst case complexities are exponential. However, they are not suited for solving more general problems, such as GIPs with additional con-straints. In particular, vertices and edges of graphs may be associated with labels that characterize them, and one may be interested in finding isomorphisms that satisfy particular constraints on these labels. This is the case, e.g., in [ ] where graphs are used to represent molecules, or in computer aided design (CAD) applications where graphs are used to represent design objects [ ].

Constraint Programming. CP is a generic tool for solving constraint satis-faction problems (CSPs), and it can be used to solve GIPs. A *CSP* [ ] is defined by a triple (X, D, C) such that :

- X is a finite set of variables,
- D is a function that maps every variable $x_i \in X$ to its domain $D(x_i)$, *i.e.*, the finite set of values that can be assigned to x_i,
- C is a set of constraints, *i.e.*, relations between some variables which restrict the set of values that can be assigned simultaneously to these variables.

Binary CSPs only have binary constraints, *i.e.*, each constraint involves two variables exactly. We shall note $C(x_i, x_j)$ the binary constraint holding between the two variables x_i and x_j, and we shall define this constraint by the set of couples $(v_i, v_j) \in D(x_i) \times D(x_j)$ that satisfy the constraint.

Solving a CSP (X, D, C) involves finding a complete assignement, which as-signs one value $v_i \in D(x_i)$ to every variable $x_i \in X$, such that all the constraints in C are satisfied.

CSPs can be solved in a generic way by using constraint programming lan-guages (such as CHOCO [ ], Ilog solver [ ], or CHIP [ ]), *i.e.*, programming languages that integrate algorithms for solving CSPs. These algorithms (called constraint solvers) are often based on a systematic exploration of the search space, until either a solution is found, or the problem is proven to have no so-lution. In order to reduce the search space, this kind of complete approach is combined with filtering techniques that narrow variables domains with respect to some partial consistencies such as Arc-Consistency [, ,].

Using CP to Solve GIPs. Graph isomorphism problems can be formulated as CSPs in a very straightforward way, so that one can use CP languages to solve them [,]. Given two graphs $G = (V, E)$ and $G' = (V', E')$, we define the CSP (X, D, C) such that :

- a variable x_u is associated with each vertex $u \in V$, *i.e.*, $X = \{x_u/u \in V\}$,
- the domain of each variable x_u is the set of vertices of G' that have the same number of entering and leaving edges than u, *i.e.*,

$$D(x_u) = \{u' \in V' \;/\; |\{(u, v) \in E\}| = |\{(u', v') \in E'\}| \text{ and}$$
$$|\{(v, u) \in E\}| = |\{(v', u') \in E'\}|\}$$

– there is one binary constraint between every pair of different variables. The constraint holding between two different variables $(x_u, x_v) \in X^2$ is denoted by $C_{edge}(x_u, x_v)$ and expresses the fact that the vertices of G' that are assigned to x_u and x_v must be connected by an edge in G' if and only if the two vertices u and v are connected by an edge in G, *i.e.*,

if $(u, v) \in E$, $C_{edge}(x_u, x_v) = E'$

otherwise $C_{edge}(x_u, x_v) = \{(u', v') \in V'^2 \mid u' \neq v' \text{ and } (u', v') \notin E'\}$

Once a GIP has been formulated as a CSP, one can use constraint programming to solve it in a generic way, and additional constraints, such as constraints on vertex and edge labels, can be added very easily.

Discussion. When formulating a GIP into a CSP, the global semantic of the problem is decomposed into a set of binary "edge" constraints, each of them expressing locally the necessity either to maintain or to forbid one edge. As a consequence, using CP to solve GIPs will often be less efficient than using a dedicated algorithm.

To improve the solution process of CSPs associated with GIPs, one can add an *allDiff* global constraint, in order to constrain all variables to be assigned to different vertices [ ]. This constraint is redundant as each binary edge constraint only contains couples of different vertices, so that it will not be possible to assign a same vertex to two different variables. However, adding this global constraint allows a constraint solver to prune the search space more efficiently, and therefore to solve GIPs quicker. Hence, with respect to the definition of globality introduced in [ ], this *allDiff* constraint is not semantically global, as it can be decomposed into a semantically equivalent set of binary constraints, but it is AC-operationally global, as an AC-filtering on the global constraint is stronger than an AC-filtering on the equivalent set of binary constraints.

In this paper, we introduce a new global constraint to define GIPs. This global constraint is not semantically global, as it can be decomposed into a set of binary edge constraints as described above. However, by considering all edges of the graphs in a global way, we can prune more efficiently the search space. Note that this GIP global constraint can be combined with an *allDiff* constraint to filter even more values.

3   Some Properties of the GIP

When looking for an isomorphism function between two given graphs, one can use vertex properties to reduce the search space. For example, one can compute the degree of each vertex, or the number of adjacent triangles to each vertex, and use these "vertex invariants" to prune every mapping which violates them. More generally, a vertex invariant is a label $l(v)$ assigned to each vertex v such that if there exists an isomorphism function which links v to v' then $l(v) = l(v')$ (but the converse is not necessary true). The most famous exemple of vertex

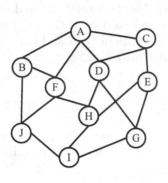

$\delta_G(u,v)$	A	B	C	D	E	F	G	H	I	J
A	0	1	1	1	2	1	2	2	3	2
B	1	0	2	2	3	1	3	2	2	1
C	1	2	0	1	1	2	2	2	3	3
D	1	2	1	0	2	2	1	1	2	3
E	2	3	1	2	0	2	1	1	2	3
F	1	1	2	2	2	0	3	1	2	1
G	2	3	2	1	1	3	0	2	1	2
H	2	2	2	1	1	1	2	0	1	2
I	3	2	3	2	2	2	1	1	0	1
J	2	1	3	3	3	1	2	2	1	0

Fig. 1. A graph $G = (V, E)$ and distances between any pair of its vertices

invariants is the *degree* of a vertex (*i.e.*, the number of incoming and outgoing edges) : if f is an isomorphism function between $G = (V, E)$ and $G' = (V', E')$, then for each vertices $v \in V$, the vertices v and $f(v)$ have the same degree.

We introduce in this section some definitions and theorems that will be used to define a new vertex invariant based on distances. We shall restrict our attention to *undirected graphs*, *i.e.*, graphs with undirected edges so that (u, v) and (v, u) are considered to be the same edge. The extension of our work to directed graphs is discussed in section   . We shall assume that graphs are connected, so that every vertex is reachable from any other vertex.

3.1   Definitions and Theorems

Definition 1. Given a graph $G = (V, E)$, a *path* between two vertices u and v is a sequence $<v_0, v_1, v_2, ..., v_k>$ of vertices such that $v_0 = u$, $v_k = v$ and for all $i \in [1, k]$, $(v_{i-1}, v_i) \in E$. The *length* of a path π, noted $|\pi|$, is the number of its edges.

Definition 2. Given a graph $G = (V, E)$, a *shortest path* between two vertices u and v is a path between u and v the length of which is minimal. The *length of the shortest path* between u and v is noted $\delta_G(u, v)$. We shall say that $\delta_G(u, v)$ is the *distance* between u and v.

Theorem 1. *Given two graphs $G = (V, E)$ and $G' = (V', E')$ such that $|V| = |V'|$, and a bijective function $f : V \rightarrow V'$, the two following properties are equivalent*

$$f \text{ is an isomorphism function, i.e., } (u, v) \in E \Leftrightarrow (f(u), f(v)) \in E' \quad (1)$$
$$\forall (u, v) \in V^2, \delta_G(u, v) = \delta_{G'}(f(u), f(v)) \quad (2)$$

Proof. (1) $\Rightarrow$ (2): *if f is an isomorphism function, then (u, v) is an edge of G iff $(f(u), f(v))$ is an edge of G' so that $< v_1, v_2, ..., v_n >$ is a path in G iff $< f(v_1), f(v_2), ..., f(v_n) >$ is a path in G', and therefore $< v_1, v_2, ..., v_n >$ is a shortest path in G_1 iff $< f(v_1), f(v_2), ..., f(v_n) >$ is a shortest path in G_2, and property (2) holds.*

(2) $\Rightarrow$ (1): *For any pair of vertices $(u, v) \in V \times V$, if (u, v) is an edge of G, then $< u, v >$ is the shortest path between u and v so that $\delta_G(u, v) = 1$, and therefore $\delta_{G'}(f(u), f(v)) = 1$, so that $(f(u), f(v))$ is an edge of G' (and vice versa).*

Theorem 1 will be used to define "distance" constraints for propagating domain reductions when solving GIPs. We now introduce some more definitions that will be used to define a partial consistency and a filtering algorithm for GIPs.

Definition 3. Given a graph $G = (V, E)$, a vertex $u \in V$, and a distance $i \in [0, |V| - 1]$, we note $\Delta_G(u, i)$ the set of vertices that are at a distance of i from u, and $\#\Delta_G(u, i)$ the number of vertices that are at a distance of i from u, i.e.,

$$\Delta_G(u, i) = \{v \in V / \delta_G(u, v) = i\} \quad \text{and} \quad \#\Delta_G(u, i) = |\Delta_G(u, i)|$$

For example, for the graph G of Fig. , we compute:

$$\begin{aligned}
\Delta_G(A, 0) &= \{A\} & \#\Delta_G(A, 0) &= 1 \\
\Delta_G(A, 1) &= \{B, C, D, F\} & \#\Delta_G(A, 1) &= 4 \\
\Delta_G(A, 2) &= \{E, G, H, J\} & \#\Delta_G(A, 2) &= 4 \\
\Delta_G(A, 3) &= \{I\} & \#\Delta_G(A, 3) &= 1 \\
\Delta_G(A, i) &= \emptyset & \#\Delta_G(A, i) &= 0, \ \forall i \geq 4
\end{aligned}$$

Definition 4. Given a graph $G = (V, E)$ and a vertex $u \in V$, we note $\#\Delta_G(u)$ the sequence composed of $|V|$ numbers respectively corresponding to the number of vertices that are at a distance of 0, 1, ... $|V| - 1$ from u, i.e.,

$$\#\Delta_G(u) = \ < \#\Delta_G(u, 0), \#\Delta_G(u, 1), ..., \#\Delta_G(u, |V| - 1) >$$

We shall omit zeros at the end of sequences.

For example, the sequences of the vertices of the graph G of Fig. are

$$\begin{aligned}
\#\Delta_G(A) &= \#\Delta_G(D) = \#\Delta_G(F) = \ < 1, 4, 4, 1 > \\
\#\Delta_G(B) &= \#\Delta_G(C) = \#\Delta_G(E) = \#\Delta_G(G) = \#\Delta_G(I) = \ < 1, 3, 4, 2 > \\
\#\Delta_G(H) &= \ < 1, 4, 5, 0 > \\
\#\Delta_G(J) &= \ < 1, 3, 3, 3 >
\end{aligned}$$

Each sequence $\#\Delta_G(u)$ characterizes the relationships of the vertex u with the other vertices of G by means of distances. Hence, when looking for a graph isomorphism, one can use these sequences as a vertex invariant to reduce the search space by pruning all mappings that associate two vertices with different sequences. However, many different vertices within a same graph may have a same sequence so that this criterion will not narrow much the search space. For example, on the graph example of Fig. , there are five different vertices the sequence of which is $< 1, 3, 4, 2 >$. Definition 6 will go one step further in order to characterize more precisely the relationships of a vertex u with the other vertices of the graph.

Definition 5. Given a graph $G = (V, E)$, we note $\#\Delta_G$ the set of all different sequences associated with the vertices of G, i.e.,

$$\#\Delta_G = \{s | \exists u \in V, s = \#\Delta_G(u)\}$$

For example, the set of all different sequences for the graph G of fig   is

$$\#\Delta_G = \{< 1, 3, 3, 3 >, < 1, 3, 4, 2 >, < 1, 4, 4, 1 >, < 1, 4, 5, 0 >\}$$

Definition 6. Given a vertex $u \in V$, we note $label_G(u)$ the set of all tuples (i, s, k) such that i is a distance, s is a sequence, and k is the number of vertices that are at a distance of i from u and the sequence of which is s, i.e.,

$$label_G(u) = \{(i, s, k) \; / \; i \in [0, |V| - 1],$$
$$s \in \#\Delta_G, \text{ and}$$
$$k = |\{v \in \Delta_G(u, i) / \#\Delta_G(v) = s\}|\}$$

We shall omit the tuples (i, s, k) such that $k = 0$.
For example, for the graph G of Fig. 1, we compute

$$label_G(A) = \{ \; (0, < 1, 4, 4, 1 >, 1),$$
$$(1, < 1, 3, 4, 2 >, 2), (1, < 1, 4, 4, 1 >, 2),$$
$$(2, < 1, 3, 3, 3 >, 1), (2, < 1, 3, 4, 2 >, 2), (2, < 1, 4, 5, 0 >, 1),$$
$$(3, < 1, 3, 4, 2 >, 1)\}$$

as there is one vertex (A) that is at a distance of 0 from A and which sequence is $< 1, 4, 4, 1 >$, two vertices $(B$ and $C)$ that are at a distance of 1 from A and which sequence is $< 1, 3, 4, 2 >$, two vertices $(D$ and $F)$ that are at a distance of 1 from A and which sequence is $< 1, 4, 4, 1 >$, etc...

Theorem 2. *Given two graphs $G = (V, E)$ and $G' = (V', E')$, if there exists an isomorphism function $f : V \rightarrow V'$ that matches the two graphs then, for each vertex $u \in V$, $label_G(u) = label_{G'}(f(u))$.*

Proof. f is a bijection, and the distance between two vertices u and v in G is equal to the distance between their associated vertices $f(u)$ and $f(v)$ in G' (see theorem). Therefore, the number of vertices of G that are at a distance of i

from u is equal to the number of vertices of G' that are a distance of i from $f(u)$, so that $\#\Delta_G(u) = \#\Delta_{G'}(f(u))$. As a consequence, the set of sequences of the two graphs are equals, i.e., $\#\Delta_G = \#\Delta_G$. Then, for each sequence $s \in \#\Delta_G$, and for each vertex $u \in V$, the number of vertices that are at a distance of i from u and which sequence is s is equal to the number of vertices that are at a distance of i from $f(u)$ and which sequence is also s, and therefore $label_G(u) = label_{G'}(f(u))$.

3.2   Algorithms and Complexities

We discuss in this section time and space complexities required to compute the different values introduced in   . These complexities are given for a non directed connected graph $G = (V, E)$ such that $|V| = n$ and $|E| = p$ with $n - 1 \leq p < n^2$.

$\delta_G(u, v)$. All definitions introduced in section 3.1 are based on distances between couples of vertices. A Breadth First Search (BFS) [ ] from each vertex of G is needed to compute them all: n BFS are needed, each of them performing $\mathcal{O}(p)$ operations, so that the time complexity is in $\mathcal{O}(np)$. The space needed to store these informations is in $\mathcal{O}(n^2)$.

$\Delta_G(u, i)$, $\#\Delta_G(u, i)$ and $\#\Delta_G(u)$. All these values can be computed in an incremental way while computing shortest paths: each time a distance $\delta_G(u, v)$ is computed, the vertex u (resp. v) is added to the set $\Delta_G(v, \delta_G(u, v))$ (resp. $\Delta_G(u, \delta_G(u, v))$), both $\#\Delta_G(u, \delta_G(u, v))$ and $\#\Delta_G(v, \delta_G(u, v))$ are incremented, and the sequences $\#\Delta_G(u)$ and $\#\Delta_G(v)$ are updated by incrementing their $\delta_G(u, v)^{\text{th}}$ component. All these operations can be done in constant time. The space needed to store these informations is in $\mathcal{O}(n^2)$.

$label_G(u)$. To compute and compare labels efficiently, we first sort the set of all sequences, so that a unique integer is associated with each different sequence. This is done in $\mathcal{O}(n^2 . log(n))$ operations as there is at most n different sequences, and the comparison of two sequences is in $\mathcal{O}(n)$. Then, the computation of all labels can be done in $\mathcal{O}(n^2)$ operations (as there are n labels to compute, each of them containing at most n different triples), and requires $\mathcal{O}(n^2)$ space. Finally, the comparison of two labels can be done in $\mathcal{O}(n)$ operations, provided that the set of triples (i, s, k) contained in each label is sorted.

As a consequence, the computation of all values introduced in section 3.1 for a graph $G = (V, E)$ requires $\mathcal{O}(|V|.|E| + |V|^2 . log(|V|))$ operations and $\mathcal{O}(|V|^2)$ space.

4   A Global Constraint for GIPs

We now introduce a new global constraint for tackling GIPs efficiently. Syntactically, this constraint is defined by the relation $gip(V, E, V', E', L)$ where

- V and V' are 2 sets of values such that $|V| = |V'|$,
- $E \subseteq V \times V$ is a set of pairs of values from V,

- $E' \subseteq V' \times V'$ is a set of pairs of values from V',
- L is a set of couples which associates one different variable of the CSP to each different value of V, $i.e.$, L is a set of $|V|$ couples of the form (x_u, u) where x_u is a variable of the CSP and u is a value of V, and such that for any pair of different couples (x_u, u) and (x_v, v) of L, both x_u and x_v are different variables and $u \neq v$.

Semantically, the global constraint $gip(V, E, V', E', L)$ is consistent if and only if there exists an isomorphism function $f : V \rightarrow V'$ such that for each couple $(x_u, u) \in L$ there exists a value $u' \in D(x_u)$ so that $u' = f(u)$.

This global constraint is not semantically global as it can be represented by a semantically equivalent set of binary constraints as described in section 2. However, the gip constraint allows us to exploit the global semantic of GIPs to solve them more efficiently. We now define a partial consistency, and an associated filtering algorithm (section   ); we shall then describe how to propagate constraints (section   ).

4.1   Label-Consistency and Label-Filtering for gip Constraints

Theorem   establishes that an isomorphism function always maps vertices that have identical labels. Hence, we can define a partial consistency for the gip constraint, called $label\text{-}consistency$, that ensures that for each couple $(x_i, v) \in L$, each value u in the domain of x_i has the same label than v.

Definition 7. The global constraint $gip(V, E, V', E', L)$ is label-consistent iff

$$\forall(x_u, u) \in L, \ \forall u' \in D(x_u), \ label_{(V,E)}(u) = label_{(V',E')}(u')$$

To achieve label-consistency, one just has to compute the label of each vertex of the two graphs, as described in section 3.2, and remove from the domain of each variable x_u associated with a vertex $u \in V$ every value $u' \in D(x_u)$ such that $label_{(V,E)}(u) \neq label_{(V',E')}(u')$.

This label-filtering often drastically reduces variable domains. Let us consider for example the graph G of Fig.   . The first three triples (sorted by increasing distance, and then by increasing sequence number) of the label of each vertex are:

$label_G(A) = \{(0, < 1, 4, 4, 1 >, 1), (1, < 1, 3, 4, 2 >, 2), (1, < 1, 4, 4, 1 >, 2), ...\}$

$label_G(B) = \{(0, < 1, 3, 4, 2 >, 1), (1, < 1, 3, 3, 3 >, 1), (1, < 1, 4, 4, 1 >, 2), ...\}$

$label_G(C) = \{(0, < 1, 3, 4, 2 >, 1), (1, < 1, 3, 4, 2 >, 1), (1, < 1, 4, 4, 1 >, 2), ...\}$

$label_G(D) = \{(0, < 1, 4, 4, 1 >, 1), (1, < 1, 3, 4, 2 >, 2), (1, < 1, 4, 4, 1 >, 1), ...\}$

$label_G(E) = \{(0, < 1, 3, 4, 2 >, 1), (1, < 1, 3, 4, 2 >, 2), (1, < 1, 4, 5, 0 >, 1), ...\}$

$label_G(F) = \{(0, < 1, 4, 4, 1 >, 1), (1, < 1, 3, 3, 3 >, 1), (1, < 1, 3, 4, 2 >, 1), ...\}$

$label_G(G) = \{(0, < 1, 3, 4, 2 >, 1), (1, < 1, 3, 4, 2 >, 2), (1, < 1, 4, 4, 1 >, 1), ...\}$

$label_G(H) = \{(0, < 1, 4, 5, 0 >, 1), (1, < 1, 3, 4, 2 >, 2), (1, < 1, 4, 4, 1 >, 2), ...\}$

$label_G(I) = \{(0, < 1, 3, 4, 2 >, 1), (1, < 1, 3, 3, 3 >, 1), (1, < 1, 3, 4, 2 >, 1), ...\}$

$label_G(J) = \{(0, < 1, 3, 3, 3 >, 1), (1, < 1, 3, 4, 2 >, 2), (1, < 1, 4, 4, 1 >, 1), ...\}$

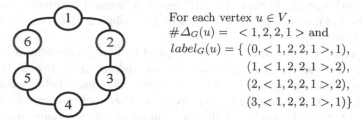

For each vertex $u \in V$,
$$\# \Delta_G(u) = \; <1,2,2,1> \text{ and}$$
$$label_G(u) = \{ \, (0, <1,2,2,1>, 1),$$
$$(1, <1,2,2,1>, 2),$$
$$(2, <1,2,2,1>, 2),$$
$$(3, <1,2,2,1>, 1) \}$$

Fig. 2. A circular graph $G = (V, E)$ and vertex sequences and labels

Actually, all vertices of G have different labels. As a consequence, for any *gip* constraint between G and another graph G', label-filtering will allow one either to detect an inconsistency (if some label of G is not in G'), or to reduce the domain of each variable to a singleton so that global consistency can be easily checked.

On this example, we can compare label-consistency of a *gip* constraint with arc consistency of the CSP defined in section 2. Let us define another graph $G' = (V', E')$ that is isomorphic to the graph $G = (V, E)$ of Fig. , and such that each vertex $u \in V$ is renamed into u' in G'. Let us consider the CSP which modelizes the problem of finding an isomorphism between these two graphs, as defined in section 2. For this CSP, the domain $D(x_u)$ of each variable x_u contains every vertex $u' \in V'$ such that u and u' have a same number of incident edges, so that :

$$D(x_A) = D(x_D) = D(x_F) = D(x_H) = \{A', D', F', G'\}$$
$$D(x_B) = D(x_C) = D(x_E) = D(x_G) = D(x_I) = D(x_J) = \{B', C', E', G', I', J'\}$$

This CSP already is arc consistent so that an AC-filtering will not reduce any domain. Note also that, on this example, adding an *allDiff* constraint does not allow to filter more domains.

4.2 Propagating Constraints

Label-filtering does not always reduce every domain to a singleton so that it may be necessary to explore the search space. Let us consider for example the graph displayed in Fig. . This graph has many symetries (it is isomorphic to any graph obtained by a circular permutation of its vertices), so that all vertices are associated with a same sequence and a same label. In this case, label-filtering does not narrow any domain.

When label-filtering does not reduce the domain of each variable to a single-ton, one has to explore the search space composed of all possible assignments by constructing a search tree. At each node of this search tree, the domain of one variable is splitted into smaller parts, and then filtering technics are applied to narrow variable domains with respect to some local consistencies. These filter-ing technics iteratively use constraints to propagate the domain reduction of one variable to other variable domains until either a domain becomes empty (the node can be cut), or a fixed-point is reached (a solution is found or the node must be splitted).

To propagate the domain reductions implied by a *gip* constraint, a first possibility is to use the set of C_{edge} constraints as defined in section 2. However, we can take advantage of results obtained while achieving label-consistency to define "tighter" constraints. By tighter, we mean that each constraint is defined by a smaller (or equal) number of allowed couples of values so that propagating them may narrow domains more strongly.

The idea is to constrain each pair of variables (x_u, x_v) associated with a pair of vertices (u, v) of the first graph to take their values within the set of pairs of vertices (u', v') of the second graph such that the distance between u and v is equal to the distance between u' and v'. Indeed, Theorem 1 proves that a bijective function between two graphs is an isomorphism function if and only if this function preserves the distances between every pair of vertices in each graph. Therefore, the global constraint $gip(V, E, V', E', L)$ is semantically equivalent to a set of "distance" constraints defined as follows: for all $((x_u, u), (x_v, v)) \in L \times L$ such that $u \neq v$,

$$C_{distance}(x_u, x_v) = \{(u', v') \in V' \times V' \mid \delta_{(V,E)}(u, v) = \delta_{(V',E')}(u', v')\}$$

One can easily show that each binary constraint $C_{distance}(x_u, x_v)$ is tighter than (or equal to) the corresponding binary constraint $C_{edge}(x_u, x_v)$ defined in section 2:

- if the vertices of G associated with the variables x_u and x_v are connected by an edge in G, then $C_{distance}(x_u, x_v) = C_{edge}(x_u, x_v) = E'$,
- otherwise, $C_{distance}(x_u, x_v) \subseteq C_{edge}(x_u, x_v)$ as $C_{edge}(x_u, x_v)$ contains all pairs of vertices of G' that are not connected by an edge whereas $C_{distance}(x_u, x_v)$ only contains the pairs of vertices of G' such that the distance between them is equal to the distance between the vertices of G associated with x_u and x_v.

As a consequence, propaging a $C_{distance}$ constraint will always remove at least as many values as propaging the corresponding C_{edge} constraint, and in some cases it will remove more values.

For example, let us consider the graph G of Fig. 2 and let us define another graph $G' = (V', E')$ that is isomorphic to G and such that each vertex $u \in V$ is renamed into u' in G'. We note x_u the variable associated with each vertex $u \in V$. The edge constraint between x_1 and x_4 contains every pair of vertices of G' that are not connected by an edge, *i.e.*,

$$C_{edge}(x_1, x_4) = \{ (1', 3'), (1', 4'), (1', 5'), (2', 4'), (2', 5'), (2', 6'),$$
$$(3', 5'), (3', 6'), (3', 1'), (4', 6'), (4', 1'), (4', 2'),$$
$$(5', 1'), (5', 2'), (5', 3'), (6', 2'), (6', 3'), (6', 4')\}$$

whereas the distance constraint between x_1 and x_4 only contains pairs of vertices of G' that are at a distance of 3 one from each other as the distance between 1 and 4 is 3, *i.e.*,

$$C_{distance}(x_1, x_4) = \{(1', 4'), (2', 5'), (3', 6'), (4', 1'), (5', 2'), (6', 3')\}$$

As the distance constraint between x_1 and x_4 is tighter than the corresponding edge constraint, it can propagate more domain reductions. For example, if x_1 is assigned to $1'$, a forward-checking propagation of $C_{distance}(x_1, x_4)$ reduces the domain of x_4 to the singleton $\{4'\}$, whereas a forward-checking propagation of $C_{edge}(x_1, x_4)$ only reduces the domain of x_4 to $\{3', 4', 5'\}$, that is, the set of vertices that are not connected to $1'$ by an edge.

Also, after the suppression of value $1'$ from the domain of x_1, an AC propagation of $C_{distance}(x_1, x_4)$ will remove the value $4'$ from the domain of x_4, whereas an AC propagation of $C_{edge}(x_1, x_4)$ will not remove any value.

5   Extensions to Directed Graphs and Subgraph Isomorphism Problems

5.1   Directed Graphs

All definitions and theorems introduced in section 3 actually hold for directed graphs, provided that paths respect edge directions. However, in this case, there may exist many couples of vertices which are not connected by a path respecting edge directions, so that the sequences describing vertices may be very short. In this case, label-filtering may not reduce much the search space.

Another way to extend our work to directed graphs is to first consider the corresponding non-directed graph (by ignoring edge directions), and to compute sequences and labels on this non-directed graph. Then, constraints can be added to express edge directions.

Finally, a last way to extend our work to directed graphs is to consider together several kinds of paths, each one having a corresponding kind of distance, e.g. directed paths, which respect edge directions, non directed paths, which ignore edge directions... We can then use these different kinds of paths to compute, for each vertex as many labels as defined distances. Two vertices can then be linked together if and only if, for each defined distance, the two vertices have the same label.

Obviously, the third possibility should allow one to narrow more tightly domains. However, it is also more expensive to achieve. Hence, we shall experimentally compare these three different possibilities.

5.2   Subgraph Isomorphism Problems

A graph $G = (V, E)$ is a *subgraph* of another graph $G' = (V', E')$, denoted by $G \subseteq G'$, if $V \subseteq V'$ and $E = E' \cap (V \times V)$. A graph $G = (V, E)$ is an *isomorphic subgraph* of another graph $G' = (V', E')$ if there exists a subgraph $G'' \subseteq G'$ that is isomorphic to G. The Subgraph Isomorphism Problem (SGIP) consists in deciding if a graph $G = (V, E)$ is an isomorphic subgraph of another graph $G' = (V', E')$.

If the theoretical complexity of the GIP is not yet completely stated, the SGIP clearly is an NP-complete problem [ ]. Actually, the SGIP is a more challenging problem for which rather small instances still cannot be solved within a reasonable amount of time.

One can modelize a SGIPs as CSPs, in a very similar way than for GIPs. However, like for GIPs, one could take benefit of the global semantic of the problem to define more powerful filtering algorithms. However, a subgraph isomorphism function $f : V \to V'$ does not preserve distances between vertices like a graph isomorphism function, as stated in Theorem 1: for every path $< v_1, v_2, ..., v_n >$ in G, there exists a path $< f(v_1), f(v_2), ..., f(v_n) >$ in G' but the opposite is not always true (f is not a bijection and it may exist some vertices of V' which are not linked to a vertex of V). As a consequence, the distance $\delta_G(u, v)$ between two vertices u and v of G may be greater than the distance $\delta_{G'}(f(u), f(v))$ between $f(u)$ and $f(v)$.

Hence, filtering technics for handling efficiently SGIPs, could be based on the following property: given two undirected graphs $G = (V, E)$ and $G' = (V', E')$, if f is an isomorphism function between G and a subgraph of G', then

$$\forall (u, v) \in V \times V, \; \delta_G(u, v) \geq \delta_{G'}(f(u), f(v))$$

This property could be used to define a partial consistency, and an associated filtering algorithm. The idea would be to check that, for every vertex u of G, the domain of the variable x_u associated with u only contains vertices u' such that, for every distance $k \in 1..|V_1| - 1$, the number of vertices $v \in G$ for which $\delta_G(u, v) \leq k$ is lower or equal to the number of vertices $v' \in G'$ for which $\delta_{G'}(u', v') \leq k$.

We could also use this property to define distance constraints for propagating domain reductions while exploring a search tree.

6 Conclusion

We have introduced in this paper a new global constraint for defining graph isomorphism problems. To tackle efficiently this global constraint, we have first defined a partial consistency, called label-consistency, and an associated filtering algorithm, that can be used to narrow variable domains before solving the CSP. This label-consistency is based on the computation, for each vertex u, of a label which characterizes the global relationship between u and the other vertices of the graph by means of shortest paths and which can be viewed as a vertex invariant. In many cases, achieving label-consistency will allow a constraint solver to either detect an inconsistency, or reduce variable domains to singletons so that the global consistency can be easily checked.

Then, for cases such that label-consistency does not allow to solve the graph isomorphism problem, we have defined a set of distance constraints that is semantically equivalent to the global GIP constraint and that can be used to propagate domain reductions. We have shown that these distance constraints are tighter than edge constraints, that simply check that edges are preserved by the mapping, so that propagating distance constraints remove more (or as many) values than propagating edge constraints. Note that this set of distance constraints can be combined with a global *allDiff* constraint to propagate even more domain reductions.

Both label-filtering and the generation of distance constraints can be done in $\mathcal{O}(np + n^2 log(n))$ operations for graphs having n vertices and p edges (such that $n - 1 \leq p \leq n^2$. As a comparison, achieving arc consistency with AC2001 on a CSP describing an isomorphism problem with edge constraints will require $\mathcal{O}(ed^2)$ operations [ ] where e is the number of constraints, *i.e.*, $e = n(n-1)/2$, and d is the size of the largest domain, *i.e.*, $d = n$. Hence, the complexity of achieving label-consistency on the global constraint is an order lower than the complexity of achieving AC-consistency on a semantically equivalent set of edge constraints. However, one should note that these two consistencies are not comparable: for some graphs, such as the graph of Fig. 1, label-consistency is stronger and actually solves the problem, whereas AC-consistency on edge constraints does not reduce any domain; for some other graphs, such as the graph of Fig. 2, label-consistency does not reduce any domain, whereas AC-consistency on edge constraints can reduce some variable domains as soon as one variable is assigned to a value.

Further work will first concern the integration of our filtering algorithm into a constraint solver (such as CHOCO [ ]), in order to experimentally validate and evaluate it. A distance constraint can be seen like an invariant of a couple of vertices : it is a label $l(u,v)$ assigned to a couple (u,v) of vertices such that, if there exists an isomorphism function which links vertices (u,v) to vertices (u',v') then $l(u,v) = l(u',v')$. One could choose stronger invariants *i.e.*, defining more "tighter" constraints than the one based on the distance between the two vertices. For example, one could choose a label $l(u,v)$ of a couple (u,v) of vertices of a graph G which describes the distances between u, v and all vertices of the graphs more than only the distance between u and v. The label $l(u,v)$ should be then a set of triples (d_u, d_v, n), each one expressing the fact that there is n vertices of G which are respectively at distance of d_u from vertex u and at distance of d_v from vertex v. Using stronger invariants can prune more efficiently the search space but can also be more expensive to compute and to compare. One has to find the best compromise between the time needed to compute it and the efficiency of the filtering, and experiments should be performed to determine this. Finally, we shall clarify relationships between different levels of partial consistencies on $C_{distance}$ and C_{edge} constraints. In particular, for all examples we have experimented, we have noticed that after the assignment of a variable, a forward checking propagation of $C_{distance}$ constraints always reduces domains as much as an AC propagation of C_{edge} constraints. Hence, we shall try to prove this property, or find a counter example to it.

References

[1] Abderrahmane Aggoun and Nicolas Beldiceanu. Extending CHIP in order to solve complex and scheduling and placement problems. In *Actes des Journées Francophones de Programmation et Logique, Lille, France,* 1992.

[2] Alfred V. Aho, John E. Hopcroft, and Jeffrey D. Ullman. *The design and analysis of computer algorithms.* Addison Wesley, 1974.

[3] Christian Bessière and Marie-Odile Cordier. Arc-consistency and arc-consistency again. In *Proceedings of the 11th National Conference on Artificial Intelligence,* pages 108–113, Menlo Park, CA, USA, July 1993. AAAI Press.

[4] Christian Bessière and Pascal Van Hentenryck. To be or not to be... a global constraint. *CP'03, Kinsale, Ireland,* pages 789–794, 2003.

[5] Christian Bessière and Jean-Charles Régin. Refining the basic constraint propagation algorithm. In Bernhard Nebel, editor, *Proceedings of the seventeenth International Conference on Artificial Intelligence (IJCAI-01),* pages 309–315, San Francisco, CA, August 4–10 2001. Morgan Kaufmann Publishers, Inc.

[6] Pierre-Antoine Champin and Christine Solnon. Measuring the similarity of labeled graphs. *5th International Conference on Case-Based Reasoning (ICCBR 2003),* Lecture Notes in Artificial Intelligence No. 2689 - Springer-Verlag:80–95, 2003.

[7] Luigi Pietro Cordella, Pasquale Foggia, Carlo Sansone, and Mario Vento. An improved algorithm for matching large graphs. In *3rd IAPR-TC15 Workshop on Graph-based Representations in Pattern Recognition,* pages 149–159. Cuen, 2001.

[8] Thomas H. Cormen, Charles E. Leiserson, and Ronald L. Rivest. *Introduction to Algorithms.* MIT Press, 1990.

[9] Pasquale Foggia, Carlo Sansone, and Mario Vento. A performance comparison of five algorithms for graph isomorphism. In *3rd IAPR-TC15 Workshop on Graph-based Representations in Pattern Recognition,* pages 188–199. Cuen, 2001.

[10] Scott Fortin. The graph isomorphism problem. Technical report, Dept of Computing Science, Univ. Alberta, Edmonton, Alberta, Canada, 1996.

[11] Michael R. Garey and David S. Johnson. *Computers and Intractability : A Guide to The Theory of NP-Completeness.* W. H. Freeman, San Francisco, 1979.

[12] John E. Hopcroft and Jin-Kue Wong. Linear time algorithm for isomorphism of planar graphs. 6th *Annu. ACM Symp. theory of Comput.,* pages 172–184, 1974.

[13] ILOG,S. A. *ILOG Solver 5.0 User's Manual and Reference Manual.* 2000.

[14] François Laburthe and the OCRE project team. CHOCO: implementing a CP kernel. In *Proc. of the CP'2000 workshop on techniques for implementing constraint programming systems, Singapore,* 2000.

[15] Eugene M. Luks. Isomorphism of graphs of bounded valence can be tested in polynomial time. *Journal of Computer System Science,* pages 42–65, 1982.

[16] James J. McGregor. Relational consistency algorithms and their applications in finding subgraph and graph isomorphisms. *Information Science,* 19:229–250, 1979.

[17] Brendan D. McKay. Practical graph isomorphism. *Congressus Numerantium,* 30:45–87, 1981.

[18] Roger Mohr and Thomas C. Henderson. Arc and path consistency revisited. *Artificial Intelligence,* 28:65–74, 1986.

[19] Jean-Charles Régin. *Développement d'Outils Algorithmiques pour l'Intelligence Artificielle. Application á la Chimie Organique.* PhD thesis, Univ. Montpellier II, 1995.

[20] Edward Tsang. *Foundations of Constraint Satisfaction*. Academic Press, 1993.

[21] Jeffrey D. Ullman. An algorithm for subgraph isomorphism. *Journal of the Association of Computing Machinery*, 23(1):31–42, 1976.   ,

Echelon Stock Formulation
of Arborescent Distribution Systems:
An Application to the Wagner-Whitin Problem

S. Armagan Tarim[1] and Ian Miguel[2]

[1] Hacettepe University
Department of Management
Ankara, Turkey
`armagan.tarim@hacettepe.edu.tr`
[2] University of York
Department of Computer Science
York, U.K.
`ianm@cs.york.ac.uk`

Abstract. An arborescent distribution system is a multi-level system in which each installation receives input from a unique immediate predecessor and supplies one or more immediate successors. In this paper, it is shown that a distribution system with an arborescent structure can also be modelled using an echelon stock concept where at any instant the total echelon holding cost is accumulated at the same rate as the total conventional holding cost. The computational efficiency of the echelon model is tested on the well-known Wagner-Whitin type dynamic inventory lot-sizing problem, which is an intractable combinatorial problem from both mixed-integer programming (MIP) and constraint programming (CP) standpoints. The computational experiments show that the echelon MIP formulation is computationally very efficient compared to the conventional one, whereas the echelon CP formulation remains intractable. A CP/LP hybrid yields a substantial improvement over the pure CP approach, solving all tested instances in a reasonable time.

1 Introduction

Inventory theory provides methods for managing and controlling inventories under different policy constraints and environmental situations. A basic distribution system consists of a supply chain of stocking points arranged in levels. Customer demands occur at the first level, and each level has its stock replenished from the one above. Typically, a *holding cost* per unit of inventory is associated with each stocking point, under the assumption that a parent stocking point has a lower holding cost than any of its children. A *procurement cost* per order is also associated with each stocking point. Given customer demands for each stocking point in the first level over some planning horizon of a number of periods, the problem is then to find an optimal *policy*: a set of decisions as to when and how much to order for each stocking point, such that cost is minimised. This is

J.-C. Régin and M. Rueher (Eds.): CPAIOR 2004, LNCS 3011, pp. 302–   , 2004.
© Springer-Verlag Berlin Heidelberg 2004

a difficult combinatorial problem, to which this paper considers mixed integer programming (MIP), constraint programming (CP) and hybrid approaches.

An important consideration in the performance of all of these approaches is the *model*, i.e. the choice of decision variables and constraints used to represent the problem. For each stocking point, at each time period in the planning horizon, the *conventional* model employs one variable for the ordering decision and another for the closing inventory level. An alternative model of inventory views the distribution structure in *echelons* [ ]. An echelon comprises a stocking point and all of its descendants, with an associated echelon inventory level corresponding to the combined inventories of the constituent stocking points. The *echelon holding cost* [ ] captures the incremental cost of holding a unit of stock at a particular stocking point instead of at its parent.

The echelon stock formulation, with an inventory variable per echelon, has been shown to be valid for serial distribution systems [ ]. This paper extends the proof of validity to *arborescent systems*, showing that the total echelon holding cost accumulates at the same rate as the total conventional holding cost.

The complexity of multi-echelon inventory problems has in the past required the use of a sequential approach to calculating the optimal policy [,]. Clark and Scarf [ ] demonstrated that, using the echelon formulation, under certain inventory control policy and cost assumptions, the optimal policy for a serial system can be determined sequentially by first determining the optimal policy at the lowest level and then proceeding sequentially to the higher levels. This paper demonstrates empirically, via a multi-echelon version of the well-known Wagner-Whitin problem [ ], that the echelon stock formulation may still improve on the conventional formulation without resorting to the sequential approach.

The paper is organised as follows. In Section , the concepts of arborescence and echelon stock, and in Section , the notation and basic definitions are given. The conventional and echelon formulations of arborescent distribution systems are presented in Section   and their equivalence is proved in Section . Section   illustrates the echelon MIP formulation of the Wagner-Whitin problem by a numerical example. Section   is devoted to numerical tests concerning the computational efficiency of echelon and conventional formulations of Wagner-Whitin problem. Conclusions are presented in the final section.

2 Multi-Echelon Systems and Echelon Stock

Figure   presents an illustrative multi-echelon inventory system. A multi-echelon inventory system can also be viewed as a directed network, where the nodes represent the stocking points and the linkages represent flows of goods. If the network has at most one incoming link for each node and flows are acyclic it is called an arborescence or inverted tree structure. More complex interconnected systems of facilities can exist; however, most of the work in multi-echelon inventory theory has been confined to arborescent structures [ ].

Consider the following distinction between an installation stock and an echelon stock. In a serial system, the stock at installation i refers only to the stock

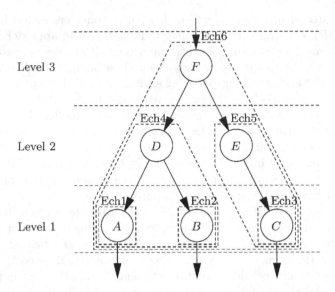

Fig. 1. An arborescent system

physically at that location; however, the echelon stock at level i refers to the sum of all the stocks at installations i, $i - 1$, ..., 2, 1 plus all the stock in transit between installations i, $i - 1$, ..., 2, 1. This definition permits useful simplifications [ ]. Under certain assumptions, a multi-stage problem can be decomposed into a set of interconnected one-stage problems, one for each echelon. In Afentakis et al. [ ], Clark and Armentano [,], Axsater and Rosling [ ], and Chen and Zheng [ ], the echelon concepts are used to reformulate production/inventory systems. Silver et al. [ ] give a full chapter treatment of the topic.

3   Notation and Definitions

Consider a level $m \in \{1, ..., L\}$ of an arborescent structure and let the number of stocking points in this level be N_m. For each stocking point, $W(i, m, n)$ is the set of descending ($n < m$) or immediate ascending ($n = m + 1$) stocking points that are in level n and connected to the ith, $i \in \{1, ..., N_m\}$, stocking point of the mth level. $G(i, m)$ is the set of all successors of stocking point i of mth level (i.e., $G(i, m) \equiv W(i, m, \{n | n < m\})$). $V(i, m)$ is the set of all stocking points that are in the first level and originate from stocking point i of the mth level (i.e., $V(i, m) \equiv W(i, m, 1)$). Each stocking point is defined by a pair of numbers (i, m), where i and m denote the stocking point and level numbers respectively.

For illustration, consider the 3-level distribution system shown in Figure , where $(i = 1, m = 1)$ refers to stocking point A, $(i = 2, m = 1)$ to stocking point B, $(i = 3, m = 1)$ to stocking point C, $(i = 1, m = 2)$ to stocking point D, $(i = 2, m = 2)$ to stocking point E, and $(i = 1, m = 3)$ to stocking point F:

- Set of immediate successors, $W(i, m, m-1)$:
  $W(1, 2, 1) = \{A, B\}$, $W(2, 2, 1) = \{C\}$, $W(1, 3, 2) = \{D, E\}$
- Set of all successors, $G(i, m)$:
  $G(1, 3) = \{A, B, C, D, E\}$, $G(1, 2) = \{A, B\}$, $G(2, 2) = \{C\}$, $G(1, 1) = G(2, 1) = G(3, 1) = \emptyset$
- Immediate predecessors, $W(i, m, m+1)$:
  $W(1, 1, 2) = D$, $W(2, 1, 2) = D$, $W(3, 1, 2) = E$, $W(1, 2, 3) = F$, $W(2, 2, 3) = F$
- Set of successors in the lowest level, $V(i, m)$:
  $V(1, 2) = \{A, B\}$, $V(2, 2) = \{C\}$, $V(1, 3) = \{A, B, C\}$

The echelon stock at stocking point i in level j at the end of period t is denoted by E_{ijt}, and e_{ij} is the echelon holding cost. The echelon holding cost at a given stocking point (i, j) is the incremental cost of holding a unit of inventory at this stocking point instead of at predecessor thereof, $W(i, j, j+1)$. The formal definitions of e_{ij} and E_{ijt} are as follows:

$$
\left.
\begin{aligned}
e_{ij} &= c_{ij} - c_{W(i,j,j+1)} \\
E_{ijt} &= I_{ijt} + \sum_{m \in G(i,j)} I_{mt}
\end{aligned}
\right\}
\quad t = 1, ..., T \quad j = 1, ..., L \quad i = 1, ..., N_j. \quad (1)
$$

where c_{ij} is the conventional unit inventory holding cost at stocking point (i, j) and I_{ijt} is the closing inventory level at the end of period t.

4   Conventional vs Echelon Formulations

In this section, inventory holding cost expressions and balance equations for multi-echelon inventory systems are presented under both conventional and echelon holding cost charging schemes. First, the conventional scheme is addressed under the assumption that a fixed holding cost c is incurred on any unit carried in inventory over from one period to the next. Under this scheme, the conventional single period total holding cost expression can be written as in Eq.( ),

$$
\sum_{j=1}^{L} \sum_{i=1}^{N_j} c_{ij} I_{ijt} \quad t = 1, ..., T. \quad (2)
$$

The pertinent inventory balance equations show that the closing inventory in any stocking point in any period is the opening inventory plus the order received minus the demand met (or the amount supplied to the other stocking points of the multi-echelon system),

$$
I_{ijt} = I_{ij(t-1)} + X_{ijt} - \sum_{m \in W(i,j,j-1)} X_{mt} \quad i = 1, ..., N_j \quad t = 1, ..., T \quad j = 1, ..., L
$$
$$(3)$$

and

$$
X_{ijt}, I_{ijt} \geq 0, \quad (4)
$$

where X_{ijt} is the stock replenishment amount received at stocking point (i,j) in period t. In Eq.( ), without loss of generality, delivery lead-time is taken as zero. This formulation, in which Eq.( ) is the cost expression and Eqs.( )–( ) are the balance equations, is called *Model I* (or the conventional model).

It is shown in Section   that, by means of the linear transformations given in Eq.( ), *Model I* of multi-echelon systems can be rewritten as Eqs.( )–( ),

$$\sum_{j=1}^{L}\sum_{i=1}^{N_j} e_{ij}E_{ijt} \qquad t=1,...,T \tag{5}$$

$$E_{ijt} = E_{ij(t-1)} + X_{ijt} - \sum_{m\in V(i,j)} d_{mt} \qquad t=1,...,T \quad j=1,...,L \quad i=1,...,N_j \tag{6}$$

$$0 \le \sum_{m\in W(i,j,j-1)} E_{mt} \le E_{ijt} \qquad t=1,...,T \quad j=1,...,L \quad i=1,...,N_j \tag{7}$$

where customer demands, d_{mt}, occur only at stocking points in level 1. By definition, $X_{m\in W(i,j,0)}$ and $d_{m\in V(i,j)}$ are equivalent. This alternative formulation is called *Model II* (or the echelon model). The concept behind this transformation is known in the MRP literature (for assembly systems) as "explosion" (see Afentakis and Gavish [ ]).

5   Equivalence of Models *I* and *II*

The validity of echelon stock and echelon holding cost concepts in arborescent structures is now addressed and the equivalence of *Models I* and *II* is proved. To serve this purpose, we show that at any instant a policy under the echelon stock charging scheme gives the same total cost as a charging scheme based on stock physically at each installation in an arborescent structure.

Lemma 1.   *Given $E_{ijt} = I_{ijt} + \sum_{m\in G(i,j)} I_{mt}$, the inventory balance equations of Models I and II are equivalent.*

Proof. Consider level j, $j \in \{1,...L\}$, of *Model I*. For all stocking points in the set of all successors, $G(i,j)$, adding up the inventory balance equations of Eq.( ),

$$\sum_{m\in W(i,j,j-1)} I_{mt} = \sum_{m\in W(i,j,j-1)} I_{m(t-1)} + \sum_{m\in W(i,j,j-1)} X_{mt} - \sum_{m\in W(i,j,j-2)} X_{mt}$$

$$\sum_{m\in W(i,j,j-2)} I_{mt} = \sum_{m\in W(i,j,j-2)} I_{m(t-1)} + \sum_{m\in W(i,j,j-2)} X_{mt} - \sum_{m\in W(i,j,j-3)} X_{mt}$$

$$\vdots$$

$$\sum_{m\in W(i,j,1)} I_{mt} = \sum_{m\in W(i,j,1)} I_{m(t-1)} + \sum_{m\in W(i,j,1)} X_{mt} - \sum_{m\in W(i,j,0)} X_{mt}$$

yields

$$\sum_{m \in G(i,j)} I_{mt} = \sum_{m \in G(i,j)} I_{m(t-1)} + \sum_{m \in W(i,j,j-1)} X_{mt} - \sum_{m \in W(i,j,0)} X_{mt}. \quad (8)$$

Thereby, using the equivalence of $X_{m \in W(i,j,0)}$ and $d_{m \in V(i,j)}$, and adding Eq.( ) to Eq.( ) give the general expression,

$$\left(I_{ijt} + \sum_{m \in G(i,j)} I_{mt} \right) = \left(I_{ij(t-1)} + \sum_{m \in G(i,j)} I_{m(t-1)} \right) + X_{ijt} - \sum_{m \in V(i,j)} d_{mt}. \quad (9)$$

Making use of Eq.( ) (i.e., the definition of E_{ijt}), Eq.( ) leads to

$$E_{ijt} = E_{ij(t-1)} + X_{ijt} - \sum_{m \in V(i,j)} d_{mt} \quad (10)$$

Hence, the equivalence of Eq.( ) and Eq.( ) is shown.

Eq.( ) can be rearranged for I_{ijt} giving

$$I_{ijt} = E_{ijt} - \sum_{m \in G(i,j)} I_{mt} \quad (11)$$

and

$$I_{ijt} = E_{ijt} - \sum_{m \in W(i,j,j-1)} E_{mt}, \quad (12)$$

from which the nonnegativity constraints of I_{ijt} lead directly to Eq.( ).

In what follows, the equivalence of conventional and echelon holding cost charging schemes is addressed. The echelon holding cost, E_{ijt}, is calculated from the conventional holding cost, c_{ij}, by the rule:

$$j = 1, ..., L \quad i = 1, ..., N_j. \quad (13)$$

Schwarz and Schrage [ ] show that in a serial system, at any instant

$$\sum_{j=1}^{L} \sum_{i=1}^{N_j=1} c_{ij} I_{ijt} = \sum_{j=1}^{L} \sum_{i=1}^{N_j=1} e_{ij} E_{ijt}, \quad t = 1, ..., T$$

that is, total echelon holding cost is accumulated at the same rate as total conventional holding cost.

Here, we show that conventional and echelon charging schemes are identical not only in serial systems, but also in arborescent systems. Hence their proof is extended to cover arborescent systems.

Lemma 2. *Given $e_{ij} = c_{ij} - \{c_m | m \in W(i, j, j+1)\}$, the cost expressions of Models I and II are equivalent.*

Proof. Assume that an arborescent system is comprised of only a single level of stocking points. In other words, the stocking points are independent. Then $E_{i1t} = I_{i1t}$ and $e_{i1} = c_{i1}$ for $i = 1, ..., N_1$, and $\sum_i e_{i1} E_{i1t} = \sum_i c_{i1} I_{i1t}$. Now assume that a new stocking point is introduced as level 2, and all independent stocking points of level 1 are connected to this new stocking point as successors. Then it follows

$$e_{12} = c_{12}$$
$$e_{i1} \leftarrow e_{i1} - c_{12} \qquad i = 1, ..., N_1,$$
$$E_{12t} = I_{12t} + \sum_{i=1}^{N_1} E_{i1t}$$

from which the conventional inventory cost increase may be calculated as

$$e_{12} E_{12t} - \sum_{i=1}^{N_1} c_{12} E_{i1t} = c_{12} \left(E_{12t} - \sum_{i=1}^{N_1} E_{i1t} \right) = c_{12} I_{12t}$$

Since this is the actual amount of increase in the conventional inventory cost, one can conclude that the echelon cost charging scheme gives the correct conventional inventory cost for a 2-echelon system.

Now assume an inventory system of N_{j-1} independent $j - 1$ level systems. A new level, level j, is introduced and all $j - 1$ level systems are connected to it as successors. Such a restructuring does not affect the echelons of the lowest $j - 2$ levels. However, the following modifications take place:

$$e_{1j} = c_{1j}$$
$$e_{i(j-1)} \leftarrow e_{i(j-1)} - c_{1j} \qquad i = 1, ..., N_{j-1},$$
$$E_{1jt} = I_{1jt} + \sum_{i=1}^{N_{j-1}} E_{i(j-1)t}.$$

Thus, the echelon holding cost expression increases by

$$e_{1j} E_{1jt} - \sum_{i=1}^{N_{j-1}} c_{1j} E_{i(j-1)t} = c_{1j} \left(E_{1jt} - \sum_{i=1}^{N_{j-1}} E_{i(j-1)t} \right) = c_{1j} I_{1jt}.$$

Since the actual amount of increase in the echelon cost is equal to the amount of increase in the conventional inventory cost for an arborescent structure with j levels, this finishes the induction step, which completes the proof.

Theorem 1. *An alternative formulation of arborescent distribution systems follows from the echelon stock and echelon holding cost definitions.*

Proof. From Lemmas ( ) and ( ).

Table 1. Period demands

Installation	P1	P2	P3	P4	P5	P6	P7	P8	P9	P10
A	100	100	100	100	100	100	100	100	100	100
B	50	200	50	50	200	250	250	100	150	150
C	250	50	350	50	250	50	250	50	350	50

6   An Illustrative Example: Wagner-Whitin Problem

The Wagner-Whitin problem [ ] describes the single stocking point planning of ordering and stocking a certain product over a discrete time planning horizon. The deterministic demand for all periods is to be satisfied, and the total sum of fixed procurement and linear holding costs is to be minimised. In this section a multi-echelon version of the Wagner-Whitin type dynamic inventory lot-sizing problem (problem 040 at www.csplib.org) is formulated using both conventional and echelon approaches. To serve this purpose, the multi-echelon structure given in Figure   is used assuming the period demands presented in Table   .

The initial inventory level is taken as zero and the replenishment lead-time is set to zero in all stocking points. A fixed procurement cost, c_o, is incurred when a replenishment order is placed, irrespective of order size. Table   gives the other parameters of the problem. Both conventional and echelon type formulations for the multi-echelon Wagner-Whitin problem are presented below. In these models, M denotes a large number and δ_{ijt} is a binary decision variable that takes the value of 1 if a replenishment order is placed in period t and 0 otherwise.

Conventional Model:

min

$$\sum_{t=1}^{T}\sum_{j=1}^{L}\sum_{i=1}^{N_j}\left(c_{ij}I_{ijt} + c_{o_{ij}}\delta_{ijt}\right)$$

subject to

$$(i = 1, ..., N_j \quad j = 1, ..., L \quad t = 1, ..., T)$$

$$I_{ijt} = I_{ij(t-1)} + X_{ijt} - \sum_{m\in W(i,j,j-1)} X_{mt}$$

$$X_{ijt} \le M\delta_{ijt}$$

$$X_{ijt}, I_{ijt} \ge 0, \quad \delta_{ijt} \in \{0,1\}$$

Echelon Model:

min

$$\sum_{t=1}^{T}\sum_{j=1}^{L}\sum_{i=1}^{N_j}\left(e_{ij}E_{ijt} + c_{o_{ij}}\delta_{ijt}\right)$$

subject to

$$(i = 1, ..., N_j \quad j = 1, ..., L \quad t = 1, ..., T)$$

$$E_{ijt} = E_{ij(t-1)} + X_{ijt} - \sum_{m\in V(i,j)} d_{mt}$$

$$X_{ijt} \le M\delta_{ijt}$$

$$X_{ijt} \ge 0, \quad \delta_{ijt} \in \{0,1\}$$

$$E_{ijt} \ge \sum_{m\in W(i,j,j-1)} E_{mt} \ge 0$$

The optimal inventory replenishment policy, with a total cost of 135,700 units, is presented in Table   . The difference between two models is significant. In the conventional case, the search tree includes 531 nodes, whereas the corresponding search tree for the echelon model has only 213 nodes.

Table 2. Problem parameters and optimal replenishment amounts, X_{ijt}

Installation /Echelon	Holding Cost, c	Echelon Cost, e	Procurement Cost, c_o	Optimal replenishment amounts, X_{ijt}									
				P1	P2	P3	P4	P5	P6	P7	P8	P9	P10
A	10	4	1,000	200	–	200	–	200	–	200	–	200	–
B	30	24	3,000	50	300	–	–	200	250	350	–	150	150
C	20	14	1,000	300	–	400	–	300	–	300	–	400	–
D	6	1	5,000	750	–	–	–	1700	–	–	–	–	–
E	6	1	7,000	700	–	–	–	1000	–	–	–	–	–
F	5	5	10,000	1450	–	–	–	2700	–	–	–	–	–

7   Computational Experiments

To get a better indication of the difference between the two models, in this section computational tests are performed on a wider set of problems, using different solution techniques. The tests are performed on a 1.2GHz Pentium-3 machine using mathematical programme solvers Xpress-MP 2003B [ ] and Ilog Cplex 8.1 and constraint solver Ilog Solver 5.3. Xpress-MP and Cplex are both used with their default settings. The models are tested on problems generated for 6-stocking point multi-level systems with three different structures, namely: arborescent systems; serial systems, and warehouse-retailer systems.

In the test problems, the structure given in Figure   is used as a design of arborescent systems. Serial systems are represented by a 6-level structure, in which A is the lowest level stocking point where the external demand is met and F is the highest level (level 6) stocking point where the external supply is received. In a warehouse-retailer system, a single second level warehouse supplies a number of first level retailers. The computational tests are performed on a 1-warehouse (denoted by F) 5-retailer (denoted by A to E) structure. The pertinent demand data for arborescent, serial and warehouse-retailer systems are presented in Table .

Three different planning horizon lengths –10, 12 and 18 periods– are used in the experiments. The number of test problems, generated for different structures and costs, totals to 35. The instances are given in Table .

7.1   Implied Constraints

During numerical experiments a number of implied constraints (IC) are incorporated into the conventional and echelon models to enhance performance. These constraints are detailed below:

IC1   All stocking points must have zero inventory at the end of the last period in an optimal solution. Hence, the corresponding inventory variables are preset.

IC2 In an optimal solution, if a parent node places an order, at least one of its children must also. Consider that, if no children make an order, the parent node incurs a holding cost that can be removed by delaying the order until a subsequent period when at least one child does place an order.

IC3 An upper bound can be derived for the inventory variables in the conventional formulation by considering that it is only worth holding stock at a node if it is cheaper than ordering it at the next period. That is: $I_{ijt}c_{ij} \leq c_{o_{ij}} + I_{ijt}c_m$, $m \in W(i,j,j+1)$, which simplifies to: $I_{ijt} \leq \frac{c_{o_{ij}}}{c_{ij}-c_m}$. This can easily be applied to the echelon model by substituting in the equality: $I_{ijt} = E_{ijt} - \sum_{m\in G(i,j)} I_{mt}$.

IC4 Similarly, an upper bound can be derived for the order variables at the leaf nodes. The principle is the same: it is only worth ordering stock not absorbed by demand at the current period if it is cheaper than waiting and ordering in a subsequent period. Consider first a bound based on deferring an order into the next period: $(X_{i1t} - d_{i1t})c_{i1} \leq c_{o_{i1}} + c_m(X_{i1t} - d_{i1t})$ which can be rearranged: $X_{i1t} \leq d_{i1t} + \frac{c_{o_{i1}}}{c_{i1}-c_m}$ This can be generalised to consider deferring an order into any of the following periods up to the planning horizon: $X_{i1t} \leq$
$$\min_{t'=t..T} \sum_{i=t}^{t'} d_{i1t} + \frac{c_{o_{i1}}}{(t'-t+1)(c_{i1}-c_m)}$$

7.2   MIP

The solution times obtained to test problems using Xpress–MP and Cplex under conventional and echelon formulations, with and without implied constraints, are given in Table . The results indicate that, irrespective of problem structure, costs involved, or MIP solver used, for the Wagner-Whitin problem the echelon formulation of a multi-echelon system is more tractable than the conventional one. The only exception to this remark is the warehouse-retailer type distribution systems which are solved by using Cplex. In these problems the computational performance of conventional and echelon models are very close and mixed.

Table 3.   Period demands (I: Arborescent; II: Serial; III: Warehouse-Retailer)

		P1	P2	P3	P4	P5	P6	P7	P8	P9	P10	P11	P12	P13	P14	P15	P16	P17	P18
	A	100	100	100	100	100	100	100	100	100	100	100	100	100	100	100	100	100	100
I.	B	50	200	50	50	200	250	250	100	150	150	50	200	50	200	50	50	200	250
	C	250	50	350	50	250	50	250	50	350	100	50	50	250	50	350	50	250	50
II.	A	50	200	250	100	150	250	50	100	150	150	100	200	50	200	250	100	150	250
	A	100	100	100	100	100	100	100	100	100	100	100	100	100	100	100	100	100	100
	B	50	200	50	50	200	250	250	100	150	150	50	200	50	50	200	250	250	100
III.	C	250	50	350	50	250	50	250	50	350	50	250	50	350	50	250	50	250	50
	D	50	50	50	50	50	50	50	50	50	50	50	50	50	50	50	50	50	50
	E	50	100	150	200	250	250	200	150	100	50	50	100	150	200	250	250	200	150

Table 4. Test Problems: Arborescent (A), Serial (S) and Warehouse-Retailer (W)

Case	Holding Costs c	Echelon Costs e	Procurement Costs c_o
	[A, B, C,D,E,F]	[A,B,C,D,E,F]	[A, B, C, D, E, F]
A-1	[3, 3, 3, 2, 2, 1]	[1, 1, 1, 1, 1, 1]	[1000,1000,1000,1000,1000,1000]
A-2	[4, 4, 4, 2, 2, 1]	[2, 2, 2, 1, 1, 1]	[1000,1000,1000,1000,1000,1000]
A-3	[4, 4, 4, 3, 3, 2]	[1, 1, 1, 1, 1, 2]	[1000,1000,1000,1000,1000,1000]
A-4	[3, 4, 5, 2, 4, 1]	[1, 2, 1, 1, 3, 1]	[1000,1000,1000,1000,1000,1000]
A-5	[3, 6, 3, 2, 2, 1]	[1, 4, 1, 1, 1, 1]	[1000,1000,1000,1000,1000,1000]
A-6	[6, 5, 4, 3, 2, 1]	[3, 2, 2, 2, 1, 1]	[1000,1000,1000,1000,1000,1000]
A-7	[3, 4, 6, 2, 5, 1]	[1, 2, 1, 1, 4, 1]	[1000,1000,1000,1000,1000,1000]
A-8	[5, 6, 3, 4, 2, 1]	[1, 2, 1, 3, 1, 1]	[1000,1000,1000,1000,1000,1000]
A-9	[4, 6, 5, 2, 3, 1]	[2, 4, 2, 1, 2, 1]	[1000,1000,1000,1000,1000,1000]
A-10	[7, 6, 5, 4, 3, 2]	[3, 2, 2, 2, 1, 2]	[1000,1000,1000,1000,1000,1000]
S-1	[6, 5, 4, 3, 2, 1]	[1, 1, 1, 1, 1, 1]	[1000,1000,1000,1000,1000,1000]
S-2	[6, 5, 4, 3, 2, 1]	[1, 1, 1, 1, 1, 1]	[1000,1500,2000,2500,3000,3500]
S-3	[6, 5, 4, 3, 2, 1]	[1, 1, 1, 1, 1, 1]	[500, 500, 500, 500, 500, 500]
S-4	[12,10, 8, 6, 4, 2]	[2, 2, 2, 2, 2, 2]	[1000,1000,1000,1000,1000,1000]
S-5	[12,10, 8, 6, 4, 2]	[2, 2, 2, 2, 2, 2]	[1000,1500,2000,2500,3000,3500]
S-6	[12,10, 8, 6, 4, 2]	[2, 2, 2, 2, 2, 2]	[2000,2000,2000,2000,2000,2000]
S-7	[18,15,12, 9, 6, 3]	[3, 3, 3, 3, 3, 3]	[1000,1000,1000,1000,1000,1000]
S-8	[18,15,12, 9, 6, 3]	[3, 3, 3, 3, 3, 3]	[1000,1500,2000,2500,3000,3500]
S-9	[18,15,12, 9, 6, 3]	[3, 3, 3, 3, 3, 3]	[2000,2000,2000,2000,2000,2000]
S-10	[30,10, 5, 3, 2, 1]	[20, 5, 2, 1, 1, 1]	[1000,1000,1000,1000,1000,1000]
S-11	[30,10, 5, 3, 2, 1]	[20, 5, 2, 1, 1, 1]	[1000,1500,2000,2500,3000,3500]
S-12	[30,10, 5, 3, 2, 1]	[20, 5, 2, 1, 1, 1]	[2000,2000,2000,2000,2000,2000]
S-13	[32,16, 8, 4, 2, 1]	[16, 8, 4, 2, 1, 1]	[1000,1000,1000,1000,1000,1000]
S-14	[32,16, 8, 4, 2, 1]	[16, 8, 4, 2, 1, 1]	[1000,1500,2000,2500,3000,3500]
S-15	[32,16, 8, 4, 2, 1]	[16, 8, 4, 2, 1, 1]	[2000,2000,2000,2000,2000,2000]
W-1	[2, 2, 2, 2, 2, 1]	[1, 1, 1, 1, 1, 1]	[1000,1000,1000,1000,1000,1000]
W-2	[3, 3, 3, 3, 3, 1]	[2, 2, 2, 2, 2, 1]	[1000,1000,1000,1000,1000,1000]
W-3	[4, 4, 4, 4, 4, 1]	[3, 3, 3, 3, 3, 1]	[1000,1000,1000,1000,1000,1000]
W-4	[2, 4, 2, 4, 2, 1]	[1, 3, 1, 3, 1, 1]	[1000,1000,1000,1000,1000,1000]
W-5	[3, 6, 3, 6, 3, 1]	[2, 5, 2, 5, 2, 1]	[1000,1000,1000,1000,1000,1000]
W-6	[2, 3, 4, 3, 2, 1]	[1, 2, 3, 2, 1, 1]	[1000,1000,1000,1000,1000,1000]
W-7	[4, 3, 2, 3, 4, 1]	[3, 2, 1, 2, 3, 1]	[1000,1000,1000,1000,1000,1000]
W-8	[6, 5, 4, 3, 2, 1]	[5, 4, 3, 2, 1, 1]	[1000,1000,1000,1000,1000,1000]
W-9	[2, 3, 4, 5, 6, 1]	[1, 2, 3, 4, 5, 1]	[1000,1000,1000,1000,1000,1000]
W-10	[3, 2, 4, 2, 3, 1]	[2, 1, 3, 1, 2, 1]	[1000,1000,1000,1000,1000,1000]

Table 5. Solution Times (in secs): Arborescent (A), Serial (S) and Warehouse-Retailer (W) cases "_" indicates no solution within 1 hour, (.) shows the total number of nodes visited

Columns 2–7: **Xpress-MP** (A,S: 12 periods; W: 10 periods). Columns 8–13: **Cplex** (A,S,W: 18 periods).

Case	No IC Conv	No IC Ech	IC1,2 Conv	IC1,2 Ech	IC1-4 Conv	IC1-4 Ech	No IC Conv	No IC Ech	IC1,2 Conv	IC1,2 Ech	IC1-4 Conv	IC1-4 Ech
A1	440 (399851)	33.5 (36507)	80.3 (97674)	1.4 (1436)	68.1 (89296)	0.9 (886)	1279 (493797)	227 (74830)	102 (34126)	29.8 (8336)	54.4 (14428)	23.7 (6203)
A2	—	28.4 (42523)	884 (5955653)	9.3 (11034)	162 (186808)	7.1 (7280)	1260 (671756)	171 (67390)	114 (54112)	33.2 (13288)	31.5 (12631)	15.3 (5399)
A3	690 (477737)	18.9 (21163)	96.8 (1236635)	2.5 (3480)	63.1 (78354)	4.2 (5359)	97.8 (33181)	72.4 (28881)	37.7 (9838)	18.5 (5812)	12.7 (3143)	9.38 (2356)
A4	—	11.7 (21011)	24.4 (36680)	1.8 (2102)	52.1 (67659)	2.7 (3478)	281 (132426)	88.0 (36922)	45.9 (16092)	24.5 (9084)	18.9 (5535)	18.0 (6823)
A5	—	17.9 (27599)	855 (531922)	4.6 (4690)	77.6 (104535)	1.7 (2060)	959 (379492)	360 (159020)	94.0 (36533)	44.1 (17678)	46.8 (16652)	33.2 (12996)
A6	—	78.9 (111988)	642 (464545)	4.6 (5997)	41.7 (61536)	3.9 (4414)	537 (275143)	50.2 (20758)	74.2 (35293)	17.3 (8227)	24.9 (11658)	9.38 (3077)
A7	115 (139905)	5.5 (8846)	14.5 (23846)	10.3 (12037)	76.2 (89690)	2.1 (2507)	134 (56468)	63.3 (22447)	23.2 (7816)	19.7 (6857)	14.9 (5219)	18.5 (5820)
A8	—	51 (58541)	380 (360470)	2.9 (4285)	111 (106018)	20.1 (28571)	502 (224897)	107 (44392)	66.6 (23565)	22.1 (7734)	45.4 (13932)	21.3 (7703)
A9	—	11.2 (19275)	169 (208609)	0.5 (473)	32.4 (49448)	1.7 (2287)	158 (8732)	57.6 (26849)	47.6 (22648)	16.9 (8169)	9.53 (3867)	9.27 (3080)
A10	1068 (702604)	2.1 (3741)	3.2 (4871)	—	3.3 (5515)	0.8 (829)	19.1 (9151)	9.15 (4795)	12.0 (4686)	4.89 (1922)	3.02 (892)	1.86 (589)
S1	858 (596027)	20.9 (26425)	3.9 (6244)	0.3 (203)	3.3 (5141)	0.2 (90)	—	175 (44185)	30.4 (18682)	3.53 (745)	56.3 (25594)	3.25 (466)
S2	363 (338530)	6.1 (8946)	1 (1105)	0.9 (147)	0.9 (988)	0.3 (134)	—	692 (389667)	20.1 (12997)	3.40 (631)	27.6 (14384)	2.87 (491)
S3	—	20.7 (30696)	7.4 (14209)	0.3 (202)	2.8 (3607)	0.2 (51)	—	15.2 (5302)	9.31 (8331)	1.53 (156)	9.24 (5154)	1.59 (163)
S4	—	6.7 (12845)	7.8 (14598)	0.3 (200)	2.7 (3623)	0.2 (51)	—	15.0 (5302)	9.27 (8331)	1.52 (156)	9.24 (5154)	1.61 (163)
S5	—	10.2 (17188)	2.3 (2686)	0.3 (185)	1.7 (2332)	0.5 (215)	—	316 (143704)	94.5 (70338)	7.58 (1510)	53.5 (33206)	4.15 (689)
S6	902 (608622)	45 (53773)	4.2 (6648)	0.2 (201)	3.2 (5028)	0.2 (90)	—	175 (44169)	29.8 (44169)	3.53 (745)	57.6 (26140)	3.24 (466)
S7	103 (186578)	3.6 (5391)	0.5 (754)	0.2 (119)	1.9 (2538)	0.3 (219)	357 (457990)	6.11 (2954)	6.72 (6135)	1.20 (198)	4.11 (1948)	1.11 (183)
S8	—	10.8 (17863)	4.9 (9015)	0.6 (569)	3.7 (5477)	0.4 (249)	—	412 (171423)	95.0 (80686)	5.21 (1161)	69.4 (53278)	5.84 (1099)
S9	—	27.7 (39264)	4.6 (7248)	0.3 (192)	21.3 (30991)	0.4 (257)	—	120 (36554)	47.8 (36047)	2.31 (454)	32.8 (20987)	2.78 (589)
S10	2800 (1163936)	11.4 (16459)	7 (15330)	0.3 (332)	9.7 (13742)	0.2 (191)	—	60.7 (36554)	33.4 (32295)	4.67 (1597)	16.5 (12169)	2.82 (940)
S11	407 (353182)	13.4 (16661)	0.9 (1449)	0.2 (95)	1.1 (1673)	0.3 (109)	—	179 (112259)	16.8 (17219)	2.89 (989)	13.2 (12017)	2.29 (747)
S12	1540 (796290)	31.3 (37976)	4 (7776)	0.5 (328)	10 (16943)	0.5 (147)	—	366 (112559)	41.3 (41326)	5.74 (2013)	36.7 (22564)	5.16 (1509)
S13	270 (296085)	1.8 (2328)	5.7 (10192)	0.2 (130)	1.7 (10932)	0.4 (334)	1460 (2613382)	31.7 (20213)	25.7 (32875)	2.13 (747)	28.0 (25268)	1.43 (535)
S14	77.3 (1236646)	.2 (2110)	2.1 (4673)	0.2 (110)	2.2 (3296)	0.4 (109)	—	88.1 (60875)	40.9 (45025)	2.95 (924)	48.3 (48538)	2.27 (830)
S15	—	6.7 (6620)	2.0 (8789)	0.9 (950)	10 (16053)	0.5 (347)	—	323 (161240)	58.9 (60760)	5.63 (1592)	—	3.47 (1144)
W1	110 (148698)	19.3 (44651)	32.1 (69549)	6.3 (14612)	29 (58612)	3 (6848)	210 (129114)	352 (234222)	71.2 (45407)	80.8 (46166)	55.8 (33547)	54.3 (33837)
W2	—	—	2.9 (6198)	1.4 (3685)	1.8 (3031)	1.1 (2448)	152 (103122)	193 (126742)	35.1 (27101)	53.3 (38514)	29.0 (20477)	37.5 (29287)
W3	164 (201504)	4.6 (6323)	0.3 (353)	0.3 (565)	0.8 (1181)	0.3 (358)	54.1 (42531)	72.1 (58695)	38.7 (30769)	32.7 (25521)	12.7 (10217)	18.7 (14362)
W4	209 (248174)	8.6 (18515)	226 (278262)	18.7 (41808)	22.1 (47635)	8.9 (21369)	138 (74797)	139 (79696)	93.4 (66715)	54.6 (30338)	25.4 (12696)	41.4 (27902)
W5	—	—	4 (7809)	1.8 (4490)	1.5 (2385)	1.8 (3689)	220 (171503)	383 (278423)	92.4 (84474)	52.9 (44689)	36.2 (30352)	50.7 (54070)
W6	1800 (336564)	81.8 (135496)	2.9 (5854)	2.9 (7057)	6.8 (13850)	2.9 (6902)	90.6 (63180)	158 (95023)	37.9 (24597)	46.2 (26848)	33.1 (23362)	34.5 (22559)
W7	—	7.6 (16921)	14.2 (30084)	2.9 (1708)	3.5 (7025)	0.9 (1379)	66.0 (43048)	76.3 (46087)	38.0 (21144)	39.4 (25396)	25.4 (13966)	18.4 (11209)
W8	87.8 (153269)	2.3 (4113)	0.6 (910)	0.4 (650)	0.3 (263)	0.3 (351)	71.1 (53948)	57.2 (36544)	30.4 (24289)	30.4 (22446)	13.6 (8801)	14.2 (10844)
W9	60.1 (112097)	5 (38850)	3.2 (6829)	5.3 (12203)	1.6 (3266)	3.4 (7755)	100 (69218)	64.6 (42328)	34.7 (27252)	56.7 (42253)	25.0 (20828)	28.7 (25977)
W10	1890 (940787)	6.3 (14956)	3.6 (7787)	1.8 (4172)	0.7 (1160)	0.7 (1420)	270 (211021)	204 (157379)	77.2 (53866)	55.4 (40103)	52.5 (40274)	54.1 (44770)

In almost half of the cases (30 out of 70) examined, the conventional model without ICs cannot produce the optimal solution in 1 hour. However, using the echelon formulation we were able to determine the optimal solution within an hour in all cases. The maximum solution time required by Xpress–MP (for 10 and 12 period problems) is less than 4 minutes, whereas this value is almost 11 minutes for Cplex (for 18 period problems).

The introduction of implied constraints IC1 and 2 dramatically improves the computational performance of both models. The improvement is especially significant in the serial systems where IC2, $\delta_{m\in W(i,j,j+1)} \leq \delta_{ijt}$ is very strong. However, when the design of distribution system exhibits a warehouse-retailer characteristic with many successors, the IC gets weaker, $\delta_{ijt} \leq \sum_{m\in W(i,j,j-1)} \delta_{mt}$. A further substantial improvement results from adding IC3 and 4. However, these constraints are observed to be weaker compared to IC1 and 2. In certain instances, it is even observed that IC3 and 4 have adverse effects on the computational performance.

The above observations clearly show that although in the Wagner-Whitin type problem an arborescent system cannot be interpreted as a nested set of echelons, the echelon formulation is still favoured to the conventional formulation. It should also be noted that the implied constraints play a significant role in the computational performance.

7.3 CP and CP/LP Hybrid

Initial experimentation with a pure constraint satisfaction model yielded disappointing results, with the solver unable to solve the arborescent problems in a reasonable time. Therefore, experiments were also performed with a hybrid CP/LP solver using Ilog Hybrid 1.3.1 to combine Solver and Cplex. The models used for the hybrid are essentially the same as the MIP models presented in Section 6. It is possible to remove the delta variables from both the conventional and echelon models by reifying the constraint $(X_{ijt} > 0)$. This reification can be used in the echelon holding cost expressions as follows (and similarly for the conventional model): $\sum_{t=1}^{T} \sum_{j=1}^{L} \sum_{i=1}^{N_j} (e_{ij}E_{ijt} + c_{o_{ij}}(X_{ijt} > 0))$. However, the delta variables are a crucial part of the linear program. Preliminary experiments show that the hybrid performs very poorly without them. Therefore the delta variables are kept in the hybrid model, but the following constraints are added: $d_{ijt} = (X_{ijt} > 0)$. By inspection, this is stronger than the big-M inequality and can be used by the constraint solver component of the hybrid for propagation.

Implied Constraints: Both formulations are enhanced through the addition of implied constraints IC1-4. Two further non-linear implied constraints are also added. The first $(IC5)$ exploits the fact that, in an optimal solution, an order is only made at a node when the inventory is 0. This principle is also known as the zero-inventory ordering policy of Wagner-Whitin solution. This can be seen by considering that if an order is made at a node with some stock at period t, the cost incurred by holding that stock from period $t - 1$ to t can be removed by increasing the size of the order at period t.

The second $(IC6)$ reduces the domains of the X (order) variables by exploiting the fact that, in an optimal solution, the sizes of all orders made are composed from the demands of the children of the associated node for a continuous stretch of time from between the current period to the end of the planning horizon [ ]. It is therefore possible to enumerate the domain elements for each X variable, replacing the simple upper/lower bounds representation. The time complexity of this process is exponential in the number of leaves beneath the order node in question, but can usefully be applied when the number of leaves is small.

Results: Experiments were performed on the same problems as those used with Xpress–MP (see Tables   and). The results are presented in Table . Due to the branching factor, it is not feasible to use $IC6$ on the warehouse problems. The delta variables were used to branch on after preliminary experimentation showed that this was more effective than using the X variables. The branching variable was selected by preferring variables with a fractional value (in the solution to the linear relaxation) as close to 0.5 as possible [ ]. Having selected a variable, branching was performed using four heuristics. The first $(H1)$ always tries 1 before 0 in order to encourage propagation based on a decision having been made to place an order. Heuristics $H2$ and $H3$ branch on the smallest and largest variation in pseudo-cost [ ] respectively. Finally, heuristic $(H4)$ branches on the value farthest away from that assigned by the relaxation. Cuts are added via Cplex at the root node.

Compared with the MIP solvers, it is immediately clear that the hybrid takes longer to solve the instances tested. One of the chief reasons for this is that the pure MIP solvers can search many more nodes per second (around five times) than the hybrid. Particularly when domain reduction $(IC6)$ is used, the search tree when using the conventional model is often smaller than that of Xpress–MP. The fact that this is not reflected in the time taken means that the reduction in search is not sufficient to compensate for the overhead of maintaining the hybrid.

The echelon model does not provide the clear advantage for the hybrid that it does for the MIP models. For the serial problems, the effect is largely positive, with a smaller search tree often explored when using the echelon model. For the arborescent problems, the echelon model is able to exploit domain reduction better than the conventional model, resulting in the echelon model performing better for a greater proportion of the instances. Performance of the echelon model versus the conventional model on the warehouse problems is dependent on the heuristic used. Sometimes an echelon model results in a smaller search tree, but longer time taken. This suggests that the use of the echelon model incurs some overhead.

Since, as confirmed in the MIP experiements, the echelon model provides a tighter relaxation, the occasions where the echelon model inhibits performance are probably due to the effects of constraint propagation. Specifically, constraint propagation will act to assign variables in the linear program, thus influencing the branching heuristic. This suggests that not only the model but the branching heuristic should be a hybrid, considering both the linear program and the constraint program.

8   Conclusion

This paper has extended Schwarz and Schrage's [ ] proof of the validity of the echelon formulation for serial distribution systems to arborescent systems. The utility of this formulation in a MIP setting was confirmed in an empirical analysis using the well known Wagner-Whitin problem. The success of the echelon formulation was less clear-cut in conjunction with the hybrid CP/LP solver. This was ascribed to the influence of constraint propagation on the branching heuristic, which considers the linear relaxation only. An important piece of future work is to develop a branching heuristic that considers both the linear program and the constraint program. Further future work will consider the echelon formulation in other arborescent structures and in different operating environments, such as backlogging of unsatisfied demand at the lowest echelon

References

[1] Afentakis, P., Gavish, B. Bill of material processor algorithms -time and storage complexity analysis. Technical report, Graduate School of Management, University of Rochester (1983).

[2] Afentakis, P., Gavish, B., Karmarkar, U. Computationally efficient optimal solutions to the lot-sizing problem in assembly systems. Management Science **30** (1984) 222–239.

[3] Axsater, S., Rosling, K. Notes: Installation vs. echelon stock policies for multilevel inventory control. Management Science **39** (1993) 1274–1279.

[4] Benichou, M., Gauthier, J. M., Hentges, G., Ribiere, G. Experiments in mixed integer linear programming. Mathematical Programming **1** (1971) 76-94.

[5] Chen, F., Zheng, Y. S. One-warehouse multiretailer systems with centralized stock information. Operations Research **45** (1997) 275–287.

[6] Clark, A. R., Armentano, V. A. Echelon stock formulations for multi-stage lot-sizing with lead-times. International Journal of Systems Science **24** (1993) 1759–1775.

[7] Clark, A. R., Armentano, V. A. A heuristic for a resource-capacitated multi-stage lot-sizing problem with lead times. Journal of the Operational Research Society **46** (1995) 1208–1222.

[8] Clark, A. J., Scarf, H. Optimal policies for a multi-echelon inventory problem. Management Science **6** (1960) 475–490.

[9] Crowston, W. B., Wagner, M. H., Williams, J. F. Economic lot size determination in multi-stage assembly systems. Management Science **19** (1973) 517–527.

[10] Optimization, D. Xpress-MP Essentials: An Introduction to Modeling and Optimization. Dash Associates, New Jersey (2002).

[11] Padberg, M. W., Rinaldi, G. A branch-and-bound algorithms for the resolution of large-scale symmetric traveling salesman problems. SIAM Review, **33**(1) (1991) 60–100.

[12] Schwarz, L. B. Multi-Level Production/Inventory Control Systems: Theory and Practice. North-Holland, Amsterdam (1981).

[13] Schwarz, L. B. Physical distribution: The anaylsis of inventory and location. AIIE Transactions **13** (1981) 138–150.

[14] Schwarz, L. B., Schrage, L. On echelon holding costs. Management Science **24** (1978) 865–866.

[15] Silver, E. A., Pyke, D. F., Peterson, R. Inventory Management and Production Planning and Scheduling. John-Wiley and Sons, New York (1998).

[16] Wagner, H. M., Whitin, T. M. Dynamic version of the economic lot size model. Management Science **5** (1958) 89–96.

[17] Zangwill, W. I. A backlogging model and a multi-echelon model of a dynamic economic lot size production system – a network approach. Management Science **15**(9) (1969) 506–527.

Table 6. Hybrid Results (secs/nodes): "–" indicates no solution within 1 hour

	Conventional Model							
	IC 1–6				IC 1–5			
	H1	H2	H3	H4	H1	H2	H3	H4
A1	347/41492	333/39447	281/34176	326/38977	994/143328	558/74616	377/45332	1650/277548
A2	644/88034	733/104322	613/80327	717/95789	736/136099	341/56385	717/130762	757/144476
A3	1030/144091	1360/201738	966/138839	1290/220931	1510/234798	1430/211116	639/74665	1390/208323
A4	137/18844	132/17806	117/14414	495/106910	308/67551	423/74196	313/56357	473/116441
A5	103/13737	155/23868	116/22690	103/14000	1260/172893	1430/206921	681/86815	1890/306806
A6	1250/176606	1150/158708	1160/182510	1510/236411	834/153700	1490/300912	810/150465	1340/278094
A7	160/24777	196/30297	177/27404	79.1/8760	287/65566	471/93533	319/66657	
A8	2770/509319	3180/608744	1600/283283	2320/428228	2010/318452	–/	1270/187083	2430/432682
A9	68.2/7312	93.2/11386	57.7/5955	68.7/7575	321/52203	205/31931	445/76195	291/45573
A10	754/106548	333/41473	820/118148	1460/159283	504/59214	565/82829	396/41762	411/49158
S1	3.81/282	3.63/298	4.12/366	4.65/468	2.85/309	2.59/244	3.02/306	4.23/505
S2	4.34/156	4.16/138	4.36/151	4.32/152	3.63/158	3.61/154	4.03/161	3.61/157
S3	11.1/1825	10.3/1650	10.8/1965	11.8/2092	7.57/1171	7.71/1013	6.13/789	9.09/1654
S4	12.9/2387	13.8/2445	15.1/2780	12.9/2327	14.1/2317	9.68/1322	15.3/2607	14.2/2326
S5	29.4/2310	39.1/2083	28.5/2173	43/3468	53.2/5134	46.5/4182	51.9/5146	44.5/3844
S6	5.19/328	4.72/278	5.19/291	6.51/490	3.88/295	3.59/267	4.14/282	5.51/509
S7	7.99/877	8.39/946	7.97/922	7.96/894	9.51/1160	9.31/1133	9.76/1141	8.81/1107
S8	31.1/3960	46.2/5028	44.1/3895	49.4/3989	60.5/4420	47.2/4186	50.7/4393	63.7/5082
S9	22.1/1930	19.7/1520	23.2/1814	25.1/1994	33.4/3326	34.1/3190	34.1/2955	27.3/2820
S10	6.72/1045	6.39/876	6.41/952	8.26/1429	5.52/998	5.78/994	5.88/980	6.72/1332
S11	4.52/280	4.12/248	5.01/331	4.34/273	4.27/268	4.33/271	5.01/360	4.41/269
S12	9.03/692	8.19/708	9.51/929	8.43/723	10.1/903	9.58/872	9.48/915	8.96/862
S13	4.27/893	4.46/882	4.26/880	4.27/903	3.28/807	3.52/825	3.71/963	3.33/855
S14	7.99/769	8.12/717	8.28/782	7.92/790	7.11/796	7.21/755	7.54/793	7.34/806
S15	8.5/1132	8.11/1065	8.37/1076	8.73/1121	6.91/926	6.65/886	7.02/973	6.64/911
W1					429/89478	438/88329	380/74273	558/129634
W2					908/187510	230/39146	963/203803	1010/226320
W3					41.1/6720	41.5/6949	40.3/6730	27.1/4098
W4					1810/425816	2140/542637	977/217213	1750/439325
W5					280/62574	136/27827	216/44857	143/28179
W6					64.8/13384	73.4/16783	74.2/17189	258/76382
W7					1130/284513	372/73895	971/239308	359/68076
W8					12.4/1664	11.6/1587	14.6/2046	14.8/2169
W9					37.6/8751	41.4/9720	37.4/8674	38.6/8999
W10					163/37997	105/21172	164/37934	108/22394

	Echelon Model							
	IC 1–6				IC 1–5			
	H1	H2	H3	H4	H1	H2	H3	H4
A1	422/50788	479/58022	292/34001	391/46499	1020/116263	408/42720	1070/131736	913/102886
A2	535/61891	551/68849	481/54697	299/29663	911/158986	581/96083	1190/231450	896/159848
A3	1630/203695	3000/511423	947/112375	1570/212956	1360/186429	1490/215757	1930/284210	1800/286333
A4	321/58663	262/41848	391/72124	243/38840	929/126296	988/132078	545/68024	1090/161570
A5	699/97698	453/61100	705/100947	575/79481	1430/193749	979/129565	822/105536	1900/277802
A6	1430/152051	1500/157195	1450/152150	1450/152150	1480/249527	2380/434608	1100/186745	1930/350496
A7	195/26560	230/33858	175/24016	230/36546	567/79609	496/65115	569/78103	1540/374719
A8	2090/318374	3160/507131	1670/223513	2130/342494	–/	–/	–/	–/
A9	78.7/8845	86.1/9279	148/20924	173/24650	286/42423	325/49181	386/73691	733/129922
A10	446/38123	551/48872	312/26625	657/99601	925/97216	1090/122752	454/49350	1130/144345
S1	3.67/310	3.15/231	3.65/308	3.41/300	3.32/404	3.65/391	3.67/449	5.71/826
S2	4.14/179	4.78/214	4.28/181	4.04/174	3.33/168	3.97/213	3.43/175	3.35/168
S3	12.3/1786	12.3/1753	14.1/1965	14.6/2252	10.4/1663	13.4/2220	11.3/1789	10.4/1689
S4	17.2/2094	17.4/1993	16.1/1791	19.1/2288	13.5/1680	13.5/1823	15.4/1777	13.4/1718
S5	36.9/2714	35.6/2633	38.3/3128	40.7/3580	21.2/1742	21.3/1726	22.5/1777	22.1/1742
S6	4.51/311	4.29/250	4.54/305	4.26/304	3.77/257	3.09/196	5.15/387	3.63/255
S7	6.79/735	6.02/503	6.53/583	6.52/611	9.35/1184	10.1/1387	8.81/1115	8.29/1026
S8	89.5/6930	68.6/5892	89.1/6846	80.3/6879	95.9/8350	108/8886	103/8718	106/9129
S9	19.6/1304	11.8/824	17.5/1257	15.3/1471	9.56/1022	7.84/648	9.78/953	11.5/1293
S10	7.41/975	7.52/960	6.88/885	7.51/1246	6.16/895	6.18/878	5.61/834	6.56/1212
S11	4.05/226	4.24/251	5.11/349	3.89/232	3.34/237	3.44/234	4.12/316	3.36/234
S12	7.21/640	7.13/595	8.42/864	7.34/630	6.55/812	6.56/750	6.38/812	6.48/727
S13	4.86/798	5.14/779	4.87/830	4.78/777	4.26/818	4.38/802	4.71/899	4.12/820
S14	6.84/719	8.88/622	5.85/507	7.05/728	7.34/804	7.36/837	6.87/793	7.18/797
S15	8.37/971	7.92/852	8.52/1094	8.21/891	4.46/728	4.68/707	4.51/695	4.61/698
W1					843/203927	649/144827	887/204364	1680/425469
W2					323/67383	386/79550	154/30806	508/103409
W3					29.9/4406	75.4/14194	29.5/4318	33.1/4902
W4					1960/469245	1870/452318	976/216520	2080/502255
W5					329/68894	113/21084	337/71589	200/38225
W6					80.6/15167	83.7/15878	86.5/16902	91.2/18139
W7					718/149184	489/92839	416/73521	253/36753
W8					16.4/2358	20.1/2943	19.3/2783	27.9/4977
W9					45.1/10149	63.3/15835	55.1/13012	42.8/9549
W10					54.1/11829	90.8/23117	56.5/13337	61.7/14019

Scheduling Abstractions for Local Search

Pascal Van Hentenryck[1] and Laurent Michel[2]

[1] Brown University
Box 1910, Providence, RI 02912
USA
[2] University of Connecticut
Storrs, CT 06269-3155
USA

Abstract. COMET is an object-oriented language supporting a constraint-based architecture for local search. This paper presents a collection of abstractions, inspired by constraint-based schedulers, to simplify scheduling algorithms by local search in COMET. The main innovation is the computational model underlying the abstractions. Its core is a precedence graph which incrementally maintains a candidate schedule at every computation step. Organized around this precedence graph are differentiable objects, e.g., resources and objective functions, which support queries to define and evaluate local moves. The abstractions enable COMET programs to feature declarative components strikingly similar to those of constraint-based schedulers and search components expressed with high-level modeling objects and control structures. Their benefits and performance are illustrated on two applications: minimizing total weighted tardiness in a job-shop and cumulative scheduling.

1 Introduction

Historically, most research on modeling and programming tools for combinatorial optimization has focused on systematic search, which is at the core of branch & bound and constraint satisfaction algorithm. It is only recently that more attention has been devoted to programming tools for local search and its variations (e.g., [, , , , ,]). Since constraint programming and local search exhibit orthogonal strengths for many classes of applications, it is important to design and implement high-level programming tools for both paradigms.

COMET [,] is a novel, object-oriented, programming language specifically designed to simplify the implementation of local search algorithms. Comet supports a constraint-based architecture for local search organized around two main components: a declarative component which models the application in terms of constraints and functions, and a search component which specifies the search heuristic and meta-heuristic. Constraints, which are a natural vehicle to express combinatorial optimization problems, are *differentiable objects* in COMET: They maintain a number of properties incrementally and they provide algorithms to

J.-C. Régin and M. Rueher (Eds.): CPAIOR 2004, LNCS 3011, pp. 319–   , 2004.

evaluate the effect of various operations on these properties. The search component then uses these functionalities to guide the local search using multidimensional, possibly randomized, selectors and other high-level control structures [ ]. The architecture enables local search algorithms to be high-level, compositional, and modular. It is possible to add new constraints and to modify or remove existing ones, without having to worry about the global effect of these changes. COMET also separates the modeling and search components, allowing programmers to experiment with different search heuristics and meta-heuristics without affecting the problem modeling. COMET has been applied to many applications and can be implemented to be competitive with tailored algorithms, primarily because of its fast incremental algorithms [ ].

This paper focuses on scheduling and aims at fostering the modeling features of COMET for this important class of applications. It is motivated by the remarkable success of constraint-based schedulers (e.g., [ ]) in modeling and solving scheduling problems using constraint programming. Constraint-based schedulers, CB-schedulers for short, provide high-level concepts such as activities and resources which considerably simplify constraint-programming algorithms. *The integration of such abstractions within* COMET *raises interesting challenges due to the fundamentally different nature of local search algorithms for scheduling.* Indeed, in constraint-based schedulers, the high-level modeling abstractions encapsulate global constraints such as the edge finder and provide support for search procedures dedicated to scheduling. In contrast, local search algorithms move from (possibly infeasible) schedules to their neighbors in order to reduce infeasibilities or to improve the objective function. Moreover, local search algorithms for scheduling typically do not perform moves which assign the value of some decision variables, as is the case in many other applications. Rather, they walk from schedules to schedules by adding and/or removing sets of precedence constraints. This is the case in algorithms for job-shop scheduling where makespan (e.g., [,   ]) or total weighted tardiness (e.g., [ ]) is minimized, flexible job-shop scheduling where activities have alternative machines on which they can be processed (e.g., [ ]), and cumulative scheduling where resources are available in multiple units (e.g., [ ]) to name only a few.

This paper addresses these challenges and shows how to support traditional scheduling abstractions in a local search architecture. Its main contribution is a novel computational model for the abstractions which captures the specificities of scheduling by local search. The core of the computational model is *an incremental precedence graph*, which specifies a candidate schedule at every computation step and can be viewed as a complex incremental variable. Once the concept of precedence graph is isolated, scheduling abstractions, such as resources and tardiness functions, become *differentiable objects* which maintain various properties and how they evolve under various local moves.

The resulting computational model has a number of benefits. From a programming standpoint, local search algorithms are short and concise, and they are expressed in terms of high-level concepts which have been shown robust

---

[1] We use *precedence constraints* in a broad sense to include distance constraints.

in the past. In fact, their declarative components closely resemble those of CB-schedulers, although their search components radically differ. From a computational standpoint, the computational model smoothly integrates with the constraint-based architecture of COMET, allows for efficient incremental algorithms, and induces a reasonable overhead. From a language standpoint, the computational model suggests novel modeling abstractions which explicit the structure of scheduling applications even more. These novel abstractions make scheduling applications more compositional and modular, fostering the main modeling benefits of COMET and synergizing with its control abstractions.

The rest of this paper first describes the computational model and provides a high-level overview of the scheduling abstractions. It then presents two scheduling applications in COMET: minimizing total weighted tardiness in a job-shop and cumulative scheduling. For space reasons, we do not include a traditional job-shop scheduling algorithm. The search component of one such algorithm [ ] was described in [ ] and its declarative component is essentially similar to the first application herein. Reference [ ] also contains experimental results. Other applications, e.g., flexible scheduling, essentially follow the same pattern.

2   The Computational Model

The main innovation underlying the scheduling abstractions is their computational model. The key insight is to recognize that most local search algorithms move from schedules to their neighbors by adding and/or removing precedence constraints. Some algorithms add precedence constraints to remove infeasibilities, while others walk in between feasible schedules by replacing one set of precedence constraints by another. Moreover, the schedules in these algorithms always satisfy the precedence constraints, but may violate other constraints. As a consequence, the core of the computational model is an *incremental precedence graph* which collects the set of precedence constraints between activities and specifies a *candidate schedule* at every computation step. The candidate schedule associates with each activity its earliest starting date consistent with the precedence constraints. It is incrementally maintained during the computation under insertion and deletion of precedence constraints using incremental algorithms such as those in [ ].

Once the precedence graph is introduced as the core concept, it is natural to view traditional scheduling abstractions (e.g., cumulative resources) as differentiable objects. A resource now maintains its violations with respect to the candidate schedule, i.e., the times where the demand for the resource exceeds its capacity. Similarly, COMET features differentiable objects for a variety of objective functions such as the makespan and the tardiness of an activity. These objective functions maintain their values, as well as a variety of additional data structures to evaluate the effect of a variety of local moves.

Although it is a significant departure from traditional local search in COMET, this computational model smoothly blends in the overall architecture of the language. Indeed, the precedence graph can simply be viewed as an incremental variable of a more complex type than integers or sets. Similarly, the scheduling abstractions are differentiable objects built on top of the precedence graph and its candidate schedule. Each differentiable object can encapsulate efficient incremental algorithms to maintain its properties and to implement its differentiable queries, exploiting the problem structure.

The overall computational model shares some important properties with CB-schedulers, including the distinguished role of precedence constraints in both architectures. Indeed, CB-schedulers can also be viewed as being implicitly organized around a precedence graph obtained by relaxing the resource constraints. (Such a precedence graph is now explicit in some CB-schedulers [ ].) The fundamental difference, of course, lies in how the precedence graph is used. In COMET, it specifies the candidate schedule and the scheduling abstractions are differentiable objects maintaining a variety of properties and how they vary under local moves. In CB-schedulers, the precedence graph reduces the domain of variables and the scheduling abstractions encapsulate global constraints, such as the edge finder, which derive various forms of precedence constraints.

3   Overview of the Scheduling Abstractions

This section briefly reviews some of the scheduling abstractions. Its goal is not to be comprehensive, but to convey the necessary concepts to approach the algorithms described in subsequent sections. As mentioned, the abstractions were inspired by CB-schedulers but differ on two main aspects. First, although the abstractions are the same, their interfaces are radically different. Second, COMET features some novel abstractions to expose the structure of scheduling applications more explicitly. These new abstractions often simplify search components, enhance compositionality, and improve performance.

Scheduling applications in COMET are organized around the traditional concepts of schedules, activities, and resources. The snippet

```
Schedule sched(mgr);
Activity a(sched,4); Activity b(sched,5);
a.precedes(b);
sched.close();
```

introduces the most basic concepts. It declares a schedule sched using the local search manager mgr, two activities a and b of duration 4 and 5, and a precedence constraint between a and b. This excerpt highlights the high-level similarity between the declarative components of COMET and constraint-based schedulers. *What is innovative in COMET is the computational model underlying these modeling objects, not the modeling concepts themselves.* In constraint-based scheduling, these instructions create domain-variables for the starting dates of the activities and the precedence constraints reduce their domains. In COMET, these instructions specify a *candidate schedule* satisfying the precedence constraints. For instance, the above snippet assigns starting dates 0 and 4 to activities a and b. The expression a.getESD() can be used to retrieve the starting date of activity a which typically vary over time.

Schedules in COMET always contain two basic activities of zero duration: the source and the sink. The source precedes all other activities, while the sink follows every other activity. The availability of the source and the sink often simplifies the design and implementation of local search algorithms.

Jobs: Many local search algorithms rely on the job structure to specify their neighborhood, which makes it natural to include jobs as a modeling object for scheduling. This abstraction is illustrated in Section   , where critical paths are computed. A job is simply a sequence of activities linked by precedence constraints. The structure of jobs is specified in COMET through precedence constraints. For instance, the snippet

```
1.    Schedule sched(mgr);
2.    Job j(sched);
3.    Activity a(sched,4); Activity b(sched,5);
4.    a.precedes(b,j);
5.    sched.close();
```

specifies a job j with two activities a and b, where a precedes b. This snippet also highlights an important feature of COMET: Precedence constraints can be associated with modeling objects such as jobs and resources (see line 4). This *polymorphic* functionality simplifies local search algorithms which may retrieve subsets of precedence constraints easily. Since each activity belongs to at most one job, COMET provides methods to access the job predecessors and successors of each job. For instance, the expression b.getJobPred() returns the job predecessor of b, while j.getFirst() returns the first activity in job j.

Cumulative Resources: Resources are traditionally used to model the processing requirements of activities. For instance, the instruction

CumulativeResource cranes(sched,5);

specifies a cumulative resource providing a pool of 5 cranes, while the instruction

a.requires(cranes,2)

specifies that activity a requires 2 cranes during its execution. Once again, COMET reuses traditional modeling concepts from CB-scheduling and the novelty is in their functionalities. Resources in COMET are not instrumental in pruning the search space: They are differentiable objects which maintain invariants and data structures to define the neighborhood. In particular, a cumulative resource maintains violations induced by the candidate schedule, where a violation is a time t where the demands of the activities executing on r at t in the candidate schedule exceeds the capacity of the resource. Cumulative resources can also be queried to return sets of tasks responsible for a given violation. As mentioned, precedence constraints can be associated with resources, e.g., a.precedes(b,crane), a functionality illustrated later in the paper.

Precedence Constraints: It should be clear at this point that precedence constraints are a central concept underlying the abstraction. In fact, precedence constraints are first-class citizens in COMET. For instance, the instruction

```
set{Precedence} P = cranes.getPrecedenceConstraints();
```

can be used to retrieve the set of precedence constraints associated with `crane`.

Disjunctive Resources: Disjunctive resources are special cases of cumulative resources with unit capacity. Activities requiring disjunctive resources cannot overlap in time and are strictly ordered in feasible schedules. Local search algorithms for applications involving only disjunctive resources (e.g., various types of jobshop and flexible scheduling problems) typically move in the space of feasible schedules by reordering activities on a disjunctive resource. As a consequence, COMET provides a number of methods to access the (current) disjunctive sequence. For instance, method `d.getFirst()` returns the first activity in the sequence of disjunctive resource `d`, while method `a.getSucc(d)` returns the successor of activity `a` on `d`. COMET also provides a number of local moves for disjunctive resources which can all be viewed as the addition and removal of precedence constraints. For instance, the move `d.moveBackward(a)` swaps activity `a` with its predecessor on disjunctive resource `d`. This move removes three precedence constraints and adds three new ones. Note that such a move does not always result in a feasible schedule: activity `a` must be chosen carefully to avoid introducing cycles in the precedence graph.

Objective Functions: One of the most innovative scheduling abstractions featured in COMET is the concept of objective functions. At the modeling level, the key idea is to specify the "global" structure of the objective function explicitly. At the computational level, objective functions are differentiable objects which incrementally maintain invariants and data structures to evaluate the impact of local moves. The ubiquitous objective function in scheduling is of course the makespan which can be specified as follows:

```
Makespan makespan(sched);
```

Once declared, an objective function can be evaluated (i.e., `makespan.eval()`) and queried to determine the impact of various local moves. For instance, the expression `makespan.evalAddPrecedenceDelta(a,b)` evaluates the makespan variation of adding the precedence $a \rightarrow b$. Similarly, the effect on the makespan of swapping activity `a` with its predecessor on disjunctive resource `d` can be queried using `makespan.evalMoveBackwardDelta(a,d)`. Note that COMET supports many other local moves and users can define their own moves using the data and control abstractions of COMET.

The makespan maintains a variety of interesting information besides the total duration of the schedule. In particular, it maintains the latest starting date of each activity, as well as the *critical* activities, which appears on a longest path from the source to the sink in the precedence graph. These information are generally fundamental in defining neighborhood search and heuristic algorithms for scheduling. They can also be used to estimate quickly the impact of a local move. For instance, the expression `makespan.estimateMoveBackwardDelta(a,d)` returns an approximation to the makespan variation when swapping activity a with its predecessor on disjunctive resource d.

Although the makespan is probably the most studied objective function in scheduling, there are many other criteria to evaluate the quality of a schedule. One such objective is the concept of tardiness which has attracted increasing attention in recent years. The instruction

```
Tardiness tardiness(sched,a,dd);
```

declares an objective function which maintains the tardiness of activity a with respect to its due date dd, i.e., $\max(0, e - dd)$ where e is the finishing date of activity a in the candidate schedule. Once again, a tardiness object is differentiable and can be queried to evaluate the effect of local moves on its value. For instance, the instruction `tardiness.evalMoveBackwardDelta(a,d)` determines the tardiness variation which would result from swapping activity a with its predecessor on disjunctive resource d.

The objective functions share the same differentiable interface, thus enhancing their compositionality and reusability. In particular, they combine naturally to build more complex optimization criteria. For instance, the snippet

```
1.   Tardiness tardiness[j in Job](sched,job[j].getLast(),dd[j]);
2.   ScheduleObjectiveSum totalTardiness(sched);
3.   forall(j in Job)
4.     totalTardiness.add(tardiness[j]);
```

defines an objective function `totalTardiness`, another differentiable function, which specifies the total tardiness of the candidate schedule. Line 1 defines the tardiness of every job j, i.e., the tardiness of the last activity of j. Line 2 defines the differentiable object `totalTardiness` as a sum of objective functions. Lines 3 and 4 adds the job tardiness functions to `totalTradiness` to specify the total tardiness of the schedule. Queries on the aggregate objective `totalTardiness`, e.g., `totalTardiness.evalMoveBackwardDelta(a,d)`, are computed by querying the individual tardiness functions and aggregating the results. It is easy to see how similar code could define maximum tardiness as an objective function.

Disjunctive Schedules: We conclude this brief overview by introducing disjunctive schedules which simplify the implementation of various classes of applications such as job-shop, flexible-shop, and open-shop scheduling problems. In disjunctive schedules, activities require at most one disjunctive resource, although they may have the choice between several such resources. Since activities are requiring at most one resource, various methods can now omit the specification of the resource which is now identified unambigously. For instance, the method `tardiness.evalMoveBackwardDelta(a)` evaluates the makespan variation of swapping activity a with its predecessor on its disjunctive resource.

```
1.    void Jobshop::state() {
2.        sched = new DisjunctiveSchedule(mgr);
3.        act = new Activity[i in ActRange](sched,duration[i]);
4.        res = new DisjunctiveResource[MachineRange](sched);
5.        job = new Job[JobRange](sched);
6.        tardiness = new Tardiness[JobRange];
7.
8.        forall(a in ActRange)
9.            act[a].requires(res[machine[a]]);
10.       forall(j in JobRange) {
11.           Activity last = sched.getSource();
12.           forall(t in TaskRange) {
13.               int a = jobAct[j,t];
14.               last.precedes(act[a]);
15.               last = act[a];
16.           }
17.           tardiness[j] = new Tardiness(sched,last,duedate[j]);
18.       }
19.       obj = new ScheduleObjectiveSum(sched);
20.       forall(j in JobRange)
21.           obj.add(weight[j] * tardiness[j]);
22.       sched.close();
23.   }
```

Fig. 1. Minimizing Total Weighted Tardiness: The Declarative Component

4   Minimizing Total Weighted Tardiness in a Job Shop

This section describes a simple, but effective, local seach algorithm for minimizing the total weighted tardiness in a job shop.

The Problem: We are given a set of jobs J, each of which being a sequence of activities linked by precedence constraints. Each activity has a fixed duration and a fixed machine on which it must be processed. No two jobs scheduled on the same machine can overlap in time. In addition, each job $j \in J$ is given a due date d_j and a weight w_j. The goal is to find a schedule satisfying the precedence and disjunctive constraints and minimizing the total weighted tardiness of the schedule, i.e., the function $\sum_{j \in J} w_j \max(0, c_j - d_j)$ where c_j is the completion of job j, i.e., the completion time of its last activity. This problem has received increasing attention in recent years. See, for instance, [, ,].

The Declarative Component: The declarative component of this application is depicted in Figure . For simplicity, it assumes that the input data is given by a number of ranges and arrays (e.g., duration) which are stored in instance variables. Lines 2-6 declare the modeling objects of the application: the schedule, the activities, the disjunctive resources, the jobs, and the tardiness array. The actual tardiness functions are created later in the method. Lines 8 and 9 specify the resource constraints. Lines 10-18 specify both the precedence constraints and the tardiness functions. Lines 11-16 declare the precedence constraints for a given job j, while line 17 creates the tardiness function associated with job j. Lines 19 defines the objective function as a summation. Lines 20-21 specify the various elements of the summation. Of particular interest is line 21 which defines the weighted tardiness of job j by multiplying its tardiness function by its weight. This multiplication creates a differentiable object which can be queried in the same way as the tardiness functions. Line 22 closes the schedule, enforcing all the precedence constraints and computing the objective function. It is worth highlighting two interesting features of the declarative statement. First, the declarative component of a traditional job-shop scheduling problem can be obtained by replacing lines 6, 17, and 19-21 by the instruction

```
obj = new Makespan(sched);
```

Second, observe the high-level nature of this declarative component and its independence with respect to the search algorithms. It is indeed conceivable to define constraint-based schedulers which would recognize this declarative specification and generate appropriate domain variables, constraints, and objective function. Of course, this strong similarity completely disappears in the search component.

The Search Component: Figure   depicts the search component of the application. It specifies a simple Metropolis algorithm which swaps activities that are critical for some tardiness function. The top-level method `localSearch` is depicted in Lines 1-13. It first creates an initial schedule (line 2) and an exponential distribution. Lines 6-9 are the core of the local search and they are executed for `maxTrials` iterations. Each such iteration selects a job j which is late (line 6), computes a set of critical activities responsible for this tardiness (line 7), and explores the neighborhood for these activities. Line 12 restores the best solution found during the local search.

Method `exploreNeighborhood` (lines 15-30) explores the moves that swap a critical activity with its machine predecessor. These moves are guaranteed to be feasible by construction of the critical path. The algorithm selects a critical activity (line 18) and evaluates the move (line 19) using the objective function. The move is executed if it improves the candidate schedule or if it is accepted by the exponential distribution (lines 20-21) and the best solution is updated if necessary (lines 22-25). These basic steps are iterated until a move is executed or for some iterations.

Method `selectCriticalPath` (lines 32-45) is the last method of the component. The key idea is to start from the activity a of the tardiness object (i.e., the last activity of its associated job) and to trace back a critical path from a to the source. Lines 37 to 43 are the core of the method. They first test if the job precedence constraint is tight (lines 37-38), in which case the path is traced back from the job predecessor. Otherwise, activity a is inserted in C as a critical

```
1.    void Jobshop::localSearch() {
2.        bestSoFar = findInitialSchedule();
3.        distr = new ExponentialDistribution(mgr);
4.        nbTrials = 0;
5.        while (nbTrials < maxTrials) {
6.            select(j in JobRange : tardiness[j].eval() > 0) {
7.                set{Activity} Criticals = selectCriticalPath(tardiness[j]);
8.                exploreNeighborhood(Criticals);
9.            }
10.           nbTrials++;
11.       }
12.       solution.restore(mgr);
13.   }
14.
15.   void Jobshop::exploreNeighborhood(set{Activity} Criticals) {
16.       int i = 0;
17.       do {
18.           select(v in Criticals) {
19.               int delta = obj.evalMoveBackwardDelta(v);
20.               if (delta < 0 || distr.accept(-delta/T)) {
21.                   v.moveBackward();
22.                   if (obj.eval() < bestSoFar) {
23.                       solution = new Solution(mgr);
24.                       bestSoFar = obj.eval();
25.                   }
26.                   break;
27.               }
28.           }
29.       } while (i++ < maxLocalIterations);
30.   }
31.
32.   set{Activity} Jobshop::selectCriticalPath(Tardiness t) {
33.       set{Activity} C();
34.       Activity a = t.getActivity();
35.       Activity source = sched.getSource();
36.       do {
37.           Activity pj = a.getJobPred();
38.           if (pj.getEFD() == a.getESD())
39.               a = pj;
40.           else {
41.               C.insert(a);
42.               a = a.getDisjPred();
43.           }
44.       } while (a != source);
45.       return C;
45.   }
```

Fig. 2. Minimizing Total Weighted Tardiness: The Local Search

Table 1. Minimizing Total Weighted Tardiness: Experimental Results

Bench	Opt	#O	BEST	AVG	WST	$\mu(TS)$	LSRW
abz5	1405	45	1405.00	**1409.92**	1464.00	7.76	1451
abz6	436	50	436.00	**436.00**	436.00	2.79	(5) 436
la16	1170	50	1170.00	**1170.00**	1170.00	5.41	(5) 1170
la17	900	41	900.00	956.62	1239.00	6.43	(5) 900
la18	929	48	929.00	929.28	936.00	5.78	(5) 929
la19	948	33	948.00	**950.58**	956.00	9.30	(3) 951
la20	809	33	809.00	821.58	846.00	7.65	(5) 809
la21	464	50	464.00	**464.00**	464.00	2.12	(5) 464
la22	1068	1	1068.00	1100.06	1185.00	11.30	1086
la23	837	1	837.00	**874.32**	879.00	5.99	875
la24	835	50	835.00	**835.00**	835.00	4.86	(5) 835
mt10	1368	44	1368.00	1393.58	1678.00	10.58	(5) 1368
orb1	2568	30	2568.00	**2600.42**	2867.00	13.79	(2) 2616
orb2	1412	5	1412.00	**1431.80**	1434.00	6.10	1434
orb3	2113	14	2113.00	**2171.56**	2254.00	12.08	2204
orb4	1623	20	1623.00	**1639.66**	1682.00	12.51	(1) 1674
orb5	1593	2	1593.00	1700.72	1756.00	12.55	1662
orb6	1792	47	1792.00	**1800.30**	2072.00	7.43	(4) 1802
orb7	590	0	611.00	675.48	766.00	13.26	618
orb8	2429	1	2429.00	**2503.04**	2624.00	11.91	2554
orb9	1316	35	1316.00	1343.80	1430.00	10.71	(3) 1334
orb10	1679	17	1679.00	**1769.90**	1840.00	7.96	1775

activity *due to the disjunctive arc $p \rightarrow a$, where p is the disjunctive predecessor* and the path is traced back from the disjunctive predecessor.

This concludes the description of the algorithm. The COMET program is concise (its core is about 70 lines of code), it is expressed in terms of high-level scheduling abstractions and control structures, and it automates many of the tedious and error-prone aspects of the implementation.

Experimental Results: Table   depicts the experimental results of the COMET algorithm (algorithm CT), on a Pentium IV (2.1mhz) and contrasts them briefly with the large step random algorithm of [ ] (algorithm LSRW) which typically dominates [ ]. These results are meant to show the practicability of the abstractions, not to compare the algorithms in great detail. The parameters were set as follows: maxIterations is set to 600,000 iterations (which roughly corresponds to the termination criteria in [ ] when machines are scaled), $T = 225$ and maxLocalIterations=5. The initial solution is a simple insertion algorithm for minimizing the makespan [ ]. Algorithm CT was evaluated on the standard benchmarks from [,   ,   ], where the deadline for job j is given by $\lceil f * \sum_{i=1}^{m} p_{ij} \rceil - 1$. ([,   ,   ] specify $\lfloor f * \sum_{i=1}^{m} p_{ij} \rfloor$ in their papers but actually use the given formula.) We used the value 1.3 for f which produces the hardest problems in [ ] and ran each benchmark 50 times. The table reports the optimal value (O), the number of times CT finds the optimum (#O), the best, average, and worst values found by CT, as well as the average CPU time to the best solution. The table also reproduces the result given in [ ], which only reports the average of LSRW over 5 runs and the number of times the optimum was found.

The results are very interesting. CT found the optimal solutions on all but one benchmark and generally with very high frequencies. Moreover, its averages generally outperform, or compare very well to, LSRW. This is quite remarkable since there are occasional outliers on these runs which may not appear over 5 runs (see, e.g., la18). The average time to the best solution is always below 14 seconds. Overall, these results clearly confirm the jobshop results [ ] as far as the practibility of the scheduling abstractions is concerned.

5   Cumulative Scheduling

This section describes the implementation of the algorithm IFLATIRELAX [ ], a simple, but effective, extension of iterative flattening [ ].

The Problem: We are given a set of jobs J, each of which consisting of a sequence of activities linked by precedence constraints. Each activity has a fixed duration, a fixed machine on which it executes, and a demand for the capacity of this machine. Each machine $c \in M$ has an available capacity $cap(c)$. The problem is to minimize the earliest completion time of the project (i.e., the makespan), while satisfying all the precedence and resource constraints.

The Declarative Component: Figure   depicts the declarative component for cumulative scheduling. The first part (lines 1-11) is essentially traditional and solver-independent. It declares the modeling objects, the precedence and resource constraints, as well as the objective function. The second part (Lines 12-13) is also declarative but it only applies to local or heuristic search. Its goal is to specify invariants which are used to guide the search. Line 12 collects the violations of each resource in an array of incremental variables, while line 13 states an invariant which maintains the total number of violations. This invariant is automatically updated whenever violations appear or disappear.

The Search Component: The search component of iterative flattening is particularly interesting and is depicted in Figure   . Starting from an infeasible schedule violating the resource constraints (i.e., the candidate schedule induced by the precedence constraints), IFLATIRELAX iterates two steps: a flattening step (line 6) which adds precedence constraints to remove infeasibilities and a relaxation step (line 7) which removes some of the added precedence constraints

```
1.    void CumulativeShop::state() {
2.        sched = new Schedule(mgr);
3.        act = new Activity[a in Activities](sched,duration[a]);
4.        resource = new CumulativeResource[m in Machines](sched,cap[m]);
5.
6.        forall(t in precedences)
7.            act[t.o].precedes(act[t.d]);
8.        forall(a in Activities)
9.            act[a].requires(resource[machine[a]],dem[a]);
10.       makespan = new Makespan(sched);
11.       sched.close();
12.       nbv = new inc{int}[m in Machines] = resource[m].getNbViolations();
13.       violations = new inc{int}(mgr) <- sum(m in Machines) nbv[m];
14.       mgr.close();
15.       bestSoFar = sum(a in Activities) duration[a];
16.   }
```

Fig. 3. Cumulative Scheduling: The Declarative Component

to provide a new starting point for flattening. These two steps are executed for a number of iterations (i.e., *maxIterations*) or until no improved feasible schedule has been found for a number of iterations (i.e., *maxStable*). The algorithm returns the best feasible schedule found after the flattening step (step 1). Note that iterative flattening can be seen as a large step local search, where each step removes (relaxation) and adds (flattening) a large set of precedence constraints.

The flattening step is performed by method flatten in lines 12-25. Flattening aims at removing all violations of the cumulative constraints by adding precedence constraints. It selects a violated cumulative constraint c (line 15) and chooses the time t with the largest violation (line 16). To remove this violation, the flattening algorithm queries the resource to obtain a number of minimal conflicts (line 18). The minimal conflicts are given by arrays of activities, potentially of different sizes. From these critical sets, the algorithm selects two activities a_i and a_j (lines 19 and 20) and inserts a precedence constraint $a_i \rightarrow a_j$ (line 21). The two activities are chosen carefully in order to minimize the impact on the makespan. More precisely, the flattening algorithm selects the two activities a_i and a_j maximizing $lsd(a_j) - efd(a_i)$ where $lds(a_j)$ denotes the latest starting date of a_j and $efd(a_i)$ is the earliest finishing date of a_i in the candidate schedule. This heuristics choice aims at keeping as much flexibility as possible in the schedule. Lines 15-23 are iterated until all violations are removed. The candidate schedule then satisfies all constraints and method updateBestSolution in line 24 possibly updates the best solution and the number of stable iterations.

After the flattening step, the algorithm has a feasible schedule and its set S of precedence constraints. Instead of restarting from scratch, the key idea behind the relaxation step is to consider the precedence constraints introduced by the flattening step and removes them with some probability. Only critical

```
1.    void CumulativeShop::search() {
3.      coin = new UniformDistribution(1..10);
4.      nbStable = 0; int it = 0;
5.      while (it++ < maxIterations && nbStable <= maxStable) {
6.        flatten();
7.        relax();
8.      }
9.      solution.restore();
10.   }
12.   void CumulativeShop::flatten() {
14.     while (violations)
15.       select(m in Machines : nbv[m] > 0) {
16.         CumulativeResource r = resource[m];
17.         selectMax(t in r.getViolations())(r.getViolationDegree(t)) {
18.           Activity c[][] = r.getMinimalConflicts(t);
19.           selectMax(k in c.rng(), i in c[k].rng(), j in c[k].rng(): i!=j)
                       (c[k][i].getLSD(makespan) - c[k][j].getEFD())
21.             c[k][j].precedes(c[k][i],r);
22.         }
23.       }
24.     updateBestSolution();
25.   }
27.   void CumulativeShop::relax() {
29.     forall(p in 1..maxRelations) {
30.       Precedence[] critical[m in Machines] =
                resource[m].getCriticalPrecedences(makespan);
31.       forall(m in Machines, i in critical[m].rng())
32.         if (coin.get() <= prRelaxation)
33.           critical[m][i].remove();
34.     }
35.   }
```

Fig. 4. Cumulative Scheduling: The Search Component

precedence constraints (i.e., constraints which correspond to critical arcs) are considered during relaxation, since only these may decrease the makespan. This relaxation is iterated for a number of iterations to avoid introducing bottlenecks early in the schedule. The relaxation step is implemented by method **relax** in lines 27-35. The algorithm collects all the critical precedence constraints introduced by flattening (line 30) and iterates over them (line 31). Line 32 flips a coin (the distribution is created in line 3) for each precedence constraint and removes the precedence constraint with some probability (lines 32-33). Line 29-33 are iterated for a small number of iterations. Once again, observe the conciseness and high-level description of algorithm IFLATIRELAX, which uses high-level modeling concepts both to state the problem and to specify the search procedure.

Table 2. Experimental Result of IFLATIRELAX on Cumulative Scheduling

	set A	set B	set MT	set C	set D
min	-0.01	-1.17	0.37	1.7	1.4
avg	1.07	0.47	3.41	4.2	2.4
T	48.55	103.96	127.84	221.3	645

Experimental Results: Algorithm IFLATIRELAX was discovered when implementing and analyzing the original iterative flattening algorithm [ ] in COMET. A detailed description of its performance results is available in [ ] and only a brief summary is presented here, since the main purpose is to illustrate the scheduling abstractions of COMET. Algorithm IFLATIRELAX found 21 new best upper bounds to standard benchmarks (with as much as 900 activities) [,] and quickly delivers solutions that are typically within 1% of the best available upper bounds. Table   summarizes the results. For each class of benchmarks, the table reports the best and average deviation (in percentage) from the upper bounds on the various classes of problems, as well as the computation times in seconds on Pentium 4 (2.4Ghz). The averages are computed over 100 runs, except for the larger classes (C and D). These results clearly indicate the practicability and benefits from the scheduling abstractions.

6   Conclusion

This paper presented a collection of scheduling abstractionsto simplify the implementation, and enhance the compositionality and reusability, of local search algorithms. The main innovation is the computational model underlying the abstractions. Its core is a precedence graph which incrementally maintains a candidate schedule at every computation step. Organized around this precedence graph are differentiable objects which are encapsulated in the scheduling abstractions such as resources and objective functions. These differentiable objects maintain various properties incrementally and support differentiable queries to evaluate the effect of local moves. The resulting abstractions and computational model nicely integrate into the COMET architecture, allows declarative components to be strikingly similar to those featured in CB-schedulers, and provides high-level concepts to specify the local search. The abstractions were illustrated with two applications, minimizing of total weighted tardiness in a jobshop and cumulative scheduling. For both applications, the COMET code is short and concise, it uses high-level scheduling concepts, and it exhibits excellent performance.

Acknowledgements

This work was partially supported by NSF ITR Awards DMI-0121495 and ACI-0121497.

References

[1] A. Cesta, A. Oddi, and S. F. Smith. Iterative flattening: A scalable method for solving multi-capacity scheduling problems. In *AAAI/IAAI*, pages 742–747, 2000.

[2] L. Di Gaspero and A. Schaerf. *Optimization Software Class Libraries*, chapter Writing Local Search Algorithms Using EasyLocal++. Kluwer, 2002.

[3] W. Kreipl. A large step random walk for minimizing total weighted tardiness in a job shop. *Journal of Scheduling*, 3:125–138, 2000.

[4] P. Laborie. Algorithms for propagating resource constraints in ai planning and scheduling: Existing approaches and new results. *AIJ*, 143(2):151–188, 2003.

[5] F. Laburthe and Y. Caseau. SALSA: A Language for Search Algorithms. In *CP'98*, Pisa, Italy, October 1998.

[6] M. Dell'Amico and M. Trubian. Applying Tabu Search to the Job-Shop Scheduling Problem. *Annals of Operations Research*, 41:231–252, 1993.

[7] M. Mastrolilli and L. Gambardella. Effective neighborhood functions for the flexible job shop problem. *Journal of Scheduling*, 3(1):3–20, 2000.

[8] L. Michel and P. Van Hentenryck. Localizer. *Constraints*, 5:41–82, 2000.

[9] L. Michel and P. Van Hentenryck. A constraint-based architecture for local search. In *OOPSLA'00*, Seattle, WA, 2002.

[10] L. Michel and P. Van Hentenryck. Maintaining longest path incrementally. In *CP'03*, Cork, Ireland, 2003.

[11] L. Michel and P. Van Hentenryck. Iterative relaxations for iterative flattening in cumulative scheduling. In *ICAPS'04*, Whistler, BC, Canada, 2004.

[12] E. Nowicki and C. Smutnicki. A fast taboo search algorithm for the job shop problem. *Management Science*, 42(6):797–813, 1996.

[13] W. Nuijten and C. Le Pape. Constraint-based job shop scheduling with ilog scheduler. *Journal of Heuristics*, 3:271–286, 1998.

[14] W. P. M. Nuijten and E. H. L. Aarts. A computational study of constraint satisfaction for multiple capacitated job shop scheduling. *EJOR*, 90(2):269–284, 1996.

[15] P. Shaw, B. De Backer, and V. Furnon. Improved local search for CP toolkits. *Annals of Operations Research*, 115:31–50, 2002.

[16] M. Singer and M. Pinedo. A computational study of branch and bound techniques for minimizing the totalweighted tardiness in job shops. *IIE Scheduling and Logistics*, 30:109–118, 1997.

[17] M. Singer and M. Pinedo. A shifting bottleneck heuristic for minimizing the totalweighted tardiness in job shops. *Naval Research Logistics*, 46(1):1–17, 1999.

[18] P. Van Hentenryck and L. Michel. Control abstractions for local search. In *CP'03*, Cork, Ireland, 2003. (Best Paper Award).

[19] S. Voss and D. Woodruff. *Optimization Software Class Libraries*. Kluwer, 2002.

[20] J. Walser. *Integer Optimization by Local Search*. Springer Verlag, 1998.

[21] F. Werner and A. Winkler. Insertion Techniques for the Heuristic Solution of the Job Shop Problem. TR, Technical Universitaet Magdebourg, 1992.

$O(n \log n)$ Filtering Algorithms
for Unary Resource Constraint

Petr Vilím

Charles University
Faculty of Mathematics and Physics
Malostranské néměstí 2/25, Praha 1
Czech Republic
vilim@kti.mff.cuni.cz

Abstract. So far, edge-finding is the only one major filtering algorithm
for unary resource constraint with time complexity $O(n \log n)$. This paper proposes $O(n \log n)$ versions of another two filtering algorithms: not-
first/not-last and propagation of detectable precedences. These two algorithms can be used together with the edge-finding to further improve
the filtering. This paper also propose new $O(n \log n)$ implementation of
fail detection (overload checking).

1 Introduction

In scheduling, *unary resource* is an often used generalization of a machine (or
a job in openshop). Unary resource models a set of non-interruptible *activities* T
which must not overlap in a schedule.

Each activity $i \in T$ has following requirements:

- earliest possible starting time est_i
- latest possible completion time lct_i
- processing time p_i

A (sub)problem is to find a schedule satisfying all these requirements. One
of the most used technique to solve this problem is a *constraint programming*.

In the constraint programming, we associate an *unary resource constraint*
with each unary resource. A purpose of such a constraint is to reduce a search
space by tightening the time bounds est_i and lct_i. This process of elimination
of infeasible values is called a *propagation*, an actual propagation algorithm is
often called a *filtering* algorithm.

Naturally, it is not efficient to remove all infeasible values. Rather we use
several fast but not complete algorithms which can find only some impossible
assignments. These filtering algorithms are repeated in every node of a search
tree, therefore their speed and filtering power are crucial.

Today, edge-finding and not-first/not-last are the mainly used filtering algorithms for the unary resource constraint. The edge-finding algorithm has time
complexity $O(n \log n)$ [ ], whereas time complexity of the not-first/not-last [,]

J.-C. Régin and M. Rueher (Eds.): CPAIOR 2004, LNCS 3011, pp. 335–   , 2004.
© Springer-Verlag Berlin Heidelberg 2004

algorithm is $O(n^2)$ (where $n = |T|$ is the number of activities). This paper introduce a new $O(n \log n)$ version of the not-first/not-last algorithm and also a third $O(n \log n)$ filtering algorithm. All these three algorithms can be used together to join their filtering powers.

Let us establish a notation concerning a subset of activities. Let $\Theta \subseteq T$ be an arbitrary non-empty subset of activities. An earliest starting time est_Θ, a latest completion time lct_Θ and a processing time p_Θ of the set Θ are:

$$\mathrm{est}_\Theta = \min \{\mathrm{est}_j, \; j \in \Theta\}$$
$$\mathrm{lct}_\Theta = \max \{\mathrm{lct}_j, \; j \in \Theta\}$$
$$\mathrm{p}_\Theta = \sum_{j \in \Theta} \mathrm{p}_j$$

Often we need to estimate an earliest completion time of a set Θ. When computing such an estimation we relax the restriction that activities are not interruptible. The resulting estimation of the earliest completion time of the set Θ is:

$$\mathrm{ECT}_\Theta = \max \{\mathrm{est}_{\Theta'} + \mathrm{p}_{\Theta'}, \; \Theta' \subseteq \Theta\} \tag{1}$$

To extend the definitions also for $\Theta = \emptyset$ let $\mathrm{est}_\emptyset = -\infty$, $\mathrm{lct}_\emptyset = \infty$, $\mathrm{p}_\emptyset = 0$ and $\mathrm{ECT}_\emptyset = -\infty$.

2 Θ-Tree

Algorithms in this paper are based on an idea of organizing a set $\Theta \subseteq T$ in a balanced binary tree. Because the set represented by the tree will always be named Θ, we will call the tree Θ-tree. The purpose of a Θ-tree is to quickly recompute ECT_Θ when an activity is inserted or removed from the set Θ.

A Θ-tree is a balanced binary search tree with respect to est_i. Each activity $i \in \Theta$ is represented by a single node. In the following we do not make a difference between an activity and the tree node representing that activity.

Notice that so far Θ-tree does not require any particular way of balancing. Any type of balanced binary tree (AVL-tree, black-red-tree *etc.*) is possible. The only requirement is a time complexity $O(\log n)$ for inserting or deleting a node, and time complexity $O(1)$ for finding a root node.

It is also possible to start with a perfect balanced tree built from all activities T with "empty" nodes . Inserting an activity means to fill an empty node reserved for that activity, deleting a node makes it empty over again. Note that building such perfect balanced tree costs $O(n \log n)$ time because activities have to be sorted first.

Let left(i) be a left son of an activity i (if it has one), similarly let right(i) be a right son of the activity i. We will also need a notation for subtrees:

---

[1] This is the implementation chosen by the author.

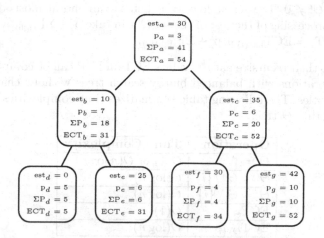

Fig. 1. An example of a Θ-tree for $\Theta = \{a, b, c, d, e, f, g\}$

let Subtree(i) be a set of all activities in the subtree rooted in i; Left(i) = Subtree(left(i)), Right(i) = Subtree(right(i)).

A Θ-tree is a balanced binary tree with respect to est_i and so:

$$\forall i \in \Theta \; \forall j \in \text{Left}(i) : \quad \text{est}_j \leq \text{est}_i$$
$$\forall i \in \Theta \; \forall j \in \text{Right}(i) : \text{est}_j \geq \text{est}_i$$

Besides the activity itself, each node i of a Θ-tree holds following two values:

$$\Sigma \text{P}_i = \sum_{j \in \text{Subtree}(i)} \text{p}_j$$

$$\text{ECT}_i = \text{ECT}_{\text{Subtree}(i)} = \max \left\{ \text{est}_{\Theta'} + \text{p}_{\Theta'}, \; \Theta' \subseteq \text{Subtree}(i) \right\}$$

Values ΣP_i and ECT_i can be computed from direct descendants of the node i:

$$\Sigma \text{P}_i = \Sigma \text{P}_{\text{left}(i)} + \text{p}_i + \Sigma \text{P}_{\text{right}(i)}$$

$$\text{ECT}_i = \max \big\{ \; \text{ECT}_{\text{right}(i)},$$
$$\text{est}_i + \text{p}_i + \Sigma \text{P}_{\text{right}(i)},$$
$$\text{ECT}_{\text{left}(i)} + \text{p}_i + \Sigma \text{P}_{\text{right}(i)} \; \big\}$$

The rule for computing ECT_i comes from the following observation. From the definition ( ) ECT_i is the maximum of $\text{est}_{\Theta'} + \text{p}_{\Theta'}$ of all $\Theta' \subseteq \text{Subtree}(i)$. With respect to the node i we will split the sets Θ' into the following three categories:

- Left(i) $\cap \, \Theta' = \emptyset$ and $i \notin \Theta'$. Clearly $\text{ECT}_i = \text{ECT}_{\text{right}(i)}$.
- Left(i) $\cap \, \Theta' = \emptyset$ and $i \in \Theta'$. In this case Θ' starts by the activity i and the maximum duration of the set Θ' is $\text{p}_i + \Sigma \text{P}_{\text{right}(i)}$. Therefore $\text{ECT}_i = \text{est}_i + \text{p}_i + \Sigma \text{P}_{\text{right}(i)}$.

- Left$(i) \cap \Theta' \neq \emptyset$. The schedule of $\Theta' \cap \text{Left}(i)$ can end at most at $\text{ECT}_{\text{left}(i)}$ and the processing of the rest of the set Θ' can take $p_i + \Sigma P_{\text{right}(i)}$ maximum. Thus $\text{ECT}_i = \text{ECT}_{\text{left}(i)} + p_i + \Sigma P_{\text{right}(i)}$.

Thanks to their recursive nature, values ECT and ΣP can be computed within the usual operations with balanced binary search trees without changing their time complexities. The following table summarizes time complexities of different operations with a Θ-tree:

Operation	Time Complexity
$\Theta := \emptyset$	$O(1)$ or $O(n \log n)$
$\Theta := \Theta \cup \{i\}$	$O(\log n)$
$\Theta := \Theta \setminus \{i\}$	$O(\log n)$
ECT_Θ	$O(1)$
$\text{ECT}_{\Theta \setminus \{i\}}$	$O(\log n)$

3   Overload Checking Using Θ-Tree

Let us consider an arbitrary set $\Omega \subseteq T$. Overload rule says that if the set Ω cannot be processed within its time bounds then no solution exists:

$$\text{lct}_\Omega - \text{est}_\Omega < p_\Omega \quad \Rightarrow \quad \text{fail}$$

Let us suppose for a while that we are given an activity $i \in T$ and we want to check this rule only for these sets $\Omega \subseteq T$ which have $\text{lct}_\Omega = \text{lct}_i$. Now consider a set Θ:

$$\Theta = \{j, \ j \in T \ \& \ \text{lct}_j \leq \text{lct}_i\}$$

Overloaded set Ω with $\text{lct}_\Omega = \text{lct}_i$ exists if and only if $\text{ECT}_\Theta > \text{lct}_i = \text{lct}_\Theta$. The idea of an algorithm is to gradually increase the set Θ by increasing the lct_Θ. For each lct_Θ we check whether $\text{ECT}_\Theta > \text{lct}_\Theta$ or not.

$\Theta := \emptyset$;
for $i \in T$ in ascending order of lct_i **do begin**
    $\Theta := \Theta \cup \{i\}$;
    if $\text{ECT}_\Theta > \text{lct}_i$ **then**
        fail; { No solution exists}
end;

Time complexity of this algorithm is $O(n \log n)$: the activities have to be sorted and n-times an activity is inserted into the set Θ.

4   Not-First/Not-Last Using Θ-Tree

Not-first and not-last are two symmetric propagation algorithms for a unary resource. From these two, we will consider only the not-last algorithm.

Let us consider a set $\Omega \subseteq T$ and an activity $i \in (T \setminus \Omega)$. The activity i cannot be scheduled after the set Ω (*i.e.* i is not last within $\Omega \cup \{i\}$) if:

$$\mathrm{est}_\Omega + \mathrm{p}_\Omega > \mathrm{lct}_i - \mathrm{p}_i \tag{2}$$

In that case, at least one activity from the set Ω must be scheduled after the activity i. Therefore the value lct_i can be updated:

$$\mathrm{lct}_i := \min \left\{ \mathrm{lct}_i, \ \max \left\{ \mathrm{lct}_j - \mathrm{p}_j, \ j \in \Omega \right\} \right\} \tag{3}$$

There are two versions of the not-first/not-last algorithms: [ ] and [ ]. Both of them have time complexity $O(n^2)$. The first algorithm [ ] finds all the reductions resulting from the previous rules in one pass. Still, after this propagation, next run of the algorithm may find more reductions (not-first and not-last rules are not idempotent). Therefore the algorithm should be repeated until no more reduction is found (*i.e.* a fixpoint is reached). The second algorithm [ ] is simpler and faster, but more iterations of the algorithm may be needed to reach a fixpoint.

The algorithm presented here can also need more iteration to reach a fixpoint then the algorithm [ ] maybe even more than the algorithm [ ]. However, time complexity is reduced from $O(n^2)$ to $O(n \log n)$.

Let us suppose that we have chosen a particular activity i and now we want to update the lct_i according to the rule not-last. To really achieve some change of lct_i using the rule ( ), the set Ω must fulfil the following property:

$$\max \left\{ \mathrm{lct}_j - \mathrm{p}_j, \ j \in \Omega \right\} < \mathrm{lct}_i$$

Therefore:

$$\Omega \subseteq \left\{ j, \ j \in T \ \& \ \mathrm{lct}_j - \mathrm{p}_j < \mathrm{lct}_i \ \& \ j \neq i \right\}$$

We will use the same trick as [ ]: lets not slow down the algorithm by searching for the *best* update of lct_i. Rather, find *some* update: if lct_i can be updated better, let it be done in the next run of the algorithm. Therefore our goal is to update lct_i to $\max \left\{ \mathrm{lct}_j - \mathrm{p}_j, \ j \in T \ \& \ \mathrm{lct}_j - \mathrm{p}_j < \mathrm{lct}_i \ \& \ j \neq i \right\}$.

Let us define the set Θ:

$$\Theta = \left\{ j, \ j \in T \ \& \ \mathrm{lct}_j - \mathrm{p}_j < \mathrm{lct}_i \right\}$$

Thus: the lct_i can be changed according to the rule not-last if and only if there is some set $\Omega \subseteq (\Theta \setminus \{i\})$ for which the inequality ( ) holds:

$$\mathrm{est}_\Omega + \mathrm{p}_\Omega > \mathrm{lct}_i - \mathrm{p}_i \tag{4}$$

The only problem is to decide whether such a set Ω exists or not.

Let us recall the definition ( ) of ECT and use it on the set $\Theta \setminus \{i\}$:

$$\mathrm{ECT}_{\Theta \setminus \{i\}} = \max \left\{ \mathrm{est}_\Omega + \mathrm{p}_\Omega, \ \Omega \subseteq \Theta \setminus \{i\} \right\}$$

Notice, that $\mathrm{ECT}_{\Theta \setminus \{i\}}$ is exactly the maximum value which can be on the left side of the inequality ( ). Therefore there is a set Ω for which the inequality ( ) holds if and only if:

$$\mathrm{ECT}_{\Theta \setminus \{i\}} > \mathrm{lct}_i - \mathrm{p}_i$$

The algorithm proceeds as follows. Activities i are taken in the ascending order of lct_i. For each one activity i the set Θ is computed using the set Θ of previous activity i. Then $\text{ECT}_{\Theta\backslash\{i\}}$ is checked and lct_i is eventually updated:

```
1   for i ∈ T do
2       lct'_i := lct_i ;
3    Θ := ∅ ;
4    Q := queue of all activities j ∈ T in ascending order of lct_j − p_j ;
5    for i ∈ T in ascending order of lct_i do begin
6        while lct_i > lct_Q.first − p_Q.first do begin
9            j := Q. first ;
10           Θ := Θ ∪ {j} ;
11           Q. dequeue ;
12       end ;
13       if ECT_{Θ\{i}} > lct_i − p_i then
14           lct'_i := min {lct_j − p_j, lct'_i} ;
15   end ;
16   for i ∈ T do
17       lct_i := lct'_i ;
```

Lines 9–11 are repeated n times maximum, because each time an activity is removed from the queue. Check on the line 13 can be done in $O(\log n)$. Therefore the time complexity of the algorithm is $O(n \log n)$.

Without changing the time complexity, the algorithm can be slightly improved: the not-last rule can be also checked for the activity j just before the insertion of the activity j into the set Θ (*i.e.* after the line 6):

```
7        if ECT_Θ > lct_Q.first − p_Q.first then
8            lct'_Q.first := lct_j − p_j ;
```

This modification can in some cases save few iterations of the algorithm.

5   Detectable Precedences

An idea of detectable precedences was introduced in [ ] for a *batch resource with sequence dependent setup times*, which is an extension of a unary resource.

The figure   is taken from [ ]. It shows a situation when neither edge-finding nor the not-first/not-last algorithm can change any time bound, but a propagation of detectable precedences can.

Edge-finding algorithm recognizes that the activity A must be processed before the activity C, i.e. $A \ll C$, and similarly $B \ll C$. Still, each of these precedences alone is weak: they do not enforce any change of any time bound. However, from the knowledge $\{A, B\} \ll C$ we can deduce $\text{est}_C \geq \text{est}_A + p_A + p_B = 21$.

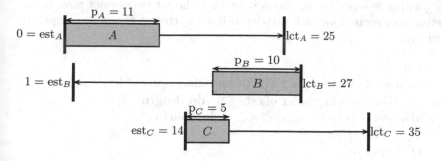

Fig. 2. A sample problem for detectable precedences

A precedence $j \ll i$ is called *detectable*, if it can be "discovered" only by comparing the time bounds of these two activities:

$$\text{est}_i + p_i > \text{lct}_j - p_j \quad \Rightarrow \quad j \ll i \tag{5}$$

Notice that both precedences $A \ll C$ and $B \ll C$ are detectable.

There is a simple quadratic algorithm, which propagates all known precedences on a resource. For each activity i build a set $\Omega = \{j \in T, \ j \ll i\}$. Note that precedences $j \ll i$ can be of any type: detectable precedences, search decisions or initial constraints. Using such set Ω, est_i can be adjusted: $\text{est}_i := \max\{\text{est}_i, \text{ECT}_\Omega\}$ because $\Omega \ll i$.

```
for  i ∈ T  do begin
    m := −∞;
    for  j ∈ T in non-decreasing order of est_j  do
        if  j ≪ i  then
            m := max{m, est_j} + p_j;
    est_i := max{m, est_i};
end;
```

A symmetric algorithm adjusts lct_i.

However, propagation of only detectable precedences can be done within $O(n \log n)$. Let Θ be the following set of activities:

$$\Theta = \{j, \ j \in T \ \& \ \text{est}_i + p_i > \text{lct}_j - p_j\}$$

Thus $\Theta \setminus \{i\}$ is a set of all activities which must be processed before the activity i because of detectable precedences. Using the set $\Theta \setminus \{i\}$ the value est_i can be adjusted:

$$\text{est}_i := \max\{\text{est}_i, \ \text{ECT}_{\Theta \setminus \{i\}}\}$$

There is also a symmetric rule for precedences $j \gg i$, but we will not consider it here, nor the resulting symmetric algorithm.

An algorithm is based on an observation that the set Θ does not have to be constructed from scratch for each activity i. Rather, the set Θ can be computed incrementally.

```
1    Θ := ∅;
2    Q := queue of all activities j ∈ T in ascending order of lctⱼ − pⱼ;
3    for i ∈ T in ascending order of estᵢ + pᵢ do begin
4        while estᵢ + pᵢ > lctQ.first − pQ.first do begin
5            Θ := Θ ∪ {Q.first};
6            Q. dequeue;
7        end;
8        est′ᵢ := max {estᵢ, ECTΘ\{i}};
9    end;
10   for i ∈ T do
11       estᵢ := est′ᵢ;
```

Initial sorts takes $O(n \log n)$. Lines 5 and 6 are repeated n times maximum, because each time an activity is removed from the queue. Line 8 can be done in $O(\log n)$. Therefore the time complexity of the algorithm is $O(n \log n)$.

6   Properties of Detectable Precedences

There is an interesting connection between edge-finding algorithm and detectable precedences:

Proposition 1. *When edge-finding is unable to find any further time bound adjustment then all precedences which edge-finding found are detectable.*

Proof. First, brief introduction of the edge-finding algorithm. Consider a set $\Omega \subseteq T$ and an activity $i \notin \Omega$. The activity i has to be scheduled after all activities from Ω, if:

$$\forall \Omega \subset T, \forall i \in (T \setminus \Omega): \quad \min(\text{est}_\Omega, \text{est}_i) + p_\Omega + p_i > \text{lct}_\Omega \implies \Omega \ll i \quad (6)$$

Once we know that the activity i must be scheduled after the set Ω, we can adjust est_i:

$$\Omega \ll i \implies \text{est}_i := \max(\text{est}_i, \max\{\text{est}_{\Omega'} + p_{\Omega'}, \Omega' \subseteq \Omega\}) \quad (7)$$

Edge-finding algorithm propagates according to this rule and its symmetric version. There are several implementations of edge-finding algorithm, two different quadratic algorithms can be found in [,], [ ] presents a $O(n \log n)$ algorithm.

Let us suppose that edge-finding proved $\Omega \ll i$. We will show that for an arbitrary activity $j \in \Omega$, edge-finding made est_i big enough to make the precedence $j \ll i$ detectable.

Edge-finding proved $\Omega \ll i$ so the condition ( ) was true before the filtering:

$$\min\left(\text{est}_\Omega,\ \text{est}_i\right) + p_\Omega + p_i > \text{lct}_\Omega$$

However, increase of any est or decrease of any lct cannot invalidate this condition, therefore it has to be valid now. And so:

$$\text{est}_\Omega > \text{lct}_\Omega - p_\Omega - p_i \tag{8}$$

Because edge-finding is unable to further change any time bound, according to ( ) we have:

$$\text{est}_i \geq \max\{\text{est}_{\Omega'} + p_{\Omega'},\ \Omega' \subseteq \Omega\}$$
$$\text{est}_i \geq \text{est}_\Omega + p_\Omega$$

In this inequality, est_Ω can be replaced by the right side of the inequality ( ):

$$\text{est}_i > \text{lct}_\Omega - p_\Omega - p_i + p_\Omega$$
$$\text{est}_i > \text{lct}_\Omega - p_i$$

And $\text{lct}_\Omega \geq \text{lct}_j$ because $j \in \Omega$:

$$\text{est}_i > \text{lct}_j - p_i$$
$$\text{est}_i + p_i > \text{lct}_j - p_j$$

So the condition ( ) holds and the precedence $j \ll i$ is detectable.

The proof for the precedences resulting from $i \ll \Omega$ is symmetrical.   □

Previous proposition has also one negative consequence. Papers [, ,] mention the following improvement of the edge-finding: whenever $\Omega \ll i$ is found, propagate also $j \ll i$ for all $j \in \Omega$ (*i.e.* change also lct_j). However, these precedences are detectable and so the second run of the edge-finding would propagate them anyway. Therefore such improvement can save some iterations of the edge-finding, but do not enforce better pruning in the end.

Several authors (*e.g.* [,]) suggest to compute a transitive closure of precedences. Detectable precedences has also an interesting property in such transitive closure.

Lets us call a precedence $i \ll j$ *propagated* iff the activities i and j fulfill following two inequalities:

$$\text{est}_j \geq \text{est}_i + p_i$$
$$\text{lct}_i \leq \text{lct}_j - p_j$$

Note that edge-finding and precedence propagation algorithm make all known precedences propagated. Thus all detectable precedences become propagated, but not all propagated precedences have to be detectable.

The following proposition has an easy consequence: detectable precedences can be skipped when computing a transitive closure.

Proposition 2. *Let $a \ll b$, $b \ll c$ and one of these precedences is detectable and the second one propagated. Then the precedence $a \ll c$ is detectable.*

Proof. We distinguish two cases:

1. **$a \ll b$ is detectable and $b \ll c$ is propagated.** Because the precedence $b \ll c$ is propagated:
$$\mathrm{est}_c \geq \mathrm{est}_b + \mathrm{p}_b$$
and because the precedence $a \ll b$ is detectable:
$$\mathrm{est}_b + \mathrm{p}_b > \mathrm{lct}_a - \mathrm{p}_a$$
$$\mathrm{est}_c > \mathrm{lct}_a - \mathrm{p}_a$$

   Thus the precedence $a \ll c$ is detectable.
2. **$a \ll b$ is propagated and $b \ll c$ is detectable.** Because the precedence $a \ll b$ is propagated:
$$\mathrm{lct}_b - \mathrm{p}_b \geq \mathrm{lct}_a$$
And because the second precedence $b \ll c$ is detectable:
$$\mathrm{est}_c + \mathrm{p}_c > \mathrm{lct}_b - \mathrm{p}_b$$
$$\mathrm{est}_c + \mathrm{p}_c > \mathrm{lct}_a$$

   Once again, the precedence $a \ll c$ is detectable. □

7 Experimental Results

A reduction of a time complexity of an algorithm is generally a "good idea". However for small n, a simple and short algorithm can outperform a complicated one with better time complexity. Therefore it is reasonable to ask whether it is the case of the new not-first/not-last algorithm. Another question is a filtering power of the detectable precedences. The following benchmark should bring answers to these questions.

The benchmark is based on a computation of destructive lower bounds for several jobshop benchmark problems taken from OR library [ ]. Destructive lower bound is a minimum length of the schedule, for which we are not able to proof infeasibility without backtracking. Lower bounds computation is a good benchmark problem because there is no influence of a search heuristic. Four different destructive lower bounds where computed. Lower bound LB1 is computed using edge-finding algorithm [ ] and new version of not-first/not-last:

```
repeat
  repeat
    edge finding;
  until no more propagation;
  not-first/not-last;
until no more propagation;
```

---

[2] Note that it is a quadratic algorithm.

Detectable precedences were used for computation of LB2:

```
repeat
  repeat
    repeat
      detectable precedences;
    until no more propagation;
    not-first/not-last;
  until no more propagation;
  edge finding;
until no more propagation;
```

Note that the order of the filtering algorithms affects total run time, however it does not influence the resulting fixpoint. The reason is that even after an arbitrary propagation, all used reduction rules remain valid and propagates the same or even better.

Another two lower bounds where computed using shaving technique as suggested in [ ]. Shaving is similar to the proof by a contradiction. We choose an activity i, limit its est_i or lct_i and propagate. If an infeasibility is found, then the limitation was invalid and so we can decrease lct_i or increase est_i. Binary search is used to find the best shave. To limit CPU time, shaving was used for each activity only once.

Often detectable precedences improve the filtering, however do not increase the lower bound. Therefore a new column R is introduced. After the propagation with LB as upper bound, domains are compared with a state when only binary precedences were propagated. The result is an average domain size in percents.

CPU time was measured only for shaving (columns T, T1–T3 in seconds). It is the time needed to proof the lower bound, $i.e.$ propagation is done twice: with the upper bound LB and LB-1. Times T1–T3 shows running time for different implementations of the algorithm not-first/not-last: T1 is the new algorithm, T2 is the algorithm [ ] and T3 is the algorithm [ ].

To improve the readability, when LB1=LB2, then a dash is reported in LB2. The same rule was applied to shaving lower bounds and columns R.

The table shows that detectable precedences improved the filtering, but not much. However there is another interesting point: detectable precedences speed up the propagation, compare T and T1 $e.g.$ for ta21. The reason is that detectable precedences are able to "steal" a lot of work from edge-finding and do it faster.

Quite surprisingly, the new not-first/not-last algorithm is about the same fast as [ ] for $n = 10$, for bigger n it begins to be be faster. Note that the most filtering is done by the detectable precedences, therefore the speed of the not-first/not-last algorithm has only minor influence to the total time. For $n = 100$, time T1 is approximately only one half of T.

---

[3] Benchmarks were performed on Intel Pentium Centrino 1300MHz.

Prob.	Size		LB1	LB2	Shaving EF+NFNL			Shaving DP+NFNL+EF				
					LB	R	T	LB	R	T1	T2	T3
abz5	10	x 10	1126	1127	1195	76.81	1.328	1196	76.95	1.395	1.392	1.447
abz6	10	x 10	889	890	940	58.76	1.526	941	66.92	1.728	1.713	1.800
orb01	10	x 10	975	-	1017	76.50	1.813	-	76.15	1.747	1.767	1.830
orb02	10	x 10	812	815	865	56.66	1.669	869	84.52	1.456	1.451	1.520
ft10	10	x 10	858	868	910	51.11	1.674	911	67.22	1.582	1.582	1.647
la21	15	x 10	1033	-	1033	72.99	0.757	-	72.93	0.749	0.761	0.814
la22	15	x 10	913	-	924	57.81	3.462	925	69.53	3.454	3.511	3.700
la36	15	x 15	1233	-	1267	87.77	5.720	-	87.66	5.321	5.398	5.709
la37	15	x 15	1397	-	1397	66.23	2.723	-	66.21	2.458	2.511	2.646
ta01	15	x 15	1190	1193	1223	73.62	9.819	1224	71.38	9.052	9.170	9.563
ta02	15	x 15	1167	-	1210	84.38	7.561	-	75.76	7.034	7.144	7.490
la26	20	x 10	1218	-	1218	97.78	0.812	-	-	0.735	0.764	0.899
la27	20	x 10	1235	-	1235	91.06	1.063	-	-	0.887	0.923	1.056
la29	20	x 10	1119	-	1119	79.57	3.552	-	79.53	3.372	3.444	3.702
abz7	20	x 15	651	-	651	62.15	3.522	-	62.08	3.204	3.288	3.537
abz8	20	x 15	608	-	621	85.88	12.34	-	85.51	11.94	12.10	12.76
ta11	20	x 15	1269	-	1295	73.74	18.58	-	70.61	14.51	14.79	15.50
ta12	20	x 15	1314	-	1336	86.63	21.06	-	86.22	17.25	17.61	18.59
ta21	20	x 20	1508	-	1546	82.62	43.60	-	80.96	37.86	38.85	40.02
ta22	20	x 20	1441	-	1499	82.84	31.03	1501	92.85	24.92	25.31	26.41
yn1	20	x 20	784	-	816	86.66	28.96	-	86.33	26.05	26.54	27.73
yn2	20	x 20	819	825	841	84.91	24.29	842	88.61	22.34	22.71	23.80
ta31	30	x 15	1764	-	1764	96.58	5.967	-	-	4.735	5.067	5.905
ta32	30	x 15	1774	-	1774	96.25	8.778	-	-	6.662	6.994	7.882
swv11	50	x 10	2983	-	2983	95.55	34.90	-	-	17.46	19.37	22.07
swv12	50	x 10	2972	-	2972	85.75	37.39	-	-	21.24	23.10	25.73
ta52	50	x 15	2756	-	2756	99.92	19.29	-	-	12.00	13.72	16.78
ta51	50	x 15	2760	-	2760	99.71	19.03	-	-	11.52	13.27	16.12
ta71	100	x 20	5464	-	5464	99.99	262.8	-	-	125.7	154.3	191.9
ta72	100	x 20	5181	-	5181	99.97	263.2	-	-	126.5	155.4	193.1

Table 1. Destructive Lower Bounds

Acknowledgements

Author would like to thank Roman Barták and to the three anonymous referees for their helpful comments and advises. The work of the author has been supported by the Grant Agency of the Czech Republic under the contract no. 201/04/1102.

References

[1] OR library. URL

[2] Philippe Baptiste and Claude Le Pape. Edge-finding constraint propagation algorithms for disjunctive and cumulative scheduling. In *Proceedings of the Fifteenth Workshop of the U. K. Planning Special Interest Group*, 1996. ,

[3] Peter Brucker. Complex scheduling problems, 1999. URL

[4] Jacques Carlier and Eric Pinson. Adjustments of head and tails for the job-shop problem. *European Journal of Operational Research*, 78:146–161, 1994. ,

[5] Yves Caseau and Francois Laburthe. Disjunctive scheduling with task intervals. In *Technical report, LIENS Technical Report 95-25*. Ecole Normale Supérieure Paris, Françe, 1995.

[6] W. Nuijten, F. Foccaci, P. Laborie. Solving scheduling problems with setup times and alternative resources. In *Proceedings of the 4th International Conference on AI Planning and Scheduling, AIPS'00*, pages 92–101, 2000.

[7] Paul Martin and David B. Shmoys. A New Approach to Computing Optimal Schedules for the Job-Shop Scheduling Problem. In W. H. Cunningham, S. T. McCormick, and M. Queyranne, editors, *Proceedings of the 5th International Conference on Integer Programming and Combinatorial Optimization, IPCO'96*, pages 389–403, Vancouver, British Columbia, Canada, 1996. , ,

[8] Claude Le Pape, Philippe Baptiste and Wim Nuijten. *Constraint-Based Scheduling: Applying Constraint Programming to Scheduling Problems.* Kluwer Academic Publishers, 2001.

[9] Philippe Torres and Pierre Lopez. On not-first/not-last conditions in disjunctive scheduling. *European Journal of Operational Research*, 1999. ,

[10] Petr Vilím. Batch processing with sequence dependent setup times: New results. In *Proceedings of the 4th Workshop of Constraint Programming for Decision and Control, CPDC'02*, Gliwice, Poland, 2002.

[11] Armin Wolf. Pruning while sweeping over task intervals. In *Principles and Practice of Constraint Programming - CP 2003*, Kinsale, Ireland, 2003.

Problem Decomposition for Traffic Diversions

Quanshi Xia[1], Andrew Eremin[1], and Mark Wallace[2]

[1] IC-Parc
Imperial College London
London SW7 2AZ, UK
{q.xia,a.eremin}@imperial.ac.uk
[2] School of Business Systems
Monash University
Clayton, Vic 3800
Australia
mark.wallace@infotech.monash.edu.au

Abstract. When a major road traffic intersection is blocked, vehicles should be diverted from the incoming roads in such a way as to avoid the roads on the diversions from also becoming over-congested. Assuming different diversions may use partly the same roads, the challenge is to satisfy the following traffic flow constraint: ensure that even in the worst case scenario, the diversions can accommodate the same volume of traffic as the blocked intersection.

The number of diversions increases quadratically with the number of roads at the intersection. Moreover any road may be used by any subset of the diversions - thus the number of *worst cases* can grow exponentially with the number of diversions.

This paper investigates two different approaches to the problem, describes their implementation on the hybrid MIP/CP software platform ECL^{i}PS^e, and presents benchmark results on a set of test cases.

1 Introduction

Cities are becoming more congested, but luckily road management technology - sensing, signs, lights *etc.* - is improving dramatically. We now have the opportunity to apply planning and optimisation techniques to road management to reduce congestion and optimise journey times.

The problem of diversions tackled in this paper is an artificial one, in that some of the assumptions do not hold on the ground. However the problem appears in the context of a larger system for traffic management, and its solution is in practical use today.

The problem focuses on planning diversions to get around a blocked junction or interchange, where a number of routes meet each other. Assuming no information about the destinations of vehicles on the road, the aim is to ensure that every incoming route is linked to every outgoing route by a diversion.

However the requirement is also to ensure that whatever traffic might have been flowing through the junction, the diversion routes are sufficiently major to

J.-C. Régin and M. Rueher (Eds.): CPAIOR 2004, LNCS 3011, pp. 348–   , 2004.
© Springer-Verlag Berlin Heidelberg 2004

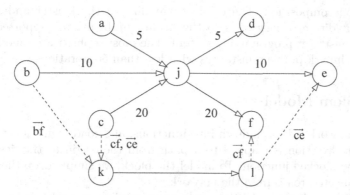

Fig. 1. A Simple Junction

cope with them. For the purposes of this problem, we ignore any traffic that happens to be using the diversion roads for other journeys, that would not have passed through the blocked junction.

The problem is scientifically interesting because, until all the diversions have been specified, it is not possible to tell what is the maximum possible traffic flow that could be generated along any given road on a diversion.

Let us illustrate this with an example: The junction **j** has three incoming roads, from **a**, **b** and **c** and three outgoing, to **d**, **e** and **f**. Each road has a grade, which determines the amount of traffic it can carry. These amounts are 5 for aj and jd, 10 for bj and je and 20 for cj and jf.

The diversion cf from **c** to **f** clearly needs to be able to carry a traffic quantity of 20. Assume that this diversion shares a road kl with the diversion bf from **b** to **f**. The total traffic on both diversions, in the worst case, is still only 20 because the diverted routes both used the road jf, which has a capacity of 20.

However if the diversion ce from **c** to **e** also uses the road kl, then in the worst case the traffic over kl goes up to 30. This case arises when there is a flow of 10 from **b** to **j**, a flow of 20 from **c** to **j**, a flow of 10 from **j** to **e**, and another flow of 20 from **j** to **f**. This means that there may potentially be end-to-end flows of 10 from **b** to **f**, from **c** to **e** and from **c** to **f**.

The total number of diversions that must be set up in this case is 9, a diversion from each incoming origin to each outgoing destination. In general, then, the number of diversions grows quadratically in the size of the junction. Moreover any subset of these diversions may intersect, so the number of worst case scenarios to be calculated is potentially exponential in the number of diversions!

The final aspect of the problem is to find actual routes for all the diversions, which satisfy all the worst case scenarios. Given all the above possibilities, a generate and test approach may need to explore a huge number of routes.

In this paper we present several approaches to solving the problem. The first is a global integer/linear model, which can solve smaller problem instances but grows in memory usage and execution time for larger problem instances which limits its scalability. The next three are increasingly sophisticated versions of a problem decomposition. The efficient handling of the master and subproblems, and addition of new rows to the master problem are supported by the ECL^iPS^e constraint programming system. The most sophisticated model solves all the benchmark problem instances with less than 50 iterations.

2   Problem Model

The road network is broken down into junctions, and road segments connecting them. Each diversion is mapped to a path from the origin to the destination, avoiding the blocked junction. To model the block, we simply drop the junction and its connected roads from the network.

The challenge is to model the capacity constraints on each road segment in each path in the network. For each road segment, the sum of the traffic flows on all the routes whose diversions pass through that road segment must be accommodated. The road segment is over-congested if, in the worst case, this sum exceeds the capacity of the road segment.

In this paper we shall write constants in lower case (e.g. e, $edge$), we shall write variables starting with an upper case letter (e.g. Q, $Quantity$), variable arrays are subscripted (e.g. with a single subscript, Q_e, or with multiple subscripts Q_{fe}), we shall write functions using brackets (e.g. dest(f)). We use **bold** identifiers to denote sets (e.g. **E**). For example, to say that edge e belongs to the set of edges **E**, we write $e \in$ **E**. Set-valued variables and functions are also written in **bold** font.

We formalism the problem in terms of a network, with *edges* representing road segments, and *nodes* representing junctions.

The network comprises a set of edges, **E** and a set of nodes, **N**. Each edge $e \in$ **E** has a capacity cap(e). Allowing for one-way traffic, we associate a direction with each edge (two way roads are therefore represented by two edges, one in each direction). The edge from origin o into the junction has capacity ocap(o) and the edge leaving the junction and entering the destination d has capacity dcap(d).

For each node n there is a set of edges **IN**(n) entering n and a set of edges **OUT**(n) leaving n.

The set of traffic flows to be diverted is **F**. Each flow $f \in$ **F** has an origin orig(f), a destination dest(f) and a maximum flow quantity quan(f).

quan(f) is limited by the size of the roads of the diverted flow, into the junction from the origin and out from the junction to the destination. Thus, quan$(f) = min\{$ocap$($orig$(f))$, dcap$($dest$(f))\}$.

The diversion for the flow f is a path **DIV**$_f$ joining its origin orig(f) to its destination dest(f). (Assuming no cycles, we model the path as a set of edges, thus **DIV**$_f$ is a set-valued variable.)

The awkward constraints are the capacity constraints. For this purpose we elicit the worst case for each edge, using an optimisation function.

Consider the total flow diverted through an edge e: for each flow f there is a non-negative flow quantity $Q_{fe} \le$ quan(f) diverted through e. For all flows f

that are not diverted through the edge e this quantity is 0, while for all flows in the set of flows diverted through an edge e, $\mathbf{F}_e = \{f : e \in \mathbf{DIV}_f\}$, there is a non-negative flow quantity.

The total diverted flow DQ_e through the edge e is therefore $DQ_e = \sum_{f \in \mathbf{F}_e} Q_{fe}$. Clearly it must be within the edge capacity: $\mathrm{cap}(e) \geq DQ_e$.

The maximum total diverted flow through an edge is in general less than the sum of the maxima, $\mathrm{quan}(f)$ of all the individual flows. Indeed the maximum quantity of the *sum* of *all* the flows which have the same origin o is constrained by $\mathrm{ocap}(o) \geq \sum_{f:\mathrm{orig}(f)=o} Q_{fe}$. Similarly for destination d: $\mathrm{dcap}(d) \geq \sum_{f:\mathrm{dest}(f)=d} Q_{fe}$.

The worst case for capacity constraint on edge e is when DQ_e is maximized, by changing the flows through the original junction. The resulting constraint is $\mathrm{cap}(e) \geq \max_{Q_{fe}} \sum_{f \in \mathbf{F}_e} Q_{fe}$.

3   Formulation as a MIP Problem

For the MIP model binary (0/1) variables X_{fe} and continuous variables Q_{fe} are introduced. For each flow f and edge e, $X_{fe} = 1$ if and only if flow f is diverted through edge e. Thus, $\mathbf{DIV}_f = \{e : X_{fe} = 1\}$

The problem is to choose diversions (by setting the values of the variables X_{fe}) such that all the worst case capacity constraints are satisfied. We introduce an optimisation expression: $\min_{X_{fe}} \sum_{f \in \mathbf{F}} \sum_{e \in \mathbf{E}} X_{fe}$, which precludes cycles in any diversion since optimisation would set the flow through any cycle to zero and minimizes the total diversion path length.

$$\min_{X_{fe}} \sum_{f \in \mathbf{F}} \sum_{e \in \mathbf{E}} X_{fe}$$

$$st. \begin{cases} \forall f \in \mathbf{F} : \begin{cases} \forall n \in \mathbf{N} \setminus \{\mathrm{orig}(f), \mathrm{dest}(f)\} : \sum_{e \in \mathbf{IN}(n)} X_{fe} = \sum_{e \in \mathbf{OUT}(n)} X_{fe} \\ n = \mathrm{orig}(f) : \sum_{e \in \mathbf{OUT}(n)} X_{fe} = 1 \\ n = \mathrm{dest}(f) : \sum_{e \in \mathbf{IN}(n)} X_{fe} = 1 \end{cases} \\ \forall e \in \mathbf{E} : \mathrm{cap}(e) \geq \begin{cases} \max_{Q_{fe}} \sum_{f \in \mathbf{F}} X_{fe} * Q_{fe} \\ st. \begin{cases} \forall o \in \mathrm{orig}(\mathbf{F}) : \mathrm{ocap}(o) \geq \sum_{f:\mathrm{orig}(f)=o} Q_{fe} \\ \forall d \in \mathrm{dest}(\mathbf{F}) : \mathrm{dcap}(d) \geq \sum_{f:\mathrm{dest}(f)=d} Q_{fe} \end{cases} \end{cases} \\ X_{fe} \in \{0,1\}, \quad \mathrm{quan}(f) \geq Q_{fe} \geq 0 \end{cases} \tag{1}$$

The embedded optimisation for each edge e can be linearized by using the Karush-Kuhn-Tucker condition [ ]. First we dualise it, introducing dual variables D_{oe} and D_{de}

$$
\text{cap}(e) \geq
\begin{cases}
\min\limits_{D_{oe}, D_{de}} \sum\limits_{o \in \text{orig}(\mathbf{F})} \text{ocap}(o) * D_{oe} + \sum\limits_{d \in \text{dest}(\mathbf{F})} \text{dcap}(d) * D_{de} \\
\text{st.} \begin{cases} \forall f \in \mathbf{F}, o = \text{orig}(f), d = \text{dest}(f): \quad D_{oe} + D_{de} \geq X_{fe} \\ D_{oe} \in \{0, 1\}, \qquad D_{de} \in \{0, 1\} \end{cases}
\end{cases}
\tag{2}
$$

Note that the upper bounds on the variables Q_{fe} are implicit from the origin and destination constraints and variable non-negativity; in forming the dual problem we have dropped these redundant bounds. Further since the coefficients of the variables D_{oe}, D_{de} in the cost function to be minimized in the dual are strictly positive and the variables non-negative an upper bound of 1 can be deduced for the value of all dual variables in any dual optimal solution, and thus in any feasible solution to the original problem, from the dual constraints and the upper bounds of X_{fe}. Moreover since the constraints of ( ) are totally unimodular any basic feasible solution and hence any basic optimal feasible solution is integral, and D_{oe}, D_{de} reduce to binary variables.

We introduce slack variables SQ_{oe} and SQ_{de} for the constraints in the primal, and dual slack variables SD_{fe} in the dual. We can now replace the embedded maximization problem for each edge e by the following constraints:

$$
\begin{cases}
\text{cap}(e) \geq \sum\limits_{f \in \mathbf{F}} Q_{fe} \\
\forall o \in \text{orig}(\mathbf{F}): \quad \text{ocap}(o) = \sum\limits_{f: \text{orig}(f) = o} Q_{fe} + SQ_{oe} \\
\forall d \in \text{dest}(\mathbf{F}): \quad \text{dcap}(d) = \sum\limits_{f: \text{dest}(f) = d} Q_{fe} + SQ_{de} \\
\forall f \in \mathbf{F}, o = \text{orig}(f), d = \text{dest}(f): \quad D_{oe} + D_{de} - SD_{fe} = X_{fe} \\
\left. \begin{array}{l} \forall f \in \mathbf{F}: \quad Q_{fe} * SD_{fe} = 0 \\ \forall o \in \text{orig}(\mathbf{F}): \quad SQ_{oe} * D_{oe} = 0 \\ \forall d \in \text{dest}(\mathbf{F}): \quad SQ_{de} * D_{de} = 0 \end{array} \right\} \text{ complementarity} \\
X_{fe} \in \{0, 1\}, \quad D_{oe} \in \{0, 1\}, \quad D_{de} \in \{0, 1\}, \quad SD_{fe} \in \{0, 1\} \\
\text{quan}(f) \geq Q_{fe} \geq 0, \quad \text{ocap}(o) \geq SQ_{oe} \geq 0, \quad \text{dcap}(d) \geq SQ_{de} \geq 0
\end{cases}
\tag{3}
$$

Since D_{oe}, D_{de}, SD_{fe} are binary (0/1) variables, the complementarity constraints can be linearized to obtain the mixed integer linear programming model which can be solved by MIP solvers:

$$\min_{X_{fe}} \sum_{f \in F} \sum_{e \in E} X_{fe}$$

$$st. \begin{cases} \forall f \in \mathbf{F}: \begin{cases} \forall n \in \mathbf{N} \setminus \{\mathbf{orig}(f), \mathbf{dest}(f)\}: \sum_{e \in \mathbf{IN}(n)} X_{fe} = \sum_{e \in \mathbf{OUT}(n)} X_{fe} \\ n = \mathbf{orig}(f): \sum_{e \in \mathbf{OUT}(n)} X_{fe} = 1 \\ n = \mathbf{dest}(f): \sum_{e \in \mathbf{IN}(n)} X_{fe} = 1 \end{cases} \\ \forall e \in \mathbf{E}: \begin{cases} \mathbf{cap}(e) \geq \sum_{f \in F} Q_{fe} \\ \forall o \in \mathbf{orig}(\mathbf{F}): \mathbf{ocap}(o) = \sum_{f:\mathbf{orig}(f)=o} Q_{fe} + SQ_{oe} \\ \forall d \in \mathbf{dest}(\mathbf{F}): \mathbf{dcap}(d) = \sum_{f:\mathbf{dest}(f)=d} Q_{fe} + SQ_{de} \\ \forall f \in \mathbf{F}, o = \mathbf{orig}(f), d = \mathbf{dest}(f): D_{oe} + D_{de} - SD_{fe} = X_{fe} \\ \forall f \in \mathbf{F}: \mathbf{quan}(f)(1 - SD_{fe}) - Q_{fe} \geq 0 \\ \forall o \in \mathbf{orig}(\mathbf{F}): \mathbf{ocap}(o)(1 - D_{oe}) - SQ_{oe} \geq 0 \\ \forall d \in \mathbf{dest}(\mathbf{F}): \mathbf{dcap}(d)(1 - D_{de}) - SQ_{de} \geq 0 \end{cases} \\ X_{fe} \in \{0, 1\}, \quad D_{oe} \in \{0, 1\}, \quad D_{de} \in \{0, 1\}, \quad SD_{fe} \in \{0, 1\} \\ \mathbf{quan}(f) \geq Q_{fe} \geq 0, \quad \mathbf{ocap}(o) \geq SQ_{oe} \geq 0, \quad \mathbf{dcap}(d) \geq SQ_{de} \geq 0 \end{cases} \quad (4)$$

The resulting performance is summarized in Table   under the column of **MIP**.

4   Formalization Using Decomposition

Most real resource optimisation problems involve different kinds of constraints, which are best handled by different kinds of algorithms. Our traffic diversions problem can be decomposed into parts which are best handled by different constraint solvers.

Both different problem decompositions and the use of different solvers for the resulting components were tried on this problem. For reasons of space, we present just one decomposition, into a master problem and a set of (similar) subproblems. The master problem is a pure integer linear programming which is best handled by a MIP solver. The subproblems are very simple linear programs and well-suited to a linear solver, although they could equally be solved by CP as in [,]. In our approach, CP provides the modelling language and the glue which enables the solvers to communicate, though not the solvers themselves.

4.1   Informal Description of the Decomposition

The original problem ( ) can be treated instead by decomposing it into a multi-commodity flow master problem, and a maximization subproblem for each edge in the network.

The master problem simply assigns a path to each flow. Initially these paths are independent. However as a result of solving the subproblems new constraints are, later, added to the master problem which preclude certain combinations of flows from being routed through the same edge.

Each subproblem takes as an input the path assigned to each flow by the latest solution of the master problem. If the edge associated with the subproblem is e, the relevant flows $\mathbf{F}_e$ are those whose paths are routed through edge e. The subproblem then maximizes the sum of the flows in $\mathbf{F}_e$. If this maximum sum exceeds the capacity of the edge, then a new constraint is created and passed back to the master problem precluding any assignment which routes all the flows in $\mathbf{F}_e$ through the edge e. Although the cuts added to the master problem are formed differently the principle behind this approach is closely related to that of classic Benders decomposition [ ] or its logic-based extension [ ].

4.2   Model Specification

The formalization of this decomposed model uses the same binary variables X_{fe} as the MIP model. Each time the master problem is solved it assigns values (0 or 1) to all these binary variables. For the assignment to X_{fe} returned by the solution of the kth master problem, we write x_{fe}^k.

The subproblems in the current model are linear maximization problems of the kind that typically occurs in production planning, which use the same continuous variables Q_{fe} as the original problem formulation in Section   above.

Accordingly the kth subproblem associated with edge e is simply:

$$\max_{Q_{fe}} \sum_{f \in F} x_{fe}^k * Q_{fe}$$

$$st. \begin{cases} \forall o \in \mathbf{orig}(\mathbf{F}): \ \mathbf{ocap}(o) \geq \sum_{f:\mathrm{orig}(f)=o} Q_{fe} \\ \forall d \in \mathbf{dest}(\mathbf{F}): \ \mathbf{dcap}(d) \geq \sum_{f:\mathrm{dest}(f)=d} Q_{fe} \\ \mathbf{quan}(f) \geq Q_{fe} \geq 0 \end{cases} \tag{5}$$

The solution to the kth subproblem associated with edge e, is a set of flow quantities, which we can write as q_{fe}^k for each flow f.

Suppose the subproblem associated with edge e indeed returns a maximum sum of flows which exceeds $\mathbf{cap}(e)$, i.e. $\sum_{f \in F} q_{fe}^k > \mathbf{cap}(e)$. Then the constraint passed to the $(k+1)^{th}$ master problem from this subproblem is

$$\sum_{f \in \mathbf{F}} x_{fe}^k * (1 - X_{fe}) \geq 1 \tag{6}$$

This constraint ensures that at least one of the flows previously routed through edge e will no longer be routed through e. Therefore, it simply rules out the previous assignment and those assignments with previous assignment as the subset [ ].

The kth master problem has the form:

$$\min_{X_{fe}} \sum_{f \in F} \sum_{e \in E} X_{fe}$$

$$st. \begin{cases} \forall f \in F : \begin{cases} \forall n \in \mathbf{N} \setminus \{\mathbf{orig}(f), \mathbf{dest}(f)\} : \sum_{e \in \mathbf{IN}(n)} X_{fe} = \sum_{e \in \mathbf{OUT}(n)} X_{fe} \\ n = \mathbf{orig}(f) : \sum_{e \in \mathbf{OUT}(n)} X_{fe} = 1 \\ n = \mathbf{dest}(f) : \sum_{e \in \mathbf{IN}(n)} X_{fe} = 1 \end{cases} \\ \text{for certain edges } e \text{ and iterations } j < k : \sum_{f \in F} x_{fe}^j * (1 - X_{fe}) \geq 1 \\ X_{fe} \in \{0, 1\} \end{cases} \quad (7)$$

This model can be solved by completing a branch and bound search at every iteration of the master problem, in order to return the shortest feasible paths, satisfying all the cuts returned from earlier subproblems. If only feasible, rather than shortest, diversions are required, optimality of the master problem solution is not necessary, and the master problem solution can be stopped as soon as an integer feasible solution is found. However, the path constraints, as they stand, admit non-optimal solutions in which there might be cyclic sub-paths in (or even disjoint from) the path. Whilst it is not incorrect to admit such irrelevant cyclic sub-paths, in fact such cycles can easily be eliminated by a pre-processing step between master and subproblem solution since the path produced by removing cycles from a feasible solution to the kth master problem remains feasible. Such a pre-processing step would make the diversion problem be solved more efficiently.

After the current master problem returns a feasible solution, it is then checked by running one or more subproblems, associated with different edges. Naturally if none of the subproblems produced a maximum flow which exceeded the capacity of its edge, then the master problem solution is indeed a solution to the original diversion problem. In this case the algorithm succeeds. If, on the other hand, after a certain number of iterations, the master problem has no feasible solution then the original diversion problem is unsatisfiable. There is no way of assigning diversions to flows that have the capacity to cope with the worst case situation.

The experimental evaluation of this algorithm is given in Table   under the column of **D(naive)**.

5   An Enhanced Decomposition

Under certain circumstances the previous decomposition leads to a very large number of iterations of the master problem, with many cuts added during the iterations. The result is that the master problem becomes bigger and more difficult to solve. Moreover, the master problem has to be solved by branch and bound search at each of a large number of iterations. This has a major impact on run times, as shown in column **D(naive)** of the experiments in Table   .

5.1 Generating Fewer Cuts

A cut that only removes the previous assignment is easy to add, but typically not very strong: a different cut has to be added for each assignment that is ruled out. This may require a very large number of cuts. Instead, for the diversion problem, one could reduce the number of cuts by considering the flow quantities of the diverted flows whose diversion is routed by the kth master problem solution through the relevant edge.

Now instead of posting the cut ( ) which simply rules out the previous assignment of diversions to edge e, we can explicitly use the flow quantities and return the constraint

$$\sum_{f \in \mathbf{F}} q_{fe}^k * X_{fe} \leq \mathsf{cap}(e) \tag{8}$$

Using this set of cuts, the kth master problem then has the form:

$$\min_{X_{fe}} \sum_{f \in \mathbf{F}} \sum_{e \in \mathbf{E}} X_{fe}$$

$$\text{st.} \begin{cases} \forall f \in \mathbf{F}: \begin{cases} \forall n \in \mathbf{N} \setminus \{\mathsf{orig}(f), \mathsf{dest}(f)\}: \sum_{e \in \mathbf{IN}(n)} X_{fe} = \sum_{e \in \mathbf{OUT}(n)} X_{fe} \\ n = \mathsf{orig}(f): \sum_{e \in \mathbf{OUT}(n)} X_{fe} = 1 \\ n = \mathsf{dest}(f): \sum_{e \in \mathbf{IN}(n)} X_{fe} = 1 \end{cases} \\ \text{for certain edges } e \text{ and iterations } j < k: \sum_{f \in \mathbf{F}} q_{fe}^j * X_{fe} \leq \mathsf{cap}(e) \\ X_{fe} \in \{0, 1\} \end{cases} \tag{9}$$

The resulting performance is summarized in Table   under the column of $\mathbf{D(0)}$.

5.2 Generating Tighter Cuts

The optimisation function in ( ) gives zero weight to any flows $f \notin \mathbf{F}_e$, for which $x_{fe}^k = 0$. For any optimal subproblem solution with $q_{fe}^k > 0$ for some $f \notin \mathbf{F}_e$ there exists an equivalent optimal solution with $q_{fe}^k = 0$. Thus the flow quantities q_{fe}^k ($\forall f \notin \mathbf{F}_e$) in optimal solutions to the subproblem may be zero rather than non-zero. The cut ( ) thus may only constrain variables X_{fe} for which $x_{fe}^k = 1$.

Instead, for the diversion problem, one could reduce the number of cuts by considering the flow quantities of *all* the diverted flows, not just the ones whose diversion is routed by the kth master problem solution through the relevant edge.

To extract the tightest cut from this subproblem, we therefore change the optimisation function so as to first optimise the flow through the relevant edge, and then, for any other flows which are still free to be non-zero, to maximize those flows too. This is achieved by simply adding a small multiplier ϵ to the other flows in the optimisation function:

$$\max_{Q_{fe}} \sum_{f \in F} (x_{fe}^k + \epsilon) * Q_{fe}$$

$$st. \begin{cases} \forall o \in \mathbf{orig}(F): \ \mathbf{ocap}(o) \geq \sum_{f:\mathbf{orig}(f)=o} Q_{fe} \\ \forall d \in \mathbf{dest}(F): \ \mathbf{dcap}(d) \geq \sum_{f:\mathbf{dest}(f)=d} Q_{fe} \\ \mathbf{quan}(f) \geq Q_{fe} \geq 0 \end{cases} \tag{10}$$

Now all the variables Q_{fe} will take their maximum possible values (denoted as $\tilde{q}_{fe}^k$), ensuring that ( ) expresses as tight a cut as possible.

$$\sum_{f \in F} \tilde{q}_{fe}^k * X_{fe} \leq \mathbf{cap}(e) \tag{11}$$

This cut not only constrains variables X_{fe} for which $x_{fe}^k = 1$, but also constrains the value of X_{fe} for other flows f which may not have used this edge in the kth subproblem.

Accordingly the kth master problem then has the form:

$$\min_{X_{fe}} \sum_{f \in F} \sum_{e \in E} X_{fe}$$

$$st. \begin{cases} \forall f \in F: \begin{cases} \forall n \in N \setminus \{\mathbf{orig}(f), \mathbf{dest}(f)\}: \ \sum_{e \in \mathbf{IN}(n)} X_{fe} = \sum_{e \in \mathbf{OUT}(n)} X_{fe} \\ n = \mathbf{orig}(f): \ \sum_{e \in \mathbf{OUT}(n)} X_{fe} = 1 \\ n = \mathbf{dest}(f): \ \sum_{e \in \mathbf{IN}(n)} X_{fe} = 1 \end{cases} \\ \text{for certain edges } e \text{ and iterations } j < k: \ \sum_{f \in F} \tilde{q}_{fe}^j * X_{fe} \leq \mathbf{cap}(e) \\ X_{fe} \in \{0, 1\} \end{cases} \tag{12}$$

The experimental results on the enhanced decomposition model are given in Table   under the column of $\mathbf{D}(\epsilon)$, where $\epsilon = 1.0e^{-5}$.

5.3 Comparison of Cuts Tightness

The 3 different cut generation formulations, ( ),( ) and ( ) have been presented. The tightness for these cuts, generated by different cut formulations are different too. For simple example, supposed that the assignment to X_{fe} returned by the solution of the kth master problem as

$$[x_{1e}^k, x_{2e}^k, x_{3e}^k, x_{4e}^k, x_{5e}^k] = [1, 1, 1, 0, 0]$$

The flow quantities Q_{fe} returned by the kth subproblem ( ) solution as

$$[q_{1e}^k, q_{2e}^k, q_{3e}^k, q_{4e}^k, q_{5e}^k] = [50, 75, 75, 50, 0]$$

and one returned by the solution of the kth subproblem ( ) as

$$[\tilde{q}_{1e}^{k}, \tilde{q}_{2e}^{k}, \tilde{q}_{3e}^{k}, \tilde{q}_{4e}^{k}, \tilde{q}_{5e}^{k}] = [50, 75, 75, 50, 75]$$

Notice that q_{4e}^{k} can also take 0 as optimal subproblem ( ) solution because $x_{4e}^{k} = 0$.

By using of the cut generation formulation ( ) we will obtain a cut

$$X_{1e} + X_{2e} + X_{3e} \leq 2$$

and the cut generated by the cut formulation ( ) is

$$50 * X_{1e} + 75 * X_{2e} + 75 * X_{3e} + 50 * X_{4e} \leq 100$$

and the cut formulation (   ) generated a cut of

$$50 * X_{1e} + 75 * X_{2e} + 75 * X_{3e} + 50 * X_{4e} + 75 * X_{5e} \leq 100$$

It is trivial to show that these cuts are getting tighter and tighter!

6   Implementation

The problem was solved using the ECL^{i}PS^e constraint programming platform [ ]. ECL^{i}PS^e provides interfaces both to CPLEX [ ] and to Xpress-MP [ ] for solving MIP problems. The current application comprises either a single MIP problem for the model presented in Section   or for the decomposed models presented in Sections   –   an MIP master problem and, in principle, LP subproblems for each edge in the network.

The MIP master problem and LP subproblems were run as separate external solver *instances* using the ECL^{i}PS^e facility to set up multiple external solver matrices. ECL^{i}PS^e enables different instances to be modified and solved at will. However the implementation does not run - or even set up - subproblems associated with edges which do not lie on any path returned by the master problem since the capacity constraints on these edges cannot, of course, be exceeded.

In the implementation we may choose how many cuts to return to the master problem after each iteration. One extreme is to run the subproblems individually and stop as soon as a new cut has been elicited while the other is to run all subproblems. These choices will respectively result in only one cut or all possible cuts being added to the master problem at each iteration. The additional cuts will tend to result in fewer iterations at the cost of more time spent per iteration on subproblem solution.

Preliminary experiments showed that it was substantially more efficient in general to run all the subproblems and add all the cuts at every iteration. This result is unsurprising for the current application since the cost of solving the master problem far outweighs that for each subproblem. The subproblems for each edge are LPs involving $|\mathbf{F}|$ variables and $|\text{orig}(\mathbf{F})| + |\text{dest}(\mathbf{F})|$ constraints, and are solved very quickly. The master problem however is a pure integer problem involving $|\mathbf{F}| \times |\mathbf{E}|$ variables and $|\mathbf{F}| \times |\mathbf{N}|$ constraints plus any cuts added so far. The initial master problem constraints are totally unimodular. As cuts are added in further iterations the unimodularity of the master problem is destroyed requiring solution by branch-and-bound search.

The ECL^{i}PS^e interface enables the user to solve a single problem instance using different optimisation functions. In the current application, the subproblem associated with each edge and each iteration has the same constraints. The difference lies only in the optimisation function. Therefore, only one subproblem needs to be set up when the algorithm is initialized: our implementation uses the same problem instance for each subproblem, simply running it with a different optimisation function for each edge.

Accordingly, in our ECL^{i}PS^e implementation, the decomposition model was set up using two different problem instances, one for the master problem and one for all the subproblems. In the following ECL^{i}PS^e code, these instances are named master and sub. First we create the master problem and the subproblem template:

```
:- eplex_instance(master).                              %1
:- eplex_instance(sub).                                 %2

diversions(Xfe_Array) :-                                %3
    % Multicommodity flow constraints written here      %4
    master:eplex_solver_setup(min(sum(Xfe_Array))),     %5

    % Subproblem constraints written here               %6
    sub:eplex_solver_setup(max(sum(Qfe_Array))),        %7

    iterate(Xfe_Array, Qfe_Array).                      %8
```

At lines 1 and 2 a master and subproblem solver instance are declared. They will later be filled with linear (and integer) constraints. Line 3 names the procedure that the user can invoke to solve the problem.

The problem constraints are entered next. These constraints are problem-instance-specific. Typically they are automatically generated from a data file holding the details of the network.

Line 5 sets up the master problem, reading in the constraints added previously, and posting a default optimisation function (minimize the total number of edges used in all the diversions).

The problem-instance-specific edge capacity constraints are now entered. Then the subproblem is set up at line 7, again with a default optimisation function. Finally at line 8 the iteration procedure is invoked.

We now write down the code which controls the iteration.

```
iterate(Xfe_Array, Qfe_Array) :-                                      %9
    master:eplex_solve(Cost),                                         %10
    master:eplex_get(typed_solution,Xfe_Values),                     %11
    % Find 'Edges' used in paths                                      %12
    ( foreach(E,Edges),                                               %13
        % Collect violated capacity constraints into 'CutList'       %14
    do                                                                %15
        % Build optimisation expression 'Expr'                        %16
        sub:eplex_probe(max(Expr),SubCost),                          %17
        ( SubCost > cap(E)  -> % Create cut                           %18
        )
    ),
    ( foreach(Cut, CutList)                                           %19
    do                                                                %20
        master:Cut  % Add cut                                         %21
    ),
    % If CutList is empty, succeed, otherwise iterate again           %22
```

At each iteration, the program solves the master problem (line 10), extracts the solution (values of all the X_{fe} variables), (line 11), finds which edges are on diversions (code not given) and then builds, for each edge (line 13), an optimisation function expression for the subproblem (code not given). The subproblem is solved (line 17), and if the maximal solution exceeds the capacity of the edge (line 18), a constraint is created.

As a result of solving the subproblem for each edge, a set of constraints are collected. Each constraint (line 19) is then added to the master problem (line 21).

If no constraints were collected (line 22), this means that no edge had its capacity exceeded, even in the worst case. Consequently the iteration terminates. If any constraints were added to the master problem, however, the iteration is repeated, with the newly tightened master problem.

The ECL^{i}PS^e language and implementation makes this decomposition very easy to state and to solve. ECL^{i}PS^e also supports many other solvers and hybridisation techniques. A description of the ECL^{i}PS^e Benders Decomposition library, for example, is available in [ ].

7   Results and Discussion

ECL^{i}PS^e implementations of the models described in Sections − were run on a number of test instances on road networks of differing sizes, the smallest involving 38 junctions and 172 road segments and the largest 365 junctions and 1526 road segments. Our data is industrial and cannot be published in detail. For each road network the choice of blocked junctions with different in- and out-degrees results in problem instances having different numbers of flows to divert.

For each test instance the first 3 columns of Table   show each example road network with the number of nodes (junctions) and edges (road segments),

Table 1. The Experimental Evaluation Result

Network (Nodes,Edges)	Flows	Obj	MIP cpu (Vars, Cstrs)	D(naive) cpu (MP)	D(0) cpu (MP)	D(ϵ) cpu (MP)
a (38, 172)	40	116	12.45 (25k,20k)	1.83 (1)	1.38 (1)	2.12 (3)
	54	132	18.77 (33k,26k)	4.05 (8)	3.38 (5)	3.52 (5)
b (38, 178)	54	132	18.90 (35k,27k)	4.27 (8)	3.39 (4)	3.85 (5)
c (38, 256)	140	328	108.97 (122k,91k)	16.07 (8)	13.63 (5)	14.15 (5)
	126	226	TO (198k,146k)	710.47 (129)	35.77 (14)	29.08 (18)
d (50, 464)	152	318	TO (236k,173k)	10060.64 (757)	48.16 (12)	72.56 (17)
	178	410	TO (274k,200k)	TO	517.60 (29)	783.94 (40)
	286	546	TO (435k,317k)	TO	459.61 (50)	99.98 (17)
e (50, 464)	350	714	TO (528k,383k)	TO	339.16 (21)	314.58 (19)
	418	890	TO (626k,453k)	TO	4033.98 (64)	2404.82 (43)
f (208, 676)	28	Fail	TO (73k,60k)	2673.38 (130)	79.66 (16)	76.63 (15)
	28	256	TO (73k,60k)	13350.71 (186)	252.44 (23)	706.21 (44)
g (208, 698)	28	76	257.64 (75k,62k)	20.83 (1)	20.90 (1)	20.76 (1)
h (208, 952)	44	108	361.84 (156k,123k)	50.79 (1)	50.69 (1)	52.87 (3)
i (208, 1186)	774	Fail	OOM (2910k,2153k)	2635.59 (22)	702.59 (3)	748.17 (3)
	88	Fail	27.55 (223k,177k)	28.59 (1)	28.28 (1)	28.39 (1)
j (212, 734)	104	Fail	33.06 (261k,207k)	TO	58.49 (4)	56.73 (3)
	106	338	TO (266k,210k)	TO	4602.80 (31)	7317.42 (40)
	154	Fail	TO (380k,300k)	TO	95.33 (5)	82.70 (2)
	142	Fail	71.22 (729k,565k)	185.76 (6)	139.47 (1)	140.31 (1)
k (365, 1526)	154	Fail	OOM (784k,606k)	TO	293.98 (4)	292.20 (3)
	178	422	OOM (900k,694k)	46930.15 (1572)	310.90 (12)	314.11 (12)

$\epsilon = 1.0e^{-5}$
cpu = seconds on Intel(R) Pentium(R) 4 CPU 2.00GHz
MP = number of iterations
OOM = Out of memory on Intel(R) Pentium(R) 4 with 1 Gig of main memory
TO = Timed out after 72000 seconds
Fail = Problem Unsatisfiable

the number of flows to reroute and optimal objective value (or Fail if no feasible solution exists). The remaining 4 columns show solution time for the four models with additionally the total number of variables and number of constraints in the original MIP model and the number of master problem iterations for the decomposed models.

While there is some variation in the difficulty of individual instances for the different methods, the decomposed models outperform the MIP by at least an order of magnitude on average. It is striking how poorly MIP scales for this problem: while the MIP is able to solve only 9 of the instances within reasonable limits on execution time and memory usage, even the initial naive decomposition solves all but 8 instances within these limits, and the improved decomposed models solve all instances within a relatively small number of master problem iterations and relatively short time periods.

MIP focuses on integer inconsistencies which do not necessarily correlate closely with the real causes of inconsistency (*i.e.* overflow on an edge). CP enables us to solve a problem relaxation with whatever scalable solver is available (*e.g.* LP, or in the current master problem, MIP), and then to select the inconsistencies/bottlenecks at the problem level rather than at the encoded linear/integer level. This yields a much more problem-focused search heuristic.

This is reflected in the results obtained: all instances solvable by the MIP approach required only a few master problem iterations in the decomposed approach. In particular the 3 infeasible instances in data sets j and k for which MIP outperforms the decomposed approach are very over-constrained. Consequently they require little search in the MIP and few iterations in the decomposed models to prove infeasibility. Similarly the 6 feasible instances in data sets a, b, c, g and h are relatively loosely constrained and easily soluble by all methods. Even here however the combination of smaller component problem size and more problem-focused search in the decomposed approach yield order of magnitude improvements. Conversely the remaining instances which are neither very loosely constrained nor very over-constrained are very difficult for both the MIP approach and the naive decomposition due to the looseness of the cuts provided, but much more amenable to solution by the problem-focused tight cuts of the improved decomposition approaches.

Although this may suggest that MIP may be preferable for problems displaying certain characteristics, it is precisely those problems which are feasible but tightly constrained or infeasible but only slightly over-constrained that are most interesting in practice.

8   Conclusion

The diversion problem is very awkward in that the constraints involve a subsidiary optimisation function. The problem can be expressed as a single MIP, using the KKT condition to transform the embedded optimisation function into a set of constraints. Nevertheless the resulting MIP problem is very large and scales poorly.

A much better approach is presented by use of the decomposition strategy. In particular, tight cuts are created to improve the efficiency and scalability. The experimental evaluation shows that decomposition can solve much larger scale problem instances.

References

[1] Nemhauser, G. L., Wolsey, L. A.: Integer and Combinatorial Optimization, John Wiley & Sons, New York. (1988)
[2] Jain, V., Grossmann, I. E.: Algorithms for Hybrid MILP/CP Models for a Class of Optimisation Problems. INFORMS Journal on Computing. **13** (2001) 258–276

[3] Thorsteinsson, E. S.: Branch-and-Check: A Hybrid Framework Integrating Mixed Integer Programming and Constraint Logic Programming. In T. Walsh, editor, Principles and Practice of Constraint Programming – CP 2001, Springer. (2001) 16–30

[4] Benders, J. F.: Partitioning Procedures for Solving Mixed-Variables Programming Problems. Numerische Mathematik. **4** (1962) 238–252

[5] Hooker, J. N., Ottosson, G.: Logic-Based Benders Decomposition. Mathematical Programming. **96** (2003) 33–60

[6] Cheadle, A. M., Harvey, W., Sadler, A., Schimpf, J., Shen, K., Wallace, M.: ECL^{i}PS^e. Technical Report 03-1, IC-Parc, Imperial College London, (2003)

[7] ILOG: ILOG CPLEX 6.5 User's Manual, http://www.cplex.com (1999)

[8] Dash Optimization: Dash Optimization Xpress–MP 14.21 User's Manual, http://www.dashoptimization.com (2003)

[9] Eremin, A., Wallace, M.: Hybrid Benders Decomposition Algorithms in Constraint Logic Programming. In T. Walsh, editor, Principles and Practice of Constraint Programming – CP 2001, Springer. (2001) 1–15

LP Relaxations
of Multiple all_different Predicates

Gautam Appa[1], Dimitris Magos[2], and Ioannis Mourtos[1]

[1] London School of Economics
London WC2A 2AE, UK.
{g.appa,j.mourtos}@lse.ac.uk
[2] Technological Educational Institute of Athens
12210 Egaleo, Greece.
dmagos@teiath.gr

Abstract. This paper examines sets of all_different predicates that appear in multidimensional assignment problems. It proposes the study of certain LP relaxations as a prerequisite of integrating CP with IP on these problems. The convex hull of vectors satisfying simultaneously two predicates is analysed and a separation algorithm for facet-defining inequalities is proposed.

1 Introduction and Motivation

The role of a useful relaxation in Constraint Programming (CP) has been extensively discussed ([ ]) and appears to be critical for problems that possibly do not accept a polynomial algorithm. A strong reason for integrating CP with Integer Programming (IP) is that IP offers a global problem view via its Linear Programming (LP) relaxation. It is common to strengthen the LP relaxation by adding implicit constraints in the form of cutting planes. This can become useful for feasibility problems, where the LP-relaxation may establish infeasibility early in the search (e.g. [ ]), but mostly for optimisation problems where LP can provide a bound on the optimal solution. For optimization models, LP is able to establish an algebraic *proof of optimality*, which can arise from CP only through enumeration.

In this context, research has focused on representing CP predicates in the form of linear inequalities. For example, the convex hull of the solutions to the *all_different* predicate and to cardinality rules has been analysed in ([ ]) and ([ ]), respectively. This paper works in the same direction by examining sets of *all_different* predicates. The *all_different* predicate arises in problems that embed an assignment structure. Hence, our motivation comes from the class of multidimensional assignment problems, which exhibit both theoretical and practical interest.

The remainder of this paper is organised as follows. Section introduces multidimensional assignment problems and presents their CP formulation. LP-relaxations of this formulation are illustrated and discussed in Section . A separation algorithm for the inequalities defining the polytope associated with the

J.-C. Régin and M. Rueher (Eds.): CPAIOR 2004, LNCS 3011, pp. 364– , 2004.
© Springer-Verlag Berlin Heidelberg 2004

all_different predicate ([ ]) is given in Section . Finally, we examine the convex hull of the solutions to two overlapping *all_different* constraints in Section and discuss further research in Section .

2 CP Models for Assignment Problems

The simplest case of an assignment problem is the well-known 2-index assignment, which is equivalent to weighted bipartite matching. Extensions of the assignment structure to k entities give rise to *multidimensional* (or *multi-index*) *assignment problems* ([ ]). A k-index assignment problem is defined on k sets, usually assumed to be of the same cardinality n. The goal is to identify a collection of n disjoint tuples, each including a single element from each set. This is the class of *axial* assignment problems ([ ]), hereafter referred to as $(k, 1)AP_n$. A different structure appears, if the aim is, instead, to identify a collection of n^2 tuples, partitioned into n disjoint sets of n disjoint tuples. These assignment problems are called *planar* and are directly linked to *Mutually Orthogonal Latin Squares (MOLS)* ([ ]). We denote k-index planar problems as $(k, 2)AP_n$ (see also [ ]).

Axial assignment structures have received substantial attention, because of their applicability in problems of data association. ([ ]). The planar problems share the diverse application fields of *MOLS*, e.g. multivariate design, error correcting codes ([ ]). Both types of assignment appear in problems of timetabling, scheduling and cryptography ([ ]).

It is easy to derive the CP formulation of $(2, 1)AP_n$:

$$all\_different(X_i : i = 0, ..., n - 1), \tag{1}$$
$$D_X = \{0, ..., n - 1\}$$

A solution to the above formulation provides n pairs (i, X_i), $i = 0, ..., n - 1$.

Recall that a solution to $(k, 1)AP_n$ is a set of n disjoint k-tuples. Hence, define variables X_i^p, $p = 1, ..., k - 1$, $i = 0, ..., n - 1$. We wish to form n tuples $(i, X_i^1, ..., X_i^{k-1})$, for $i = 0, ..., n - 1$. Evidently, variables $\{X_0^p, ..., X_{n-1}^p\}$ must be pairwise different for each $p \in \{0, ..., k - 1\}$. The formulation follows.

$$all\_different(X_i^p : i = 0, ..., n - 1), \text{for } p = 1, ..., k - 1 \tag{2}$$
$$D_X = \{0, ..., n - 1\}$$

This formulation includes $(k-1) \cdot n$ variables and $(k-1)$ *all_different* constraints.

Concerning $(k, 2)AP_n$, we are looking for n^2 $(k - 1)$-tuples that can be partitioned into n sets of n tuples. Each set of n disjoint tuples is denoted as $(i, X_{ij}^1, ..., X_{ij}^{k-2})$, $i = 0, ..., n - 1$, $j \in \{0, ..., n - 1\}$. Each of the n distinct values of index j corresponds to a different set of n disjoint tuples. Clearly, variables $\{X_{ij}^p : i = 0, ..., n - 1\}$ and $\{X_{ij}^p : j = 0, ..., n - 1\}$ must be pairwise different for each of the n sets to contain n disjoint tuples.

In addition, define the auxilliary variables Z_{ij}^{qr}, where $q, r = 1, ..., k - 2$ and $q < r$, including also the constraints $Z_{ij}^{rq} = X_{ij}^r + n \cdot X_{ij}^q$. Variables X_{ij}^p and Z_{ij}^{rq}

have domains of cardinality n and n^2, respectively. Observe that an *all_different* constraint must be imposed on all Z_{ij}^{rq} variables to ensure that no two tuples can contain the same value for both indices i and j. The following model involves $(k-1) \cdot n^2$ variables and $\binom{k-2}{2}(1+n^2) + 2 \cdot (k-2) \cdot n$ constraints.

$$all\_different(X_{ij}^p : i = 0, ..., n-1), \text{ for } j = 0, ..., n-1, \quad p = 1, ..., k-2$$
$$all\_different(X_{ij}^p : j = 0, ..., n-1), \text{ for } i = 0, ..., n-1, \quad p = 1, ..., k-2$$
$$all\_different(Z_{ij}^{rq} : i, j = 0, ..., n-1), \text{ for } q, r = 1, .., k-2, \ q < r \qquad (3)$$
$$Z_{ij}^{rq} = X_{ij}^r + n \cdot X_{ij}^q, \text{ for } i, j, = 0, ..., n-1, \ q, r = 1, .., k-2, q < r$$
$$D_X = 0, ..., n-1 \ , \quad D_Z = 0, ..., n^2-1$$

Notice that the CP formulation of $(k, 2)AP_n$ contains *all_different* constraints with one variable in common, in contrast to the CP formulation of $(k, 1)AP_n$ that contains non-overlapping *all_different* constraints.

3   LP Relaxations

A relaxation can be derived from the IP models of $(k, 1)AP_n$ and $(k, 2)AP_n$. To illustrate these IP models, some notation should be introduced. Let $K = \{1, ..., k\}$ and assume sets M_i, $i \in K$, each of cardinality n. For $S \subseteq K$, assume $S = \{i_1, ..., i_s\}$ and let $M^S = M_{i_1} \times \cdots \times M_{i_s}$. Also let $m^S \in M^S$, where $m^S = (m^{i_1}, ..., m^{i_s})$. A k-tuple is denoted as $(m^1, m^2, ..., m^k)$.

Define the binary variable $x_{m^1m^2...m^k}$, written also as x_{m^K} according to the above notation. Clearly, $x_{m^1m^2...m^k} = 1$ iff tuple $(m^1, m^2, ..., m^k)$ is included in the solution. It is not difficult to see that the IP formulation for $(k, 1)AP_n$ is obtained by summing over all possible subsets of $k-1$ out of k indices (consider as an example the 2-index assignment).

$$\sum \{x_{m^K} : m^{K \setminus \{i_1\}} \in M^{K \setminus \{i_1\}}\} = 1, \forall m^{i_1} \in M_{i_1}, \forall i_1 \in K$$
$$x_{m^K} \in \{0, 1\}, \forall m^K \in M^K$$

Similarly, the IP formulation for $(k, 2)AP_n$ is obtained by summing over all possible subsets of $k-2$ out of k indices. As an example, consider the formulation of the Orthogonal Latin Squares problem, which is the $(4, 2)AP_n$ ([ ]).

$$\sum \{x_{m^K} : m^{K \setminus \{i_1, i_2\}} \in M^{K \setminus \{i_1, i_2\}}\} = 1, \forall m^{\{i_1, i_2\}} \in M^{\{i_1, i_2\}}$$
$$\forall (i_1, i_2) \in K^2, i_1 < i_2, x_{m^K} \in \{0, 1\}, \forall m^K \in M^K$$

Both models are special cases of the model discussed in [ ]. The LP-relaxation arises from these models by simply replacing the integrality constraints with non-negativities. This relaxation, however, involves a large number of binary variables, therefore becoming impractical as n grows. The relaxation can be further strengthened by introducing facet-defining inequalities, although known families of facets remain limited, except for small orders, e.g. $(3, 1)AP_n$, $(3, 2)AP_n$ and

$(4, 2)AP_n$. A family of facets for all axial problems, i.e. $(k, 1)AP_n$, is presented in [ ].

A different relaxation is implied by the CP formulations, if each *all-different* predicate is replaced with an equality constraint. It is easy to see that the predicate *all-different*$(y_1, ..., y_n)$, where $D_y = \{0, .., n - 1\}$, implies the equality $\sum_{i=1}^{n} y_i = \frac{n(n-1)}{2}$.

For example, constraints ( ) imply the validity of the equalities:

$$\sum_{i \in \{0, ..., n-1\}} X_i^p = \frac{n(n - 1)}{2} \text{ for } p = 1, ..., k - 1$$

This second relaxation is weaker in the sense that it allows for infeasible integer values to be assigned to variables X_i^p. Given however that the convex hull of all vectors satisfying an *all-different* predicate is known ([ ]), the above relaxation can be tightened by adding valid inequalities, which are facets of this convex hull. If all facets are included, the solution to the augmented LP will be integer feasible. The problem lies in the exponential number of these facets and can be resolved only via an efficient separation algorithm, i.e. a polynomial algorithm that adds these inequalities only when violated by the current LP solution. Such an algorithm is presented next.

4   A Separation Algorithm

Let $N = \{0, ..., n - 1\}$ and consider the constraint *all-different*$\{y_i : i \in N\}$, where $y_i \in \{0, ..., h - 1\}$, $h \geq n$. Let all indices $i_1, ..., i_l$ belong to N and be pairwise different. Let $P = \{y \in \mathbb{R}^n : \sum_{i=1}^{n} y_i = \frac{n(n-1)}{2}, y \geq 0\}$. The convex hull of all non-negative integer n-vectors that contain pairwise different values is the polytope $P_{alldiff} = conv\{y \in P \cap \mathbb{Z}^n : \text{all-different}\{y_i : i \in N\}\}$. It was proved in [ ] that $P_{alldiff}$ is the intersection of the following facet-defining inequalities:

$$\sum_{S \subseteq N : |S| = l} \{y_t : t \in S\} \geq \frac{l(l - 1)}{2}, \forall S \subseteq N, \forall l \in \{1, ..., n\} \tag{3}$$

$$\sum_{S \subseteq N : |S| = l} \{y_t : t \in S\} \leq \frac{l(2h - l - 1)}{2}, \forall S \subseteq N, \forall l \in \{1, ..., n\} \tag{4}$$

Although the number of these inequalities is exponential in n, i.e. equals $2(2^n - 1)$, the separation problem for these inequalities can be solved in polynomial time.

Algorithm 1 *Let* $y \in P_{alldiff} \backslash P$.
Step I: Order $\{y_1, ..., y_n\}$ *in an ascending order* $\{\bar{y}_{i_1}, ..., \bar{y}_{i_n}\}$;
Step II: For $l = 1, ..., n$, *if* $\sum_{t \leq l} \bar{y}_{i_t} < \frac{l(l-1)}{2}$
   return;
Step III: For $i = n, ..., 1$, *if* $\sum_{t \geq l} \bar{y}_{i_t} > \frac{l(2k-l-1)}{2}$
   return;

Proposition 1. *Algorithm   determines in $O(n^2)$ steps whether a facet of $P_{alldiff}$ is violated.*

Proof. Because of the ordering, it holds that $\sum_{t \leq l} \bar{y}_{i_t} \leq \sum_{S \subseteq N : |S|=l} \{y_t : t \in S\}$. Therefore, if no inequality of the form $\sum_{t \leq l} \bar{y}_{i_t} \geq \frac{l(l-1)}{2}$ is violated, none of the inequalities ( ) will be either. The case is similar for inequalities ( ). This proves the correctness of the algorithm. Furher, it is not difficult to establish that the algorithm runs in $O(n^2)$ steps in the worst case.

This algorithm is a necessary step towards solving an *all_different* constraint via Linear Programming, without any enumeration. One should solve an initial LP containing a subset of the facet-defining inequalities and then repetitively add violated cuts and re-optimise until a feasible integer solution is identified. Evidently, any linear objective function could be included, therefore this method is also appropriate for the optimisation version of the problem.

5   The Convex Hull
of Two Overlapping all_different Constraints

Since multidimensional assignment problems include more than one *all_different* predicates, it is important to analyse the convex hull of vectors satisfying simultaneously multiple *all_different* constraints. The simplest case to consider is that of two *all_different* constraints of cardinality n having t common variables, where $t \in \{0, ..., n-1\}$. For $t = 0$, this reduces to $(3,1)AP_n$, discussed in [ ].

Define the variables $X_1, ..., X_n, ..., X_{2n-t}$ with domain $D_X = \{0, ..., n-1\}$. The constraints are:

$$all\_different(X_1, X_2, ..., X_n) \tag{5}$$
$$all\_different(X_{n-t+1}, X_{n-t+2}, ..., X_{2n-t})$$

Let $P_L = \{y \in \mathbb{R}^{2n-t} : \sum_{i=1}^{n} y_i = \frac{n(n-1)}{2}, \sum_{i=n-t+1}^{2n-t} y_i = \frac{n(n-1)}{2}, y \geq 0\}$ and $P_I = conv\{y \in P_L \cap \mathbb{Z}^{2n-t} : all\_different(y_1, ..., y_n), all\_different(y_{n-t+1}, ..., y_{2n-t})\}$. The following results are illustrated without giving extensive proofs.

Theorem 2. $\dim P_I = 2(n-1) - t$

Proof. There are two implied equalities, satisfied by all $y \in P$:

$$y_1 + ... + y_n = \frac{n(n-1)}{2}$$
$$y_1 + .. + y_{n-t} = y_{n+1} + ... + y_{2n-t}$$

Given that the above two equalities are linearly independent, the minimum equality system for P_L has rank 2. Therefore, $\dim P_L = 2n - t - 2 = 2(n-1) - t$. Since $P_I \subset P_L$ it holds that $\dim P_I \leq \dim P_L$. Equality is proved by illustrating $2(n-1) - t + 1$ affinely independent points.

It can be proved that the inequalities defining facets of the polytope associated with a single *all_different* constraint, i.e. ( ) and ( ), remain facet-defining for P_I. Whether these are the only facets of P_I, i.e. they define a minimal polyhedral description of the problem, remains an open question. Experiments with the PORTA software ([ ]) show that no other facets of P_I exist for $n \leq 6$ and $t \leq 3$ (PORTA provides the convex hull of an input set of vectors).

Given these polyhedral aspects, the applicability of the separation algorithm of Section   to the case of two overlapping *all_different* constraints is direct. CP and IP techniques can also be combined with CP handling a certain subset of the variables and applying a tightened LP relaxation thereafter.

6   Further Research

One obvious extension of the presented work is to examine 2 *all_different* constraints having (a) different cardinalities and (b) variable domains of cardinality larger than the number of variables. Further, dealing with $(k,1)AP_n$ and $(k,2)AP_n$ implies solving sets of more than 2 *all_different* constraints. Computational results are also required in order to assess the efficiency of our approach in comparison to existing methods for the $(k,1)AP_n$.

References

[1] Appa G., Mourtos I., Magos D.: Integrating Constraint and Integer Programming for the Orthogonal Latin Squares Problem. In van Hentenryck P. (ed.), Principles and Practice of Constraint Programming (CP2002), *Lecture Notes in Computer Science* **2470**, Springer-Verlag, 17-32 (2002).   ,

[2] Appa   G.,   Magos   D.,   Mourtos   I.:   Polyhedral   resuls   for   assignment   problems.   LSE   CDAM   Workin   Paper   Series   (URL: http://www.cdam.lse.ac.uk/Reports/Files/cdam-2002-01.pdf).   ,   ,

[3] Balas E., Saltzman M. J.: Facets of the three-index assignment polytope. *Discrete Applied Mathematics* **23**, 201-229 (1989).   ,

[4] Christoff T., Löbel A.: Polyhedral Representation Transformation Algorithm v.1.4.0 (2003) (URL: http://www.zib.de/Optimization/Software/Porta).

[5] Dénes J., Keedwell A. D.: *Latin Squares: New developments in the Theory and Applications.* North-Holland (1991).

[6] Hooker, J. N.: *Logic Based Methods for Optimization*, Wiley, NY (2000).

[7] Pierskalla W. P.: The multidimensional assignment problem. *Operations Research* **16**, 422-431 (1968).

[8] Spieksma F. C. R.: Multi-index assignment problems: complexity, approximation, applications. In Pitsoulis L., Pardalos P. (eds.), *Nonlinear Assignment Problems, Algorithms and Applications*, 1-12, Kluwer Academic Publishers (2000).

[9] Yan H., Hooker J. N. : Tight Representation of Logic Constraints as Cardinality Rules. *Mathematical Programming* **85** (2), 363-377 (1999).

[10] Williams H. P., Yang H.: Representations of the all-different Predicate of Constraint Satisfaction in Integer Programming. *INFORMS Journal on Computing* **13**, 96-103 (2001).   ,   ,

Dispatching and Conflict-Free Routing of Automated Guided Vehicles: A Hybrid Approach Combining Constraint Programming and Mixed Integer Programming

Ayoub Insa Corréa, André Langevin, and Louis Martin Rousseau

Département de Mathématiques et de Génie Industriel
École Polytechnique de Montréal and GERAD
C.P 6079, Succ. Centre-ville
Montréal, Canada, H3C 2A7
{iacorrea, andre.langevin, louis-martin.rousseau}@polymtl.ca

Abstract: This paper reports on the on-going development of a hybrid approach for dispatching and conflict-free routing of automated guided vehicles used for material handling in manufacturing. The approach combines Constraint Programming for scheduling and Mixed Integer Programming for routing without conflict. The objective of this work is to provide a reliable method for solving instances with a large number of vehicles. The proposed approach can also be used heuristically to obtain very good solution quickly.

Keywords: Automated guided vehicles, hybrid model, constraint programming, material handling systems, routing.

1 Introduction

This study focuses on automated guided vehicles (Agvs) in a flexible manufacturing system (FMS). An Agv is a material handling equipment that travels on a network of paths. The FMS is composed of various cells, also called working stations, each with a specific function such as milling, washing, or assembly. Each cell is connected to the guide path network by a pick-up / delivery (P/D) station where the pick-ups and deliveries are made. The guide path is composed of aisle segments with intersection nodes. The vehicles are assumed to travel at a constant speed and can stop only at the ends of the guide path segments. The guide path network is bidirectional and the vehicles can travel forward or backward. The unit load is one pallet. The number of available vehicles and the duration of loading and unloading at the P/D stations are known. As many vehicles travel on the guide path simultaneously, collisions must be avoided. There are two types of collisions: the first type may appear when two vehicles are moving toward the same node. The second type of collision occurs when two vehicles are traveling on a segment in opposite directions. Every day, a list of orders is given, each order corresponding to a specific product to manufacture (here, product means one or many units of the same product). Each order determines a

J.-C. Régin and M. Rueher (Eds.): CPAIOR 2004, LNCS 3011, pp. 370-379, 2004.
© Springer-Verlag Berlin Heidelberg 2004

sequence of operations on the various cells of the FMS. Then, the production is scheduled. This production scheduling sets the starting time for each order. Pallets of products are moved between the cells by the Agvs. Hence, each material handling request is composed of a pickup and a delivery with their associated earliest times. At each period, the position of each vehicle must be known. Time is in fifteen second periods. A production delay is incurred when a load is picked up or delivered after its planned earliest time. The problem is thus defined as follows: *Given a number of Agvs and a set of transportation requests, find the assignment of the requests to the vehicles and conflict-free routes for the vehicles in order to minimize the sum of the production delays.*

2   Literature Review

For a recent general review on Agvs problems and issues, the reader is referred to the survey of Qiu *et al.* (2002). These authors identified three types of algorithms for Agvs problems: (1) algorithms for general path topology, (2) path layout optimization and (3) algorithms for specific path topologies. In our study, we work on algorithms for general path topology. Methods of this type can be divided in three categories: (1) static methods, where an entire path remains occupied until a vehicle completes its route; (2) time-window based methods, where a path segment may be used by different vehicles during different time-windows; and (3) dynamic methods, where the utilization of any segment of path is dynamically determined during routing rather than before routing as with categories (1) and (2). Our method belongs to the third category and we focus on bidirectional networks and conflict-free routing problems with an optimization approach. Furthermore we have a static job set, i.e., all jobs are known *a priori*. Krishnamurthy *et al.* (1993) proposed first an optimization approach to solve a conflict-free routing problem. Their objective was to minimize the makespan. They assumed that the assignment of tasks to Agvs is given and they solved the routing problem by column generation. Their method generated very good solutions in spite of the fact that it was not optimal (column generation was performed at the root node of the search tree only). Langevin *et al.* (1996) proposed a dynamic programming based method to solve exactly instances with two vehicles. They solved the combined problem of dispatching and conflict-free routing. Désaulniers *et al.* (2003) proposed an exact method that enabled them to solve instances with up to four vehicles. They used slightly the same data set as Langevin *et al.* Their approach combined a greedy search heuristic (to find a feasible solution and set bound on delays), column generation and a branch and cut procedure. Their method presents however some limits since its efficiency depends highly on the performance of the starting heuristic. If no feasible solution is found by the search heuristic, then no optimal solution can be found. The search heuristic performs poorly when the level of congestion increases and the system considers at most four Agvs.

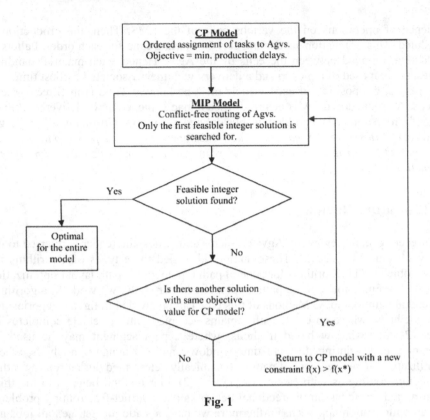

Fig. 1

3   A Constraint Programming/Mixed Integer Programming Approach

The decisions for dispatching and conflict-free routing of automated guided vehicles can be decomposed into two parts: first, the assignment of requests to vehicles with the associated schedule, then, the simultaneous routing of every vehicle. The hybrid approach presented herein combines a Constraint Programming (CP) model for the assignment of requests to the vehicles with their actual pick-up or delivery times (in order to minimize the delays) and a Mixed Integer Programming (MIP) model for the conflict-free routing. The two models are imbedded in an iterative procedure as shown in Fig. 1. For each assignment and schedule found by the CP model, the MIP model tries to find conflict-free routes satisfying the schedule. CP is used to deal with the first part because it is very efficient for scheduling and, in the present case, it allows identifying easily all optimal solutions. Here optimal solutions that are equivalent in terms of value but represent different assignment might yield very different routing solution. The routing part is addressed with MIP since it can be modeled with a time-space network with some interesting sub-structures that allow fast solutions.

The method can be described in three steps:

- Step 1: find an optimal solution x^* (i.e., an assignment of requests to vehicles) to the CP model. Let z^* be the optimal objective function value (the total delay).
- Step 2: use x^* in the MIP model to find a conflict-free routing. If there exists any, the optimal solution to the entire model is found. Otherwise (no feasible solution found), go to step 3.
- Step 3: find another optimal solution to the CP model different from x^* but with the same objective function value. If there exists any, return to step 2. If no feasible solution has been found with any of the optimal solutions of the CP model, go to step 1 and add a lower bound to the objective function ($f(x) > z^*$) before solving anew the CP model. This lower bound is set to z^* and is always updated when returning to step 1.

3.1 The CP Model

The model answers the question "Which vehicle is processing what material handling task and when?" by yielding an ordered assignment of tasks to Agvs. The total amount of delays is measured by summing the difference between the actual start time and the earliest start time of all deliveries. In this model, the distance (time) matrix is obtained by using shortest paths between nodes. Thus, the delays calculated (which don't take into account the possible conflicts) are an approximation (a lower bound) of the actual delays.

Sets and Parameters Used:

DummyStartTasks: set of dummy starting tasks. Each of them is in fact the starting node of a vehicle corresponding to the last delivery node of a vehicle in the previous planning horizon.

Start[k]: starting node of Agv k.

Pickups: set of pick-up tasks.

SP [·,·]: length of the shortest path between a couple of nodes

Node (p): node for task p. It is used here to alleviate the notation.

nbRequests: number of requests to perform.

nbChar: number of vehicles available.

Requests: set of requests. Each request contains two fields: the pick-up task and the associated delivery task.

DummyStartRequests: set of dummy starting requests.

Inrequest: set of dummy start requests and real requests.

Pick [·]: pick-up field of a request.

Del [·]: delivery field of a request. Each task (dummy or not) is defined by three fields: the node where the task is to be performed, the processing time at this work station and the earliest starting time of the task.

Duration [·]: duration of a task.

Priorities: set of couples of tasks linked by a precedence relationship (the first task is to be performed before the other).

Tasks: set of all tasks with a (mandatory) successor. This model uses the three following variables:

- **Alloc[i]** = k if task i is performed by vehicle k. The index lower than 1 represent dummy requests.

- **Succ[u]** = v if request v is the successor of request u on the same vehicle.

- **Starttime[j]** is the start time of task j.

For each vehicle a couple of dummy tasks are created, a starting task and an end task. The starting task has the following characteristics: its node is the starting node of the Agv, its duration and earliest starting time are set to zero. We define the set **Tasks** by the set of dummy start tasks and real pickups and deliveries tasks. A request consists of a pickup and a delivery tasks. The constraints used in the model are the following:

---

(1) $\forall\, k \in$ Char

Alloc[1 - k] = Alloc[nbRequests + k] = k

(2) $\forall\, r \in$ DummyStart Requests

$\sum_{s\ \text{in Inrequest}} (\text{Succ}[r] = s) = 1$

(3) $\forall\, o \in$ Tasks

Alloc[o] = Alloc[Succ[o]]

(4) $\forall\, d \in$ DummyStart Tasks

Starttime[d] = 0

(5) $\forall\, k \in$ Char, $\forall\, d \in$ DummyStart Requests, $\forall\, r \in$ Requests

(Alloc[d] = k) $\wedge$ (Succ[d] = r) $\Rightarrow$ Starttime[pick[r]] $\geq$ SP[Start[k], node(pick[r])]

(6) Alldifferent(Succ)

(7) $\forall\, r \in$ Requests

Startime[pick[r]] + 1 + SP[node(pick[r]), node(del[r])] $\leq$ Starttime[del[r]]

(8) $\forall\, r1, r2 \in$ Requests

Succ[del[r1]] = pick[r2] $\Rightarrow$ (Starttime[del[r1]] + 1 + SP[node(del[r1]), node(pick[r2])] $\leq$ Starttime[pick[r2]])

(9) $\forall\, u \in$ Priorities

Starttime[before[u]] + duration(before[u]) $\leq$ Starttime[after[u]]

(10) $\forall\, i, j \in$ Tasks : $(i \neq j)$ and node(i) = node(j)

(Starttime[i] $\geq$ Starttime[j] + 1) $\vee$ (Starttime[i] + 1 $\leq$ Starttime[j])

Constraints (1) ensure that a dummy starting task and its dummy end task are performed by the same Agv. Constraints (2) ensure that the successor of a dummy start request is either a real request or a dummy end request (in this case the vehicle is idle during the entire horizon but can move to avoid collisions). Constraints (3) ensure that every request and its successor must be performed by the same Agv. Constraints (4) ensure that, at the beginning of the horizon (period zero), each vehicle is located at its starting node. Constraints (5) specify that each vehicle must have enough time to reach its first pick-up node. Constraints (6) imply that the successor of each request is unique. Constraints (7) specify that each vehicle processing a request must have enough time to go from the pick-up node to the delivery node of the request. Constraints (8) ensure that if one request is the successor of another request on the same vehicle, the Agv must have enough time to make the trip from the delivery node of the first request to the pick-up node of the second request. They link the tasks that must be processed at the same nodes so that there is no overlapping. Constraints (9) enforce that, for every couple of tasks linked by precedence relationship, the first task must start and be processed before the beginning of the second task. Constraints (10) ensure that for each couple of tasks that must be performed on the same node, one must start one period after the beginning of the other.

3.2 The MIP Model

For a given schedule obtained from the CP model, the MIP model allows to find whether there exists a feasible routing without conflict. This could be seen as a Constraint Satisfaction Problem since we only search for a feasible routing without conflict. However, the inherent network structure of the routing problem allows using a MIP model where only the first feasible integer solution is searched for, thus preventing a potentially time consuming search for the optimal solution of the MIP. The MIP corresponds to a time-space network which defines the position of every vehicle at anytime (see Fig. 2.). The original guide path network is composed of segments of length 1, 2 and 3. This network has been transformed into a directed network where all arcs are of length 1 by incorporating dummy nodes on segments of length 2 or 3. At every time period, there is a node for each intersection node (including the dummy nodes) of the guide paths. An arc is defined between two nodes of two successive time periods if the corresponding intersection nodes are adjacent on the guide path layout. Each vehicle enters the network at a given node at period 0. The time-space network model has the following characteristics:

- One unit of flow is equivalent to one vehicle present in the system.
- The total amount of entering flow at each period is equal to the total number of Agvs (busy or idle).
- At most one unit of flow can enter in a node (no collision can occur at a node).
- There is flow conservation at each node.
- An arc whose origin and destination are the same node at two successive periods corresponds to waiting at that node.
- A vehicle can move without having a task to perform, just for avoiding conflicts.

See Figure 2 for time-space network description.

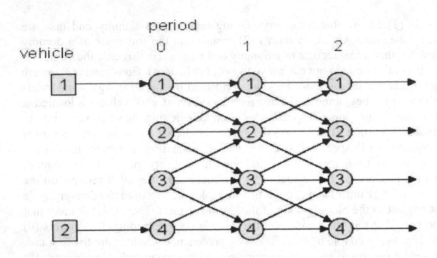

Fig. 2. Description of the time-space network (MIP)

Several versions of MIPs are presently under investigation. Here, we present one that has given interesting results up to now.

Sets and Parameters of the MIP Model:

Char: the set of agvs

Nodes: the set of nodes

Periods: the set of periods.

ArcsPlus: the set of all arcs (including those with dummy nodes), represented as an interval of integers.

M is the length of the horizon (number of periods). The variables **Alloc** [·] and **Starttime** [·] obtained from the CP model are used as input.

Segment[a] is a record having two fields. The first field (**Segment[a].orig**) is the origin of the arc whereas the second field (**Segment[a].dest**) is the destination of a.

The variables of the MIP model:

Y [k, t, p] = 1 if vehicle k ∈ Char is on node p ∈ Nodes at period t ∈ Periods.

Z [k, t, a] = 1 if vehicle k ∈ Char starts visiting arc a ∈ ArcsPlus at period t ∈ [0 ... M-1].

The MIP model is defined as follows:

$$\text{Min} \sum_{k \in Char} \sum_{t \in Periods} \sum_{p \in Nodes} Y[k, t, p]$$

s.t

(1) $\forall\, k \in$ Char

$\quad Y[k, \text{Starttime}[\,2 - k], \text{node}[\text{Task}[\,2 - k]]] = 1$

(2) $\forall\, r \in$ Requests

$\quad Y[\text{Alloc}[r], \text{Starttime}[\text{pick}[r]], \text{node}[\text{Task}[\, r]]] = 1;$

(3) $\forall\, r \in$ Requests

$\quad Y[\text{Alloc}[r], \text{Starttime}[\text{del}[r]], \text{node}[\text{Task}[\, r]]] = 1;$

(4) $\forall\, r \in$ Requests

$\quad Y[\text{Alloc}[r], \text{Starttime}[\text{pick}[r]] + 1, \text{node}[\text{Task}[\, r]]] = 1;$

(5) $\forall\, r \in$ Requests

$\quad Y[\text{Alloc}[r], \text{Starttime}[\text{del}[r]] + 1, \text{node}[\text{Task}[\, r]]] = 1;$

(6) $\forall\, t$ in Periods, $\forall\, k$ in Char

$\quad \sum_{p\ in\ Nodes} Y[k, t, p] = 1;$

(7) $\forall\, k \in$ Char, $\forall\, t \in [0 .. M - 1], \forall\, a \in$ Arcs

$\quad Y[k, t, \text{Segment}[a].\text{orig}] + Y[k, t + 1, \text{Segment}[a].\text{dest}] - Z[k, t, a] <= 1;$

(8) $\forall\, k \in$ Char, $\forall\, t \in [0 .. M - 1], \forall\, a \in$ Arcs

$\quad Z[k, t, a] <= Y[k, t, \text{Segment}[a].\text{orig}];$

(9) $\forall\, k \in$ Char, $\forall\, t \in [0 .. M - 1], \forall\, a \in$ Arcs

$\quad Z[k, t, a] <= Y[k, t + 1, \text{Segment}[a].\text{dest}];$

(10) $\forall\, t \in [0 .. M - 1], \forall\, k \in$ Char

$\quad \sum_{a\ \in\ ArcsPlus} Z[k, t, a] = 1;$

(11) $\forall\, t \in$ Periods, $\forall\, p \in$ Nodes

$$\sum_{p\ \in\ Nodes} Y[k, t, p] \le 1 + \left(\sum_{\substack{r \in \text{Realtasks}: \\ t = \text{Starttime}[\, r] \wedge p = \text{node}(\text{Task}[\, r])}} 1 \right) * \left(\sum_{\substack{r \in \text{RealTasks}: \\ t = \text{Starttime}[\, r] - 1 \wedge p = \text{node}(\text{Task}[\, r])}} 1 \right);$$

(12) $\forall\, t \in [0 .. M - 1], \forall\, a \in$ ArcPlus

$\quad \sum_{k\ \in\ Char} Z[k, t, a] + \sum_{\substack{k \in Char, \\ b \in Opp[a]}} Z[k, t, b] \le 1;$

Constraints (1) specify that every vehicle must be present at its starting node at period 0. Constraints (2-3) enforce the presence of vehicles at their task node in due time. Constraints (4-5) ensure that every vehicle stays at least one period at its task node to load or unload. Constraints (6) ensure that every vehicle has a unique position at each period. Constraints (7) imply that if a vehicle starts visiting the origin of an arc at period t, it will visit the destination at period t+1. Constraints (8) enforce that if a vehicle is on a node at period t, it means that it has started visiting an arc (waiting arc or not) at period t-1. Constraints (9) enforce that if a vehicle is on a node at period t+1, it means that it has started visiting an incoming arc (waiting arc or not) at period t. Constraints (10) ensure that every vehicle starts running on a unique arc (real or waiting arc) at each period. Constraints (11) forbid the presence of two vehicles on the same node except the case where one vehicle is finishing its task while another is starting its task on a work station. In a certain sense, these are anti-collision constraints on nodes. Constraints (10) are anti-collision constraints on arcs: no two vehicles can travel at the same time on the same arc in opposite directions.

4   Preliminary Results

The method has been implemented in OPL Script. We compared our method to the approach of Desaulniers *et al.* and we not only gain on flexibility by using CP but we were able to solve formerly unsolved cases (their algorithm failed in two instances). We also solved some new cases with five and six Agvs (the maximum number of Agvs in Desaulniers et al. was four). The number of Agvs used was limited to six since it didn't make sense to increase the number of Agvs with regards to the number of requests. However, larger applications like container terminal operations use dozens of agvs and no optimization automated solutions exist Presently, the size of the MIP model for the routing part is very large. It depends largely on the size of the horizon. Then larger time horizons will likely be more difficult to handle. We need to test our method on problems of larger number of tasks or Agvs with the idea of rolling horizons.

Our method took more computing time than that of Desaulniers et al. even though the computation times found are below the limit of ten minutes that we set. Our tests were done on a Pentium 4, 2.5 GHz. Desaulniers et al. did their tests on a SUNFIRE 4800 workstation (900 MHz).

Our approach can be transformed into a heuristic version by limiting the time of each scheduling solution to 30 seconds. Experiments are planned to see if this technique can yield quickly very good solutions.

5   Conclusion

This article reports on the development of a flexible hybrid algorithm based on the decomposition into CP and MIP components. We were able to solve some formerly unsolved cases of a complex AGV dispatching and routing problem. Tests on problems with a greater number of tasks or Agvs are needed to fully evaluate the effectiveness of the proposed method. It would be interesting to analyze the impact of the number and the diversity of precedence relationships between tasks. This research is an ongoing project as we are now working on refining the presented models and the iterative loop that guides them. As future work, an adaptation of our approach to a mining context should be very interesting due to the high level of congestion present in these problems.

References

[1]   Desaulniers, G., Langevin, A. and Riopel, D. 2003. Dispatching and conflict-free routing of automated guided vehicles: an exact approach. *International Journal of Flexible Manufacturing Systems. To appear.*

[2]   Krishnamurthy, N.N., Batta, R. and Karwan, M.H. 1993. Developing Conflict-free Routes for Automated Guided Vehicles. *Operations Research,* vol. 41, no. 6, 1077-10-90.

[3]   Langevin, A., Lauzon, D. and Riopel, D., 1996 Dispatching, routing and scheduling of two automated guided vehicles in a flexible manufacturing system. *International Journal of Flexible manufacturing Systems,* 8, 246-262.

[4]   Qiu, L., Hsu, W-J., Huang, S-Y. and Wang, H. 2002 Scheduling and routing algorithms for Agvs: a survey. *International Journal of Production Research,* vol. 40, no 3, 745-760.

Making Choices
Using Structure at the Instance Level
within a Case Based Reasoning Framework*

Cormac Gebruers[1], Alessio Guerri[2], Brahim Hnich[1], and Michela Milano[2]

[1] Cork Constraint Computation Centre
University College Cork, Cork, Ireland
{c.gebruers,b.hnich}@4c.ucc.ie
[2] DEIS
University of Bologna
Viale Risorgimento 2, 40136 Bologna, Italy.
{aguerri,mmilano}@deis.unibo.it

Abstract. We describe using Case Based Reasoning to explore *structure at the instance level* as a means to distinguish whether to use CP or IP to solve instances of the Bid Evaluation Problem.

1 Introduction

Constraint programming (CP) and Integer Linear Programming (IP) are both highly successful *technologies* for solving a wide variety of combinatorial optimization problems. When modelling a combinatorial optimization problem, there is often a choice about what technology to use to solve that problem. For instance in the Bid Evaluation problem (BEP) –a combinatorial optimization problem arising in combinatorial auctions– both CP and IP can successfully be used [ ]. While there exist domains where one can easily predict what technology will excel, in many domains it is not clear which will be more effective. The BEP falls into the latter case.

How do we choose among technologies when all instances share the same *problem structure*? We are currently investigating machine learning methodologies to explore *structure at the instance level* as a means to distinguish whether to use CP or IP to solve instances of the BEP.

In [ ] a Decision Tree technique is used to select the best algorithm for a BEP instance. In this paper, we use another Artificial Intelligence technique, Case Based Reasoning (CBR), as a framework within which to carry out this same exploration. CBR utilises similarity between problems to determine when to reuse past experiences to solve new problems. An experience in this context is what technology to use to solve an instance of the BEP. Using a representation

---

* This work has received support from Science Foundation Ireland under Grant 00/PI.1/C075. This work was partially supported by the SOCS project, funded by the CEC, contract IST-2001-32530.

that distinguishes problems at the instance level, and a similarity function to compare problems expressed in this representation, enables us to determine what technology to use on a new problem instance. We present a preliminary result showing this methodology can predict 80% of the time whether to use CP or IP to solve the BEP.

2   Bid Evaluation Problem

Combinatorial auctions are an important e-commerce application where bidders can bid on combination of items. The *Winner Determination Problem WDP* is the task of choosing, from among $\mathcal{M}$ bids, the best bids that cover all items at a minimum cost or maximum revenue. The winner determination problem is NP-hard. The *Bid Evaluation Problem BEP*, is a time constrained version of the *WDP* and consequently involves temporal and precedence constraints. Items in a bid are associated with a time window (and hence duration) and are interconnected by precedence constraints. A solution to the *BEP* involves solving the *WDP* and additionally satisfying the temporal and precedence constraints. In [ ], different approaches based on pure CP, pure IP and hybrid approaches mixing the two have been developed and tested. Furthermore, variants of these technologies are considered involving different parameter settings for each technology. In this work we concern ourselves with a sub-problem of deciding whether to use CP or IP as a solution technology for the BEP. Due to lack of space, we refer the reader to [ ] for detailed information about the CP and IP models developed for the BEP.

3   What Technology?

As a consequence of dominant structure apparent in a problem domain, there are situations where clear predictions about whether to use CP or IP can be made. For instance, when side constraints complicate the problem, CP can take advantage of them. When the problem is highly structured, polyhedral analysis can be highly effective. When the problem has a loose continuous relaxation, CP can overcome this weakness. If, instead, relaxations are tight and linear constraints tidily represent the problem, IP should most probably be used. However deciding whether to use IP or CP to solve a particular combinatorial optimisation problem is often an onerous task. Experimental results evaluating the performance of the two strategies considered for the BEP, show that neither CP nor IP is a clear winner [ ]. These results further indicate that there isn't any simplistic structure or problem feature, that correlates with choice of suitable technology. Hence we propose to explore whether structure at the instance level can be used to discriminate between CP and IP for the BEP.

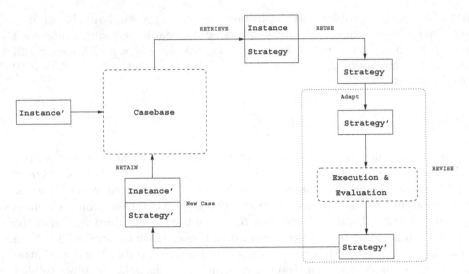

Fig. 1.   Methodology Overview

4   CBR Framework

CBR enables past problem solving experiences to be reused to solve new problems [ ]. CBR has been successfully used in the context of e.g. diagnosis and decision support [ ], design and configuration [ ], etc.

Experiences are stored along with the problems they solve as *cases*. A case is a representative example of a cluster of instances (a cluster can contain from 1 to n instances) that are similar to one another, but different from other clusters of instances. A particular technology (CP or IP in this work), is associated with one or more clusters of instances i.e. cases.

A CBR system consists of a four step cycle; *retrieve*, *reuse*, *revise*, and *retain*. To solve a new problem, we *retrieve* a case from the casebase, whose problem part is most similar to the new problem. We then *reuse* the experience part of the case to solve the new problem. The casebase may be *revised* in light of what has been learned during this most recent problem solving episode and if necessary the retrieved experience, and the new problem, may be *retained* as a new case. Every time a new problem instance is presented to the system, this cycle enables a CBR system to both learn new experiences and to maintain or improve the quality of the cases.

We now describe a CBR system called SELECTOR that enables us explore the use of structure at the instance level to predict whether to use CP or IP for BEP instances. Firstly, consider a case in SELECTOR. For now we will simply refer to the problem part of a case in abstract terms. We do this because the choice of how to represent a problem instance is one aspect to be explored. The experience part of a case corresponds to the appropriate technology for that instance (determined by experimentation) i.e. CP or IP. A case is thus a tuple composed of an abstract representation of an instance, and an appropriate technology (either CP or IP) that efficiently solves that instance. The final element of the system is a function f_{sim} to compute the similarity between two problem instances. The choice of f_{sim} is inextricably linked to the problem representation and hence is the other aspect we wish to explore in this work.

SELECTOR has two modes of operation; a training mode and a testing mode. A dataset is randomly divided into two sub-sets for training and testing purposes. Initially, the casebase is seeded with a random instance. In training mode, a casebase is assembled using the training problems. We expect that the instances retained in the casebase constitute examples of when to use the appropriate technology. The training mode consists of the following activities:

Retrieval: The current training instance is compared with every case in the casebase and the most similar case (established using an f_{sim}) is returned;

Reuse: The current training instance is solved using the technology identified by the case retrieved during the retrieval step;

Training Evaluation: The current training instance is solved using the other technologies that could have been chosen. The results of this step and the reuse step are recorded in preparation for the next step.

Revise & Retain: If the retrieved case has predicted incorrectly, then a new case is assembled consisting of the current training instance and the most appropriate technology (rather than the retrieved technology). This case is then saved to the casebase.

Once the casebase is non-empty, we can enter testing mode. For each testing problem, test mode does the following:

Retrieval: As for training mode;

Reuse: As for training mode;

Testing Evaluation: if the system correctly predicts the appropriate technology, we increment the good prediction score. Otherwise we increment the failure score. No new cases are added to the casebase in testing mode.

The training and testing phases can be intertwined at an interval n of the user's choosing. The system continues in training mode until n new cases have been added, then switches to testing mode. Once the testing set has been exhausted, the system reverts to training mode until a further n cases have been added... and so on until there are no instances left in the training set. This approach of intertwining training and testing enables us to identify learning behavior as casebases grow, and to consider the impact of an individual or group of cases on casebase performance. These factors are important for examining the impact of structure at the instance level.

This methodology is based on the intuition that if two instances are similar, then it follows that the same technology should be appropriate for both problems. Whether this approach works or not depends on two critical factors; how to represent problem instances and how we decide they are similar. In the next section, we give an example of representations and similarity measures.

5   Case Study

5.1   Problem Representations and Similarity Measures

Feature Based Approaches. We consider representations based on four features of the BEP that fully describe the problem; the number of bids (b), the number of tasks (t), the number of includes constraints (i), and the number of precedence constraints (p). All these features can be determined from the problem instance in polynomial time.

We explore different similarity measures between two *BEP* instances p1 and p2 with feature vectors p_1 and p_2, respectively:

Weighted Block City: If $p_1 = \langle b_1, t_1, i_1, p_1 \rangle$, $p_2 = \langle b_2, t_2, i_2, p_2 \rangle$, then we refer to following family of block city similarity measures as *weighted similarity functions*:

$$sim(p_1, p_2) =$$

$$= w_b \left(\frac{|b_1 - b_2|}{b_{max} - b_{min}} \right) \bowtie w_t \left(\frac{|t_1 - t_2|}{t_{max} - t_{min}} \right) \bowtie w_i \left(\frac{|i_1 - i_2|}{i_{max} - i_{min}} \right) \bowtie w_p \left(\frac{|p_1 - p_2|}{p_{max} - p_{min}} \right)$$

where $\bowtie \in \{+, *\}$, and f_{min} and f_{max} are the minimum and the maximum values for feature f, respectively. Note that we use normalization to remove the effects of different problem sizes.

Euclidean Distance: The similarity measure between two feature vectors is defined as the Euclidean distance between these two vectors.

Structural Approach: Should the feature based representations prove insufficient to distinguish between IP and CP, we also propose to examine structural representations such as graphs. The principle graph representation we adopt for this purpose is a minor natural extension of the bid-good graph , where we add precedent edges between task vertices to represent the precedence constraints that exist between tasks. We compute similarity between problem instances using a graph matching technology based on an incomplete branch and bound approach. However, due to space limitations, the details of the algorithms are not shown.

5.2   Preliminary Results

We generated a large variety of instances, spacing from easiest with 5 tasks and 15 bids to hardest with 30 tasks and 1000 bids, and with variable tasks-per-bid values and precedence graph structures. In this work, we applied CBR only to harder instances, where the difference between search times of the algorithms becomes considerable. This considered data set is composed of 90 instances, the easiest of which has 15 tasks, about 400 bids and a mean tasks-per-bid value little higher than 1.

---

[1] a common way to represent the BEP, where vertices represent bids and tasks, and edges between bids and tasks represent the inclusion of tasks in bids. See [ ].

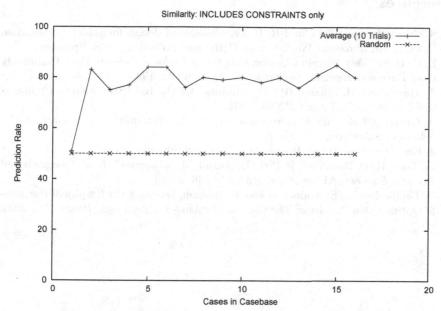

Fig. 2. Prediction using Includes constraints (i); Block City Similarity Measure

The 90 problems were randomly divided 10 times into pairs of training and testing sets. The ratio of training to testing problems was $\frac{2}{3}$ training and $\frac{1}{3}$ testing. Each experiment was repeated 10 times using the 10 pairs of training and testing sets and results averaged over these 10 trials.

We observe it is possible to achieve an average prediction rate across 10 trials of 80% by using the number of Includes constraints (i). However, it is very misleading to draw any further conclusions based on such elementary experiments. No significant difference in prediction ability was observed between using a similarity measure based on Block-City or Euclidean Distance, hence we show the Block City result only. The representations and similarity measures considered thus far may appear quite basic, however it is informative to see how effective relatively simple measures can be [ ].

6   Conclusion

In this paper, we propose an investigation of instance structure as a discriminating factor among solution technologies for the BEP within a CBR framework. In our future work, we plan to perform extensive exploration of both feature based and more complex structure based representations.

References

[1] S. Craw, N. Wiratunga, and R. Rowe. Case-based design for tablet formulation. In Proc. 4th European Workshop on CBR, pages 358–369, 1998. Springer.

[2] R. C. Holte. Very Simple Classification Rules Perform Well on Most Commonly Used Datasets. *Machine Learning*, vol. 3, pp. 63-91, 1993.

[3] A. Guerri and M. Milano, IP-CP techniques for the Bid Evaluation in Combinatorial Auctions, in Proc. CP2003, 2003.   ,

[4] A. Guerri and M. Milano, Learning techniques for Automatic Algorithm Portfolio Selection, Submitted.

[5] J. Kolodner, Case-Based Reasoning, Morgan Kaufmann, 1993.

[6] M. Lenz, H.-D. Burkhard, P. Pirk, E. Auriol, M. Manago: CBR for Diagnosis and Decision Support. AI Commun. 9(3): 138-146 (1996)

[7] K. Leyton-Brown, E. Nudelman and Y. Shoham, Learning the Empirical Hardness of Optimization Problems: The Case of Combinaorial Auctions, *Proc CP02*, 2002.

The Challenge of Generating Spatially Balanced Scientific Experiment Designs*

Carla Gomes, Meinolf Sellmann, Cindy van Es, and Harold van Es

Cornell University
{gomes,sello}@cs.cornell.edu
{clv1,hmv1}@cornell.edu

The development of the theory and construction of combinatorial designs originated with the work of Euler on Latin squares. A Latin square on n symbols is an $n \times n$ matrix (n is the order of the Latin square), in which each symbol occurs precisely once in each row and in each column. Several interesting research questions posed by Euler with respect to Latin squares, namely regarding orthogonality properties, were only solved in 1959 [ ]. Many other questions concerning Latin squares constructions still remain open today.

From the perspective of the Constraint Programing (CP), Artificial Intelligence (AI), and Operations Research (OR) communities, combinatorial design problems are interesting since they possess rich structural properties that are also observed in real-world applications such as scheduling, timetabling, and error correcting codes. Thus, the area of combinatorial designs has been a good source of challenge problems for these research communities. In fact, the study of combinatorial design problem instances has pushed the development of new search methods both in terms of systematic and stochastic procedures. For example, the question of the existence and non-existence of certain quasigroups (Latin squares) with intricate mathematical properties gives rise to some of the most challenging search problems in the context of automated theorem proving [ ]. So-called general purpose model generation programs, used to prove theorems in finite domains, or to produce counterexamples to false conjectures, have been used to solve numerous previously open problems about the existence of Latin squares with specific mathematical properties. Considerable progress has also been made in the understanding of symmetry breaking procedures using benchmark problems based on combinatorial designs [, , ,]. More recently, the study of search procedures on benchmarks based on Latin squares has led to the discovery of the non-standard probability distributions that characterize complete (randomized) backtrack search methods, so-called heavy-tailed distributions [ ].

In this paper we study search procedures for the generation of *spatially balanced Latin squares*. This problem arises in the design of scientific experiments. For example, in agronomic field experiments, one has to test and compare different soil treatments. Two different soil treatments may correspond to two different fertilizers or two different ways of preparing the soil. Most agronomic

---

* This research was partially supported by AFOSR grants F49620-01-1-0076 (Intelligent Information Systems Institute) and F49620-01-1-0361 (MURI).

J.-C. Régin and M. Rueher (Eds.): CPAIOR 2004, LNCS 3011, pp. 387–   , 2004.
© Springer-Verlag Berlin Heidelberg 2004

field experiments are implemented through randomized complete block designs (RCBD) where each block has as many experimental units as treatments [ ]. Use of blocks is in most cases justified by spatial variability in fields, and this layout is an attractive way to organize replications. This approach to experimental design uses random allocation of treatments to plots, which is used to ensure that a treatment is not continually favored or handicapped in successive replications by user bias or some extraneous source of variation [ ]. Although this randomization approach is intuitively attractive, it has been shown to cause biases and imprecision under most field conditions [   ]. The reason for this is that underlying soil characteristics are typically non-random and show field trends, spatial autocorrelation, or periodicity [   ]. For example, fertility patterns in fields often exhibit high and low areas due to, among others, erosion, drainage variability, and management history. The classical randomization process does not explicitly account for such field patterns, and many realizations of such RCBD designs may result in undesirable outcomes.

To address the limitations of the traditional RCBD designs, van Es and van Es [   ] proposed *spatially balanced* experimental designs that are inherently robust to non-random field variability. This approach uses dummy indicators and the treatments are randomly assigned to the indicators. In other words, treatments are randomly allocated to optimized designs, rather than to plots. Such designs may be spatially-balanced complete block designs or spatially-balanced Latin squares, the latter being a special case of the former where the number of treatments equal the number of replications.

We report our preliminary results concerning the generation of spatially-balanced Latin squares. In a spatially-balanced Latin square all pairs of symbols (treatments, in agronomic terms) have the same total distance in the Latin square. The distance of two symbols in a given row is the difference between the column indices of the symbols. The existence of spatially-balanced Latin squares is an open question in combinatorics, and no polynomial time constructions for the generation of spatially-balanced Latin squares have been found yet. Therefore, in order to get some insights into the structure of spatially-balanced Latin squares we used general local and complete search methods. We discovered that local search methods do not scale well on this domain, failing to find the global optimum for instances larger than order 6. This result was somehow surprising, especially given that local search methods perform well on generating (regular) Latin squares. Note that generating spatially-balanced Latin squares is considerably more difficult than generating regular Latin squares. On the other hand, our results with a CP based approach were very promising and we could generate totally spatially-balanced Latin squares up to order 18. Furthermore

---

[1] While the current state of the art of local search and backtrack search methods can easily generate Latin squares of order 100 or larger, the largest spatially-balanced Latin squares that we can generate is 18, using considerably more sophisticated techniques. There are constructions for generating Latin squares of arbitrary order. Our comparison considers only the generation of Latin squares using local search or backtrack search methods.

the CP based models provided us with interesting insights about the structure of this problem that allowed us to conjecture the existence of polynomial time constructions for generating spatially-balanced Latin squares. We are currently working on finding such efficient constructions.

The structure of the paper is as follows. In the next section we provide basic definitions. In section   we describe our simulated annealing approach and we present our main CP based model. In section   we provide empirical results.

1   Preliminaries

Definition 1. [**Latin square and conjugates**] *Given a natural number $n \in$ $\mathbb{N}$, a Latin square L on n symbols is an $n \times n$ matrix in which each of the n symbols occurs exactly once in each row and in each column. We denote each element of L by l_{ij}, $i, j \in \{1, 2, \cdots, n\}$. n is the order of the Latin square.*

Given a Latin square L of order n, its row (column) conjugate R (C) is also a Latin square of order n, with symbols, $1, 2, \cdots, n$. Each element r_{ij} (c_{ij}) of R (C) corresponds to the row (column) index of L in which the symbol i occurs in column (row) j.

Definition 2. [**Row distance of a pair of symbols**] *Given a Latin square L, the distance of a pair of symbols (k, l) in row i, denoted by $d_i(k, l)$, is the absolute difference of the column indices in which the symbols k and l appear in row i.*

Definition 3. [**Average distance of a pair of symbols in a Latin square**] *Given a Latin square L, the average distance of a pair of symbols (k, l) in L is $\bar{d}(k, l) = \sum_{i=1}^{n} d_i(k, l)/n$.*

We make the following important observation:

Remark 1. Given a Latin square L of order $n \in \mathbb{N}$, the expected distance of any pair in any row is $\frac{n+1}{3}$. (Proof omitted due to lack of space. See also [ ].)

Proof. (See [ ]) When we denote the probability that a random pair has distance h with $P(h)$, the expected distance is $\mu = \sum_{h=1}^{n-1} hP(h)$. It holds $P(h) = \frac{2(n-h)}{n(n-1)}$, and therefore, $\mu = \frac{2}{n(n-1)} \left(n \sum_{h=1}^{n-1} h - \sum_{h=1}^{n-1} h^2 \right)$. Simplification yields $\mu = \frac{n+1}{3}$.

Clearly, a Latin square L of order $n \in \mathbb{N}$ is totally spatially balanced if every pair of symbols $1 \le k < l \le n$ has an average distance $\bar{d}(k, l) = \frac{n+1}{3}$. Consequently, we define:

Definition 4. *Given a natural number $n \in \mathbb{N}$, a Totally Spatially Balanced Latin Square (TBLS) is a Latin square in which $\bar{d}(k, l) = \frac{n+1}{3}$ $\forall 1 \le k < l \le n$.*

Since $n\bar{\mathrm{d}}(k,l) = \sum_i \mathrm{d}_i(k,l) \in \mathbb{N}$, it follows that:

Remark 2. If there exists a TBLS of order n, then $n \bmod 3 \neq 1$.

In the following section, we present different computational approaches for the generation of Totally Spatially Balanced Latin Squares. We refer to this problem as the TBLS problem.

2   Totally Spatially Balanced Latin Square Models

2.1   Simulated Annealing

We started by developing a simulated annealing approach for the TBLS problem. Our work borrows ideas from a successful simulated annealing approach for the Traveling Tournament Problem [ ].

Objective Function: Given that we are interested in the research question of the existence of Totally Spatially Balanced Latin Squares, our problem becomes a decision problem. Therefore, we first relax some of its constraints and try to minimize the constraint violation. We relax both the balancedness and the Latin square constraints. In case we cannot find a totally balanced Latin square, we would like to balance both the worst case pair of symbols as well as the average over all pairs. Therefore, we penalize the unbalancedness of a square with the term, $b := \sum_{1 \le k < l \le n} \left(\bar{\mathrm{d}}(k,l) - \frac{n+1}{3}\right)^4$.

The second term of the objective function penalizes a symbol if it occurs more than once in the same column. Denoting with $f_k(i)$ the number of times that symbol k occurs in column i, the Latin square penalty is defined as $ls := \sum_{1 \le i, k \le n} \max\{f_k(i) - 1, 0\}$.

The overall objective then is to minimize $b + \beta\, ls$, whereby β is a variable factor that oscillates during the optimization. It allows us to guide the search towards or away from the search region containing Latin squares. For more details on strategic oscillation we refer the reader to [,].

Neighborhood: As with every local search technique, the other fundamental design decision regards the neighborhood. We experimented with different neighborhoods, finding that a rather simple type of moves gives smoother walks and results in better performance than more complicated neighborhoods. In our approach, there is just one simple move allowed: Swap a random pair of symbols in a random row, whereby we only consider such pairs that will result in a change in the number of Latin square constraints that are fulfilled.

2.2   Constraint Programming Approach

In our basic CP model every cell of our square is represented by a variable that takes the symbols as values. We use an AllDifferent constraint [ ] over all cells in the same column as well as all cells in the same row to ensure the Latin

---

[2] We would like to thank an anonymous reviewer for suggesting this neighborhood!

square requirement. We also keep a dual model in form of the row conjugate that is connected to the primal model via channeling constraints for the quasi-group completion problem [,]. This formulation is particularly advantageous given that by having the dual variables at hand it becomes easier to select a "good" branching variable as well as to perform symmetry checks, as we will see shortly. In order to enforce the balancedness of the Latin squares, we introduce variables for the values $\bar{d}(k, l)$ and enforce that they are equal to $\frac{(n+1)}{3}$.

Variable Selection: As with many discrete problems, it turns out that the selection of the branching variable has a severe impact on the performance of our algorithm. For Latin square type problems it has been suggested to use a strategy that minimizes the options both in terms of the position as well as the value that is chosen. In our problem, however, we must also be careful that we can detect unbalancedness very early in the search. Therefore, we traverse the search space symbol by symbol by assigning a whole column in the row conjugate before moving on to the next symbol. For a given symbol, we then choose a row in which the chosen symbol has the fewest possible cells that it can still be assigned to. Finally, we first choose the cell in the chosen row that belongs to the column in which the symbol has the fewest possible cells left.

Symmetry Breaking: In order to avoid that symmetric search regions are explored repeatedly, we implemented Symmetry Breaking by Dominance Detection (SBDD) (see [,]). According to our strategy for the variable selection, we try all different mappings of symbols in the current search node to the symbols of the previously explored nodes. Given that mapping, using the dual model we can easily check in linear time whether there exists a permutation of the rows such that the current search node is dominated. However, since there exist $n!$ different permutations of the symbols, this symmetry check is rather costly. In order to reduce the computational effort, it is important to treat unassigned symbols implicitly, which gives great advantages especially when comparing against previously expanded search nodes that are located high up in the search tree. Still, symmetry breaking is expensive. In the following section we will therefore evaluate whether this enhanced symmetry breaking procedure pays off.

Composition of TBLS: We also developed a very promising strategy for generating TBLS instances using as building blocks TBLS instances of smaller orders. Given a TBLS instance of order n, our model generates TBLS instances of orders $2n$ (and $3n$) by making 1 (or 2) copies of the initial TBLS instance of order n, and appropriately renaming the symbols of the copy (or 2 copies). In this approach we only manipulate *entire* columns of the building blocks and therefore the number of variables is reduced to the number of columns of the composed TBLS. The domain of each variable in this model corresponds to the different columns of the building blocks.

3 Computational Results

We now present preliminary computational results obtained with our implementations of the local search as well as the constraint programming algorithm. The

Table 1. Comparison of constraint programming (CP-X) based and local search (LS-X) based approaches for the TBLS. Time given in CPU seconds. All values are medians over 10 runs

	Order						
	3	5	6	8	9	11	12
CPI - time	0.01	0.02	0.06	16.14	241	-	-
CPS - time	0.01	0.02	0.09	¿300	-	-	-
LSO - time	0.01	0.01	0.36	79.56	153	334	883
LSO - max. dev.	opt	opt	opt	0.25	0.22	0.36	0.5
LSO - av. dev.	opt	opt	opt	0.04	0.11	0.16	0.15
LSN - time	0.01	0.01	0.03	32.03	60.95	246	no LS
LSN - max. dev.	opt	opt	opt	opt	opt	0.36	no LS
LSN - av. dev.	opt	opt	opt	opt	opt	0.14	no LS

simulated annealing approaches were implemented in C++ compiled with the gnu g++ compiler version 3.2.2 on an Intel XEON 2.0 GHz CPU and 1.0 GB RAM. The CP approaches were implemented using ILOG Solver 5.1 and the gnu g++ compiler version 2.91 on an Intel Pentium III 550 MHz CPU and 4.0 GB RAM.

Table   shows our results for the two approaches. Comparing the two CP variants, the first (CPI) using an initialization of the first row and the second (CPS) using SBDD, we find that sophisticated symmetry breaking does not pay off for the problem sizes that we can tackle so far. Note that we cannot initialize the first row when using the traversal strategy without spending a lot of time in the symmetry checks since then all symbol permutations must be checked from the beginning. Instead, we can initialize the first column in the row dual, thus fixing the traversal of symbol 0.

Next, we see that the local search approach can also find optimal solutions for orders up to 9 if we do not use strategic oscillation (LSN). However, strategic oscillation (LSO) helps us to obtain feasible Latin square solutions (see order 12 for example), which is why we favor this approach for higher orders for which we are unable to provide totally balanced Latin squares so far. Table   also gives the maximum and the average deviation from the perfect balance in these cases.

As mentioned in the previous section, we also developed a CP based model for the generation of spatially-balanced Latin squares by means of *composition of columns of spatially-balanced Latin squares*. In this approach we use spatially-balanced Latin squares of order n as building blocks to produce spatially-balanced Latin squares of order $2n$ or order $3n$. Using such a strategy we were able to generate, for example, a spatially-balanced Latin square of order 18, by composing spatially-balanced Latin squares of order 9. We are currently working on streamlining this approach in order to produce efficient constructions for the generation of totally spatially-balanced Latin square instances of order n, n mod 2 = 0 or n mod 3 = 0. While we do not see how the local search approach could be further improved so that it can provide optimal solutions for much larger orders, the idea of composing squares can be stated naturally as a constraint program.

4 Conclusions and Future Work

We present several models for the generation of totally spatially-balanced Latin squares. While it is unclear at this stage how the local search approach could be tuned to give optimal solutions for orders greater than 9, our results with CP based models were very encouraging; We could find totally spatially-balanced Latin instances up to order 18. Moreover, our different CP based models provided us with good insights about the structure of the problem. In fact, we conjecture that totally spatially-balanced Latin squares can be generated using a polynomial time construction, based on a representation that exploits the underlying traversal structure of Latin squares corresponding to matchings in bipartite graphs, as well as the duality between rows, columns, and symbols in a balanced Latin square. We also conjecture that, for certain orders, spatially-balanced Latin squares can be generated by means of composition, in polynomial time. If some symbols are pre-assigned to specific cells of the Latin square, our conjecture is that the problem of deciding if a partially filled Latin square can be completed into a balanced Latin square is an NP-complete problem. We hope that our results will further stimulate research on this interesting and challenging problem.

References

[1] A. Anagnostopoulos, L. Michel, P. van Hentenryck, and Y. Vergados. A Simulated Annealing Approach to the Traveling Tournament Problem. In *Proc. CPAIOR'03*, 2003.

[2] A. Atkinson and R. Bailey. One hundred years of design of expriements on and off the pages of biometrika. *Biometrika*, 88:53–97, 2001.

[3] R. Bose and S. Shrikhande. On the falsity of euler's conjecture about the nonexistence of two orthogonal latin squares of order 4t+2. In *Proc. Nat. Ac. of Sc. 45*. 1959.

[4] W. Cochran and G. Cox. *Experimental design*. John Wiley and Sons, Inc., 1950.

[5] T. Fahle, S. Schamberger, and M. Sellmann. Symmetry Breaking. In *Principles and practice of Constraint Programming (CP01) Lecture Notes in Computer Science*, pages 93–107. Springer-Verlag, 2001.

[6] F. Focacci and M. Milano. Global Cut Framework for Removing Symmetries. In *Principles and practice of Constraint Programming (CP01) Lecture Notes in Computer Science*, pages 77–92. Springer-Verlag, 2001.

[7] F. Glover and M. Laguna. *Tabu Search*. Kluwer Academic Publishers, 1997.

[8] C.P. Gomes, B. Selman, N. Crato, and H. Kautz. Heavy-tailed phenomena in satisfiability and constraint satisfaction problems. *J. of Automated Reasoning*, 24(1–2):67–100, 2000.

[9] W. Harvey. Symmetry Breaking and the Social Golfer Problem. In *Proc. SymCon'01*, 2001.

[10] B. Hnich, B. M. Smith, and T. Walsh. Dual modelling of permutation and injection problems. 2003.

[11] J. Regin. A Filtering Algorithm for Constraints of Difference in CSPs. In *Proc. of AAAI*, pages 362–367, 1994.

[12] I. J. D. Rodriguez, A. del Val, and M. CebriÃín. Redundant Modeling for the QUasigroup Completion Problem. In F. Rossi, editor, *Principles and practice of Constraint Programming (CP03) Lecture Notes in Computer Science*, pages 288–302. Springer-Verlag, 2003.

[13] B. Smith. Reducing Symmetry in a Combinatorial Design Problem. In *Proc. CPAIOR'01*, pages 351–360, 2001.

[14] H. van Es. Sources of soil variability. In *Methods of Soil Analysis, Part 4: Physical Properties*. Soil Sci. Soc. Am, 2002.

[15] H. van Es and C. van Es. The spatial nature of randomization and its effects on outcome of field experiments. *Agron. J.*, (85):420–428, 1993.

[16] H. Zhang. Specifying latin square problems in propositional logic. In *Automated Reasoning and Its Applications*. MIT Press, 1997.

Building Models through Formal Specification

Gerrit Renker and Hatem Ahriz

School of Computing
The Robert Gordon University
Aberdeen. Scotland, UK.
{gr,ha}@comp.rgu.ac.uk

Abstract. Over the past years, a number of increasingly expressive languages for modelling constraint and optimisation problems have evolved. In developing a strategy to ease the complexity of building models for constraint and optimisation problems, we have asked ourselves whether, for modelling purposes, it is really necessary to introduce more new languages and notations. We have analyzed several emerging languages and formal notations and found (to our surprise) that the already existing Z notation, although not previously used in this context, proves to a high degree expressive, adaptable, and useful for the construction of problem models. To substantiate these claims, we have both compiled a large number of constraint and optimisation problems as formal Z specifications and translated models from a variety of constraint languages into Z. The results are available as an online library of model specifications, which we make openly available to the modelling community.

1 Motivation

Formal methods and notations are most commonly associated with software development in procedural and object-oriented implementation languages. We are developing a strategic software engineering approach for modelling constraint and optimisation problems (CSOPs); one of the main underlying objectives is to integrate the notion of such problems into the standard software design cycle [ ]. For this purpose, we have been investigating the use of formal notation in general and of Z in particular, coming to the conclusion that advantages are to be had in at least four areas.

The first concerns the inception phase of building an initial or conceptual model. A modeller must first come up with an understanding of the problem requirements before being able to exploit its specific features. Quoting Smith, a recognized expert in the area of modelling: "*Hence, although constraint programming does require an understanding of search and constraint propagation, it is by understanding the problem and building in that understanding that we can develop a successful model.*" [, sec. 13]

Secondly, as larger-scale software is mostly developed in a (possibly distributed) team context and problem-solving strategies are shared across the modelling community, we see the importance of formal notation as a means

J.-C. Régin and M. Rueher (Eds.): CPAIOR 2004, LNCS 3011, pp. 395–   , 2004.
© Springer-Verlag Berlin Heidelberg 2004

of communication which is not constrained by and tied to the specifics of a particular implementation language. Specifications of constraint and optimisation problems in the scientific literature are often either based on the use of non-standardized (sometimes informal) mathematical notation, or in form of source-code descriptions at implementation level. The importance of not committing to a certain implementation format was also one of the crucial insights realized by the founders of CSPLib, as *"representation remains a major issue in the success or failure to solve constraint satisfaction problems. All problems are therefore specified in natural language ..."* [ ]

The third aspect lies in proving and verifying that a constraint problem at hand is indeed syntactically and semantically covered by a given model. Formal specification languages here allow the interaction with computer tools (figure) for simplifying, reducing and rewriting statements, thus allowing to generate (canonical) forms of expressions which are either more general or more suitable for the problem at hand. The notion of debugging in constraint programming fundamentally differs from that in procedural programming, a verified model specification reduces the need for debugging by highlighting conceptual errors at an early stage of development. We support the argumentation of Law and Lee in [ ] in that we would like to reason about properties of CSP models without actually having to solve these. Modelling in constraint programming is further, like mathematical modelling, a rather abstract mental activity, and so the verification of a model can provide concrete evidence, reassuring the modeller that a chosen concept is indeed correct.

Finally, and in keeping with evolving concepts of constraint problem modelling, formal specifications allow higher-level abstractions of model formulations. Over the past years increasingly more expressive and abstract modelling languages have evolved. OPL [ ] innovated a uniform abstraction to deal with both CP and OR problems at the same language level. The $\mathcal{F}$ language [ ] introduced useful model abstractions based on function variables, which is being developed further in the ESRA language for relational modelling of constraint problems [ ]. Work on automated model refinement [ ] has provided substantial support for the conjecture that constraint problem models can be constructed by compositional refinement of abstract specifications. Such compositionality is also at the heart of introducing algebraic CSP model operators to support a modular design of constraint problem formulations [ ]. Furthermore, Law and Lee speak in that study of *"reusable model components"* and *"model patterns"* [, sec. 5]. The latter recently stirred interest in form of an invited lecture [ ].

From the above considerations we have chosen Z [,   ], due to the fact that its style is generic and not geared towards a certain programming paradigm. The schema format, as introduced in section , proved a natural match for expressing the main bodies of constraint and optimisation problems. To substantiate our claims and to evaluate Z, we have compiled a large number of well-known constraint and optimisation problems, which we make openly available as online

---

1

[2] A formal specification can also prototypically be verified through animation.

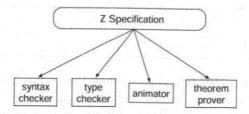

Fig. 1. Further processing of model specifications

library of specifications to the modelling community (cf. section   ). The remainder of this document is structured as follows. After a brief summary of relevant Z features in section   , we show how to use Z for the specification of CSOPs in section   , followed by an example in section   and conclusions in section   .

2   A Brief Recapitulation of Z Features

Z is a typed formal specification language based on first-order logic and Zermelo-Fraenkel set theory. It provides a precise syntax and a semantics based on classical mathematics for the abstract specification of systems in a model-oriented way. The language has been standardized as ISO/IEC standard 13568:2002, and its reference manual [   ] comes with a mathematical toolkit of common operations on sets and numbers. Main elements of a Z specification are given sets, axiomatic definitions and schemas. *Given sets* are introduced as further unspecified global names within square brackets, e.g.

[*Warehouses*]

This allows to reference the set *Warehouses* as type throughout the specification. *Axiomatic definitions* also have global scope and are often used to introduce constants or constant mappings. An axiomatic definition consists of a declaration part and an optional predicate part, separated by a horizontal line.

$$square : \mathsf{N} \longrightarrow \mathsf{N}$$
$$\forall n : \mathsf{N} \bullet square(n) = n * n$$

The example   introduces a total function *square* on N. Several type constructors, e.g. tuples, Cartesian product and (finite) power-sets, are provided by default, as well as common mathematical data types such as relations, functions, sequences and bags. Composite and heterogeneous data types can be introduced using schemas, which are one of the most powerful features of Z. A *schema* is an elementary building block of a Z specification. Like axiomatic definitions, schemas divide into a declaration and optional predicate part, the difference being that all declared constants and variables are locally-scoped. For example,

---

[3] This example first appeared in [, pp. 123/24].

```
┌─ SQPAIR ──────────────────────────────────────────────
│  x, y : N
│ ─────────────────────────────────────────────────────
│  y = square(x)
└───────────────────────────────────────────────────────
```

Here, x, y are local to $SQPAIR$ and y is assigned the value of applying the global function $square$. The elements in the declaration part are called *components* of the schema. A schema can therefore be viewed as a set of named components that are constrained by predicates. Schemas can be combined into new ones using the operations of the schema calculus such as inclusion, composition, projection, conjunction, disjunction, negation and hiding. A schema can also be seen as a mere abbreviation for the text it contains. Instead of

$$\exists\, x, y : N \bullet y = square(x) \wedge x > 100$$

we can equivalently write $\exists\, SQPAIR \bullet x > 100$. The *type* of a schema is the signature of its components, where the order of appearances is irrelevant. The type of the above schema is $\langle\!\langle x : N;\, y : N \rangle\!\rangle$. Likewise, the term $\{SQPAIR\}$ is the set of all schema bindings which have the type $\langle\!\langle x, y : N \rangle\!\rangle$ and contain exactly those values for x, y such that $y = square(x)$. More sophisticated variants of schemas in Z allow generic and parameterised definitions that specify entire families of schemas rather than sets of complying objects [ ].

3   Adapting Z for Constraint and Optimisation Problems

Constraint satisfaction problems are usually defined as a triple $\langle X, D, C \rangle$ of variables X, domains D and constraints C formulated over X. In the majority of constraint (logic) programming languages, the constraints in C can be expressed as quantified formulae of first-order logic. This allows a representation of constraint satisfaction problems in Z by single schemata, named e.g. CSP, where the elements of X and D are contained in the declaration part and the constraints in C in the predicate part. In cases of complex domains d_i, the base type (e.g. $\mathbb{Z}$) appears in the declaration part in combination with additional unary constraints on x_i in the predicate block. These concepts are illustrated by the example in section . Following the semantics of Z [ ], the solution set of constraint problems defined in the aforementioned way is simply the set $\{CSP\}$ (wrt. the above schema name), since it is the set containing all objects of type $\langle\!\langle x_1 : d_1 \ldots, x_n : d_n \rangle\!\rangle$ such that the constraints C of the predicate block hold. This permits the definition of a template for specifying optimisation problems. In *constrained optimisation problems*, we are interested in selecting the 'best' out of a set of solutions to a problem, where the evaluation criterion is determined by an objective function mapping solutions into numerical values. As is customary in IP and many CP languages, we will here assume that objective functions range over $\mathbb{Z}$. Using the above format for expressing constraint satisfaction problems, let the constraints of the problem be given as a schema $CSOP$.

---

[4] Besides, we can also make modular use of other and auxiliary schemata.

We can then define the objective function in a separate schema, as a function from *CSOP* to ℤ, and express the solution of the problem in terms of optimising the value of this function. This is also illustrated in section . The procedure for *unconstrained optimisation problems* is the same as for the constrained variants, the difference being that the predicate block of the main constraint schema remains empty. As ℤ itself does not make restrictions on the domains to use, we can in principle extend the concept also to the domain of Real (or even complex) numbers, although we would need to supply an appropriate toolkit.

4   An Example Specification

We now illustrate the main concepts of the last section on a small example, the bus driver scheduling problem (prob022 in CSPLib). We are a given set of tasks (pieces of work) and a set of shifts, each covering a subset of the tasks and each with an associated, uniform cost. The shifts need to be partitioned such that each task is covered exactly once, the objective is to minimise the total number of shifts required to cover the tasks. The sets of interest are *pieces* and *shifts*,

$$\mid \; pieces, shifts : \mathbb{P}\, \mathbb{N}$$

defined here as sets of natural numbers. The function *coverage* is part of the instance data and denotes the possible subsets of *pieces*. The only decision

───
 Driver_Schedule _____

  coverage : *shifts* ⟶ 𝔽 *pieces*
  allocate : iseq *shifts*
 ─────────────────────────────
  coverage ∘ *allocate* **partition** *pieces*
───

variable is *allocate*, an injective sequence of *shifts*. Composition of *allocate* with *coverage* yields a sequence of subsets of *pieces*. The built-in **partition** operator of ℤ [, p. 122] asserts that this sequence of subsets is a partition of the set *pieces*. We continue with the optimisation part, which illustrates the template format for modelling optimisation problems we mentioned in section . The *objective* function maps each element from the solution set *Driver_Schedule* into a natural number, in this case the number of shifts represented as the cardinality of the *allocate* variable.

───
 Optimisation_Part _____

  objective : *Driver_Schedule* ⟶ ℕ
  solution : *Driver_Schedule*
 ─────────────────────────────
  ∀ *ds* : *Driver_Schedule* • *objective*(*ds*) = #(*ds.allocate*)
  objective(*solution*) = min(*objective*⦅*Driver_Schedule*⦆)
───

─────────────
[5] As one reviewer rightly pointed out, the set-based nature of this small example is not very indicative of ℤ's abstraction facilities, but this example shows how succinct a formulation is possible. We refer to the more than 50 online examples.
[6] Shifts can appear at most once.

The last expression states that the element *solution* of the solution set must have a minimal value of the *objective* function. For this purpose, we use relational image of the entire solution set *Driver_Schedule* through *objective*.

5   Conclusion and Further Work

In this paper we have summarized the successful use of Z as a precise modelling notation for CSOPs. With regards to expressivity, we had positive results in mapping constructs and models from OPL, ESRA and $\mathcal{F}$ [ ]. We initially wrote the specifications without any tool support, subsequent verification (using the *f*uzz type checker and the Z-Eves prover) however proved so helpful in ironing out inconsistencies and improving the understanding of the problems that now all models are electronically verified prior to documentation. Our main focus at the moment is the modelling strategy and formal analysis of models. Aspects of further investigation are the translation of Z models into an implementation language, model animation and further tool support. The online repository is at

.

Acknowlededgment

We kindly thank the reviewers for their constructive criticism.

References

[1] A. Bakewell, A. M. Frisch, and I. Miguel. Towards Automatic Modelling of Constraint Satisfaction Problems: A System Based on Compositional Refinement. In *Proceedings of the Reform-03 workshop, co-located with CP-03*, pages 2–17, 2003.

[2] P. Flener, J. Pearson, and M. Ågren. Introducing ESRA, a Relational Language for Modelling Combinatorial Problems. In *Proc. Reform-03*, pages 63–77, 2003.

[3] I. P. Gent and T. Walsh. CSPLib: A Benchmark Library for Constraints. In J. Jaffar, editor, *Proceedings of CP'99*, pages 480–481. Springer, 1999.

[4] P. V. Hentenryck. *The OPL Optimization Programming Language*. MIT, 1999.

[5] B. Hnich. *Function Variables for Constraint Programming*. PhD thesis, Department of Information Science, Uppsala University, Sweden, 2003.

[6] Y. C. Law and J. H. M. Lee. Algebraic Properties of CSP Model Operators. In *Proceedings of Reform-02, co-located with CP'02*, pages pp. 57–71, 2002.

[7] G. Renker. A comparison between the F language and the Z notation. Technical report, Constraints Group, Robert Gordon University, Aberdeen, November 2003.

7

8

[8] G. Renker, H. Ahriz, and I. Arana. A Synergy of Modelling for Constraint Problems. In *Proc. KES'03*, volume 2773 of *LNAI*, pages 1030–1038. Springer, 2003.

[9] B. Smith. Constraint Programming in Practice: Scheduling a Rehearsal. Technical Report APES-67-2003, APES Research Group, September 2003.

[10] J. M. Spivey. *Understanding Z: A specification language and its formal semantics*, volume 3 of *Cambridge tracts in theoretical computer science*. CUP, 1988.

[11] J. M. Spivey. *The Z Notation: A Reference Manual*. Oriel College, Oxford, 1998.

[12] T. Walsh. Constraint Patterns. In *Proc. CP'03*, pages 53–64. Springer, 2003.

Stabilization Issues
for Constraint Programming Based
Column Generation

École Polytechnique de Montréal
C.P. 6079, succ. centre-ville, Montréal
Canada H3C 3A7
louism@crt.umontreal.ca

Abstract. Constraint programming based column generation is a hybrid optimization framework recently proposed that uses constraint programming (CP) to solve column generation subproblems. In the past, this framework has been successfully used to solve both scheduling and routing problems. Unfortunately the stabilization problems well known with column generation can be significantly worse when CP, rather than Dynamic Programming (DP), is used at the subproblem level. Since DP can only be used to model subproblem with special structures, there has been strong motivation to develop efficient CP based column generation in the last five years. The aim of this short paper is to point out potential traps for these new methods and to propose very simple means of avoiding them.

Introduction

Column generation was introduced by Dantzig and Wolfe [ ] to solve linear programs with decomposable structures. It has been applied to many problems with success and has become a leading optimization technique to solve Routing and Scheduling Problems [,]. However column generation methods often show very slow convergence due to heavy degeneracy problems.

Constraint programming based column generation can be particulary affected by convergence problems since its natural exploration mechanisms, such as Depth First Search (DFS), do not tend to generate sufficient information for the master problem. The main contribution of this paper is to explain why constraint programming can yield more unstable column generation processes than dynamic programming and to discuss how the use of known techniques, such as LDS [ ], can improve the behavior this hybrid decomposition method.

The next section, which gives a brief overview of the column generation framework, is followed by a description of the stabilization problem. Section 3 then discusses constraint programming based column generation with respect to convergence and stabilization. Some preliminary results on the Vehicle Routing with Time Windows are reported in section 4.

J.-C. Régin and M. Rueher (Eds.): CPAIOR 2004, LNCS 3011, pp. 402– , 2004.
© Springer-Verlag Berlin Heidelberg 2004

1   Column Generation

Column generation is a general framework that can be applied to numerous problems. However, since an example is often useful to give a clear explanation, all results in this paper will be presented with respect to the Vehicle Routing Problem with Time Windows (VRPTW). This is done without loss of generality nor restrictions to deal with this particular problem. The VRP can be described as follows: given a set of customers, a set of vehicles, and a depot, find a set of routes of minimal length, starting and ending at the depot, such that each customer is visited by exactly one vehicle. Each customer having a specific demand, there are usually capacity constraints on the load that can be carried by a vehicle. In addition, there is a maximum amount of time that can be spent on the road. The time window variant of the problem (VRPTW) imposes the additional constraint that each customer must be visited during a specified time interval. One can wait in case of early arrival, but late arrival is not permitted.

In the first application of column generation to the field of Vehicle Routing Problems with Time Windows, presented by Desrochers et al. [ ], the basic idea was to decompose the problem into sets of customers visited by the same vehicle (routes) and to select the optimal set of routes between all possible ones. Let r be a feasible route in the original graph (which contains N customers), R be the set of all possible routes r, c_r be the cost of visiting all the customers in r, $A = (a_{ir})$ be a Boolean matrix expressing the presence of a particular customer (denoted by index $i \in \{1..N\}$) in route r, and x_r be a Boolean variable specifying whether the route r is chosen ($x_r = 1$) or not ($x_r = 0$). The full Set Partitioning Problem, denoted S, is usually insolvable since it is impractical to construct and to store the set R (due to its very large size. It is thus usual to work with a partial set R' that is enriched iteratively by solving a subproblem. This is why, in general, the following relaxed Set Covering formulation is used as a Master Problem (M):

$$\min \sum_{r \in R'} c_r x_r \tag{1}$$

$$s.t \sum_{r \in R'} a_{ir} x_r \geq 1 \; \forall i \in \{1..N\} \tag{2}$$

$$x \geq 0 \tag{3}$$

To enrich R', it is necessary to find new routes that offer a better way to visit the customers they contain, that is, routes with a negative reduced cost. The reduced cost of a route is calculated by replacing the cost of an arc (the distance between two customers) d_{ij} by the reduced cost of that arc $c_{ij} = d_{ij} - \lambda_i$, where λ_i is the dual value associated with the covering constraint ( ) of customer i. The dual value associated with a customer can be interpreted as the marginal cost of visiting that customer in the current optimal solution (given R'). The objective of the subproblem is then the identification of a negative reduced cost path, i.e., a path for which the sum of the travelled distance is inferior to the

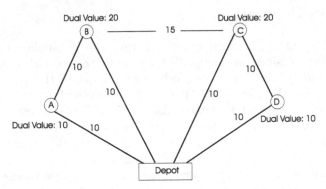

Fig. 1. Path $Depot - B - C - Depot$ has a negative reduced cost of -5

sum of the marginal costs (dual values). Such a path represents a new and better way to visit the customers it serves.

The optimal solution of (M) has been identified when there exists no more negative reduced cost path. This solution can however be fractional, since (M) is a relaxation of (S), and thus does not represent the optimal solution to (S) but rather a lower bound on it. If this is the case, it is necessary to start a branching scheme in order to identify an integer solution.

2   Stabilization Problems

Most column generation frameworks depend heavily on marginal costs to guide the search at the subproblem level. In the first iterations of the optimization process, it is possible that the marginal costs associated with each customer are not appropriately estimated by the dual values. For instance, it is possible that in some routes some customers pick up most of the total dual value. If this is the case (as illustrated in figure), then in the subproblem a path that visits each of those overweighed customers (B, C) will be considered a good route (with reduced cost of -5), even though it is not the case. With a more realistic distribution of dual values, all nodes would have been given a value of 15 and no more reduced cost path would have been found, thus saving the need for a last iteration.

This problem occurs because the Set Covering problem is degenerate and thus its dual has an infinite number of optimal solutions. The standard function that returns dual values in most LP codes usually returns an extreme point of the dual polyhedron. This behavior encourages dual variables to take either very large or very small values. This behavior led to the introduction of a stabilization method by du Merle et al. [ ] which defines a box around the dual values to prevent them from taking extreme values. Rousseau, Gendreau and Feillet [ ] recently proposed an Interior Point Stabilization (IPS) method that allows to select dual values inside their interval rather than on their boundaries. Stabilization methods attempt to accelerate convergence by limiting the negative effects of degeneracy.

During column generation the bounds on the dual values are tightened at each iteration when columns are added to R'. These columns, in the dual representation of M, represent cuts that limit and define the marginal costs of each nodes. That is why most column generation frameworks will try to generate a large number of columns at each iteration in order to refine the marginal costs intervals and get as much precise values as possible for the next iteration.

3   CP Based Column Generation

There have been few attempts to combine column generation and constraint programming. Chabrier [ ] presented a framework that uses constraint programming search goals to guide the column generation process that was recently packaged as a software library called Maestro [ ]. Junker et al. [   ] have proposed a framework they call *constraint programming based column generation* (which led to Fahle et al. [ ]) which uses CP to solve constrained shortest path subproblems in column generation. Fahle and Sellmann [ ] later enriched the CP based column generation framework with a *Knapsack* constraint for problems which present knapsack subproblems. This framework however required that the underlying graph be acyclic. Rousseau et al. [   ] then proposed a shortest path algorithm, based on CP and dynamic programming, that allowed the solution of problems presenting a cyclic structure like the routing problem.

The general idea of CP based column generation is to use specialized operations research (OR) algorithms, such as non elementary resource constrained shortest paths or knapsacks, both as relaxation and branching heuristics for a CP model. The subproblem model defined with constraint programming thus uses these OR techniques to achieve efficient pruning and search with respect to the optimization component of the subproblem (identifying the best negative reduced costs paths). Traditional CP mechanisms are used to deal with the other more messy side constraint that can be present in the model. The columns that should be added to R' are thus identified by solving the CP model with traditional techniques such as DFS branching.

That is the trap that can lead to unstable behavior and slow convergence. Depending on the branching strategy, DFS can tend to generate solutions that are very similar to one another i.e. paths with a lot of nodes in common. Furthermore, when the total number of columns needed for one iteration have been identified, it is possible that a significant portion of the nodes remained uncovered with the new columns. This means that no new information (cuts) has been provided to the master problem M about these nodes and that their marginal cost can be left unchanged for another iteration.

In contrast, when dynamic programming traverses its state space graph to identify negative reduced cost paths, it does so by traversing the whole graph simultaneously. This means that, in general, it tends to generate columns that are more diverse in terms of the nodes they cover. The set of possible marginal cost is then reduced for a larger number of node at each iterations.

Although it was not presented form the stand point of stabilization, this notion of diversity was raised by Sellman *et al.*in [ ] as technique which would allow more profitable combination in the master problem. The authors present a simple modification to the value selection heuristic that forbids each variable from taking the same value too many times.

Particular care must thus be taken to make sure that the search strategy used to solve the CP model will tend to cover the largest possible set of nodes. Limited Discrepancy Search (LDS) [ ] is an interesting alternative to DFS since it limits the search effort that is spent in one particular region of the solution space. Other techniques can also be used, for instances rejecting columns that are too similar to the ones already generated or devising branching heuristics that maximize the scope of the search.

4   Preliminary Results

We compare the number of iterations needed to reach optimality using DFS, LDS and Interior Point Stabilization (IPS) [ ] a very simple and efficient stabilization method. The idea behind Interior Point Stabilization is to generate a dual solution that is an interior point of the optimal dual space rather than an extreme point. The proposed way to achieve this goal is to generate several extreme points of the optimal dual polyhedron and to generate an interior point corresponding to a convex combination of all these extreme points. We report in table   the figures averaged by problem class.

The observed CPU times were larger for LDS than DFS due to the naive implementation of LDS. Each time the LDS level was augmented, all columns previously generated at a lower LDS level were regenerated but ignored (not added to R'). Results being encouraging, a more efficient implementation should be devised. However the IPS method is considerably faster then DFS, which is encouraging for a better implemented LDS. Since at this point CPU times do not provide valid information to perform a real efficiency comparison, we preferred not to include them in the results.

Since the aim of these first experiments was to assess the impact of LDS as a stabilization method and that improvements can be well observed at the root node of M, we did not perform the Branch and Price search for integer solutions.

The column generation framework used for comparison is the one presented in [ ] and the experiment were run on the 56 VRPTW instances of Solomon of size 25. These problems are classified into three categories: randomly (R) distributed customers, clustered (C) customers, and a mix of the two (RC). At each iteration we ask the subproblem to generate 100 columns and the maximum allowed time to prove optimality was set to 3600 seconds.

LDS does seem to improve the method convergence speed since it reduces the needed number of iteration to reach optimality. LDS seems to be sufficient

Table 1. Average number of iterations by problem class

Class	DFS	LDS	IPS
R	12.8	10.5	10.75
C	30.5	19.8	17.7
RC	22.75	20.25	8.5

in stabilizing the R and C classes of the Solomon Problems, however it is not as efficient as a true stabilization method (IPS) on the RC instances.

5   Conclusion

Slow convergence is a well known problem to OR researchers working with column generation and a number of methods to address this problem have been developed. The objective of this short paper was to stress that, by nature, CP based column generation can be even more affected by slow convergence than traditional column generation methods that use dynamic programming. Preliminary experiments tend to show that LDS can serve as a remedy to this problem but that a dedicated stabilization technique is still needed to accelerate convergence.

References

[1] C. Barnhart, E. L. Johnson, G. L. Nemhauser, M. W. P. Savelsbergh, and P. H. Vance. Branch-and-Price: Column Generation for Huge Integer Programs. *Operations Research*, 46:316–329, 1998.

[2] A. Chabrier. Using Constraint Programming Search Goals to Define Column Generation Search Procedures. In *Integration of AI and OR Techniques in Constraint Programming for Combinatorial Optimisation Problems*, pages 19–27, 2000.

[3] A. Chabrier. Maestro: A column and cut generation modeling and search framework. Technical report, ILOG SA, 2002. http://chabrier.free.fr/work.html.

[4] G. B. Dantzig and P. Wolfe. Decomposition principles for linear programs. *Operations Research*, 8:101–111, 1960.

[5] J. Desrosiers, Y. Dumas, M. M. Solomon, and F. Soumis. Time Constrained Routing and Scheduling. In M. O. Ball, T. L. Magnanti, C. L. Monma, and Nemhauser G. L., editors, *Network Routing*, volume 8 of *Handbooks in Operations Research and Management Science*, pages 35–139. North-Holland, Amsterdam, 1995.

[6] O. du Merle, D. Villeneuve, J. Desrosiers, and P. Hansen. Stabilized Column Generation. *Discrete Mathematics*, 194:229–237, 1999.

[7] T. Fahle, U. Junker, S. E. Karisch, N. Kohl, M. Sellmann, and B. Vaaben. Constraint Programming Based Column Generation for Crew Assignment. *Journal of Heuristics*, (8):59–81, 2002.

[8] T. Fahle and M. Sellmann. Constraint Programming Based Column Generation with Knapsack Subproblems. *Annals of Operations Research*, 115:73–94, September 2002.

[9] W. Harvey and M. Ginsberg. Limited Discrepancy Search. In *Proc. of the Fourteenth International Joint Conference on Artificial Intelligence (IJCAI-95)*, pages 607–615, Montréal, Canada, 1995. Morgan Kaufmann.   ,

[10] U. Junker, S. E. Karisch, N. Kohl, B. Vaaben, T. Fahle, and M. Sellmann. A Framework for Constraint Programming Based Column Generation. In *Principles and Practice of Constraint Programming*, Lecture Notes in Computer Science, pages 261–274, 1999.

[11] L.-M. Rousseau, Focacci F., M. Gendreau, and G. Pesant. Solving VRPTWs with Constraint Programming Based Column Generation. *Annals of Operations Research*, 2004. to appear.   ,

[12] L.-M. Rousseau, M. Gendreau, and D. Feillet. Interior Point Stabilization for Column Generation. Technical report, Centre de recherche sur les transports, 2003.   ,

[13] M. Sellmann, K. Zervoudakis, P Stamatopoulos, and T Fahle. Crew Assignment via Constraint Programmging: Integrating Column Generation and Heuristic Tree Search. *Annals of Operations Research*, 115:207–225, 2002.

A Hybrid Branch-And-Cut Algorithm for the One-Machine Scheduling Problem

Ruslan Sadykov

CORE
Université Catholique de Louvain
34 voie du Roman Pays, 1348 Louvain-la-Neuve
Belgium
sadykov@core.ucl.ac.be

Abstract. We consider the scheduling problem of minimizing the weighted sum of late jobs on a single machine ($1 \mid r_j \mid \sum w_j U_j$). A hybrid Branch-and-Cut algorithm is proposed, where infeasibility cuts are generated using CP. Two ways are suggested to increase the strength of cuts. The proposed approach has been implemented in the *Mosel* modelling and optimization language. Numerical experiments showed that the algorithm performs at least as well as the best to our knowledge exact approach [ ] on sets of public test instances.

1 Introduction

A set of jobs $N = \{1, \ldots, n\}$ has to be processed on a single machine. The machine can only process one job at a time and preemptions are not allowed. Each job $j \in N$, has a release date r_j, a processing time p_j, a due date d_j and a weight w_j. For a given schedule π, job j is *on time* if $c_j(\pi) \leq d_j$, where $c_j(\pi)$ is the completion time of j in the schedule π, and otherwise job j is *late*. The objective is to minimize the sum of the weights of the late jobs. Using standard scheduling notation, this problem is denoted as $1 \mid r_j \mid \sum w_j U_j$. The problem is $\mathcal{NP}$-hard in the strong sense.

Recently several approaches to solve this problem have appeared in the literature. First, the method, based on the notion of *master sequence*, i.e. a sequence that contains at least one optimal sequence of jobs on time, was suggested by Dauzère-Pérès and Sevaux [ ]. A Lagrangean relaxation algorithm is used to solve an original MIP formulation, derived from the master sequence. Another exact approach, based on a time-indexed Lagrangean relaxation of the problem, has been proposed by Péridy, Pinson and Rivreau [ ]. This method is able to solve to optimality 84.4% of a set of 100-job test instances in 1 hour. To tackle larger instances, some genetic algorithms have been suggested by Sevaux and Dauzère-Pérès [ ].

2 A Hybrid Branch-And-Cut Method

Recently some hybrid IP/CP algorithms have been applied to the multi-machine assignment scheduling problem (MMASP). In this problem jobs with release

J.-C. Régin and M. Rueher (Eds.): CPAIOR 2004, LNCS 3011, pp. 409–   , 2004.
© Springer-Verlag Berlin Heidelberg 2004

dates and deadlines are processed on unrelated machines and the total assignment cost is minimized (the cost of each job depends on the machine, to which the job is assigned). Jain and Grossmann [ ] have proposed an iterative hybrid algorithm for the MMASP. Later Bockmayr and Pisaruk [ ] have improved it using a Branch-and-Cut approach. Finally, Sadykov and Wolsey [ ] have introduced valid inequalities to tighten the "machine assignment" IP relaxation. This allowed them to solve to optimality significantly larger instances.

The problem $1 \mid r_j \mid \sum w_j U_j$ is a special case of the MMASP. Therefore we can apply the algorithm described above to it. Now we give some details about the algorithm. We use a binary variable x_j, which takes value one if job j is on time, and value zero if it is late. Let $\alpha_{li} = (r_i - r_l)^+$ and $\beta_{jl} = (d_l - d_j)^+$. Then we can write the following formulation for the problem:

$$\min \sum_{j \in N} w_j (1 - x_j) \tag{1}$$

$$s.t. \sum_{l \in N} \min \left[d_j - r_i, (p_l - \max\{\alpha_{li}, \beta_{jl}\})^+ \right] x_l \le d_j - r_i$$

$$\forall i, j \in N : r_i < d_j \tag{2}$$

$$x \in X \tag{3}$$

$$x_j \in \{0,1\}^n \quad \forall j \in N \tag{4}$$

Here $x \in X$ if and only if the corresponding set of jobs $J = \{j \in N : x_j = 1\}$ is *feasible*. This means that the set J of jobs can be processed on the machine without violating release and due dates. The tightened inequalities ( ) were introduced in [ ].

The hybrid Branch-and-Cut algorithm works in the following way. First, the formulation ( ), ( ) and ( ) is fed to a MIP solver. Whenever an integer solution $\bar{x}$ is found at some node of the global search tree, it is checked for feasibility. This can be done either by using a CP algorithm, typically the so-called "disjunctive" global constraint. Otherwise one can use any algorithm that checks if $\bar{x} \in X$ or not. If $\bar{x}$ is infeasible, we cut it off by adding a so called "no-good" cut:

$$\sum_{j \in J} x_j \le \mid J \mid -1, \tag{5}$$

where $J = \{j \in N : \bar{x}_j = 1\}$.

However, the direct application of this hybrid Branch-and-Cut algorithm to the $1 \mid r_j \mid \sum w_j U_j$ problem has its limits. The method does not work well when the number of jobs per machine is sufficiently large (about 50 jobs and more). This is because a large number of cuts are required to cut off infeasible solutions, produced by the IP part of the algorithm. This fact can be partly explained by the weakness of the "no-good" cuts.

The goal of this paper is to look for ways to overcome this weakness, which one can also expect to encounter when tackling other similar problems.

3 The Ways to Improve Infeasibility Cuts

Here we examine ways to improve the hybrid Branch-and-Cut algorithm by suggesting two approaches to produce stronger infeasibility cuts.

First, the following fact can be observed. When the solution $\bar{x}$ produced by the IP part of the algorithm is infeasible (i.e. the set of jobs $J = \{j \in N : \bar{x}_j = 1\}$ is infeasible), it is usually possible to isolate a smaller infeasible subset of jobs S, $S \subset J$, which is also infeasible. Obviously, a "no-good" cut, based on such a smaller subset of variables S, is stronger than a standard cut, which involves all the variables from the set J.

However, isolating a minimum, or even a minimal infeasible subset of jobs seems to be a very hard task, as even checking feasibility is NP-hard. Thus our aim is not necessarily to find a minimal infeasible subset, but as small one as possible. To do this we suggest a modification of the Carlier algorithm [ ]. This algorithm is used to find a schedule with minimum maximum lateness (or tardiness) for a set of jobs with release and due dates. The lateness of job j in schedule π is the difference between its completion time in π and its due date: $L_j(\pi) = c_j(\pi) - d_j$. Clearly, there exists schedule π such that $\max_{j \in J} L_j(\pi) \leq 0$ if and only if the set J of jobs is feasible.

Here we present the main ideas of the modified Carlier algorithm. At each node of the search tree, it constructs a schedule π_S using the Schrage heuristic in $O(n \log n)$ time. If π_S is *feasible* (i.e. $\max_{j \in J} L_j \leq 0$), then the set J of jobs is feasible and the algorithm terminates. If $\max_{j \in J} L_j > 0$, then in certain cases one can determine in linear time that it is impossible to reduce maximum lateness below zero for a certain subset $S \subset J$. In this case the current node is pruned. Otherwise we proceed as in the Carlier algorithm. One has available job c and a set $P \subset J$, such that c cannot be processed inside P in a feasible schedule. Thus two descendant nodes are created in one of which c is processed before P and after P in the other. To introduce these additional precedence relations, respectively, the due date or the release date of job c is changed.

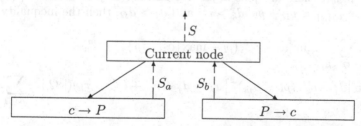

If Carlier's algorithm fails to find a feasible schedule for the set J of jobs, each leaf of the tree has associated with it an infeasible subset. Then we begin to ascent sequentially from the leaves to the root of the tree. Each node receives the infeasible subsets S_a and S_b from its descendants. Notice, that if job c does not belong to the subset S_a or S_b, then the corresponding subset is also infeasible for the current node, as only the parameters of job c are changed in the descendant nodes. The infeasible subset for the current node is determined the following way.

- If $c \notin S_a$ and $c \notin S_b$ then if $|S_a| < |S_b|$ then $S = S_a$, else $S = S_b$.
- If $c \notin S_a$ and $c \in S_b$ then $S = S_a$.
- If $c \in S_a$ and $c \notin S_b$ then set $S = S_b$.
- If $c \in S_a$ and $c \in S_b$ then set $S = P \cup S_a \cup S_b$. Job c can be scheduled before, inside or after P. In these cases, respectively, S_a, $P \cup \{c\}$ or S_b is infeasible, and all three of them are included in S.

After returning to the top node, the modified algorithm returns a required infeasible subset $S \subseteq J$.

Now we consider another way to generate infeasibility cuts. The tightened inequalities ( ) can be generalized in the following manner. Notice, that constraints ( ) exist for each pair of release and due dates (r_i, d_j), $i, j \in N$, for which $r_i < d_j$. The basic idea is to try to use one or more CP propagation algorithms to reduce certain time windows and adjust respective release and due dates according to some precedence relations, which can be deduced by propagation. Then after such adjustments more inequalities similar to ( ) can be obtained.

We propose to use the "Edge-Finding" constraint propagation technique to develop such cuts. For some set of jobs S, let $d_S = \max_{j \in S} d_j$, $r_S = \min_{j \in S} r_j$ and $p_S = \sum_{j \in S} p_j$. The "Edge-Finding" propagation rule for release dates is the following [, p. 22]:

$$\forall \Omega \in J, \forall k \in J \setminus \Omega: \text{ if } d_\Omega - r_{\Omega \cup \{k\}} < p_\Omega + p_k \qquad (6)$$

$$\text{then } \Omega \to k \text{ and } r_k^\Omega = \max_{\Omega' \in \Omega}\{r_{\Omega'} + p_{\Omega'}\} \qquad (7)$$

The rule for due dates is symmetric.

Let $\alpha_{lk}^\Omega = (r_k^\Omega - r_l)^+$ and $\beta_{jl} = (d_l - d_j)^+$. From ( ) and ( ) we can obtain the following. If there are jobs $k, j \in N$, and set of jobs $\Omega \subset N \setminus \{k, j\}$ such that $d_\Omega - r_{\Omega \cup \{k\}} < p_\Omega + p_k$, $d_k > r_k^\Omega$ and $d_j > d_\Omega$, then the inequality

$$\sum_{l \in N \setminus \Omega \setminus \{k\}} \min\left[d_j - r_k^\Omega, (p_l - \max\{\alpha_{lk}^\Omega, \beta_{jl}\})^+\right] x_l^m +$$

$$\min\left[d_j - r_k^\Omega, (p_k - \beta_{jk})^+\right] x_k \le d_j - r_k^\Omega + (r_k^\Omega - r_k)(|\Omega| - \sum_{o \in \Omega} x_o) \qquad (8)$$

is valid for the formulation ( )-( ).

This can be easily shown as follows. When $|\Omega| - \sum_{o \in \Omega} x_o \ge 1$, then the inequality ( ) is a relaxation of the inequality ( ) for the pair (r_k, d_j). When $|\Omega| - \sum_{o \in \Omega} x_o = 0$, all the jobs from Ω are on time and from ( ) job k can be executed only from the time moment r_k^Ω. Inequalities for due dates are similar.

Of course, we cannot add all the inequalities of type ( ) to the initial formulation because of their exponential number. But we can add them as cuts. Thus we need a separation algorithm to check whether a solution $\bar{x}$ violates any

Table 1. Comparison of the results for the sets of test instances

Inst.	Results in [ ]			Our results			
	solved in 1 h.	av. CPU	max. CPU	solved in 1 h.	solved in 1000 s.	av. CPU	max. CPU
b50	100%	47.1	409.1	100%	100%	1.7	55.7
b60	100%	157.9	2147.0	100%	100%	3.3	34.4
b70	97.8%	168.6	1745.3	100%	100%	7.0	105.8
b80	97.8%	294.6	3567.6	100%	100%	14.5	365.5
b90	88.9%	383.0	3542.7	98.9%	98.9%	36.8	670.4
b100	83.3%	515.7	3581.6	100%	97.8%	104.2	1871.5
s40	100%	26.6	448.6	100%	100%	0.6	8.6
s60	99.4%	178.2	2127.5	100%	100%	3.3	238.9
s80	95.0%	496.5	3552.2	100%	99.7%	15.3	1141.8
s100	84.7%	1049.9	3560.0	99.7%	99.1%	46.0	1505.4

inequality of type ( ). Such a separation algorithm can be obtained by adapting the "Edge-Finding" propagation algorithm. We use the algorithm of complexity $O(n^2)$ described in the book by Baptiste et al. [, p. 24]. The propagation algorithm considers the set of jobs $J = \{j \in N : \bar{x}_j = 1\}$ and whenever it finds a pair (Ω, k), which satisfies ( ), the inequalities of type ( ) for this pair (Ω, k) and all jobs $j \in J \setminus \Omega$, such that $d_j > r_k^\Omega$, are checked whether one of them is violated by $\bar{x}$. The complete separation algorithm has complexity $O(n^3)$.

So, the improved hybrid Branch-and-Cut algorithm for the $1 \mid r_j \mid \sum w_j U_j$ problem can be described in the following way. First, the formulation ( ),( ),( ) is fed to the IP solver. Whenever an integer solution $\bar{x}$ is found on some node of the global search tree, we check if $\bar{x}$ violates an inequality of type ( ) using the separation algorithm. If a violated inequality is found, then it is added to the formulation as a cut. If not, then we run the modified Carlier algorithm for the set of jobs $J = \{j \in N : \bar{x}_j = 1\}$. We interrupt it after a certain number of iterations (1000 in the experiments), because in some cases it takes a lot of time. If the algorithm terminates before the limit, a strengthened "no-good" cut is added in case of infeasibility, else we run the standard Carlier algorithm to check the feasibility of J and add a standard "no-good" cut if J is infeasible.

4   Numerical Experiments

Numerical experiments have been carried out on the same sets of test instances, as in the paper by Péridy, Pinson and Rivraux [ ]. To our knowledge the best results for the $1 \mid r_j \mid \sum w_j U_j$ problem are presented there, and we compare our results with theirs.

The hybrid Branch-and-bound method described here was implemented in the *Mosel* modelling and optimization language [ ], using *XPress-MP* as the IP solver. In Table   we present our results, as well as results from [ ] for comparison.

In the first column there are the sets of instances with number of jobs ('b' - the sets from the paper by Baptiste et al. [ ] with 90 instances for each dimension, 's' - the sets from the paper by Dauzère-Pérès and Sevaux [ ] with 320 instances for each dimension). The next three columns display results from [ ]: percentage of instances solved to optimality in 1 hour, average and maximum computing times (excluding not solved instances), respectively. In the last four columns we present our results: percentage of instances solved to optimality in 1 hour, 1000 seconds, average and maximum computing times (also excluding not solved instances). Notice, that the experiments in [ ] have been carried out done on a 450 MHz computer, whereas we have used a 2 GHz computer, so we included a column with time limit of 1000 seconds to provide a reasonable comparison.

It appeared that our method is always superior or has the same efficiency, even with taking into account the difference in speed of the computers. The suggested algorithm hasn't solved to optimality only 2 instances from 1820 considered in 1 hour and only 7 instances in 1000 seconds.

Notice, that the method presented here can be extended to the multi-machine case. For this the column generation approach is suggested, where the subproblem is exactly the one-machine problem considered here. For details see [ ].

References

[1] Baptiste, Ph., A. Jouglet, C. Le Pape, W. Nuijten. 2000. A constraint-based approach to minimize the weighted number of late jobs on parallel machines. *Research Report UTC* 2000/288.

[2] Baptiste, Ph., C. Le Pape, W. Nuijten. *Constraint-based scheduling: applying constraint programming to scheduling problems.* Kluwer Academic Publishers (2001).

[3] Bockmayr, A., N. Pisaruk. 2003. Detecting infeasibility and generating cuts for MIP using CP. In *5th International Workshop on Integration of AI and OR techniques in Constraint Programming for Combinatorial Optimization Problems, CP-AI-OR'03, Montreal, Canada.*

[4] Carlier, J. 1982. The one machine sequencing problem. *European J. of Oper. Res.* **11**:42-47.

[5] Colombani, Y., T. Heipcke. 2002. Mosel: an extensible environment for modeling and programming solutions. *4th International Workshop on Integration of AI and OR techniques in Constraint Programming for Combinatorial Optimization Problems, CP-AI-OR'02, Le Croisic, France,* 277-290.

[6] Dauzère-Pérès, S., M. Sevaux. 2003. Using Lagrangean relaxation to minimize the weighted number of late jobs on a single machine. *Naval Res. Logistics.* **50**(3):273-288.

[7] Jain, V., I. E. Grossman. 2001. Algorithms for hybrid MILP/CLP models for a class of optimization problems. *INFORMS Journal on Computing.* **13**(4):258-276.

[8] Péridy L., E. Pinson, D. Rivraux. 2003. Using short-term memory to minimize the weighted number of late jobs on a single machine. *European J. of Oper. Res.* **148**:591-603.

[9] Sadykov, R., L. Wolsey. 2003. Integer Programming and Constraint Programming in Solving a Multi-Machine Assignment Scheduling Problem with Deadlines and Release Dates. *CORE Discussion Paper* 2003/81.

[10] Sevaux M., S. Dauzère-Pérès. 2003. Genetic algorithms to minimize the weighted number of late jobs on a single machine. *European J. of Oper. Res.* **151**:296-306.

Author Index